# Lead Toxicity

# Lead Toxicity

Edited by

**R.L. Singhal, Ph.D.**
Professor and Chairman
Department of Pharmacology
Faculty of Health Sciences
School of Medicine
University of Ottawa, Ottawa, Canada

**J.A. Thomas, Ph.D.**
Professor of Pharmacology and Associate Dean
West Virginia University Medical Center
Morgantown, West Virginia, U.S.A.



**Urban & Schwarzenberg** • Baltimore-Munich 1980

Urban & Schwarzenberg, Inc.
7 E. Redwood Street
Baltimore, Maryland 21202
U.S.A.

Urban & Schwarzenberg
Pettenkoferstrasse 18
D-8000 München 2
GERMANY



**Library of Congress Cataloging in Publication Data**

Main entry under title:
Lead toxicity.

Includes index.
1. Lead-poisoning. 2. Lead—Physiological effect. 3. Neurotoxic agents. I. Singhal, Radhey Lal, 1940– II. Thomas, John A., 1933– [DNLM: 1. Lead poisoning. QV292 L4342]
RA1231.L4L43 616.8′62 79-16784

ISBN 0-8067-1801-3 (Baltimore)
ISBN 3-541-71801-3 (München)

Printed in the United States of America

# Contents

# Contributors

**J.U. Bell**
Department of Pharmacology
West Virginia University
School of Medicine
Morgantown, West Virginia 26506
U.S.A.

**R.J. Bull**
Toxicological Assessment Branch
Health Effects Research Laboratory
U.S. Environmental Protection Agency
Cincinnati, Ohio 45268
U.S.A.

**J.J. Chisolm**
Department of Pediatrics
Baltimore City Hospital
4940 Eastern Ave.
Baltimore, Maryland 21224
U.S.A.

**D.D. Choie**
Laboratory of Toxicology
National Cancer Institute
National Institutes of Health
Bethesda, Maryland 20205
U.S.A.

**G.S. Cooper**
Department of Environmental Health
University of Cincinnati
Cincinnati, Ohio 45267
U.S.A.

**T.C. Dubas**
Drug Toxicology Division
Health Protection Branch
Health and Welfare
Ottawa, Ontario K1A OL2
Canada

**A. Goldberg**
Department of Medicine
University of Glasgow
Gardiner Institute of Medicine
Glasgow G11 Scotland

**A.M. Goldberg**
Johns Hopkins University
School of Hygiene and Public Health
Dept. of Environmental Health Sciences
Baltimore, Maryland 21205
U.S.A.

**I. Hanin**
Department of Psychiatry
Western Psychiatric Institute and Clinic
University of Pittsburgh
School of Medicine
Pittsburgh, Pennsylvania 15261
U.S.A.

**S. Hernberg**
Institute of Occupational Health
Haartmanin Katu 1
SF 00290 Helsinki 29
Finland

**P.D. Hrdina**
Department of Pharmacology
Faculty of Health Sciences
School of Medicine
University of Ottawa
Ottawa, Ontario K1N 9A9
Canada

**K.M. Jason**
Neuropsychopharmacology Research
Unit
Department of Psychiatry
New York University Medical Center
New York, New York 10016
U.S.A.

**S. Kacew**
Department of Pharmacology
Faculty of Health Sciences
School of Medicine
University of Ottawa
Ottawa, Ontario K1N 9A9
Canada

**C.K. Kellogg**
Department of Psychology
University of Rochester
Rochester, New York 14627
U.S.A.

**K.R. Mahaffey**
Division of Nutrition
Food and Drug Administration
1090 Tusculum Avenue
Cincinnati, Ohio 45267
U.S.A.

**P.A. Meredith**
Department of Materia Medica
Stobhill General Hospital
Glasgow G21 3UW Scotland

**I.A. Michaelson**
Division of Toxicology
University of Cincinnati
College of Medicine
3223 Eden Avenue
Cincinnati, Ohio 45267
U.S.A.

**M.R. Moore**
University of Glasgow
Department of Medicine
Gardiner Institute of Medicine
Glasgow G11 Scotland

**H.L. Needleman**
Children's Hospital Medical Center
Harvard Medical School
300 Longwood Avenue
Boston, Massachusetts 02115
U.S.A.

**J.O. Nriagu**
National Water Research Institute
Canada Centre for Inland Waters
Burlington, Ontario L7R 4A6
Canada

**D.C. Rice**
Toxicology Section
Toxicology Research Division
Food Directorate
Health Protection Branch
Tunney's Pasture
Health & Welfare Canada
Ottawa, Ontario K1A OL2
Canada

**G.W. Richter**
Department of Pathology
University of Rochester
School of Medicine & Dentistry
Rochester, New York 14642
U.S.A.

**C.D. Sigwart**
Dept. of Environmental Health
University of Cincinnati
Cincinnati, Ohio 45267
U.S.A.

**E.K. Silbergeld**
Section of Neurotoxicology
National Institutes of Health
National Institute of Neurological and Communicative Disorders and Stroke
Bethesda, Maryland 20205
U.S.A.

**R.L. Singhal**
Department of Pharmacology
Faculty of Health Sciences, School of Medicine
University of Ottawa
Ottawa, Ontario K1N 9A9
Canada

**J.A. Thomas**
Department of Pharmacology
West Virginia University
Medical Center
Morgantown, West Virginia 26506
U.S.A.

**J.F. Truelove**
Toxicology Section
Toxicology Research Division
Food Directorate
Health Protection Branch
Tunney's Pasture
Health & Welfare Canada
Ottawa, Ontario K1A OL2
Canada

**R.F. Willes**
Toxicology Section
Toxicology Research Division
Food Directorate
Food Directorate
Health Protection Branch
Tunney's Pasture
Health & Welfare Canada
Ottawa, Ontario K1A OL2
Canada

# Preface

Ancient history has recorded many toxicological actions of lead. In recent years, the dangers of human exposure to lead from industrial sources and consequently the environment has sparked considerable debate that emphasizes the need for a more complete understanding of the physiological and metabolic consequences of lead toxicity. This monograph is therefore designed to provide a comprehensive reference and an up-to-date summary of the recent advances in this field and to stimulate new ideas on problems related to lead toxicity.

This volume contains a selection of authoritative papers dealing with clinical and experimental studies of lead poisoning and is intended to assist clinicians and researchers in pediatrics, toxicology, internal medicine, nephrology, hematology, neurology and environmental pharmacology, biochemistry and physiology. The topics selected range from chapters devoted to the problems in clinical and experimental studies of lead intoxication and treatment to the critical exposure levels and the effects of lead on a variety of organ systems. This volume also deals with the neurotoxic, behavioral and neurological aspects of lead exposure as well as the importance of nutritional interactions of lead-induced CNS effects. Topics such as the influence of lead on tissue growth processes and glucose homeostatic mechanisms are also included. Finally, the interrelationships of lead with other metals and trace elements have been reviewed, and the health effects and toxicological potential of this heavy metal are placed in perspective with regard to both human and animal species including infrahuman primates.

Each chapter is organized for ease of comprehension as well as for rapid retrieval of progress and essential information concerning lead toxicity. Attention has been focused on aspects of chemical techniques, experimental design, clinical significance, unresolved problems and on the limitations of current experimental and clinical approaches to the subject. The contributing authors were chosen because of their widely recognized expertise in the specific field of lead toxicology covered in which major advances have been made through their own research endeavours. The editorial policy of imposing as few restrictions as possible has enabled contributors to encompass a wide range of material and to express their ideas freely. The responsibility for precision and accuracy of data and references, allocations of priority, expressions of judgment and evaluation, therefore lies with the individual authors.

The editors join the contributing authors in hoping that this monograph will provide the physician, scientist and student with a series of timely reviews of discrete areas of lead research, stimulate new ideas and suggest directions for future investigation. We remain grateful to Dr. E.K. Silbergeld of the National Institutes of Health for her valuable suggestions concerning various possible topics and contributors for this monograph. The editors would also like to express their sincere gratitude to Mr. Braxton D. Mitchell and his staff at Urban & Schwarzenberg for all of their help and advice and to Mrs. Diane McNeil for her tireless assistance with assembling this volume.

*R. L. Singhal*
*J. A. Thomas*

# Lead Toxicity

# Human Lead Exposure: Difficulties and Strategies in the Assessment of Neuropsychological Impact

*Herbert L. Needleman*

TABLE OF CONTENTS



## I. INTRODUCTION

> You will see by it, that the opinion of this mischievous effect from lead is at least above 60 years old; and you will observe with concern how long a useful truth may be known and exist, before it is generally received and practis'd on.
>
> *Benjamin Franklin, 1786*

### A. Historical Overview

The useful properties and the toxic potential of lead have been known to man since antiquity. Artifacts of lead and warnings about hazards attending its use have been found dating back at least as far as the second century B.C. Nicander, in the Alexipharmaca, vividly described the colic, anemia and neurotoxic sequelae of lead ingestion. In Rome large quantities of lead were used in the construction of aqueducts, domestic plumbing and cookware. While Pliny advocated lining copper pots with lead to avoid the unpleasant taste and toxic effects of copper, he was at the same time aware that drinking wine from such vessels could produce "dangling, paralytic hands." Pliny was less ambiguous about the dangers of lead inhalation: "Whilst it (lead) is being melted the breathing passages should be protected . . . otherwise the noxious and deadly vapour of the lead furnace is inhaled" (Waldron, 1973).

Until relatively recently, lead poisoning was thought to be almost exclusively a disease of industrial exposure. The risks for children from exposure to lead became apparent only at the turn of the twentieth century. In 1904, J. Gibson, an Australian ophthalmologist described lead-induced oculo neuritis in children and identified lead paint as the major source (Gibson, 1917). In 1924, Ruddock described pica as a cause of lead poisoning in children (Ruddock, 1924).

For many years it was generally assumed that, upon recovery from the acute phase of intoxication, children were left without significant residual deficit. Byers and Lord (1943), however, followed 20 children who were assumed to have recovered from lead poisoning and found that 19 out of 20 were failing in school or were severely behavior disordered. In their paper, they raised for the first time the question of whether some idiopathic neuropsychologic deficit in school children was the result of unidentified lead intoxication.

## B. The Threshold Question

Two questions dominate modern discourse about human lead exposure: What threshold, if any, exists for the effects of lead on human health? What is the earliest change that may be regarded as a health effect? Because one molecule of lead entering a cell will bind a ligand and thus cause an effect, the significance of the word "threshold" is elusive. Considerable controversy has revolved around this issue in testimony given before federal agencies. The question of whether slight degrees of enzyme inhibition or porphyrin elevations are adverse health effects or "merely" biochemical changes have practical consequences for regulatory decisions. Hernberg (1972) presented a rather elegant heuristic model which clarifies this question. Since individual differences with respect to both absorption and physiologic response to lead are universal, a given population exposed to varying concentrations of lead may be expected to display responses as depicted in Figure 1.

This figure portrays four specific responses to lead (A, B, C, and D) in relation to increasing dose, plotted on the abscissa. Response A represents a

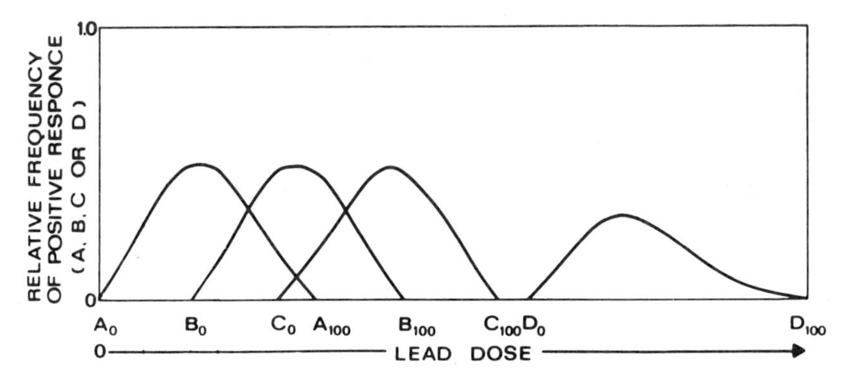


**Figure 1.** A model of the spectrum of responses to lead at increasing dose. Each letter represents a separate biological response. Each response has a threshold ($A_0$, $B_0$, $C_0$, $D_0$) and a dose of lead at which the entire sample is affected ($A_{100}$, $B_{100}$, $C_{100}$, $D_{100}$) (Hernberg, 1972).

low dose response of debatable consequence, *e.g.*, a sensitive enzyme which is not rate-limiting. Response D represents death. Between A and D is a spectrum of adverse effects of differing severity, each with a different position on the abscissa. $A_0$, $B_0$, $C_0$, and $D_0$ represent the levels at which no member of a population displays the response, and $A_{100}$, $B_{100}$, $C_{100}$, and $D_{100}$ are the levels at which all members of the population respond. In addition, the thresholds for some events are easily recognized, while the onset of others is not readily detected.

The onset dose for neuropsychologic dysfunction is not readily measured. This fact has resulted in considerable conflict with regard to the precise blood lead level which should be defined as hazardous, the level at which treatment should begin and the standards for allowable concentrations in air, paint and water. It is the principle thesis of this chapter that brain function, as expressed by neuropsychologic performance and behavior, is quite sensitive to lead; therefore, the threshold belongs at the left side of the graph. It is also the thesis of this chapter that the position on the graph at which the threshold will be placed in an individual study will depend on the sensitivity of the methods brought to bear upon measurement of performance and the rigor of the epidemiologic techniques applied.

## II. RECENT POPULATION STUDIES IN CHILDREN

In a study supported by the Center for Disease Control, Landrigan et al. (1975) studied children who lived close to a major lead smelter and compared the performance of those subjects with blood lead levels greater that 40 $\mu$g/dl to subjects with blood lead levels below 40 $\mu$g/dl. The outcome measures chosen were the Weschler Intelligence Scale and some quantitative neurologic measures. High lead children were found to have significantly lower scores on the performance scales of the IQ test. In an industry sponsored study of children from the same area, McNeil et al. (1974) were unable to find any significant difference between high and low lead subjects on intelligence tests. McNeil's study used proximity to the smelter as the major criterion to separate exposed from unexposed children rather than a tissue level of lead. Of 206 children who resided near the smelter and were invited to participate, 138 joined the study. No data comparing participating and excluded children were given. Lead exposed children were judged as more likely to be behaviorally disturbed on the California Test of Personality.

Kotok (1972) found no difference in outcome as measured by the Denver Developmental Screening Test between high and low lead children matched on socioeconomic status. However, blood lead levels in his control group ranged as high as 39 μg/dl. The Denver Developmental Test is a somewhat coarse screening tool and may not be sensitive to small changes in behavior. In a subsequent paper, Kotok et al. (1977) compared children with blood lead levels between 61 and 200 μg/dl to controls believed to have concentrations below 40 μg/dl. Blood lead concentrations were not available, however, on all control children. The children ranged in age from 20 to 67 months at the time of testing. No differences were found between groups. Perino and Ernhart (1974) compared performance on the McCarthy Scales of childrens abilities in black preschoolers who had blood lead levels between 40 and 70 μg/dl and a control group with levels below 30 μg/dl. Lead level was not found to be related to parental intelligence, socioeconomic status, birthweight, birth order or family size. Lead level, however, was inversely related to parental education level. The study found that as lead levels increased, general cognitive, verbal and perceptual abilities were significantly decreased. The authors interpret the inverse relationship between parental education and lead level to indicate that less well educated parents were not as assiduous in protecting their children from exposure. Others have suggested that parental education in this study could confound the effects of lead.

Hebel et al. (1976) compared performance on a group intelligence test (11-plus examination) in children from three areas. Those were, first, an industrial area in the proximity of a lead smelter, second, an industrial area removed from any lead source and, third, a residential area. In the lead smelter area, outcome was compared among children at three radial distances from the point source. No tissue lead levels were obtained. Outcome was compared after standardizing scores for maternal age, social class and birthrank. Children in the lead smelter area were found to have somewhat higher scores than children from the other two areas. In the lead smelter area, children living closer to the point source, especially males, were found to have lower scores than their counterparts at greater distance. Lansdown et al. (1974) compared intelligence, reading ability and behavioral rating in children who lived near a smelter. Children were categorized on the basis of the distance of their home from the smelter or on blood level. The age of subjects in the study was between 6 and 16 years. While blood lead level did vary somewhat in relation to distance from source, the correlation was quite low ($r = 0.176$). The authors state that the sample was relatively homogeneous, but made no measurements of other covariates known to affect development (*e.g.*, family size, socioeconomic status, birthweight).

## III. METHODOLOGIC DIFFICULTIES THAT ATTEND HUMAN POPULATION STUDIES

Most of the methodologic difficulties encountered in the studies briefly reviewed above can be classified into the following four types.

1. Problems in exposure classification
2. Problems due to sampling bias
3. Problems due to confounding variables
4. Problems due to insensitive performance measures

Because lead cannot be systematically manipulated in human studies, investigators are forced to examine groups which differ in their exposure to lead and resultant body burden. Markers of exposure (generally the level of blood lead) are chosen to classify the lead burden of subjects. However, lead in the blood is not the concentration of main concern. In addition, the blood lead level is a single static datum which is the result of a number of forces, such as intake, excretion and sequestration in tissue. It measures only recent exposure and cannot be used to classify subjects accurately with respect to exposure long past.

After subjects have been classified as to exposure or lead burden, they are generally invited to participate in studies of neuropsychological performance. Two types of methodologic difficulties may arise at this point. The selection strategy chosen by the investigator can affect both the results obtained and the ability to generalize conclusions drawn from the study to the population at large. Studies which draw their sample from schools for the retarded or psychiatric clinics, for example, may not apply to other groups. It may also be true that retarded subjects are more likely to eat paint. In addition, all subjects recruited usually do not choose to participate; it is possible that those who are excluded or those who exclude themselves may differ significantly from participants, either with respect to the independent variable (lead exposure) or the dependent variable (behavior). There is no easy way of determining the existence and extent of selection bias without some empirical evaluation. For example, parents who are worried about their children's abilities may either seek out the study or avoid it, depending upon how the study is perceived. If the investigation is seen as being an early step in the remedy of their child's state, parents of children with the deficit may be more likely to respond to an invitation than parents of children without the deficit. If the study is seen as a potential interference in the lives of the family, parents of children with the deficit (real or perceived) may avoid the study. For this reason, epidemiologic studies which are perceived as originating with the industry that delivered the pollutant may be less likely to recruit affected subjects. It is important, then, that in population studies attempts be made to compare some measure of independent and dependent variables in participant and nonparticipant subjects in order to evaluate the possibilitiy and extent of ascertainment bias.

Exposure to lead often segregates with poverty. The circumstances of being poor entail multiple assaults on the complex and easily perturbed tasks of successful cognitive, motor and social development. Among these other assaults are poor nutrition, poor prenatal care, exposure to other environmental hazards, lack of stimulation and an increased incidence of infection and trauma. The matrix for the confounding of lead's effect is illustrated in Figure 2. It is necessary to measure and control these variables in order to avoid the confounding of the effect of lead.

Finally, the effects of frank lead toxicity on children are vague, less than dramatic and often missed. It is likely then that the effects of lead at lesser doses are even more subtle. If this is the case, an investigation which hopes to evaluate successfully the effect of lower doses must use sensitive measures of performance. Group tests or clinical screening tests cannot be expected to discover mild performance deficits.

## IV. A SAMPLE STRATEGY TO DEAL WITH THESE DESIGN ISSUES

In a recently completed population study (Needleman et al., 1979) the Lead Exposure Study group at the Children's Hospital Medical Center attempted to address the four major design difficulties alluded to earlier while measuring neuropsychologic function in a group of children who were considered asymptomatic with respect to lead. The exposure of children in this study was classified by measuring the concentration of lead in shed deciduous

**Figure 2.** The matrix for confounding lead's effects on development. Since lead exposure often accompanies poverty, and poverty is associated with other hazards to development, the possible confounding of other impacts with lead is always present. In addition, the children with developmental deficit may have more pica (as indicated by the dotted arrow).

teeth. Because the tooth is a long term storage site for lead, it provides a reliable means for classifying past exposure (Altschuller et al., 1962; Needleman et al.; 1972 and 1974; Strehlow, 1969). High and low lead children were then evaluated in a blind study using a large battery of sensitive neuropsychological instruments. Children who participated and those who were excluded were compared with respect to both lead burden and behavior in an attempt to evaluate the degree of sampling bias. A large number of nonlead covariates known to affect development were measured and their effect on outcome was evaluated.

### A. Methods

From a total population of 3329 first and second grade children in two towns near Boston, 2335 children gave one shed deciduous tooth or more. Children who shed teeth brought them to their teachers, who then verified the presence of an empty socket. The teachers, after at least 2 months of classroom experience with the child, evaluated each child with an 11 item forced-choice questionnaire rating behavior in the class (Table 1). Dentine slices were analyzed for lead (Needleman et al., 1974), the children were ranked and those in the highest and lowest 10 percentiles were then invited to participate in the study. Only children who had two or more concordant dentine lead values were included. The following types of children were excluded: 1) children of bilingual parents, 2) children whose birthweights were below 2500 g, 3) children who had a history of significant head injury and 4) children who had a history of plumbism. A large number of nonlead covariates important to development were measured for each child and family (Tables 2, 3, and 4). These covariates were compared for high and low lead subjects; any which differed by t test at the $p < 0.1$ level were then controlled for using analysis of covariance. One hundred low lead subjects and 58 high lead subjects were then evaluated under strict blind conditions with a panel of neuropsychologic tests. The neuropsychologic evaluation of one child was accomplished in 4 hours by two examiners. The outcome measures employed consisted of the following evaluations:

*Psychometric Intelligence:* WISC-R
*Concrete Operational Intelligence:* Piagetian Conservation
*Academic Achievement:* Peabody Achievement Tests of Mathematics, Reading Recognition, Reading Comprehension
*Auditory and Language Processing:* Sentence Repetition Test, Token Test, Seashore Rhythm Test, Wepman Test of Auditory Discrimination
*Visual Motor Coordination:* Visual Motor Integration Test, Frostig Test
*Attentional Performance:* Reaction Time Under Intervals of Varying Delay
*Motor Coordination:* Elements of Halstead-Reitan Battery

**Table 1.** Teacher Questionnaire

1. Is this child easily distracted during his work?
2. Can he persist with a task for a reasonable amount of time?
3. Can this child work independently and complete assigned tasks with minimal assistance?
4. Is his approach to tasks disorganized (constantly misplacing pencils, books, etc.)?
5. Do you consider this child hyperactive?
6. Is he over-excitable and impulsive?
7. Is he easily frustrated by difficulties?
8. Is he a daydreamer?
9. Can he follow simple directions?
10. Can he follow a sequence of directions?
11. In general, is this child functioning as well in the classroom as other children *his own age*?

## B. Results

Lead exposure as indicated by dentine lead levels is shown in Figure 3. The distribution of dentine lead levels was log normal ($\bar{x} \pm$ S.D. = 14.0 ± 9.2 ppm). Twenty-four high dentine lead subjects and 54 low dentine lead subjects had blood lead determinations done 4 to 5 years prior to shedding teeth. These values were compared (Table 5). The range of levels in the high lead group indicates that these children can truly be considered to have low level lead exposure by current definitions.

## C. Evaluation of Sampling Bias

The distribution of dentine lead levels in children who participated was similar to that found in those who did not (Table 6). When teachers' reports of excluded children were compared to those included in the study, the children were not found to differ in any important respect (Table 6).

## D. Control Variables

High and low lead level children were remarkably alike on most personal nonlead variables. The children were primarily white, from intact working class homes. Boys and girls were equally represented. High lead level children tended to be older at the time of testing, but this difference was controlled by age-normalizing those tests not previously standardized for age. High lead level children came from slightly larger families. The fathers of high lead subjects were of significantly lower socioeconomic status as calculated by the 2 Factor (education, occupation) Index of Hollingshead and Redlich (1958). Mothers of high lead children tended to be younger at the time they gave birth to the subject and to be less well educated. The IQ of

**Table 2.** Comparison of subject control variables in high and low lead groups

| | Low dentine lead (%) | High dentine lead (%) |
|---|---|---|
| Male | 49.5 | 55.9 |
| White | 97.0 | 98.3 |
| Vaginal Delivery | 87.9 | 94.9 |
| Father Head of Household | 77.2 | 67.8 |

| Physical variables at date of testing | Low dentine lead ($\bar{x} \pm$ S.D.) | High dentine lead ($\bar{x} \pm$ S.D.) | P Value (2 tail) |
|---|---|---|---|
| Age (mo) | 87.2 ± 7.7 | 90.7 ± 8.4 | 0.009 |
| Height (cm) | 126.6 ± 6.3 | 126.4 ± 6.3 | 0.85 |
| Weight (kg) | 25.8 ± 4.9 | 26.5 ± 4.6 | 0.37 |
| Head circumference (cm) | 51.8 ± 1.6 | 51.7 ± 1.5 | 0.54 |
| Skinfold: right arm (cm) | 9.5 ± 3.5 | 9.8 ± 4.2 | 0.57 |
| left arm (cm) | 9.5 ± 3.4 | 9.7 ± 4.2 | 0.75 |
| Past Medical History | | | |
| Birthweight (g) | 3400.0 ± 448.6 | 3346.0 ± 514.0 | 0.45 |
| Length infant hospital stay (days) | 4.9 ± 1.8 | 4.4 ± 1.5 | 0.11 |
| Birth order | 2.4 ± 1.7 | 2.7 ± 2.0 | 0.25 |
| Number of hospital admissions | 0.47 ± 1.2 | 0.42 ± 1.6 | 0.86 |

**Table 3.** Comparison of family control variables in high and low lead groups

| | Low dentine lead ($\bar{x} \pm$ S.D.) | High dentine lead ($\bar{x} \pm$ S.D.) | P Value (2 tail) |
|---|---|---|---|
| Number of pregnancies | 3.3 ± 1.8 | 3.8 ± 2.3 | 0.10 |
| Mother's age at subject's birth (yrs.) | 26.2 ± 5.5 | 24.5 ± 5.8 | 0.07 |
| Mother's social class (2 Factor Hollingshead) | 4.1 ± 0.8 | 4.2 ± 0.8 | N.S. |
| Mother's education (grade) | 11.9 ± 2.0 | 11.4 ± 1.7 | 0.08 |
| Mother's occupation | 5.5 ± 1.1 | 5.5 ± 1.3 | N.S. |
| Father's age at subject's birth (yr) | 28.8 ± 7.1 | 27.5 ± 7.9 | N.S. |
| Father's social class (2 Factor Hollingshead) | 3.8 ± 1.0 | 4.1 ± 0.8 | 0.02 |
| Father's education (grade) | 12.7 ± 2.8 | 11.1 ± 2.3 | 0.001 |
| Father's occupation | 4.7 ± 1.6 | 5.0 ± 1.2 | N.S. |
| Parent I.Q. | 111.8 ± 14.0 | 108.7 ± 14.5 | N.S. |

**Table 4.** Control variables: parental attitude scores in high and low lead subjects

| | Low dentine lead ($\bar{x} \pm$ S.D.) | High dentine lead ($\bar{x} \pm$ S.D.) |
|---|---|---|
| Parental aspirations for child | 19.7 ± 5.6 | 19.5 ± 4.6 |
| Home learning environment | 37.6 ± 6.3 | 37.1 ± 5.4 |
| Parental attitude toward school | | |
| Resignation | 17.4 ± 3.4 | 17.1 ± 2.9 |
| Futility | 17.8 ± 2.7 | 17.7 ± 2.5 |
| Conservatism | 20.3 ± 2.2 | 20.4 ± 1.9 |
| Parental attitude toward child | 34.4 ± 4.3 | 34.5 ± 4.8 |
| Parental restrictiveness | 19.1 ± 2.1 | 19.4 ± 2.2 |

These scores were compiled from a 58-item paper and pencil questionnaire evaluating parental attitude in the seven areas listed above. No significant differences were found.

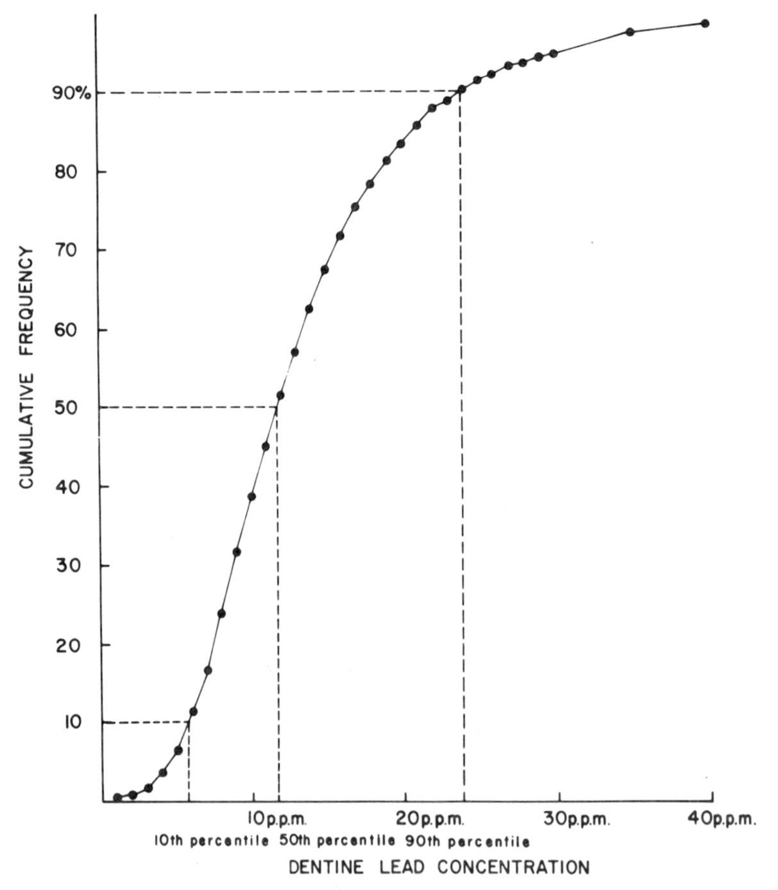


**Figure 3.** Cumulative frequency distribution of dentine lead concentrations (N = 3211 teeth). The points plotted are actual (unsmoothed) values (Needleman et al., 1979).

**Table 5.** The relationship between earlier blood lead level and current dentine lead level

| Dentine lead level (1977) | N | Mean Blood lead level | Range (1973–1974) |
|---|---|---|---|
| High (>20 ppm) | 24 | 35 μg/dl | 18–54 μg/dl |
| Low (<8.5 ppm) | 54 | 24 μg/dl | 12–37 μg/dl |

the parents of high lead subjects tended to be lower. These variables (number of pregnancies, maternal age and education, father's socioeconomic status and parental IQ) were then controlled as covariates in analysis of covariance.

## E. Performance Differences in High and Low Lead Subjects

Significant differences in high lead subjects were found in three of the seven neuropsychologic performance areas evaluated. High lead subjects performed significantly less well on the WISC-R, particularly on the verbal subtests (Table 7), on auditory and language processing (Table 8) and on attentional performance (Fig. 4). Teachers' ratings showed that high lead subjects had a significantly higher incidence of negative reports on 9 of the 11 classroom behavioral items (Table 9).

## F. Teachers' Evaluations of the Entire First and Second Grade Sample

Of those children who submitted at least one tooth (2335), 2146 were rated by the teachers. Students were classified into six groups according to dentine lead level and the incidence of negative reports on each classroom behavior was compared across groups (Fig. 5). The incidence of negative reports on each behavior is seen to increase in a dose-related fashion.

**Table 6.** Comparison of tested and excluded subjects with respect to the independent variable (lead burden) and one dependent variable (classroom behavioral rating)

| | Low dentine lead (<10 ppm) (%) | High dentine lead (>20 ppm) (%) |
|---|---|---|
| Tested | 61 | 39 |
| Excluded | 56 | 44 |

| | Sum Scores of Teachers' Ratings | | | |
|---|---|---|---|---|
| | ≤3 (%) | 4–6 (%) | 7–9 (%) | 10–11 (%) |
| Tested | 8.0 | 7.0 | 15 | 70 |
| Excluded | 7.5 | 11.5 | 21 | 60 |

**Table 7.** Outcome variables in high and low lead subjects: WISC-R

| WISC-R | Low lead x̄ | High lead x̄ | P value (2 tail) |
|---|---|---|---|
| Full scale IQ | 106.6 | 102.1 | 0.03 |
| Verbal IQ | 103.9 | 99.3 | 0.03 |
| Information | 10.5 | 9.4 | 0.04 |
| Vocabulary | 11.0 | 10.0 | 0.05 |
| Digit Span | 10.6 | 9.3 | 0.02 |
| Arithmetic | 10.4 | 10.1 | 0.49 |
| Comprehension | 11.0 | 10.2 | 0.08 |
| Similarities | 10.8 | 10.3 | 0.36 |
| Performance IQ | 108.7 | 104.9 | 0.08 |
| Picture Completion | 12.2 | 11.3 | 0.03 |
| Picture Arrangement | 11.3 | 10.8 | 0.38 |
| Block Design | 11.0 | 10.3 | 0.15 |
| Object Assembly | 10.9 | 10.6 | 0.54 |
| Coding | 11.0 | 10.9 | 0.90 |
| Mazes | 10.6 | 10.1 | 0.37 |

**Table 8.** Outcome variables in high and low lead subjects: verbal processing and reaction time

| | Low lead x̄ values | High lead x̄ values | P value (2 tail) |
|---|---|---|---|
| Seashore Rhythm Test | | | |
| Subtest A | 8.17 | 7.1 | 0.002 |
| Subtest B | 7.5 | 6.8 | 0.03 |
| Subtest C | 6.0 | 5.4 | 0.07 |
| Sum | 21.6 | 19.4 | 0.002 |
| Token Test | | | |
| Block 1 | 2.9 | 2.8 | 0.37 |
| Block 2 | 3.7 | 3.7 | 0.90 |
| Block 3 | 4.1 | 4.0 | 0.42 |
| Block 4 | 14.1 | 13.1 | 0.05 |
| Sum | 24.8 | 23.6 | 0.09 |
| Sentence Repetition Test | 12.6 | 11.3 | 0.04 |
| Reaction time under varying intervals of delay | (x̄ ± S.D.) | (x̄ ± S.D.) | |
| Block 1 (3 sec) | 0.35 ± 0.08 sec. | 0.37 ± 0.09 sec. | 0.32 |
| Block 2 (12 sec) | 0.41 ± 0.09 | 0.47 ± 0.12 | 0.001 |
| Block 3 (12 sec) | 0.41 ± 0.09 | 0.48 ± 0.11 | 0.001 |
| Block 4 (3 sec) | 0.38 ± 0.10 | 0.41 ± 0.12 | 0.01 |

**Table 9.** Comparison of high and low lead subjects on teachers' behavioral rating scale. Percent of students in each group receiving a negative response on each item is reported

| Item | Low lead % | High lead % | P value |
|---|---|---|---|
| Distractible | 14 | 36 | 0.003 |
| Not persistent | 9 | 21 | 0.05 |
| Dependent | 10 | 23 | 0.05 |
| Disorganized | 10 | 20 | 0.14 |
| Hyperactive | 6 | 16 | 0.08 |
| Impulsive | 9 | 25 | 0.01 |
| Easily frustrated | 11 | 25 | 0.04 |
| Daydreamer | 15 | 34 | 0.01 |
| Does not follow: | | | |
| Simple directions | 4 | 14 | 0.05 |
| Sequence of directions | 12 | 34 | 0.003 |
| Low overall functioning | 8 | 26 | 0.003 |
| | Mean | Mean | |
| Sum score | 9.5 | 8.2 | 0.02 |

## G. Summary

These data demonstrate a statistically significant impairment in verbal performance, auditory discrimination, language processing and attention, when measured in the neuropsychologic laboratory, after controlling for a number of nonlead varibles known to be important to neuropsychological development. These performance deficits appear to be reflected in the real

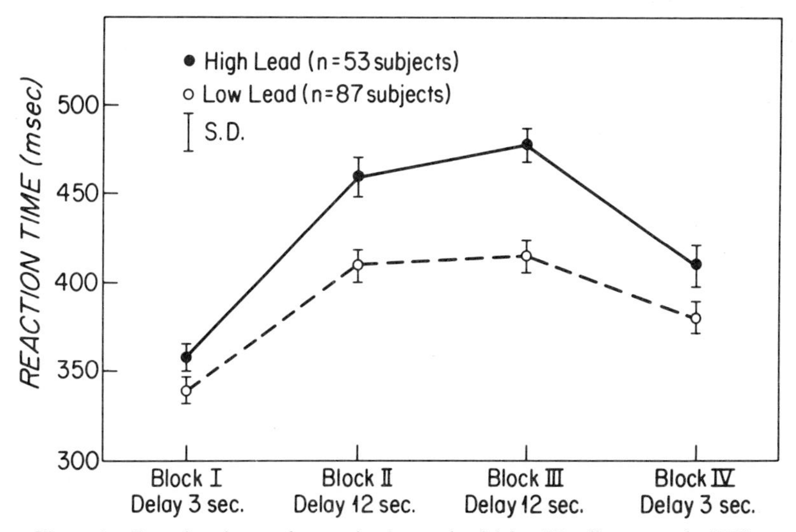


**Figure 4.** Reaction time under varying intervals of delay (Needleman et al., 1979).

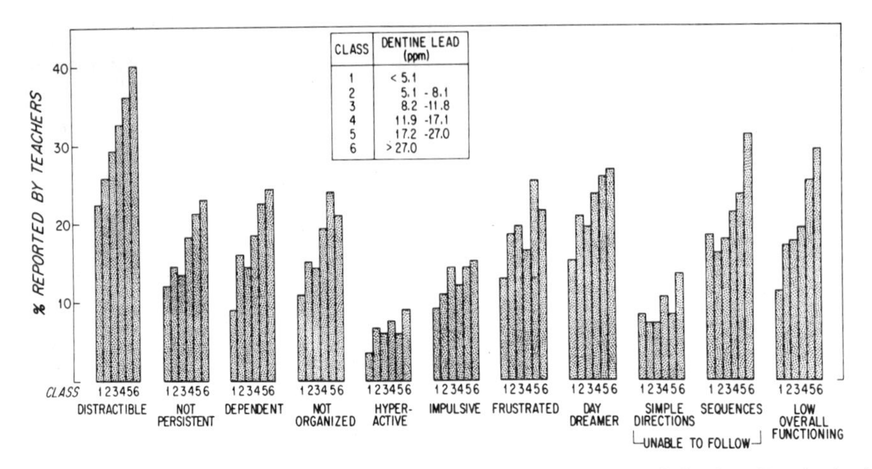


**Figure 5.** Distribution of negative ratings by teachers on 11 classroom behaviors. Dentine lead levels and teachers' ratings were completed on 2146 students. Students were classified into six groups according to dentine lead level. The relationship of negative classroom behavior to dentine lead level is seen for each behavior evaluated (Needleman et al., 1979).

life classroom behavior of children. Teachers' ratings of children demonstrate a dose-response relationship for the entire sample, not simply a difference between highest and lowest exposure groups.

## V. FUTURE DIRECTIONS FOR HUMAN STUDIES

It appears that the perceived threshold for lead's effect on the central nervous system of children varies inversely with the sensitivity and rigor of the methods brought to bear on measuring the performance of the subjects under investigation. It is reasonable to expect that with increasingly sensitive experimental design, this threshold will be further reduced in future studies.

An almost infinite number of questions can be asked of future population studies. The following two topics are selected as among the most important.

### A. Prospective Studies from Early Pregnancy

At high doses, lead has devastating effects on the fetus. This was known to British factory inspectors at the turn of the century when they reported that women exposed to lead in the manufacture of pottery were likely to be barren or to bear children who were destined to be short-lived (Rom, 1976). The increased vulnerability to lead of the developing nervous system has been demonstrated with respect to anatomical change (Pentschew and Garro, 1966), transport across the blood-brain barrier (Lorenzo and Gewirtz, 1977) and behavior (Brown, 1975). Lead crosses the placenta and is found in the human neonate's cord blood (Scanlon, 1971; Barltrop, 1969).

In a case control study in Glasgow, Beattie et al. (1975) compared the concentration of residential water lead of mothers who bore retarded children to controls matched on geography and socioeconomic status. Living in a home with a high concentration of lead in the drinking water was associated with a relative risk of 1.7 of bearing a retarded offspring. Moore et al. (1977) retrospectively examined blood lead in some of these cases and controls from filter paper blood specimens on file from a Phenylketonuria survey. Retardates had significantly high blood lead levels during the first week of life than did controls.

Data from prospective population studies of children, tracking exposure from early pregnancy and controlling for covariates such as nutrition, smoking, alcohol and drugs, are needed with some urgency. Such studies would evaluate children at birth and follow them forward in time at intervals close enough to detect any postnatal exposure. Appropriate outcomes for this study would be physical, biochemical and developmental variables.

## B. Lead and Aging

Because of the vulnerability of young organisms to lead, and the propensity of children to mouth foreign substances, most of the consideration of lead's health effects other than those due to occupational exposure has been directed at children. Considerable, although far from complete, evidence exists to warrant study of the effects of lead in older populations and, indeed, upon the aging process itself.

Schroeder (1971) demonstrated decreased life span in rodents exposed to lead during the fetal period. A 2.5 increase in the incidence of death due to the vascular disease in lead workers has been reported (Dingwall-Fordyce and Lane, 1963) and an increased incidence in death from nephropathy and hypertension was reported by Cooper and Gaffey (1975). A few authors have suggested that lead may produce a motor neuron disease similar to amyotrophic lateral sclerosis (ALS) or may in fact be one etiologic agent in the production of this disease. Petkau et al. (1974) found elevations in lead levels in nerves, spinal cord, cardiac and skeletal muscle in ALS patients, only some of whom had an occupational history of lead exposure. Felmus et al. (1976) compared, in a case control study, the occupational history of ALS patients with age and sex matched controls. ALS patients had a higher incidence of lead and mercury exposure. Conradi et al. (1976) described elevated cerebrospinal fluid and plasma lead levels in ALS patients. One investigator (Niklowitz, 1975) has reported increased neurofibrillary tangles in rabbits given both organic and inorganic lead and suggests that lead's role in the production of Alzheimer's Disease be considered.

By far the greatest pool of lead in the body is in bone. Because this pool has a biologic half-life of about $10^4$ days (Rabinowitz et al., 1976), most people tend to regard lead in bone as inert and of little biologic consequence. However, with advancing age, bone demineralization begins and it is likely that some lead is mobilized at this time. This mobilization is synchronous in some older people with restricted intake of proteins, calories and other trace elements, raising the intriguing possibility that some of the cognitive changes in older individuals are an effect of lead. The behavioral and biochemical status of older subjects with respect to both lead exposure and lead mobilization could well be a fertile area for investigation.

## VI. SUMMARY AND CONCLUSIONS

The toxic effects of lead at high doses have been known since antiquity and are no longer a subject of controversy. Considerable debate continues to revolve around the question of whether a threshold for adverse health effects exists and, if it does, at what internal dose it is reached.

Because of individual differences, responses to lead in population studies can be expected to follow predictable probability distributions with respect to external dose. The onset dose for a measured effect in a given sample will depend upon the sensitivity of the detection methods for the response variable (effect) and the rigor of the design.

A number of recent studies of behavioral effects of lead at low doses in a vulnerable subgroup, children, were reviewed. Some investigators found neuropsychological deficit and others did not. The major problem in experimental design and execution in these studies are found to fall into four categories: classification of exposure, ascertainment bias, confounding with other variables, and insensitive measures of dependent variables.

A population study in first and second grade children who were considered asymptomatic for lead, designed to address these issues, was presented. High lead level children were found to have significant deficits in IQ, auditory and language processing and attention. These behaviors, measured in the laboratory, were supported by scaled behavioral assessments evaluated by the teachers who were unaware of the lead status of the children.

The child in the first 6 years of life is in one of the more vulnerable subgroups in the population. Evidence is presented to suggest that two other subgroups may also be vulnerable. The effects of lead exposure prior to birth deserve systematic evaluation in prospective studies. Because lead is stored in bone, and because this bone demineralizes as part of the aging process, it is suggested that lead mobilization may occur with age. This subgroup also deserves study.

The chapters that follow provide ample testimony to the diversity of lead's effects. Many important and fascinating questions remain to be answered. Enough is known at present of lead's mischief, its sources and passages to argue for the orderly but rapid removal of this hazard from the human environment.

**Acknowledgments**

The author's studies cited here were supported by a Program Project Grant (DHEW) HD 08945 and by EPA Contract #68-02-2217.

## REFERENCES

Altschuller, L.F., Halak, D., Landing, B., and Kehoe, R. (1962). *J. Pediatr. 60:*224.

Barltrop, D. (1969). D. Barltrop and W.L. Burland (Eds.), In "Mineral Metabolism in Pediatrics," p. 135. F.A. Davis Co., Philadelphia.

Beattie, A.D., Moore, M.R., and Goldberg, A. (1975). *Lancet:*589.

Brown, D.R. (1975). *Toxicol and Appl. Pharmacol. 32:*628.

Byers, R.K., and Lord, E. (1943). *Am. J. Dis. Childr. 66:*471.

Conradi, S., Ronneni, L.A., and Vesterberg, O. (1976). *J. Neurol. Sci. 29:*259.

Cooper, W.C., and Gaffey, W.R. (1975). *J. Occup. Med. 17:*100.

Dingwall-Fordyce, I., and Lane, R.E. (1963). *Brit. J. Industr. Med. 20:*313.

Felmus, M.T., Patten, B.M., and Swanke, L. (1976). *Neurol. 26:*167.

Gibson, J.L. (1917). *Med. J. Australia 2:*201.

Hebel, J.R., Kinch, D., and Armstrong, E. (1976). *Brit. J. Prev. Med. 30:*170.

Hernberg, S. (1972). D. Barth, A. Berlin, R. Engel, P. Recht, J. Smeets (Eds.), In "Environmental Health Aspects of Lead," Commission of the European Communities, Luxembourg, p. 617.

Hollingshead, A., and Redlich, F. (1958). "Social Class and Mental Illness," p. 387. John Wiley & Sons, New York.

Kotok, D. (1972). *J. Pediatr. 80:*57.

Kotok, D., Kotok, R., and Heriot, T. (1977). *Am. J. Dis. Childr. 131:*791.

Landrigan, P.J., Whitworth, R.H., Baloh, R.W., Staehling, N.W., Barthel, W.H., and Rosenblum, B.T. (1975). *Lancet:*708.

Lansdowne, R.G., Shepherd, J., Clayton, B., Delves, H., Graham, P., and Turner, W. (1974). *Lancet:*538.

Lorenzo, A.V., and Gewirtz, M. (1977). *Brain Res. 132:*386.

McNeil, J., and Ptasnik, J. (1974). H. Berlin, R. Engel, V. Vouk (Eds.), In "Recent Advances in the Assessment of the Health Effects of Environmental Pollution," Vol. II, Commission of the European Communities, Luxembourg, p. 571.

Moore, M.R., Meredith, P.A., and Goldberg, A. (1977). *Lancet:*717.

Needleman, H.L., Davidson, I., Sewell, E.M., and Shapiro, I.M. (1974). *New Engl. J. Med. 290:*245.

Needleman, H.L., Gunnoe, C., Leviton, A., Reed, R., Peresie, H., Maher, C., and Barrett, P. (1979). *New Engl. J. Med.* 300:689.

Needleman, H.L., Tuncay, O.C., and Shapiro, I.M. (1972). *Nature 235:*111.

Niklowitz, W.J. (1975). *Neurology 25:*927.

Pentschew, A., and Garro, F. (1966). *Acta Neuropath. 6:*266.

Perino, J., and Ernhart, C.B. (1974). *J. Learn. Disabl. 7:*26.

Petkau, A., Sawatzky, A., Hillier, C.R., and Hoogstraten, J. (1974). *Br. J. Indus. Med. 31:*275.

Rabinowitz, M.G., Wetherill, G.W., and Kopple, J.D. (1976). *J. Clin Invest. 58:*260.

Rom, W.N. (1976). *Mt. Sinai J. Med. 43:*542.

Ruddock, J.C. (1924). *JAMA 82:*1682.

Scanlon, J. (1971). *Am. J. Dis. Childr. 121:*325.

Schroeder, H.A., and Mitchener, M. (1971). *Arch. Environ. Hlth 23:*102.

Strehlow, C.D., and Kneip, T. (1969). *Am. Ind. Hygiene Assoc. J. 30:*372.

Waldron, H.A. (1973). *Med. History 17:*391.

# Problems in Experimental Studies of Lead Poisoning

*Ellen K. Silbergeld and Alan M. Goldberg*

**TABLE OF CONTENTS**



## I. INTRODUCTION

Lead has been recognized as toxic to humans since the times of ancient medicine and there have been many experimental studies to confirm both its toxicity and its multiplicity of toxic signs. The more recently designed studies have been directed at understanding the cellular and molecular mechanisms of lead toxicity. While these studies have increased knowledge of many specific aspects of lead toxicity, as detailed in this monograph, many controversies have also occurred. This chapter is intended to provide an overview of the problems encountered in experimental lead studies. Some of these problems may, upon examination, turn out not to be real problems; others may be associated with particular methods of animal experimentation, while other problems may be inherent in the definition and measurement of toxic effects. The existence of these controversies attests to the vigor of the field. However, they should not obscure the appreciation of areas of consensus on basic mechanisms involved in the toxic actions of lead.

Experimental research in lead poisoning shares with research in any area of toxicology and pharmacology problems of a general and fundamental nature. These include the definition of toxic effects, as distinct (if possible) from statistically significant alteration in a given parameter; extrapolation from one experimental animal to another, and, most importantly, from animals to humans; methods of design or exposure, and the equivalence of exposure (duration and dose) between relatively short-lived species and humans; separation of systemic effects from actions on specific target organs; and interaction of the general state of the organism with the specific effect of the compound under study. At another level, there are several problems which may be considered specifically related to research on lead, such as possible differences in effect related to the chemical species of lead; appropriate means of measuring internal doses of lead at various cellular sites; unintended or inadvertent effects of lead on nutritional state, resulting from administering lead directly or indirectly through the diet; and the relation of effect to dose, that is, consideration of possible nonlinearity of toxic effects and the evidence for thresholds for these effects. These issues have been considered in relation to human toxicity studies (see Hernberg, this volume), and they are even more important in assessing experimental studies, where parameters of dose and intersubject variability appear to be controllable.

In addition, it is important to consider the objectives for experimental research on lead toxicity. First, the toxicity of lead has been known since ancient times; it is not necessary to demonstrate that lead is toxic. Beyond information on toxicity, other objectives for toxicology research are more complex. Animal experimentation can be used to provide models, or homologues of human disease, by the production of similar pathology at similar levels of exposure. For example, cerebellar hemorrhage is known to be a consequence of acute, high level lead exposure in humans (Browder et al., 1973). A similar pathologic process has been produced in the rat (Pentschew and Garro, 1966) and guinea pig (Bouldin and Krigman, 1975) by exposure of neonates to relatively high levels of lead. In such studies, the relationship between dose and effect is of great importance, since the effect, within limits of species differences, should be the same. As another example, recent research has attempted to develop animals with impaired behavioral function as models of lead-associated subencephalopathic neurological dysfunction described clinically (see Needleman, this book). In this case, the problems in homology are more difficult, since much less is known of the most appropriate parallels in behavioral and neutopsychological function between humans and experimental animals. Some experimental studies (Sauerhoff and Michaelson, 1973; Silbergeld and Goldberg, 1975; Sobotka et al., 1975; Kostas et al., 1978) have selected for homologous modeling symptoms of minimal brain dysfunction based on clinical reports associating aspects of hyperkinetic behavior with asymptomatic lead poisoning in man (Byers and Lord, 1943; David et al., 1972; de la Burdee and Choate,

1975). However, the clinical definition of hyperkinesis itself is under considerable discussion (Rapoport et al., 1978) and the aspects selected for modeling by lead research, which are increased motility and altered response to stimulants, may be debatable as indices of dysfunction. As this example suggests, the utility of animal models as homologues of clinical disease depends in large part upon the extent to which the clinical entity is defined and understood.

Animal models are also useful for studies of basic mechanisms of toxic action. In such cases, comparability with clinical conditions of dose and effect is of relatively less importance. In pharmacology and toxicology, it is often the case that significantly larger doses of compounds are used to produce effects specifically for investigation at the molecular level. It is thus of great importance to distinguish the rationale for this type of toxicological research from that for modeling clinical effects, since there are fundamental differences in the types of questions they are designed to answer.

An additional objective of research on lead toxicity is the development of environmental and public health standards based on experimental research (see Nriagu, this book). This use of experimental research involves at least three processes not involved in the prior two types of models. These are 1) extrapolation of experimental data to human situations by means of assumptions concerning dosage and exposure; 2) the determination of the significance of biochemical or functional changes for the health of an individual or of a population and 3) the application of political and economic judgements, such as risk-benefit analysis, to evaluate the social significance of these biological effects. These processes involve the use of experimental data by others with different objectives and standards of judgement. Inherent in such extrapolative jumps is a confusion of criteria, since the scientific research may have been undertaken with quite different rationales than those involved in the uses to which the experimental data are ultimately put in environmental policy making. An example of this may be taken from recent litigation and legislation in the United States concerning allowable concentrations of lead in paint. The statute was written in such a way that a lower level of allowable lead in paints would be automatically imposed unless the regulatory agency found that a higher level was safe. The findings of no effect of a substance at a specific dose does not necessarily imply lack of any other effects at a lower dose. It is therefore critical that the objectives of the original research be considered in trying to assess the utility of the data for other purposes.

This chapter will specifically focus on those areas of experimental studies where one can identify problems in interpretation of results arising from the differences in the questions being asked. It will attempt to distinguish experimental studies designed to model aspects of clinical lead poisoning from studies designed to understand mechanisms of toxicity. In addition, it will consider the impact of such issues as species selectivity and sensitivity, dosage (internal and external), age, sex and other interacting variables. This chapter will address the questions of model selection with regard to appropriateness of the experimental model to answer the questions being asked. These issues will be considered both from the standpoint of toxicology in general and of lead toxicity specifically.

## II. SPECIFIC PROBLEMS IN EXPERIMENTAL STUDIES

### A. Inter- and Intraspecies Variability

Experimental studies of lead toxicity have used, among other species, fish, chicks, pigeons, mice, rats, guinea pigs, cats, beagle dogs, rhesus monkeys, baboons, horses, and sheep. Such a multiplicity of subjects produces great problems for comparison of data, since there are vast differences among these species in target organ sensitivity, as well as in absorption and distribution of administered lead. As an example, pigeons show little or no encephalopathic signs at blood lead concentrations which exceed those found at death in humans, dogs, and rodents (Barthalmus et al., 1977). In terms of target organ sensitivity, among rodents the rat appears highly susceptible to cerebellar hemorrhage (Pentschew and Garro, 1966; Goldstein et al., 1974), while the mouse appears relatively resistant to this effect (Rosenblum and Johnson, 1968; Silbergeld and Goldberg, 1974). Also, the rat appears uniquely sensitive to lead as a carcinogen (Boyland et al., 1962). Comparisons of experimental lead research must consider fundamental differences among species. This is important not only in determining dose-effect relationships, but also in investigating mechanisms of toxic action, which in some species may be expressed in significantly different ways. The choice of a specific animal depends, therefore, upon the particular effect of lead under study.

Within a species, the toxic effects of lead appear to vary with age, sex and specific organ system. These three sources of variance can produce apparent contradictions in the experimental literature.

**1. Age** The dependency of toxic effects upon age reflects maturational changes in several physiological parameters. Generally, young animals respond more markedly to xenobiotics than do older animals (Dobbing, 1968). However, in the case of lead, at least two specific factors related to age are involved in the greater sensitivity of the young. First, younger animals and children appear to absorb and retain significantly more lead per dose than older animals and humans (Momcilovic and Kostial, 1974). The mechanisms for this age-related difference are not completely understood, but they appear to involve changes in renal function, gastro-intestinal absorption and calcium metabolism. There may also be age-related differences in excretion of lead by the kidney (Brown, 1975); this has been less extensively studied. Differences in retention may be associated with differences in apparent dose response between studies. This factor is probably the most significant cause of the apparent variability in the results of experimental studies.

The second major factor related to age is the developmental state of specific target organs. For the central nervous system, this has been reviewed by Dobbing (1968); its relevance to lead poisoning has been discussed by Silbergeld et al. (1978). Pentschew and Garro (1966) described a method for producing lead intoxication in the neonatal rodent by noninvasive techniques. This great advance has been found to be a useful and relevant model for the study of infant and juvenile lead poisoning in humans, as the chapters in this book demonstrate.

Age-related differences in target organ response to lead may be very precise, *i.e.*, small differences in age of experimental subjects may be associated with significant differences in toxic effects. Adverse effects of lead on learning behavior have been associated with small increases in blood lead concentrations when these increases occur in neonatal rats exposed during the first 10 days of life and not when exposure takes place from 11 to 21 days of life (Brown, 1975). Vasculopathies in the central nervous system (CNS) are produced in neonates only when exposure occurs before 14 days of age and not when exposure is initiated in mature rats or in sucklings older than 14 days (Bouldin et al., 1975). The age-dependent sensitivity of the CNS may depend not only upon the maturational state of the brain (synaptogenesis, myelination, rate of synthetic processes, etc.) but also upon the functional state of transport systems at the blood-brain barrier, which vary with age not only in functional uptake of essential amino acids but also in response to lead (Lorenzo and Gewirtz, 1977; O'Tuama et al., 1976).

Other organ systems also show age-related differences in toxic effects of lead. It has been observed clinically that children younger than 2 years do not show the expected relationship between blood lead concentrations and increases in erythrocyte protoporphyrin (Chisolm, personal communication). This can be explained by the fact that, postnatally, the hematopoietic system increases in capacity (Marver and Schmid, 1972). The accumulation of the heme precursor erythrocyte protoporphyrin is dependent upon the normal rate of heme synthesis; thus, in infants and young children with a lower basal synthesis rate for heme, the amount of lead-induced accumulation of propoporphyrin would be lower, even though the extent of functional inhibition may be the same as in adults. Another age-dependent toxic effect of lead may be on hepatic drug metabolism, since the drug-metabolizing enzymes are induced after birth. The reports that acute lead exposure inhibits drug metabolism in mature rodents (Goldberg et al., 1978) have not been replicated in immature humans (Alvarez et al., 1975). This discrepancy may reflect age-related differences in the functional capacity of hepatic metabolism and its response to inhibition, as well as age-related differences in heme synthesis supplying cytochrome P-450 as discussed above.

These examples serve only to indicate age-related variables important to experimental studies of lead poisoning. It should be emphasized, however, that age-related differences in the absorption, retention, and distribution of lead are of primary importance, since these factors determine the concentration of lead at sites of action within the organism.

**2. Sex** Sex-related differences in response to lead have not been well studied. Experimentally, the preference for using males has prevented gathering information on sex-related effects. There appear to be effects of lead on sexual functions and on implantation or fertility in the female (Hildebrand et al., 1973). The question has not been answered as to differences in susceptibility to lead between males and females to various effects. Sex-related differences have been reported in hematological alterations in populations exposed occupationally to lead, but these may reflect greater incidence of slight iron deficiency among women, as compared to men. In an experimental study of neurotoxic effects apparently transmitted genetically, exposure of either the father or the mother, or both, was sufficient to alter learning in offspring (Brady et al., 1975). No difference has been found between male and female mice in the production of hyperactivity by neonatal lead exposure (Silbergeld and Goldberg, 1974). There are considerable differences between males and females in the effects of chronic lead exposure on drug metabolism. Using hexobarbital sleeping time as an index, lead-exposed male rats were found to sleep slightly but not significantly less than control male rats, while lead-exposed female rats slept significantly longer than female controls (Silbergeld, et al, 1979b). Possible sex-related differences in lead toxicity do not obscure the experimental literature, which, as noted above, tends to utilize male subjects exclusively, but these differences may decrease the relevance of this research to clinical medicine, which must deal with both males and females exposed to lead.

### B. Target Organ Sensitivity and Specificity

The sensitivity of specific organs and systems to lead varies among species. Within an animal, there are also differences in organ sensitivity to lead, as is apparent in the chapters dealing with specific organ systems in this monograph. The relative sensitivity of different systems is not known, although it is generally thought that either the hematopoietic or nervous system is the most sensitive, in terms of dose and response. In addition, however, differences in sensitivity raise one of the general toxicological problems, *i.e.*, separation of systemic, or other, effects from the effects on a specific system under study.

Problems related to organ specificity particularly complicate interpretations of mechanisms of toxic action, since the effects under study may really be secondarily caused by effects of lead on another, unexamined system. An example of this is the role of the heme precursor delta-aminolevulinic acid (ALA) in lead toxicity on CNS function. ALA is significantly increased in the blood of lead-exposed animals and humans and it is able to cross the blood-brain barrier (Silbergeld et al., 1979). ALA appears to alter neurochemical function of $\gamma$-aminobutyric acid (GABAergic) synapses (Mueller and Snyder, 1977; Becker et al., 1976). Altered heme metabolism has been proposed to directly affect neurological function and, as a consequence, behavior in lead poisoning (Dagg et al., 1960). However, most studies of lead-induced neurotoxicity have concentrated only on the actions of lead itself as a neurotoxin and not on the possible role played by concomitant elevations in brain ALA concentrations (see Hrdina et al., this book). It is difficult to study lead and ALA independently of each other, since experimental paradigms of lead exposure also produce elevations in ALA, and the paradigms for producing increased ALA involve acute administration of compounds (such as allylisopropyl acetamide and barbiturates) which have a broad spectrum of neurochemical and metabolic effects.

Another example of such unexplored interactions among target systems may be found in the mechanisms involved in altered response to psychoactive drugs which have been reported for lead exposed rats, rabbits and mice (see Hrdina, et al., this book). These differences in drug response are usually ascribed to lead-induced changes in CNS neurochemistry (Silbergeld and Goldberg, 1974; Shih et al., 1976; Kostas et al., 1978), but they may also result from lead-induced changes in uptake of drugs by the brain and in metabolism and clearance of drugs from circulation. These variables, which clearly influence the duration and nature of drug effects on the CNS, have not been taken into account in the interpretations of neuropharmacological studies of lead poisoning.

### C. Dosage

In many toxicological and pharmacological studies, methods of exposure or dosage produce many problems, particularly when it is difficult to use the same conditions under which exposure occurs clinically. This is particularly true for experimental studies of lead, few of which expose animals to airborne particulates, a route of substantial clinical importance. In addition, it can be difficult to devise methods which do not produce artifactual or unexpected results, or side-effects, in themselves. Inadvertent administration-related results have been reported in pharmacology. For example, there are differing behavioral effects of the dopamine agonist apomorphine apparently associated with the site of injection (Ljungberg and Ungerstedt, 1977). When the drug was administered subcutaneously in the flank, rats showed predominantly sniffing responses; when apomorphine was administered subcutaneously in the neck region, rats responded predominantly with increased locomotion.

In lead research, animals have been exposed both directly and indirectly. One of the most indirect routes is the paradigm where lead is administered to the nursing dam, serving as the means of exposing the sucklings which are the subjects of study. Other routes of administration include intraperitoneal, subcutaneous or intravenous injection, gastrointestinal gavage, addition of lead to food or drinking water or inhalation of suspended particulates. In some paradigms, it can be difficult to determine actual dose as distinct from the concentration presented to the animal. Further, uptake and retention depend greatly upon the method of dosage employed. Intravenous, subcutaneous or intraperitoneal injection of lead acetate in solution results in the rapid deposition of insoluble lead compounds, mostly carbonates, from which lead is only slowly leached. Insoluble deposits can cause nonspecific trauma (Bischoff and Bryson, 1977) which may be hard to distinguish from lead-related effects.

In experimental toxicology, it is difficult to determine equivalence in exposure among experimental animals or between animals and humans. Many experimental designs assume some sort of constant relationship between dose and duration, such that long exposure at a low dose is equated (in terms of effect or body burden) to short exposure at a high dose. In part, this assumption has to be invoked because of the different life spans of most experimental animals (notably rodents) and humans, as well as the practical constraints of biomedical research. However, in the case of lead, such an equation is not true clinically. Acute exposure to high doses of lead is associated with effects which are not seen even after very prolonged exposure to low levels of lead; these are encephalopathy, seizures, coma and death (Browder et al., 1973). In determining dose, or in classifying type of exposure (low-level, acute, etc.), there is a need to define the parameters which will determine equivalence of exposure. It may be assumed that external dose is the most important defining parameter for equivalence definition. However, this would appear to invite error, since different species appear to absorb and retain different amounts of lead. Varying dosage schedules may be required to produce equivalent concentrations of lead at sites of toxic action. More importantly, species-, age-, and sex-dependent variability in target organ sensitivity may require different external doses to reach equivalent effects. A reasonable definition of equivalence, in animal models of toxicant exposure, may be equivalence of *effect*, which takes into account the variance both in absorption and in doses of lead required to produce specific signs or alterations associated with intoxication. Under this definition, the differences required in external dose should not be the basis of criticism of animals studies as unrealistic or irrelevant to clinical lead poisoning.

Another complicating factor in experimental studies is the use of different chemical compounds of lead. Lead exists in the environment as organic compounds, relatively insoluble salts and some soluble salts. It is important to note here that lead may be administered in one form, but as the result of physical-chemical actions, may reach the organism or its target organs in another chemical form. As indicated above, injections of soluble lead acetate results in the precipitation of lead as insoluble carbonates.

No comprehensive study has been done to compare the toxicity of lead compounds on different organ systems. Absorption and retention are influenced by chemical species, as found in studies of different lead-containing pigments and driers (Zook et al., 1976). In a study of the effects on rats of intraperitoneally injected lead species, the two-day $LD_{50}$ was found to be 215 mg/kg/day for lead acetate and 65.9 mg/kg/day for lead nitrate (Adler and Adler, 1977).

*In vitro* studies have also suggested that there are significant differences in toxic effects related to chemical species of lead, particularly when organic and inorganic compounds are compared. Synaptosomal uptake of dopamine is inhibited by lead chloride *in vitro* (Silbergeld, 1977); however, the organolead compound is more potent, by orders of magnitude, than lead acetate in producing this effect (Bondy et al., 1979). In addition, lead is relatively insoluble; therefore, actual concentrations in experimental media should be measured. Further, lead salts added in high concentrations may affect pH or the solubility of other compounds, proteins, etc.

### D. Measurement of Dose

The problems arising from varying methods of exposure are compounded further by problems in determining the means for appropriate measurement of lead at internal sites. Calculation of the dose itself presents problems. In clinical and experimental studies, dosage of lead has been presented as mg/day, mg/kg/day, $mg/m^2$ body surface/day or total dose (mg) over time. It is not clear which method is most appropriate, particularly for determining subsequent relationships between external and internal doses of lead (see Hernberg, this book). Clinical research is undecided on how to measure internal doses of lead or how to define exposure (Needleman and Hernberg, this book). It follows that experimental studies have also relied on variable methods for obtaining data on internal dose. The correlation coefficients among various parameters used to measure lead exposure exhibit no consistent pattern in children (Chisolm et al., 1977). In animal studies under controlled conditions of constant dosing, such parameters as blood lead concentrations vary considerably both for one animal and as the mean of a similarly exposed group of animals over time. Such variability can result from several interacting factors, including lack of a steady-state condition of distribution among various pools for lead within the organism stress and rapid, day-to-day changes in metabolism and nutrition. It is not clear under what conditions a steady-state condition for lead distribution can be achieved. Goldberg and Cohen (1977) have shown that prior chronic exposure to lead alters body distribution of subsequent acute doses.

Blood lead concentrations and increased heme precursors are the most commonly used indices of internal dose. However, their relationship to the concentrations of lead at tissue sites, other than those related to red cell heme metabolism, is unknown. Bornschein et al. (1975) have proposed that blood lead concentrations are directly related to brain lead concentrations in neonatal rodents, but this has not been found by other investigators (Goldstein et al., 1974). When tissue and blood lead concentrations have been determined at the same time in the same animals, it appears that measurements of lead in blood do not provide useful information on lead concentrations at sites of toxic action, other than to confirm a parallelism of increasing concentrations. Both clinical and experimental research on lead toxicity has been hampered by the lack of good indicators of lead concentration at sites other than the biosynthesis of heme.

Even if tissue concentrations of lead were known, there remain problems of dose and effect related to the partitioning of lead between inactive and metabolically active fractions and between extracellular and intracellular binding sites. The relative enrichment of lead in the nuclear fraction of kidney glomerular cells is well documented; it is not known if the uptake of lead into these intranuclear inclusion bodies is a sequestering and detoxifying mechanism or a toxic effect on the cell nucleus. A similar possibility exists for the nucleated red blood cell of the pigeon (Barthalmus et al., 1977). Lead is taken up and concentrated within mitochondria of neuronal and brain capillary tissue; only by indirect evidence is this assumed to be associated with toxic effects expressed in the functions of these cells (Goldstein et al., 1977; Silbergeld et al., 1977). The unequal distribution of lead among various subcellular compartments may explain apparent discrepancies between studies of the effects of *in vivo* and *in vitro* lead on the mitochondria which use tissue homogenates, as opposed to purified mitochondrial fractions (Bull et al., 1975; Holtzman and Shen Hse, 1976; Silbergeld and Adler, 1978).

## E. Controls

A specific problem in animal studies of lead toxicity concerns the development of control populations. Lead exposure *in vivo* appears to produce effects other than those considered to be specific to lead itself. This creates difficulties in defining appropriate controls. As discussed at length by Michaelson this volume several models of lead poisoning in neonates produced animals which grew less rapidly than littermate or age-matched controls. However, several reports indicate that within 1 week after cessation of exposure, pups reach the weights of age-matched or littermate controls. The signficance of a decreased rate of growth during the suckling or exposure period is debated. The long term meaning of undernutrition restricted to early neonatal periods on later brain development and behavior is not clear. Early undernutrition in experimental animals has been associated with a wide spectrum of behavioral and functional changes.

The nature of decreased rate of growth in pups indirectly exposed to lead is not clear; it could be the result of decreased food consumption (suckling by the neonates or decreased dietary intake by mothers) or associated with changes in the nutritional composition of milk produced by leaded mothers. It is important, for purposes of modeling lead poisoning, to determine if it is a lead-specific effect. One possible control may be found in studies of manganese poisoning in rats conducted by a similar paradigm of neonatal exposure. Such exposure reduces growth rate to a much greater extent than exposure to lead (Aylmer and Silbergeld, in preparation). Pair-fed controls were generated as controls in this study. Manganese-exposed rats were found to be significantly less active, measured in open-field crossings and by electronic actometers, than unexposed, normal controls. However, they were not different from rats restricted in access to food to produce a similarly depressed growth rate (Fig. 1). Other parameters (dopamine metabolism, response to amphetamine and apomorphine) which were altered in manganese-exposed rats were similarly altered in the food-restricted control group, as compared to normal growth controls. These results suggest that the effects of undernutrition on behavior and neurochemistry are not completely known and may vary at least as much as the effects of a toxin such as lead. This does not rule out the possibility that the reduced growth rate associated with lead exposure may be specific to the effects of lead on nutritional parameters and hence properly should be considered as one of the effects of lead, rather than as a confounding variable.

If reduced growth rate is considered a complicating variable, then strategies can be devised to factor it out of experimental studies. There are many studies in which lead-exposed suckling rodents are reported not to show decreases in body weight gain, while other effects of lead remain observable (Silbergeld et al., 1978). Reduced litter size appears to produce control and exposed pups with equal growth rates (Tennekoon et al., 1979). Administration of lead directly to pups by gavage or injection appears to avoid effects on growth (Brown, 1975; Sobotka et al., 1975, Louis-Ferdinand et al., 1978). Experimentally controlling for reduced growth rate is

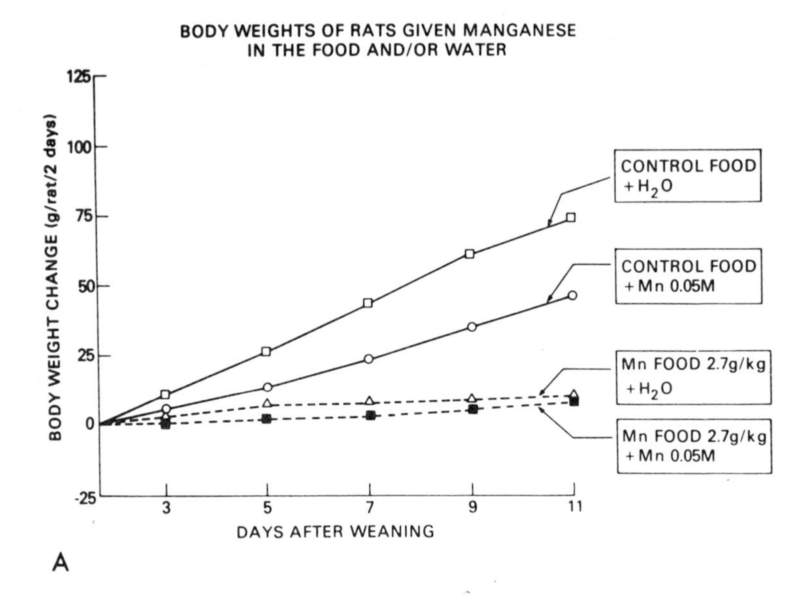


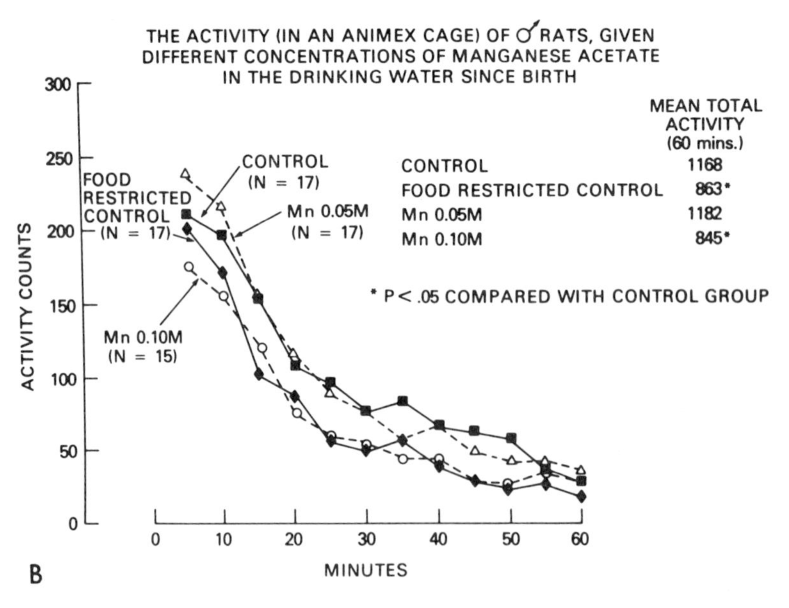


**Figure 1A, B.** The effects of manganese feeding on body weight; spontaneous motor activity in manganese-fed and pair-fed controls (Aylmer and Silbergeld, unpublished data).

more difficult, since the nature of this reduction is not known. Pair feeding has been used. This assumes that the cause of reduced growth rate is reduced food intake, rather than possible changes in nutrition such as constitution of dams' milk associated with lead exposure of the nursing mother. It is important to approach the problem of the effects of lead on growth rate as a lead-related effect, which needs further definition, rather than as a purely confounding variable, unrelated to lead exposure.

Another problem in development of controls involves problems of dose control. Since lead is so ubiquitous in the environment, lead contamination is difficult to control. Detectable amounts of lead are present in many commercially available animals chows, in the air, tap water and dust fallout of laboratories, in the solder used to make animal cages, in the air of animal rooms and in many parts of the environment of both exposed and so-called control populations of animals used in experimental research. Thus, the preparation of truly lead-free controls is very difficult even when gnotobiotic techniques are used. The inadvertent exposure of lead-dosed animals to environmental lead also complicates measurement of dose, since it will add to the intentionally administered dose. Although there are no data on the subject, it is possible that altered behavioral patterns in animals may increase their exposure from these environmental sources, *i.e.*, coprophagy and excessive grooming among either controls or lead-exposed animals increases the intake of lead found in feces or on body surfaces. Humans also exist in an environment widely contaminated with lead. The lowest blood lead concentration reported in the Three Cities Study in 1965 was 11 $\mu$g/100 mg, found in suburban nonsmokers near Philadelphia, and the highest value was 34 $\mu$g/100 ml, for parking lot attendants in Cincinnati. Thus, unlike some toxicologic studies, the experimental and clinical study of lead poisoning cannot be based on the comparison of unexposed with exposed subjects, but rather must be based on the comparison of subjects at opposite ends of a continuum of exposure. The concept of a continuum of lead exposure, in which exposed and control populations are defined on a relative scale of exposure, has important implications. Determination of significance of effect on measures of outcome depends on the definition of exposed and control populations. Inappropriate grouping of subjects can particularly confound epidemiologic studies.

Another problem in developing controls results when certain types of outcome measures are used. Under conditions where chronic treatment paradigms are used in pharmacology or toxicology, it is not possible to use each animal as its own control since, over the time required for exposure, a great deal of change in function may also occur associated with maturation (or senescence). Many tests of behavioral function, particularly operant measures, are difficult to interpret under such conditions, since animals cannot be trained to a specific criterion of performance. The subsequent effects of lead on maintenance of that level of performance, or extinction of behavior, cannot be used as indices of toxic effects. Similar methodological problems arise in clinical studies (see Needleman, this volume). In addition, when exposure is chronic, all other conditions need to be maintained in as similar a manner as possible between exposed and control groups. The effects of environmental parameters on such phenomena as aggression and activity are well known.

## III. MEASUREMENT OF TOXIC EFFECTS

The definition of toxic effect is a fundamental problem in all toxicologic research. In the case of experimental lead research, that lead is toxic is not disputed; the nature of its toxic effects, particularly those occurring before overt physical signs, has been the subject of much of the recent research reviewed in this monograph. The application of standard biochemical, functional and behavioral measures has shown that lead can affect many aspects of biochemistry and behavior. However, the recent literature has presented inconsistent findings. Variability in results may reflect differences in the many factors discussed above, such as exposure, species and conditions under which effects are measured. These variances are considerable and their consequences for outcome are undetermined. It may, therefore, be more fruitful to develop strategies based on the development of useful measures to describe the observed differences in lead-exposed animals. In neurobehavioral studies, these anecdotally reported differences include increased aggression and irritability (Sauerhoff and Michaelson, 1973; Silbergeld and Goldberg, 1973; Hopkins and Dayan, 1974; Allen et al., 1975; Williams et al., 1977), abnormal startle response (Bouldin and Krigman, 1975; Wince et al., 1976; O'Tuama et al., 1976), increased activity (Maker et al., 1975; O'Tuama et al., 1976; Wince et al., 1976), restlessness (Winneke et al., 1977) and hyperexcitability, usually in response to manipulations by the experimenter (Silbergeld and Miller, 1977). These behaviors have been observed under many conditions of exposure in rats, mice, guinea pigs and baboons. However, it has been difficult to translate these observations into quantifiable measures demonstrating significant changes due to lead exposure. In a study of mice exposed according to the same paradigm in which increased aggression was anecdotally reported (Silbergeld and Goldberg, 1973), Reiter et al. (1977) were unable to find quantitative evidence of increased aggressive behavior. Such discrepancies emphasize the need to develop sensitive, and inventive, measures capable of detecting the changes which upon observation are so striking.

Interacting multiple effects of lead may also interfere with detection and measurement of a specific effect. Lead-induced changes in brain transport of amino acids might be reflected in altered blood levels of these amino acids (Lorenzo and Gewirtz, 1978). However, the aminoaciduria produced by lead acting on the kidney would also affect blood levels of amino acids, thus making it difficult to determine the actual effect of lead or amino acid metabolism. As another example, some tests of behavioral function rely upon reinforcement of performance by providing water or food under specific schedules to acutely deprived animals. However, exposure to lead via the diet (such as indirect exposure through the mother in neonatal studies) may alter the response of animals to food or water. This was shown for mice exposed indirectly to lead; when presented with one- or two-bottle preference tests to determine response to leaded drinking water or tap water, lead-exposed mice were found to drink more tap water than controls (Morrison etal., 1974). Such a change in drinking behavior could interact with performance in those behavioral tests which use water as a reinforcement.

Another type of methodological problem producing inconsistencies may arise from inappropriate measures of outcome. Hyperactivity was reported as a behavioral effect of lead in rodents (Sauerhoff and Michaelson, 1973; Silbergeld and Goldberg, 1973). However, other experiments which measured activity have produced different results, including hypoactivity and no change in activity after lead exposure. In large part, these discrepancies may be related to differences in methodologies for measuring locomotion or activity in animals, as discussed by Goldberg and Silbergeld (1977). Since the effects of lead on behavior are probably complex, simple approaches to behavioral measurement may produce confusing results. It may be that the primary effect of lead on behavior is to increase reactivity, rather than motility, as indicated by several anecdotal reports cited above. In that case, increases in motility may be found in studies using methods of measurement which involve exposure of the animal to a novel environment (Silbergeld and Goldberg, 1973; Winneke et al., 1977), rather than, for example, residential mazes (Reiter et al., 1977). There is a need to devise more accurate measures of outcome in order to translate into quantitative data the observations made over many studies. Measurement of the effects of lead is complicated by an increased variability of response, which appears characteristic of animals exposed to a variety of toxins, including lead. Part of this variability may be associated with inability to provide consistent doses among animals in the same exposure group. As an example, the results of the effects of *in vivo* lead exposure on aspects of GABA metabolism in rats are shown in Figure 2 (data from Silbergeld et al., 1979c). Similar reports of increased variability in lead-exposed rats have been made concerning catecholaminergic function (see Hrdina, et al., this volume). If increased variability *per se* is a consequence of lead exposure, after factoring out differences in tissue concentrations among animals, then appropriate statistical measures must be used to compare data

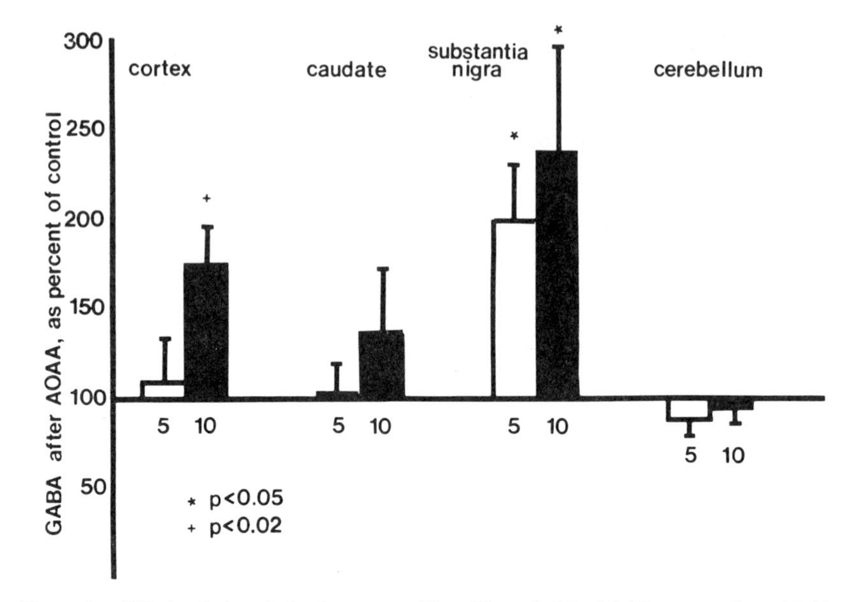


**Figure 2.** Effects of chronic lead exposure (5 or 10 mg/ml in drinking water from birth) on GABA turnover in cortex, caudate, substantia nigra, and cerebellum of rats. Turnover is measured as the increase in GABA levels after inhibition of GABA transaminase by systemic amino-oxyacetic acid (AOAA) (Silbergeld et al, 1979c).

from lead-exposed and control animals. Most commonly, Student's t test of sample means is used to determine significance of effect on a statistical basis. This statistical test is not appropriate under conditions where variability of the sample population is expected to differ from that of the control population. Increased variability of one group will by itself confound findings of significance by this test. These methodologic problems in statistical analysis have been approached only in some clincal studies, where distribution frequencies of intellectual function and mental retardation have shown effects of environmental lead exposure when comparison of mean IQ scores shows no significant difference (see Needleman, this volume). Similarly, more meaningful statistical procedures and experimental designs need to be applied in experimental studies of lead poisoning.

In the measurement of toxic effects, it is important to distinguish initial biologic effects from long term functional consequences. In making this distinction, it can be difficult to decide when to measure long term consequences. Brown (1975) has shown that effects on learning in rats after early neonatal exposure may not be expressed until the organism has a repertoire capable of more complex behaviors and, therefore, of reflecting more subtle deficits. Scotopic vision defects were observed in rhesus monkeys 18 months after cessation of lead exposure, when blood lead levels were no longer elevated (Bushnell et al., 1977). Some effects may be amplified over generations (Brady et al., 1975), which raises questions of teratological and genetically transmitted effects.

The possible nonlinearity of effects is another important factor in understanding dose-effect relationships. All dose-effect studies depend on the validity of the assumed relationship between dose and effect. In toxicology, such assumptions are usually based on one or two theoretical relationships; either there is a proportional, monotonic (though not necessarily linear or directly proportional) relationship between dose and effect or there is a threshold relationship between dose and effect. The first assumption implies that at every dose of lead, there is some effect; the second implies that there is a level of lead exposure below which no effects occur. The empirical data and ramifications of such assumptions have been most fully discussed in clinical and experimental studies of the effects of lead on heme metabolism. It is not presently clear which relationship provides the most valid model for the relationship between lead and ALA or lead and erythrocyte protoporphyrin (see Hernberg, and Moore et al., this volume; also Chisolm et al., 1977).

Apparent nonlinearity of effects can result from the interaction of various effects of lead. Interactions occur under conditions of continuous exposure to lead, which obscure detection of early or low level effects of lead by the induction of other effects. For example, lead decreases peripheral neuromuscular function by interfering with presynaptic neurotransmitter release (Manalis and Cooper, 1973; Silbergeld et al., 1974). Decrements in neuromuscular performance may interfere with the expressed behaviors used to measure learning performance, such as lever-pressing or maze-running, both of which are reported to be altered by lead exposure (see Jason and Kellogg, this volume). Similarly, effects of lead on the kidney are not continuous over a range of exposure; rather, distinctly different effects are associated with different levels of exposure and concentrations of lead in the kidney. The effects of lead to increase mitochondrial uptake of calcium are observed over a discrete range of lead concentrations in vitro as shown in Figure 3 (Silbergeld and Adler, 1978). The apparent reversal of this effect at concentrations of lead greater than 1 $\mu$M may relate to the induction of other effects by lead at this concentration, such as chelation of ATP or inhibition of Na-K ATPase, which would functionally counteract the hypothesized effects of lead on calcium efflux. However, this should not be taken as absence of toxic effects at higher lead concentrations, but is rather the failure to detect a specific change in the presence of many changes. Nonlinearity may exist for many specific effects of lead, while other effects appear to show (at least *in vitro*) completely linear dose dependence. This has been reported for the inhibitory effects of lead *in vitro* on brain adenylate cylcase (Nathanson and Bloom, 1975).

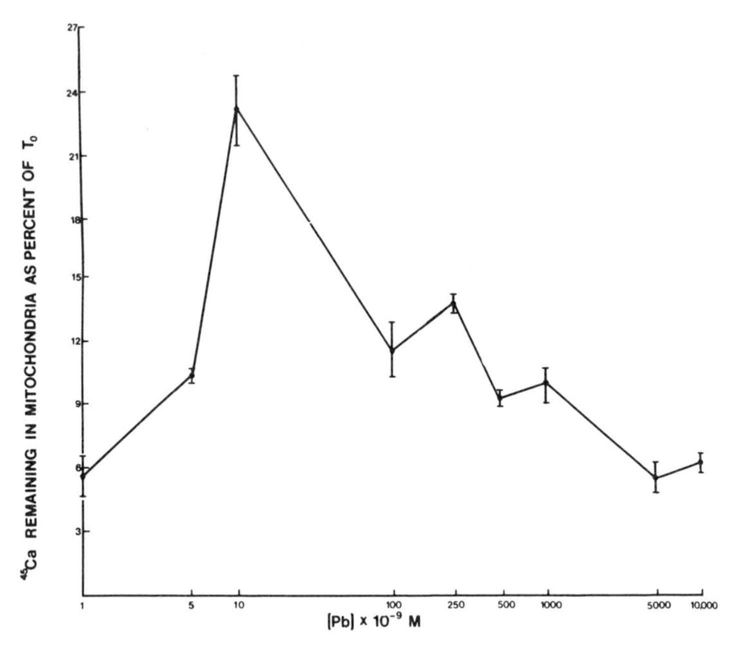


**Figure 3.** Effects of lead on mitochondrial $^{45}$Ca release *in vitro* (Silbergeld and Adler, 1978).

However, *in vivo*, there is an apparent lack of dose-dependent effects of lead on brain cholinergic function (Carroll et al., 1977). In summary, therefore, it is probable that lead induces a spectrum of effects, each of which may be associated with a different internal dose. Some of these effects are nugatory of others, while others may obscure or prevent detection of previously observed symptoms.

## IV. *IN VITRO* SYSTEMS

One method for dealing with many of the problems associated with the lack of correspondence between external and internal doses is to use *in vitro* systems. The fundamental problem with this approach is, of course, extrapolation since many *in vitro* assay systems are methodologically isolated from expression of function. For studies of toxic mechanisms, however, *in vitro* systems can provide information which would be difficult to obtain otherwise. The processes involved in the formation of renal intranuclear inclusion bodies and in the effects of lead on blood-brain barrier function have been explicated primarily on the basis of *in vitro* studies (Goldstein et al., 1977). *In vitro* systems have also been useful in studies of lead neurotoxicity, possibly because of the particular difficulties in developing *in vivo* models. The effects of lead on peripheral cholinergic synapses have been extensively studied with *in vitro* preparations of the neuromuscular junction (Manalis and Cooper, 1973; Silbergeld et al., 1974). The synaptosome, or pinched-off nerve terminal derived from homogenization and partial purification of brain tissue, has also been used in *in vitro* studies of lead and its neurotoxic mechanisms of action (Carroll et al., 1975; Silbergeld, 1977; Silbergeld and Adler, 1978; Bondy et al., 1978). However, these studies are limited to providing information on one parameter of function, electrophysiology or neurochemistry. However, a recent preparation of rat hemidiaphragm phrenic nerve permits not only monitoring of neuromuscular function but also collection of released acetylcholine and other substances associated with synaptic chemical transmission (Bierkamper and Goldberg, 1978). The use of such systems, which link *in vitro* biochemistry with functional parameters, will be very important in research on lead and other toxins.

Some of the problems raised by *in vitro* studies are partly met by techniques which combine *in vitro* methods of exposure with measurement of function in intact animals or preparations. The effects of lead on transmission in the superior cervical ganglion of the cat have been studied by applying lead *in vitro* to the exposed ganglion and measuring functional change in two ways, response of the nictitating membrane to pre- and post-ganglionic stimulation and bioassay measurement of released acetylcholine into the perfusing bath (Kostial and Vouk, 1957). An elaborate preparation of rat cerebellar explants grown in the chamber of the eye has also been used to study the effects of lead on electrophysiological aspects of norepinephrine-mediated neurotransmission in Purkinje cells in the explants (Taylor et al., 1979).

## V. SUMMARY AND CONCLUSIONS

Recent clinical and experimental studies on lead poisoning have interacted with each other to reproduce considerable increases in our current knowledge about lead. Correlations between clinical and experimental data are extensive, in terms of the production of similar or homologous effects in animals to those reported in humans. Such effects include, for example, hyperactivity, impaired learning behavior, altered catecholaminergic metabolism, neuromuscular deficits, cardiac arrhythmias, decreased renal function and changes in heme synthesis; all have been reported in humans and confirmed in experimental animals. In this sense, the recent experimental literature has provided many models for the study of the clinical effects of lead; many of these are discussed specifically in other chapters in this monograph. At present, the most significant problems in experimental lead research related to dose and effect, concern determination of dose. Currently used paradigms of exposure increase the probability of wide variability in internal dose; differences in species and age of experimental subjects also complicate comparability. Methods for determining internal dose are not standard; clinical data continues to provide reasons to question experimental results. Until internal doses are more reliably determined and controlled, it is premature to criticize the variability introduced into the experimental literature by different outcome measures in experimental studies. It is also important to remember, as a separate issue, that the results of such measures are determined by the interaction of the toxic effect produced in the animal with the instrument used to measure this effect.

In terms of mechanisms, recent experimental studies have significantly advanced knowledge of the toxic actions of lead. However, in the evaluation of these types of studies, it is important to use relevant critical standards. Correlations with clinical data, in terms of dose, are of less importance than the production of verifiable effects at any dose. It is not possible to study mechanisms of effect in the absence of effect.

The third use of animal models, for the development of environmental standards for "safe exposure," depends upon assumptions beyond the data gathered from recent experimental studies. The questions of linearity of effect and the existence of thresholds for dose-effect relationships remain largely unanswered. However, the experimental literature warns against simplistic interpretation of either clinical or experimental data for this purpose. The many effects of lead are produced along a continuum of exposure; some of these effects interact, in some cases to obscure detection of each other and in other cases possible to reverse each other. These interactions particularly obscure measurements of functional changes in whole, intact organisms, where any outcome is variously affected by many parameters, each subject to the effects of lead. An effect of lead which increases variability of response also obscures outcome measures.

Finally, it is fair to ask if the problems in experimental studies of lead poisoning are different, qualitatively or qunatitatively, from those encountered in other experimental studies. It may be argued that many of these problems (definitions of controls, regulation of dose, determination of concentration at site of action, variability of effects, translation of observed effects into quantifiable measurements) have existed in other areas of research. In considering the problems associated with experimental models of lead poisoning, it is important to define those which relate generally to problems in toxicological research and those which arise specifically from lead poisoning. From the latter, it is then imperative to distinguish those which represent actual confounding variables, such as inadvertent contamination of the environment, from those which are the results of effects of lead on systems other than those under study in specific experimental design. The separation of specific effects of lead, or any toxic compound, from the whole range of its effects presents a difficult problem; however, it is important to understand that all the effects, however confusing and difficult to sort out, are in fact effects of lead upon an organism or subcellular system. In this way, experimental research on lead poisoning fulfills its primary purpose, which is to understand and predict the effects of exposure to lead and to understand the normal functioning of an organism through the study of perturbation of function.

## REFERENCES


Adler, M.W., Adler, C.H. (1977) *Clin. Pharmacol. Ther. 22:*774.

Allen, J.R., McWey, P.J., and Suomi, S.J. (1975). *Envir. Health Persp. 7:*239.

Alvarez, A., Kapelner, S., Sassa, S., and Kappas, A. (1975). *Clin. Pharmacol. Ther. 17:*179.

Barthalmus, G.T., Leander, J.D., McMillan, D.E., Mushak, P., and Krigman, M.R. (1977). *Toxicol. App. Pharmacol. 42:*271.

Becker, D.M., Viljoen, J.D., and Kramer, S. (1976). M. Doss (Ed.), In "Porphyrins in Human Diseases," p. 163. Karger, Basel.

Bierkamper, G., and Goldberg, A.M. (1978). *J.E.P.T. 6:*40.

Bischoff, F., and Bryson, G. (1977). *Res. Comm. Chem. Pathol. Pharmacol. 18:*201.

Bondy, S.C., Anderson, C.L., Harrington, M.E., and Prasad, K.N. (1979). *Envir. Research. 19:*102–111.

Bornschein, R.L., Michaelson, I.A., and Fox, D. (1975). *Pharmacol. 17:*212.

Bouldin, T.W., and Krigman, M.R. (1975). *Acta Neuropath.* (*Berlin*) *33:*185.

Bouldin, T.W., Mushak, P., O'Tuama, L.A., and Krigman, M.R. (1975). *Envir. Health Perspect 12:*81.

Boyland, E., Dukes, C.E., Grover, P.L., and Mitchley, B.C.V. (1962). *Brit. J. Cancer 16:*283.

Brady, K., Herrera, Y., and Zenick, H. (1975). *Pharmacol. Biochem. Behav. 3:*561.

Browder, A.A., Joselow, M.M., and Louria, D.C. (1973). *Medicine 52:*121.

Brown, D.R. (1975). *Toxicol. Appl. Pharmacol. 32:*628.

Bull, R.J., Stanaszek, P.M., O'Neill, J.J., and Lutkenhoff, S.D. (1975). *Environ. Health Perspect. 12:*89.

Bushnell, P.J., Bowman, R.E., Allen J.R., and Marlan, R.J. (1977). *Science 197:*333.

Byers, R.K., and Lord, E.E. (1943) *Am. J. Dis. Child. 66:*471.

Carroll, P.T., Silbergeld, E.K., and Goldberg, A.M. (1977). *Biochem. Pharmacol. 26:*397.

Chisolm, J.J., Barrett, M.B., and Mellitts, E.D. (1975). *J. Pediat. 87:*1152.

Dagg, J.H., Goldberg, A., Tochlead, A., and Smith, J.A. (1965). *Quart. J. Med. XXXIV* (*134*)*:*163.

David, O., Clark, J., and Voeller, K. (1972). *Lancet II:*900.

de la Burde, B., and Choate, M.S. (1972). *J. Pediat. 81:*1088.

Dobbing, J. (1968). A.M. Davison and J. Dobbing (Eds.), In "Applied Neurochemistry," p. 287. Davis Publishing, Philadelphia.

Goldberg, A.M., and Cohen, S. (1977). *Proc. 1st Internat. Cong. Toxicol.* (*Toronto*) p. 11.

Goldberg, A., Meredith, P.A., Miller, S., Moore, M.R., and Thompson, G.G. (1978). *Por. J. Pharmac. 62:*529.

Goldberg, A.M., and Silbergeld, E.K. (1977). E. Usdin and I. Hanin (Eds.), In "Animal Models in Neurclogy and Psychiatry," p. 371. Pergamon Press, Oxford.

Goldstein, G.W., Asbury, A.K., and Diamond, I. (1974). *Arch. Neurol. 31:*382.
Goldstein, G.W., Wolinsky, J.S., and Csejtey, J. (1977). *Ann. Neurol. 1:*235.
Hilderbrand, D.C., Der, R., Fahim, Z., and Fahim, M.S. (1974). *Res. Commun. Chem. Pathol. Pharmacol. 9:*723.
Holtzman, D., and Shen Hse, J. (1976). *Pediat. Res. 10:*70.
Hopkins, A.P., and Dayan, A.D. (1974). *Brit. J. Ind. Med. 31:*128.
Kostas, J., McFarland, D.J., and Drew, W.G. (1978). *Parmacology 16:*226.
Kostial, K., and Vouk, V.B. (1977). *Brit. J. Pharmacol Chemother. 12:*219.
Ljungberg, T., and Ungerstedt, U. (1977). *Eur. J. Pharmacol. 46:*45.
Lorenzo, A.V., and Gewirtz, M. (1977). *Brain Res. 132:*386.
Louis-Ferdinand, R.T., Brown, D.R., Fiddler, S.F., Daugherty, W.C., and Klein, A.W. (1978). *Toxicol. Appl. Pharmacol. 43:*351.
Maker, H.S., Lehrer, G.M., Silides, D.J., Weissbarth, S., and Weiss, C. (1975). *Environ. Res. 10:*76.
Manalis, R.S., and Cooper, G.P. (1973). *Nature 243:*354.
Marver, H.S., and Schmid, R. (1972). J.B. Stanbury, J.B. Wyngaarden and D.S. Frederickson (Eds.), In "The Porphyrias," 3rd edition, p. 1087. McGraw-Hill, New York.
Momcilovic, B., and Kostial, L. (1974). *Environ. Res. 8:*214.
Morrison, J., Olton, D.S., Goldberg, A.M., and Silbergeld, E.K. (1974). *Dev. Psychobiol. 8:*389.
Mueller, W. and Snyder, S.H. (1977). *Am. Neurol. 2:*340.
Nathanson, J.A., and Bloom, F. (1976). *Nature 255:*419.
O'Tuama, L.A., Kim, C.S., Gatzy, Krigman, M.R., and Mushak, P. (1976). *Toxicol. Appl. Pharmacol. 35:*1.
Pentschew, A., and Garro, F. (1966). *Acta. Neuropathol. 6:*266.
Rapoport, J.L., Buchsbaum, M.S., Zahn, T.P., Weingartner, H., Ludlow, C., and Mikkelsen, E.J. (1978). *Science 199:*560.
Reiter, L., Anderson, G.E., Lashey, J.W., and Catrill, D.F. (1972). *Envir. Health Persp. 12:*119.
Rosenblum, W.I., and Johnson, M.G. (1968). *Arch. Pathol. 85:*640.
Sauerhoff, M., and Michaelson, I.A. (1973). *Science 182:*1022.
Shih, T.M., Khachaturian, Z.S., and Hanin, I. (1977). *Psychopharmacology 55:*187.
Silbergeld, E.K. (1977). *Life Sci. 20:*309.
Silbergeld, E.K., and Adler, H.S. (1978). *Brain Res. 148:*451.
Silbergeld, E.K., Adler, H.S., and Costa, J. (1977). *Res. Commun. Chem. Pathol. Pharmacol. 17:*715.
Silbergeld, E.K., Carroll, P.T., and Goldberg, A.M. (1978). In "Proc. International Conference on Heavy Metals in the Environment," p. 213. Univ. of Toronto Press, Toronto.
Silbergeld, E.K., Fales, J.T., and Goldberg, A.M. (1974). *Neuropharmacol. 13:*795.
Silbergeld, E.K., and Goldberg, A.M. (1973). *Life Sci. 13:*1275.
Silbergeld, E.K., and Goldberg, A.M. (1974). *Env. Health Persp. 7:*227.
Silbergeld, E.K., and Goldberg, A.M. (1975). *Neuropharmacol. 14:*431.
Silbergeld, E.K., Hruska, R.E., Lamon, J.E., Frykholm, B.F., and Hess, R. (1979a). *Clin. Res. 27:*A307.
Silbergeld, E.K., Hrnska, R.E., Lamon, J.E., Frykholm, B.F., and Hess, R. (1979b). *Pharmacol. 21:*209.
Silbergeld, E.K., and Miller, L.P. (1977). *Neurol. 4:*405.
Silbergeld, E.K., Miller, L.P., Kennedy, S., and Eng, N. (1979c). *Envir. Res. 19:*371–382.

Sobotka, T.J., Brodie, R.E., and Cook, M.P. (1975). *Toxicology* 5.175.
Taylor, D., Nathanson, J., Hoffer, B., Olson, L., and Seiger, A. (1978). *J. Pharmacol. Exp. Ther. 206:*371–381.
Tennekoon, G., Aitcheson, C.S., Frangia, J., Price, D.L., and Goldberg, A.M. (1979). Ann. Neurol. In press.
Williams, B.J., Griffity, W.H., Albrecht, C.M., Pirch, J.H., and Heitmancik, M.R. (1977).
Wince, L.C., Donovan, C.H., and Azzaro, A.J. (1976). *Pharmacol. 18:*473.
Winneke, G., Brockhause, A., and Baltissen, R. (1977). *Arch. Toxicol. 37:*247.
Zook, B.C., London, W.T., Sever, J.L., and Sauer, R.M. (1976). *J. Med. Primatol. 5:*23.

# Aspects of Molecular Mechanisms Underlying the Biochemical Toxicology of Lead

*Sam Kacew and Radhey L. Singhal*

**TABLE OF CONTENTS**



## I. INTRODUCTION

The continuous emission of lead into the environment from industrial sources and automobile exhaust and the accumulation and persistence of this heavy metal in the atmosphere, as well as the high affinity of lead to remain bound to mammalian tissues, emphasizes the need for a more complete understanding of the metabolic consequences of heavy metal toxicity. The ubiquitous nature of lead and its potential hazard to man has stimulated studies designed to elucidate the mechanisms of certain physiologic and toxicologic reactions to this heavy metal. Because young children may be more susceptible than adults to the effects of prolonged low-level lead exposure (Klein, 1974), it is important to determine the effects of this heavy metal on neonatal animals. The present chapter is concerned primarily with a review of acute, subacute and chronic effects of lead on glucose homeostasis and certain biochemical parameters associated with growth in rats.

Since lead has been found to produce an elevation in the concentration of blood and urinary glucose (Chisolm, 1962; Goyer, 1971b), the influence of this heavy metal on the capacity of liver and kidney to synthesize glucose was examined as a possible source of sugar in blood and urine. In particular, the effects of lead on the activities of the four key gluconeogenic enzymes (pyruvate carboxylase, phosphoenolpyruvate carboxykinase, fructose 1,6-diphosphatase and glucose 6-phosphatase) as well as on blood glucose were investigated. Since the concentration of blood glucose is also elevated by hepatic glycogenolysis, the effect of lead treatment on hepatic glycogen levels was determined. Although the exact mechanism(s) underlying hormonal modulation of hepatic and renal carbohydrate metabolism is still unknown, it has been postulated that stimulation of membranal adenylate cyclase and the consequent elevation of endogenous cyclic 3′, 5′-adenosine monophosphate (cAMP) may be involved in the control of glucose synthesis. An apparent similarity between the effects of cyclic AMP and lead on carbohydrate metabolism prompted studies on the influence of this metal on the renal and hepatic cyclic AMP-adenylate cyclase system.

Chronic lead poisoning in man is known to result from inhalation or ingestion, a process which is rather slow but continuous. One of the known consequences is the excretion of small amounts of lead through the mother's milk to the neonate. We therefore examined whether the chronic exposure of neonatal rats to lead resulted in any functional or biochemical abnormalities in pancreatic, renal and hepatic tissue. In this regard, newborn rats were given relatively low amounts of lead either indirectly through lactating mothers or directly by gastric intubation; changes in renal gluconeogenesis and endogenous cyclic AMP levels were measured. In addition, nucleic acids, protein levels and the incorporation of thymidine into deoxyribonucleic acid (DNA) were investigated in renal, hepatic and pulmonary tissue as an index of growth processes. In order to investigate the mechanisms involved in lead-induced alterations in macromolecular synthesis, the endogenous concentrations of putrescine, spermidine and spermine, as well as adenylate cyclase activity and cyclic AMP levels, were examined in kidney, liver and lung tissue.

## A. Lead Exposure and Glucose Homeostasis: Consequences in Liver, Kidney and Pancreas

Although considerable work has been carried out on the morphologic changes produced by lead in kidney and liver, few studies have been done on the metabolic consequences induced in these tissues as a result of exposure to this heavy metal. In the present review, an attempt is made to relate the ability of lead to influence carbohydrate metabolism with modulations in the adenylate cyclase-cyclic AMP system. In mammals, the full enzymatic potential is present for both glucose synthesis from noncarbohydrate precursors and for glucose degradation via the glycolytic pathway in both hepatic and renal cortical tissue. Gluconeogenesis comprises the synthesis of glucose from lactate, pyruvate, glycerol and certain amino acids, while the reverse pathway (*i.e.*, glycolysis) results in the formation of pyruvate from glucose (Fig. 1). The four key rate-limiting enzymes of gluconeogenesis shown in hatched boxes in Fig. 1 are pyruvate carboxylase, phosphoenolpyruvate carboxykinase, fructose 1,6-diphosphatase and glucose 6-phosphatase; the circles denote glucokinase, phosphofructokinase and pyruvate kinase, the triad of rate-limiting glycolytic enzymes. The quartet of gluconeogenic enzymes catalyze irreversible reactions in the formation of glucose from noncarbohydrate precursors and are predominantly located in liver and kidney cortex, although minor amounts of pyruvate carboxylase, phosphoenolpyruvate carboxykinase and fructose 1,6-diphosphatase have been demonstrated in skeletal muscle (Crabtree et al., 1972). Pyruvate carboxylase is the enzyme involved in the synthesis of oxaloacetate from pyruvate; phosphoenolpyruvate carboxykinase biotransforms oxaloacetate into phosphoenolpyruvate. In addition to pyruvate, malate and aspartate can also be converted into phosphoenolpyruvate. The other two important gluconeogenic enzymes are fructose 1,6-diphosphatase and glucose 6-phosphatase. Fructose 1,6-diphosphate is converted to fructose 6-phosphate by fructose 1,6-diphosphatase, while glucose is the end product of the reaction catalyzed by glucose 6-phosphatase.

### 1. Subacute lead treatment

***a. Changes in blood glucose, hepatic glycogen and renal gluconeogenic enzymes*** Some of the most prominent toxic manifestations noted in plumbic animals are interstitial nephritis and contracted kidney tumors (Morgan et al., 1966; Van Esch and Kroes, 1969). These lead-induced morphologic alterations are associated with functional disturbances such as enhanced capacity of renal tissue to biosynthesize nucleic acids (Choie and Richter, 1974a and 1974b) and proximal tubular dysfunction as evidenced by aminoaciduria, glycosuria, hyperphosphatemia and hyperuricemia (Chisolm, 1962; Goyer, 1971a). The kidneys therefore appear to be a primary target for heavy metal inflicted toxicity. Experiments were undertaken to examine the influence of a large dose of lead given subacutely on the ability of renal tissue to maintain glucose homeostasis. Male rats, initially weighing approximately 125 g, were injected intraperitoneally with lead chloride (5 mg/kg/day $\times$ 2) for 7 days. Although treatment with lead for 7 days significantly lowered the body weight, kidney size remained unaltered. Data in Table 1 show that exposure to lead for 7 days produced an increase in the activities of renal pyruvate carboxylase to 122%, fructose 1,6-diphosphatase to 189% and glucose 6-phosphatase to 130%. Whereas the activity of kidney phosphoenolpyruvate carboxykinase displayed a tendency toward a rise in plumbic rats (125%), enzymic activity remained insignificantly different from controls (Stevenson et al., 1976). The observed increase in gluconeogenic enzyme activities was accompanied by slight but insignificant changes in blood glucose, liver glycogen and serum urea.

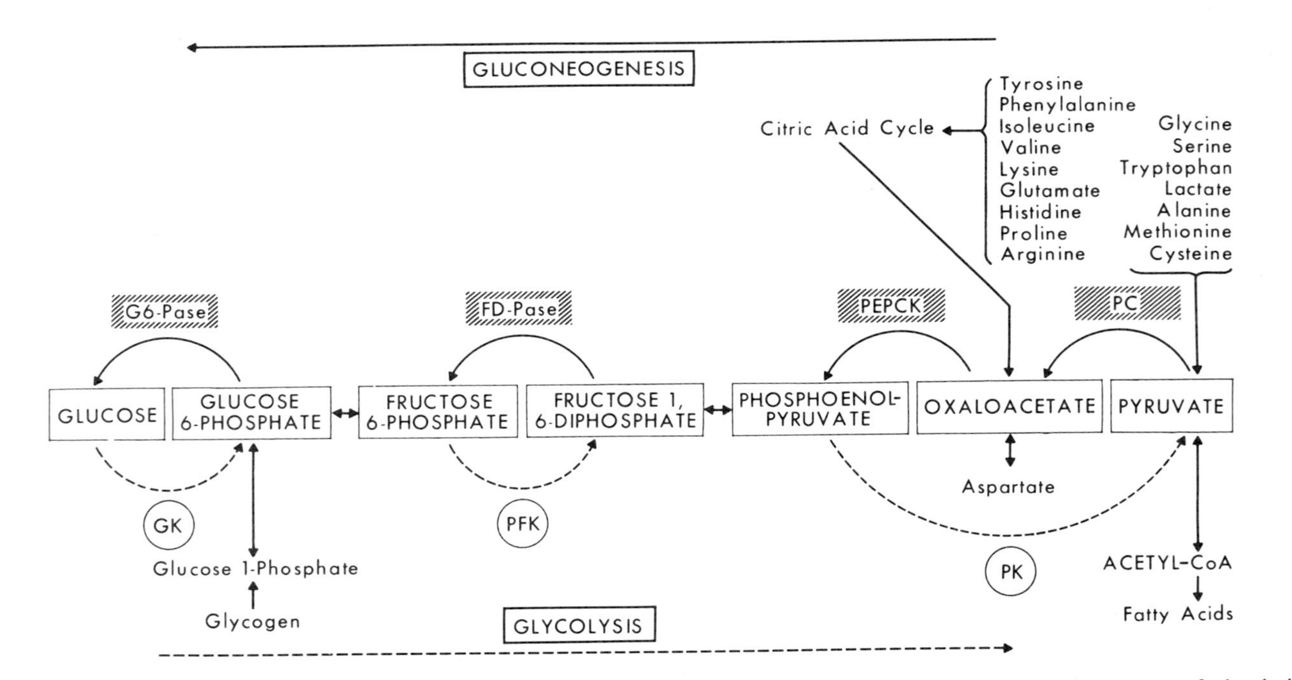


**Figure 1.** Pathways of hepatic and renal carbohydrate metabolism showing the important enzymes involved in the process of glycolysis and gluconeogenesis. PFK, phosphofructokinase; PK, pyruvate kinase; PC, pyruvate carboxylase; PEPCK, phosphoenolpyruvate carboxykinase; FD-Pase, fructose 1,6-diphosphatase; G6-Pase, glucose 6-phosphatase.

**Table 1.** Gluconeogenic enzymes, cyclic AMP and adenylate cyclase in rat kidney cortex after subacute administration of lead

| Parameter | Control | Lead treated | Percent of control |
|---|---|---|---|
| PC | 238 ± 17 | 289 ± 9 | 122* |
| PEPCK | 21.5 ± 1.4 | 27.0 ± 3.2 | 125 |
| FD-Pase | 5.2 ± 0.5 | 9.7 ± 0.8 | 189* |
| G6-Pase | 7.5 ± 1.2 | 9.7 ± 0.9 | 130* |
| Cyclic AMP | 4.92 ± 0.36 | 3.94 ± 0.17 | 80* |
| Adenylate cyclase | | | |
| Basal | 3.48 ± 0.47 | 3.50 ± 0.38 | 101 |
| NaF | 14.98 ± 2.06 (430)† | 15.64 ± 1.93 (447)† | 104 |
| EPI | 5.18 ± 0.77 (149)† | 4.96 ± 0.41 (142)† | 96 |

Means ± SEM represent at least five animals in each group. Lead (5 mg/kg/day ×2) was administered intraperitoneally to rats for 7 days and animals were sacrificed 24 hr after the last injection. PC, PEPCK, FD-Pase and G6-Pase are expressed as micromoles of substrate metabolized per hr per mg protein, while adenylate cyclase is expressed as picomoles cyclic AMP formed per 10 min per mg protein. Cyclic AMP levels are expressed as picomoles per mg tissue. The concentrations of NaF and EPI are 10 mM and 0.05 mM, respectively.

* Significantly different from control ($p < 0.05$).

† Statistically significant difference when compared with respective basal adenylate cyclase activity ($p < 0.05$).

***b. Role of pancreas in glucose homeostasis*** The maintenance of blood glucose levels is also dependent on the hormones secreted by the pancreas. In order to determine whether changes in pancreatic function contributed to the lead-induced glycosuria, the effects of this metal were examined on serum insulin and the insulinogenic index. Administration of lead (5 mg/kg/day × 2) failed to significantly alter the immunoreactive insulin levels in serum (Stevenson et al., 1976). The ability of subacute lead to tolerate high levels of blood glucose was studied after a single intraperitoneal injection of dextrose (2 g/kg). Although lead failed to alter the basal concentration of blood glucose, a marked rise was seen at various time intervals in animals given dextrose (Fig. 2). The observed increase was maintained at 15 min (147%), 30 min (135%) and 60 min (151%) after glucose administration, suggesting that the capacity of lead-exposed rats to compensate for high levels of blood sugar may have been disturbed. In response to a rise in blood glucose, increased amounts of insulin are usually released into the circulation (Henning et al., 1963), suggesting that the serum immunoreactive insulin levels may serve as an important index of glucose tolerance in mammals. Lead did not alter the basal levels of serum immunoreactive insulin; the concentration of this hormone fell to 58% at 30 min and returned to control levels after 1 hr. A similar pattern of respon-

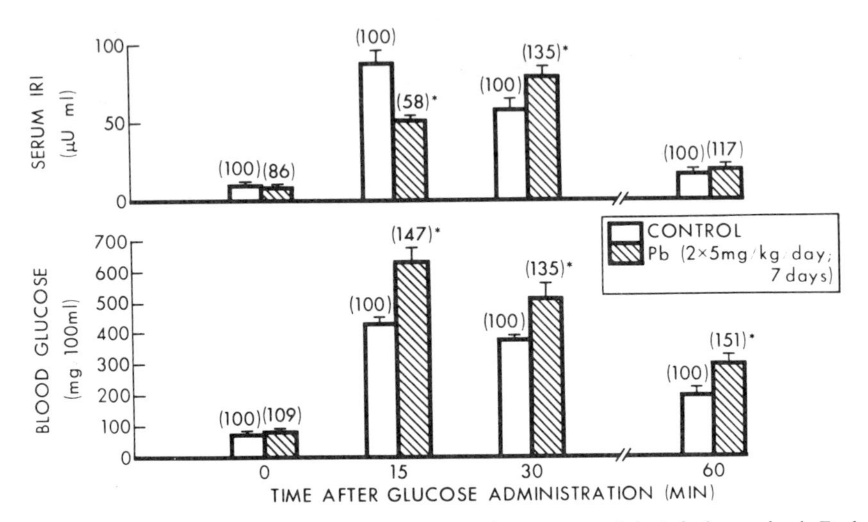


**Figure 2.** Effects of subacute lead treatment on tolerance to an injected glucose load. Each bar represents the mean ± SEM of at least five animals in each group. Rats were injected intraperitoneally with lead chloride (5 mg/kg) twice daily for a 1-week period. Following a 16-hr fast, animals were given glucose (2 g/kg, ip) and killed at 0, 15, 30 and 60 min following the glucose load. Blood glucose levels are expressed as mg per 100 ml, and serum immunoreactive insulin (IRI) as μU per ml. Data are also given in percentages (in parentheses), with the values for control animals taken as 100%. Asterisk indicates statistically significant difference when compared with the control values ($p < 0.05$) (Stevenson et al., 1976).

siveness was noted in serum immunoreactive insulin levels following a glucose load in control rats; however, the changes in normal animals preceded those seen in rats exposed to lead, indicating that this metal delays the release of insulin from pancreas.

Since lead impaired the ability of rats to tolerate a glucose load, the influence of this metal on the insulinogenic index, a measure of pancreatic secretory activity, was examined. Table 2 shows that the insulinogenic index was markedly decreased in heavy metal treated rats 15 min after glucose administration (52%). At none of the other time periods studied was there a significant difference in insulinogenic indices of lead-treated animals as compared to controls. Sixty min after the injection of glucose, the insulinogenic index was decreased in both control (54%) and treated animals (49%).

***c. Relationship between renal gluconeogenesis and the adenylate cyclase-cyclic AMP system*** Modulation in the cyclic AMP-adenylate cyclase system has been suggested to play a role in the control of renal gluconeogenesis. An increase in the incorporation of gluconeogenic precursors into glucose was found in kidneys of rats treated with cyclic AMP. Weiss et al. (1972) demonstrated that cyclic AMP augmented the entry and cumulative uptake of several amino acids into rat kidney cortex slices. From studies with gluconeogenic precursors, Guder et al. (1971) suggested that the rate-limiting step in cyclic AMP-stimulated renal gluconeogenesis involved an enhanced formation of phosphoenolpyruvate by phosphoenolpyruvate carboxykinase. Kacew and Singhal (1974) also demonstrated that administration of exogenous cyclic AMP elevated the activities of renal pyruvate carboxylase, phosphoenolpyruvate carboxykinase, fructose 1,6-diphosphatase and glucose 6-phosphatase. Since lead increased the activities of various gluconeogenic enzymes in kidney cortex, the ability of heavy metal treatment to exert an effect on the adenylate cyclase-cyclic AMP system was explored. Results in Table 1 illustrate that lead decreased the endogenous concentration of renal cyclic AMP to 3.94 ± 0.17 from a control value of 4.92 ± 0.36 picomoles per mg tissue. In contrast, lead failed to alter markedly the basal activity of kidney adenylate cyclase. Table 1 also shows that fluoride and epinephrine produced enhancement in the activity of basal adenylate cyclase. However, the fluoride- and epinephrine-stimulated forms of adenylate cyclase remained unchanged in kidneys obtained from plumbic rats.

### 2. Chronic lead injection

***a. Changes in liver glycogen and certain blood parameters*** Since long term exposure to small amounts of lead presents a potentially greater health hazard than subacute treatment, studies were undertaken to examine the effects of prolonged metal administration on carbohydrate metabolism. Male rats, initially weighing about 100 g, were injected daily by the intraperitoneal route with lead chloride (0.2 or 1.0 mg/kg) for 45 days. The weight of kidney, testes and heart remained unaltered in rats treated with either dose of lead (Singhal et al., 1973). However, liver weight was reduced

**Table 2.** Insulinogenic index following subacute lead administration in rats given an intraperitoneal glucose load

| | Time following glucose injection (Min) | | | |
|---|---|---|---|---|
| Treatment | 0 | 15 | 30 | 60 |
| Control | 13.2 ± 1.8 | 17.4 ± 1.6 | 15.3 ± 1.7 | 6.1 ± 0.7 |
| Percent of initial value | 100 | 132 | 116 | 46† |
| Lead | 12.1 ± 1.3 | 8.3 ± 1.0 | 17.3 ± 0.9 | 6.7 ± 1.5 |
| Percent of control | 92 | 48* | 113 | 109 |
| Percent of initial value | 100 | 63† | 131 | 51† |

Means ± SEM represent at least five animals in each group. Lead (5 mg/kg/day ×2 for 1 week) was administered intraperitoneally and rats were sacrificed 24 hr after the last injection. Insulinogenic index represents a ratio of serum immunoreactive insulin to blood glucose at various times following an injected glucose load of 2 g/kg.

* Significantly different from controls at a given time ($p < 0.05$)

† Statistically significant difference from control values at zero time ($p < 0.05$)

by approximately 35% in both treatment groups. In contrast, adrenal and thymus weights were significantly increased in rats receiving the heavy metal. Data presented in Table 3 illustrate the effects of lead injection on liver glycogen, serum urea and blood glucose. In contrast to subacute treatment, administration of either 0.2 or 1.0 mg/kg lead markedly reduced liver glycogen and increased the concentration of serum urea and blood glucose. The observed alterations were related to the dose; greater changes were seen in rats receiving the larger amount of the metal. A group of rats treated with 1.0 mg/kg lead was maintained for an additional period of 28 days in order to determine the effects of heavy metal withdrawal. Withdrawal for 28 days in rats previously given the metal for 45 days resulted in partial restoration of these metabolic alterations; however, the decrease in liver glycogen, as well as the elevation in serum urea and glucose, still remained significantly different from controls.

Because chronic lead treatment produced alterations in glucose homeostasis, the effects of this metal on serum immunoreactive insulin was examined. Results presented in Table 3 also demonstrate that a dose of 0.2 mg/kg of lead failed to significantly alter serum immunoreactive insulin. In

**Table 3.** Changes in liver glycogen, blood glucose, serum urea, serum immunoreactive insulin and insulinogenic index following chronic injection of lead and subsequent withdrawal

| | | Lead chloride | | |
|---|---|---|---|---|
| Parameter | Control | 0.2 mg/kg, 45 days | 1.0 mg/kg, 45 days | 1.0 mg/kg, 45 days; 45 days withdrawal |
| Liver Glycogen (g/100 g) | 2.2 ± 0 (100) | 1.3 ± 1 (59)* | 0.4 ± 0.1 (18)* | 1.2 ± 0.2 (53)* |
| Serum Urea (mg/100 ml) | 26 ± 1 (100) | 37 ± 1 (142)* | 47 ± 1 (181)* | 39 ± 1 (151)* |
| Blood Glucose (mg/100 ml) | 81 ± 2 (100) | 124 ± 2 (153)* | 158 ± 3 (195)* | 144 ± 2 (178)* |
| Serum IRI ($\mu$U IRI/ml serum) | 58.5 ± 3.8 (100) | 53.2 ± 2.2 (90) | 48.4 ± 1.1 (82)* | 58.6 ± 2.4 (100) |
| Insulinogenic Index | 72.6 ± 6.4 (100) | 42.9 ± 2.4 (59)* | 30.6 ± 1.3 (42)* | 40.7 ± 2.2 (56)* |

Means ± SEM represent at least five animals in each group. Rats were treated daily with lead chloride (0.2 or 1.0 mg/kg) injected intraperitoneally for 45 days and killed 24 hr after the last injection. Some rats treated with the 1.0 mg/kg dose were withdrawn from the injections for an additional 28 days before sacrifice. Data are also expressed as percentages (in parentheses), with the values of control animals taken as 100%.

* Statistically significant difference when compared with the values of control animals ($p < 0.05$)

contrast, the larger amount of the metal (1.0 mg/kg) significantly decreased the concentration of immunoreactive insulin by 18%. Withdrawal of lead for 28 days in rats previously given the metal for 2 months produced a return in the level of serum immunoreactive insulin to control amounts. The insulinogenic index was also depressed by either dose of the heavy metal. Withdrawal of the treatment regimen partially restored the insulinogenic index, although the values were still significantly lower than those of controls.

***b. Response of hepatic and renal gluconeogenic enzymes*** Administration of either 0.2 or 1.0 mg/kg lead produced a rise in the activities of hepatic pyruvate carboxylase, phosphoenolpyruvate carboxykinase, fructose 1,6-diphosphatase and glucose 6-phosphatase (Fig. 3). Withdrawal from lead treatment for 28 days partially restored the activities of phosphoenolpyruvate carboxykinase, fructose 1,6-diphosphatase and glucose 6-phosphatase to control values, although the observed alterations still remained significantly higher than controls. In contrast, the metal-induced enhancement in the activity of pyruvate carboxylase in "withdrawn" rats remained elevated to the same extent as that seen in animals injected for 45 days (Singhal et al., 1973). Since the kidney is another site of glucose production, it was determined whether lead treatment also affected the activities of renal gluconeogenic enzymes. Results presented in Figure 3 show that injection of either amount of lead for 45 days elevated the activities of pyruvate carboxylase, phosphoenolpyruvate carboxykinase, fructose 1,6-diphosphatase and glucose 6-phosphatase in rat kidney cortex. As in the case of liver, withdrawal from treatment for 28 days in rats previously given lead for 45 days restored the activities of phosphoenolpyruvate carboxykinase, fructose 1,6-diphosphatase and glucose 6-phosphatase to the levels seen in controls. In contrast, the activity of pyruvate carboxylase still remained significantly elevated (181%), although it was lower than that seen in animals treated for 45 days (307%).

***c. Changes in hepatic and renal adenylate cyclase-cyclic AMP system*** Daily administration of lead (1.0 mg/kg for 45 days) produced significant elevation in cyclic AMP levels and a rise in the activity of adenylate cyclase in rat liver (Table 4). However, the smaller amount (0.2 mg/kg) of heavy metal failed to alter markedly hepatic cyclic AMP content and the activity of adenylate cyclase. Table 4 also shows that withdrawal from lead treatment for 28 days in animals previously given the metal for 45 days restored both cyclic AMP levels and the activity of adenylate cyclase to approximately control levels. In kidney, lead at a dose of 0.2 or 1.0 mg/kg failed to alter significantly the concentration of cyclic AMP (Table 4). Although basal adenylate cyclase activity was insignificantly different from control in rats receiving 0.2 mg/kg lead, the larger amount of the

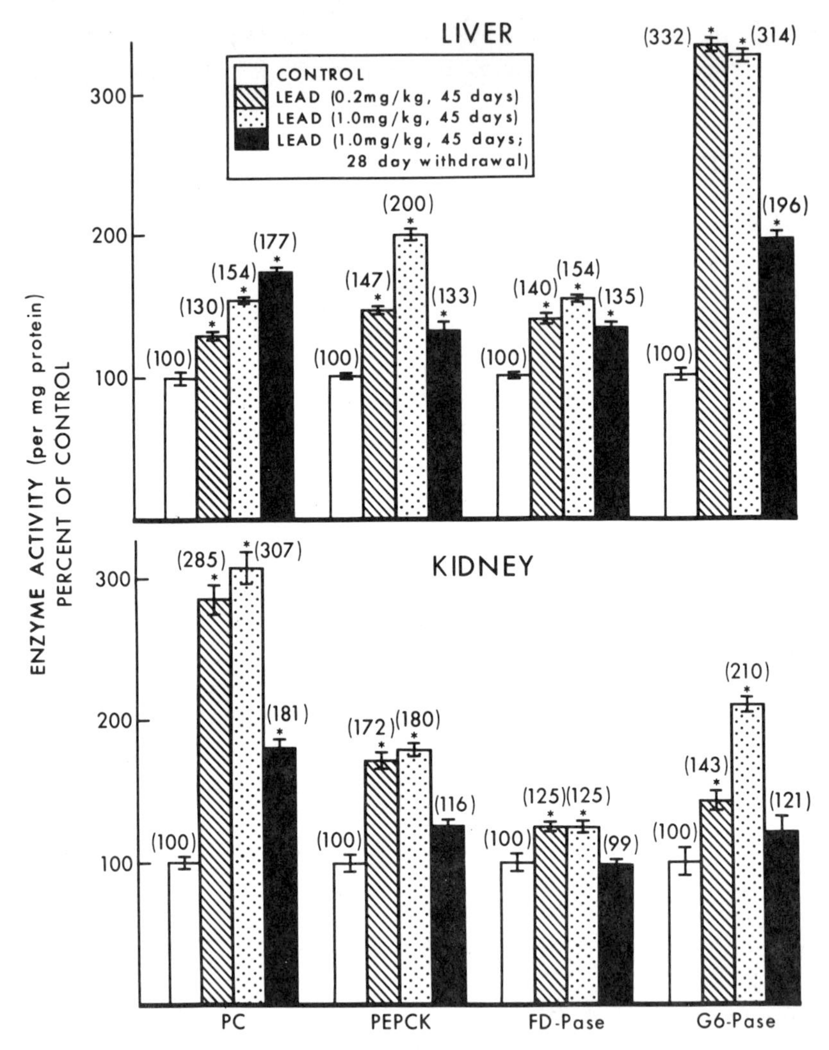


**Figure 3.** Influence of chronic lead and subsequent withdrawal on hepatic and renal gluconeogenic enzymes. Each bar represents the mean value ± SEM of five animals in each group. Animals were given lead daily by the intraperitoneal route (0.2 or 1.0 mg/kg $PbCl_2$) for 45 days and killed 24 hours after the last injection. Five rats pretreated with 1.0 mg/kg lead were maintained for an additional period of 28 days without any treatment. Enzyme activities were calculated as micromoles of substrate metabolized per hour per mg protein. Data are given in percentages, taking the values of control animals as 100%. Asterisk indicates statistically significant difference when compared with the values of control rats ($p < 0.05$).

**Table 4.** Effects of chronic lead injection on hepatic and renal cyclic AMP-adenylate cyclase system

| Tissue examined | Parameter | | Control | Lead: 0.2 mg/kg | Lead: 1.0 mg/kg | Lead: 1.0 mg/kg; 28 day withdrawal |
|---|---|---|---|---|---|---|
| Liver | Cyclic AMP | | 0.93 ± 0.05 | 1.01 ± 0.06 | 1.26 ± 0.14* | 1.02 ± 0.22 |
| | Adenylate cyclase | Basal | 1.74 ± 0.27 | 2.19 ± 0.15 | 3.04 ± 0.31† | 2.20 ± 0.22 |
| | | NaF | 13.44 ± 1.80† | 14.75 ± 1.47† | 14.50 ± 0.87† | 14.89 ± 0.70† |
| | | EPI | 3.37 ± 0.35† | 3.76 ± 0.25† | 3.67 ± 0.59† | 2.43 ± 0.25*† |
| | | Glucagon | 8.43 ± 0.41† | 18.59 ± 1.89† | 13.06 ± 1.75*† | 15.21 ± 1.13*† |
| Kidney cortex | Cyclic AMP | | 4.41 ± 0.11 | 4.49 ± 0.32 | 4.06 ± 0.15 | 4.47 ± 0.13 |
| | Adenylate cyclase | Basal | 1.39 ± 0.04 | 1.41 ± 0.24 | 0.54 ± 0.07* | 0.66 ± 0.13* |
| | | NaF | 9.63 ± 0.91† | 4.86 ± 0.32*† | 9.75 ± 1.76† | 18.24 ± 1.85*† |
| | | EPI | 3.04 ± 0.15 | 3.12 ± 0.38† | 2.45 ± 0.29† | 3.39 ± 0.30† |

Means ± SEM represent at least five animals in each group. Rats were treated daily with lead (0.2 or 1.0 mg/kg) intraperitoneally for 45 days and killed 24 hr after the last injection. Some rats treated with 1.0 mg/kg were withdrawn from injections for an additional 28 days before sacrifice. Cyclic AMP levels are given as picomoles per mg tissue, while adenylate cyclase activity is expressed as picomoles cyclic AMP formed per 10 min per mg protein.

* Significantly different from controls ($p < 0.05$)

† Statistically significant difference as compared to basal values in respective treatment groups ($p < 0.05$)

metal produced a reduction in the activity of adenylate cyclase. Indeed, the lead-induced decrease in the activity of adenylate cyclase was maintained even 28 days after withdrawal from lead treatment (52%). In contrast, termination of lead treatment for 28 days restored the endogenous concentration of cyclic AMP to control amounts. A similar difference in the responsiveness of liver and kidney cyclic AMP-adenylate cyclase system was reported by Kacew et al. (1976a and 1977) in animals given cadmium. As in the case of lead, cadmium decreased renal cyclic AMP levels, accompanied by elevation in hepatic adenylate cyclase activity and cyclic AMP content.

***d. Changes in hormonal and ionic stimulation of adenylate cyclase*** One of the most striking features of adenylate cyclase in mammalian cells is its ability to respond to hormones and fluoride ions. Hepatic adenylate cyclase can be stimulated by catecholamines (Murad et al., 1962; Pohl et al., 1969), glucagon (Makman and Sutherland, 1964) and fluoride (Perkins, 1973). Glucagon is a much more potent stimulator than any of the catecholamines, not only in the sense that smaller concentrations of glucagon are required, but also in that the maximal response seen is much greater than that observed with catecholamines (Bitensky et al., 1968). In order to assess the effect of chronic lead exposure on the hormonal responsiveness of hepatic and renal adenylate cyclase, the enzyme activity was measured in the presence of fluoride, epinephrine and glucagon. Chronic lead treatment and subsequent withdrawal did not appear to affect the five- to eight-fold increase in the activity of adenylate cyclase produced by fluoride ions (Table 4). Similarly, chronic heavy metal administration failed to alter the activity of the epinephrine-stimulated form of the enzyme; however, upon termination of heavy metal administration for 28 days, a fall in the activity of this form of adenylate cyclase was noted. It is of interest that epinephrine produced no additional stimulation in adenylate cyclase activity over the basal levels in livers of rats administered 1.0 mg/kg lead chloride. Chronic exposure to lead produced a marked enhancement in the glucagon-stimulated form of adenylate cyclase. The rise in activity of the glucagon-stimulated form of adenylate cyclase was maintained even after lead was withdrawn for 28 days. As in the case of glucagon, chronic lead treatment had no significant effect on the activity of the epinephrine-stimulated form of adenylate cyclase. Although treatment with 0.2 mg/kg lead produced a reduction in the fluoride-stimulated form of adenylate cyclase (50%), enzymic activity remained unchanged in kidney cortices obtained from 1.0 mg/kg treated rats. Withdrawal for 28 days markedly enhanced the fluoride-stimulated form of renal adenylate cyclase (189%). Unlike the basal activity of adenylate cyclase in lead-treated animals, a marked enhancement in the fluoride-stimulated form of adenylate cyclase was noted in rats given 1.0 mg/kg of lead chloride and also after withdrawal from metal injection (Table 4).

### 3. Chronic oral administration

***a. Changes in hepatic glycogen and certain blood parameters*** Since chronic oral ingestion of lead by children has been identified as a major environmental hazard, the effects of this mode of metal administration on carbohydrate metabolism in newborn rats was examined. Lead was administered as a 2% lead acetate solution to dams for a period of 8 weeks. This dietary dose of lead has been reported to result in lead concentrations ranging from 25 to 40 ppm in the milk (Michaelson and Sauerhoff, 1974). When the pups reached 21 days of age, they were weaned and separated randomly into groups receiving 20, 40 or 80 ppm lead in drinking water for a further 35 day period. Data demonstrate that lead failed to alter both body and kidney weights. In contrast, liver glycogen was markedly reduced by all three amounts of heavy metal; the lowest value (19% of control) was obtained with the 40 ppm concentration. Serum urea was decreased in animals given 20 ppm lead (22%), but an increase was seen in the 80 ppm group (133%); there was no change in rats receiving 40 ppm. The concentration of blood glucose was elevated in rats given either 20, 40 or 80 ppm lead for 8 weeks in a dose-dependent manner and greater changes were seen as the dose was increased (Table 5). The observed hyperglycemia was associated with a fall in the concentration of serum immunoreactive insulin in animals receiving the highest amount of the metal. Exposure to 40 or 80 ppm lead also significantly decreased the insulinogenic index (Table 5).

***b. Response of renal gluconeogenic enzymes*** Data presented in Figure 4 illustrate that the 20 ppm dose of lead significantly enhanced only the activities of renal phosphoenolpyruvate carboxykinase and pyruvate carboxylase. When the concentration of lead was increased to 40 ppm, a resultant elevation in the quartet of kidney gluconeogenic enzymes occurred. Treatment with 80 ppm metal produced a rise in phosphoenolpyruvate carboxykinase to 273%, pyruvate carboxylase to 227%, glucose 6-phosphatase to 180% and fructose 1,6-diphosphatase to 260% of control values, respectively. Figure 4 also shows that the lead-induced increase in gluconeogenic enzyme activity was related to the dose, with maximal changes produced by the highest metal concentration used.

***c. Changes in renal cyclic AMP and adenylate cyclase*** Administration of all three doses of lead increased the concentration of cyclic AMP in kidney cortex (Table 5). The metal-induced rise in cyclic nucleotide levels was related to the amount of lead ingested. Under the experimental regimen used, maximal stimulation to 195% of the control values was seen with the 80 ppm concentration. Treatment with either 20 or 40 ppm lead failed to

**Table 5.** Summary of effects of chronic oral lead

| Parameter | | Control | Lead 20 ppm | 40 ppm | 80 ppm |
|---|---|---|---|---|---|
| Body weight (g) | | 214 ± 18 | 211 ± 15 | 223 ± 9 | 206 ± 11 |
| Kidney weight (g) | | 1.85 ± 0.14 | 1.92 ± 0.13 | 1.77 ± 0.08 | 1.84 ± 0.11 |
| Liver glycogen (g/100 g) | | 3.6 ± 0.6 | 1.2 ± 0.4* | 0.7 ± 0.3* | 0.8 ± 0.2* |
| Serum urea (mg/100 ml) | | 19.2 ± 1.5 | 15.0 ± 1.1* | 19.6 ± 1.7 | 25.6 ± 1.5* |
| Blood glucose (mg/100 ml) | | 82.0 ± 4.1 | 109.1 ± 6.7* | 147.1 ± 9.2* | 179.3 ± 4.8* |
| Serum IRI ($\mu$U IRI/ml) | | 57.3 ± 1.9 | 58.7 ± 3.2 | 53.8 ± 3.6 | 48.9 ± 3.6* |
| Insulinogenic index | | 65.8 ± 3.4 | 65.3 ± 3.1 | 39.5 ± 3.3* | 27.5 ± 2.6* |
| Cyclic AMP (pmoles/mg tissue) | | 1.15 ± 0.03 | 1.29 ± 0.05* | 1.55 ± 0.06* | 2.25 ± 0.15* |
| Adenylate cyclase (pmoles cAMP formed/10 min/mg protein) | Basal | 2.53 ± 0.12 | 2.73 ± 0.15 | 2.94 ± 0.31 | 3.30 ± 0.22* |
| | +NaF | 12.89 ± 0.61† | 13.17 ± 0.79† | 14.65 ± 0.90† | 16.17 ± 1.99† |
| | +EPI | 6.69 ± 0.24† | 6.80 ± 0.39† | 7.68 ± 0.52† | 8.19 ± 1.20† |

Means ± SEM represent at least five animals in each group. Lead was administered for 8 weeks from birth. Dams received 2% lead (as lead acetate) in the drinking water; 21 days after parturition, pups were separated into random groups receiving 20, 40 or 80 ppm metal in the drinking water for the remaining 35 days.

* Significantly different from control values ($p < 0.05$)

† Statistically significant difference when compared with the respective basal adenylate cyclase activity ($p < 0.05$)

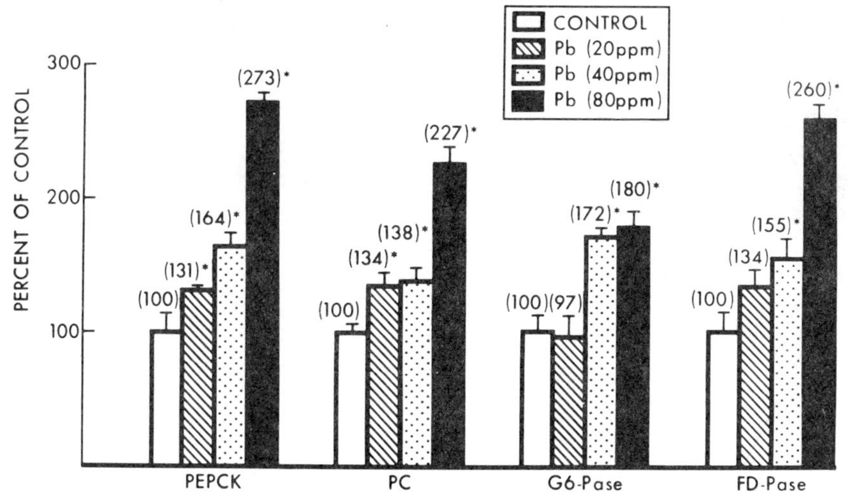


**Figure 4.** Effect of chronic lead ingestion on the quartet of key gluconeogenic enzymes in kidney cortex. Each bar represents the mean value ± SEM of at least five animals in each group. Mothers received 2% lead (as lead acetate) in their water supply for the first 21 days and, subsequently, groups of infants had access to either 20, 40 or 80 ppm heavy metal in the drinking water for an additional period of 35 days. Enzyme activities were calculated as micromoles of substrate metabolized per hour per mg protein. Data are expressed in percentages, taking the values of control animals as 100%. Asterisk indicates statistically significant difference when compared with the values of control animals ($p < 0.05$) (Stevenson et al., (1976).

alter basal adenylate cyclase, but a significant elevation in enzymic activity was noted with the 80 ppm dose. It is of interest that the activities of the fluoride- and epinephrine-stimulated forms of renal adenylate cyclase remained unchanged in metal-treated rats (Stevenson et al., 1976).

**4. Comparison between varying lead treatment regimens on glucose homeostasis** Irrespective of the duration of experiment or the mode of administration, lead was found to exert marked effects on carbohydrate metabolism. Although intraperitoneal injection of lead for 7 days failed to alter the concentration of blood glucose and liver glycogen, this treatment significantly increased the activity of renal pyruvate carboxylase, fructose 1,6-diphosphatase and glucose 6-phosphatase. When the daily injection period was prolonged over 45 days, lead elevated the concentration of blood glucose and urea, as well as reduced hepatic glycogen levels. In addition, daily administration of lead over a 45-day period produced augmentation in the activities of pyruvate carboxylase, phosphoenolpyruvate carboxykinase, fructose 1,6-diphosphatase and glucose 6-phosphatase in liver and kidney, the two major tissues which possess the complete enzymatic potential for glucose synthesis from noncarbohydrate precursors (Weber et al., 1964).

Abstinence from treatment for 28 days in rats previously given lead for 45 days reduced the heavy metal inflicted biochemical alterations. However, certain changes observed in the "withdrawn" group were still significantly different from controls, indicating the persistent nature of some of the lead-induced alterations in liver and kidney metabolism. It is interesting that chronic lead treatment increased the urinary excretion of glucose and protein (30 to 60 mg% above control), as measured by using the Dextrostix and Combistix tapes, respectively; observed glycosuria and proteinuria also persisted 4 weeks after the end of treatment (Singhal et al., 1973). Recently, Goyer (1971b) found that, whereas chronic lead treatment also increased the urinary excretion of protein, glucose and phosphate, the excretion of uric acid was decreased and the observed functional abnormalities were associated with kidney damage. Two other heavy metals, mercury and cadmium, which exert similar toxic actions on the kidney (Vallee and Ulmer, 1972) also produced marked proteinuria and glucosuria (Nomiyama et al., 1973; Friberg and Vostal, 1972). In recent reports, Singhal et al. (1974) and Merali et al. (1975) demonstrated that the organomercurial- and cadmium-induced rise in urinary excretion of glucose and protein may also have been related to stimulation in the gluconeogenic capacity of liver and kidney cortex. In contrast to the findings with lead, withdrawal from organomercurial or cadmium treatment in rats which had been given the heavy metal for 45 days was found to produce persistent changes in the quartet of renal and hepatic gluconeogenic enzymes, liver glycogen, serum urea and glucose (Singhal et al., 1974; Merali et al., 1975). Since the metabolic alterations produced by lead persisted only partially after 4 weeks of withdrawal, it is possible that the immobilization of lead in bone stores results in a lower effective toxicity of a large body burden of the heavy metal.

A more relevant experimental model of heavy metal intoxication was one in which animals were given either 20, 40 or 80 ppm lead for a period of 56 days. Chronic oral administration of lead at a concentration of 40 ppm in the drinking water produced significant reduction of liver glycogen and elevation of blood glucose, as well as augmentation of the activities of renal pyruvate carboxylase, phosphoenolpyruvate carboxykinase, fructose 1,6-diphosphatase and glucose 6-phosphatase. These lead-induced alterations in glucose synthesis were dose related; significant enhancement in kidney gluconeogenesis could be seen with a concentration as low as 20 ppm given for 56 days. Changes in glucose homeostasis as evidenced by glycosuria also have been reported by Chisolm (1962) and Goyer (1971b) following exposure to lead. Berteloot and Hugon (1975) found that incubation of rat liver homogenates with less than 0.5 mM lead nitrate enhanced the activity of glucose 6-phosphatase, although increasing the concentration of the heavy metal depressed enzymic activity. In acute studies (Filkins, 1973; Cornell and Filkins, 1974), decreased capacity of hepatic tissue slices to biotransform alanine or pyruvate to glucose was noted in animals given a 5 mg/kg dose of lead acetate. In contrast to the present study, in which lead was administered at a level as low as 20 ppm for 56 days, Hirsch (1973) observed a decrease in glucose production from pyruvate in rat kidney cortex slices, using a dietary concentration of heavy metal as high as 2%, an amount known to exceed the level to which humans may be exposed environmentally.

One of the prominent toxic manifestations of plumbism is renal damage associated with an increased urinary excretion of glucose (Chisolm, 1962). In the present study, the observed augmentation in kidney glucose synthesis may have contributed to enhanced excretion of glucose in the urine. The reduction in hepatic glycogen stores following lead treatment suggests that the process of glycogenolysis may also have provided a source for the increased levels of glucose present in blood and urine. Chronic lead exposure resulted in a dose-related increase in blood glucose and a decrease in insulinogenic index. Further, in rats subacutely exposed to lead and challenged with a glucose load, the insulin secretory response was delayed and suppressed and the ability to tolerate a glucose load was significantly impaired. These data suggest that alterations in glucose homeostasis may be related to the action of lead on the pancreas. A similar dysfunction in pancreatic activity, characterized by inhibition of the glucose-stimulated release of insulin, has been reported in animals given other heavy metals such as nickel and cadmium (Clary, 1975; Ghafghazi and Mennear, 1975). The precise mechanisms by which lead affects pancreatic function still remains to be elucidated; however, it is conceivable that this heavy metal may interfere with certain divalent ions such as zinc and calcium, known to be essential for normal pancreatic activity. Indeed, Merali and Singhal (1976) demonstrated that the concurrent administration of zinc produced a partial protection against the pancreatotoxic actions of cadmium.

Our results also demonstrate that chronic administration of lead produces an elevation in the concentration of serum urea. Since urea represents the chief metabolic product of protein and amino acid catabolism, the present data seem to be consistent with the suggestion that lead alters protein metabolism. Indeed, ultrastructural studies revealed that administration of lead produced characteristic lead-protein intranuclear inclusion bodies in kidney as well as reduced protein synthesis in rabbit reticulocytes (Farkas, 1975; Moore et al., 1973). The observed elevation of serum urea in rats chronically exposed to lead may thus be related to enhanced glucose synthesis from proteins and amino acids. Other heavy metals such as mercury and cadmium also produced an elevation in serum urea and gluconeogenesis in hepatic and renal tissue; these metal-inflicted biochemical alterations persisted even after withdrawal from the treatment for 4 weeks (Singhal et al., 1974; Merali et al., 1975).

At present, the mechanism(s) by which the kidney responds to various nephrotoxic agents such as lead remains obscure. Kirschbaum et al. (1973) noted that lead poisoning increased the activities of renal glycoprotein:glycosyltransferase, protease, acid phosphatase, N-acetyl glucosaminidase, $\alpha$-galactosidase and $\beta$-galactosidase, suggesting that acid hydrolases might participate in cellular damage to the kidney. Cellular oxidation processes and mitochondrial structures also were found to be altered in livers of lead-treated rats and rabbits (Carson et al., 1974; Vallee and Ulmer, 1972). Since energy production appears to be altered by this heavy metal (Vallee and Ulmer, 1972), it is possible that lead enhances the glucose-synthesizing enzymatic potential of both renal and hepatic tissue in order to meet the increased demand for energy. In addition, the glycosuria observed after chronic lead treatment might be related to a decreased reabsorption of glucose as a result of kidney damage (Morgan et al., 1966).

In the present study, chronic ingestion of lead via drinking water elevated the endogenous levels of cyclic AMP, augmented the activity of adenylate cyclase and enhanced the gluconeogenic capacity of the renal cortex, indicating that the observed alterations in renal carbohydrate metabolism may be related to modulation in the adenylate cyclase-cyclic AMP system. Although the ingestion of lower amounts of heavy metal tended to produce only a slight rise in adenylate cyclase activity, significant increases in endogenous cyclic AMP levels were noted in kidney cortices of rats receiving either 20 or 40 ppm lead for 56 days. A similar elevation in the endogenous concentration of hepatic cyclic AMP and adenylate cyclase activity was noted in rats chronically injected with 1.0 mg/kg of lead for 45 days. In contrast, prolonged treatment with lead via the intraperitoneal route produced a significant decrease in cyclic AMP levels of kidney cortex. In addition, subacute exposure to this metal failed to alter markedly the activity of adenylate cyclase and only slightly reduced the concentration of renal cyclic AMP. It is difficult to ascertain the reason for the observed decrease in renal cyclic AMP levels after intraperitoneal lead administration, since both lead and cyclic AMP were previously found to produce marked enhancement in renal gluconeogenesis (Sutherland and Robison, 1969). However, the possibility exists that the observed fall in kidney cortex cyclic AMP may be due to enhanced breakdown by cyclic nucleotide phosphodiesterase and/or increased urinary excretion of cyclic AMP due to tubular damage. Indeed, chronic exposure to lead in man and experimental animals has been found to produce renal tubular dysfunction (Fanconi syndrome) including glycosuria, aminoaciduria and fructosuria (Chisolm, 1962). It is of interest that treatment with cadmium, which produced distubances in kidney function characterized by glycosuria and proteinuria, also resulted in a significant depression in the concentration of renal cyclic AMP (Merali and Singhal, 1975).

## B. Influence of Lead on Parameters Associated with Tissue Growth

**1. Concentration of lead in various tissues** In recent years, it has been shown (Hardy et al., 1971; Klein, 1974) that a correlation exists between the incidence of lead poisoning and the prevalence of pica (ingestion of nonfood items) in young children. According to Klein (1974), a single chip of paint of approximately 1 square centimeter surface area contains 1.5 to 3.0 mg lead (provided the chip initially contained one coat of paint which was 10% lead by weight). Since ingestion of 150 μg of lead in paint is already in excess of an individual's maximal permissible daily intake of metal, it is not too surprising that lead tends to accumulate and induce toxicity (Klein, 1974). Since children rather than adults exhibit a tendency to consume paper, paint chips, solder from cans and dirt, investigators have recently focused their attention on the effects of exposure to low levels of this heavy metal on certain metabolic aspects of tissue growth and development in young mammals.

In any consideration of the interpolation and comparison of animal data to human lead poisoning, it is essential first to determine the concentration of this heavy metal in various tissues of both species. In the experimental model employed, Stevenson et al. (1977a) administered 50 μg of lead as lead acetate daily via gastric intubation to 1-day-old rat pups for the first 21 days after birth. These animals were subsequently weaned on Master Laboratory Chow and water containing 80 ppm lead for the next 35 days. Control rats received corresponding amounts of the vehicle (sodium acetate). The tissue lead levels at various time intervals during the treatment period are presented in Table 6. Treatment of rat pups from birth with lead revealed that the kidney retained the metal during the entire exposure period, with the highest amount being noted after 2 weeks. Although the liver contained higher levels of lead 2 weeks after intubation was started, continuation of treatment resulted in a decreased retention as compared to 2 week old animals. However, the concentration of metal retained by this tissue at 4 and 8 weeks was still significantly greater than controls. It is of interest that, in comparison to liver tissue, kidneys contained twice as much lead in 4- and 8-week-treated pups. In contrast, at none of the time intervals studied did pulmonary lead values exceed control values. Azar et al. (1973) found that treatment of rats with 60 ppm of lead for 1 year produced a threefold greater accumulation of the metal in kidney in comparison to hepatic tissue. In rats treated for 1 year with 60 ppm of dietary lead, significant amounts of metal were retained in the kidney (0.85 ppm), liver (0.31ppm), brain (0.28 ppm) and bone (17.5 ppm), while the metal content

**Table 6.** Tissue lead levels after chronic exposure

| Treatment period | Experimental group | Tissue lead content ($\mu g/g$) | | |
|---|---|---|---|---|
| | | Kidney | Liver | Lung |
| Two weeks | Control | 0.080 ± 0.010 | 0.102 ± 0.003 | 0.088 ± 0.015 |
| | Treated | 0.463 ± 0.031* | 0.402 ± 0.038* | 0.080 ± 0.005 |
| Four weeks | Control | 0.082 ± 0.010 | 0.101 ± 0.003 | 0.080 ± 0.015 |
| | Treated | 0.334 ± 0.039* | 0.169 ± 0.034* | 0.091 ± 0.007 |
| Eight weeks | Control | 0.083 ± 0.010 | 0.104 ± 0.003 | 0.088 ± 0.015 |
| | Treated | 0.362 ± 0.036* | 0.169 ± 0.007* | 0.090 ± 0.010 |

Means ± SEM represent at least eight animals in each group. Neonatal rats were given daily oral doses of 50 $\mu g$ lead for 20 days from birth. Thereafter, for the next 35 days, groups of rats were given 80 ppm lead in drinking water.

* Statistically significant difference as compared with control values ($p < 0.05$)

of pulmonary tissue was not reported. Hubermont et al. (1976) demonstrated that, in 3-week-old newborn rats which received lead via the mother 3 weeks prior to (*in utero*) and 3 weeks after birth (lactation), this metal accumulated in the kidney, while no significant amounts of lead were present in the liver. These results together with our findings suggest that, unlike liver, kidney tissue possesses a greater capacity to accumulate and store administered lead. In contrast to results in rodents, Gross et al. (1975) found that the concentration of renal and hepatic lead decreased with age in humans, yet the overall body burden of metal increased due to skeletal deposition. Although Barry (1975) failed to detect elevated levels of lead in kidney and liver with increasing age in humans, male adults were reported to have higher amounts of this metal in these tissues than females and children, suggesting that sex as well as age may play an important role in tissue lead retention. Similarly, Kostial et al. (1974) found that the toxicity to rats of lead acetate was related to sex and age, with adult males being the most susceptible.

The present investigation (Table 6) shows that, although the lung is the only organ which receives the entire output of the right heart and is thus highly susceptible to ingested chemicals, no detectable amounts of lead were present in pulmonary tissue. Examination of children with no known exposure to lead revealed the presence of trace concentrations of metal in the lung (0.1 ppm); the levels observed were similar to those obtained in other studies (Barry, 1975). Barry (1975) showed that the concentration of lead in pulmonary tissue of human infants remained unchanged, with increasing age confirming the finding (Stevenson et al., 1977a) that prolongation of exposure need not necessarily elevate the amount of metal in the lung. Why only trace amounts of lead were detected in the lung as compared to other tissues is unknown; however, it is conceivable that the mechanism(s) which clears this metal from pulmonary tissue may be more efficient than that found in kidney and liver. In contrast to lead, Kacew and Singhal (unpublished data) observed that intubation of neonates with cadmium (0.003 μg/g/day) for 2 weeks resulted in an accumulation of this metal in the lung; pulmonary cadmium content returned to control values after 8 weeks. Since the levels of lead noted by Stevenson et al. (1977a) closely resemble those previously seen in unexposed children, the possibility remains that data obtained with the use of this experimental model may indeed reflect metabolic and functional alterations which occur among humans suffering from lead intoxication.

**2. Factors involved in lead-induced changes in growth** Children are most prone to ingest lead-containing materials (Klein, 1974) and display a greater susceptibility to the toxic effects of the metal as evidenced by hematopoietic disorders, altered bone formation and hyperexcitability (Scharding and Oehme, 1973; Sassa et al., 1975; de la Burde and Choate, 1975). This knowledge prompted studies on metabolic consequences of lead in young developing animals. Pentschew and Garro (1966) reported substantially lower body weights in pups nursed by dams fed 4% lead carbonate. McLellan et al. (1974) also demonstrated a decreased body weight, abnormal motor activity and increased mortality in fetuses of mouse dams given 360 nM/g of lead. In an early study, Sauerhoff and Michaelson (1973) found that dietary administration of lead to lactating mothers resulted in a reduction of rat pup body weight and brain dopamine content, accompanied by enhanced motor activity. However, on repetition of the identical experimental protocol, lead failed to affect brain dopamine levels, but increased the concentration of norepinephrine in this tissue, suggesting that diet, rather than the nonessential element, played an important role in observed responses (Golter and Michaelson, 1975). Stevenson et al. (1976) found that daily intraperitoneal administration of lead (10 mg/kg for 7 days) resulted in enhanced kidney gluconeogenesis accompanied by a significant fall in body weight, indicating that both the metal and diet may play a role in the observed alteration in renal function. Indeed, Barltrop and Khoo (1975) demonstrated that the ability of ingested lead to reach target tissues such as bone, kidney or liver is highly dependent on diet, with greatest absorption seen in rats on a low mineral diet. It is thus not surprising that parathyroid hormone, which exerts an effect on mammalian mineral levels, was recently found to increase the concentration of renal lead (Mouw et al., 1978). The parathyroid hormone-induced enhancement in kidney lead levels appeared to be specific, as the metal content of liver, red cells and plasma was not affected by this hormone. An improper diet and lead also resulted in a dramatic decrease in mouse body growth and brain development, as well as sexual and behavioral maturation (Maker et al., 1975). Similarly, Brown (1975) suggested that effects of lead on rat behavior may be associated with failure of animals to grow due to metal-induced malnutrition, rather than being a direct effect of lead. Thus, it is not too surprising that lead poisoning occurs more frequently among children from poor families in which malnourishment is more common (Klein, 1974; Adebonojo and Strahs, 1974). Mahaffey et al. (1974) found that the lead-induced increase in kidney weight and metal accumulation was enhanced in the presence of ethanol; however, this synergistic change was attributed to a defect in nutrition. Based on these findings, it has been suggested that nutrition may be responsible for the toxicologic synergism observed between lead and alcohol seen among industrial workmen (Mahaffey et al., 1974). Unexpectedly, vitamin E, which has a stimulatory effect on heme synthesis and is the reverse of lead, was reported by Bartlett et al. (1974) to increase the concentration of lead in rabbit liver. In rats, the addition of lead acetate (1000 ppm) to drinking water for 35 days was reported to be harmless, while toxicity characterized by anemia and decreased growth rate appeared only when Purina Rat Chow was substituted for a semipurified diet, which prevented normal growth (Mylroie et al., 1978). The lead-induced decrease in mouse body growth could be reversed simply by the addition of 2% tannic acid to the diet (Peaslee and Einhellig, 1977). Singhal et al. (1973) demonstrated that the daily intraperitoneal administration of lead (1.0 mg/kg) for 45 days produced alterations in renal and hepatic metabolism without any associated significant changes in body weight. This finding was later confirmed by Brown (1975), who found that an intraperitoneal dose as high as 5 mg/kg was the maximum amount which could be administered daily consistent with a normal growth rate. In experiments using the oral route of administration, Brown (1975) established that the effects of malnutrition on body growth were minimized, provided the concentration of lead did not exceed 10 mg/100 ml in the drinking water. In agreement with these observations, Stevenson et al. (1977a) found no change in body weight of rats receiving lead from lactating mothers for the first 21 days and then subjected to 80 ppm of lead in their drinking water for the following 35 days. Since dams tend to eat less when the diet contains lead, the influence of nutrition on suckling neonates was further minimized when the lactating mother was subjected to less lead in her diet, so that consumption of metal-treated food by dams did not differ from that seen in controls (Michaelson and Sauerhoff, 1974).

However, it is conceivable that pups do not suckle at the same rate; thus, the amount of lead ingested during the lactation period could vary greatly. In an attempt to eliminate the problems associated with suckling, Golter and Michaelson (1975) administered lead directly to rat pups. They reported that, although this metal increased motor activity, no significant change was seen in body weight gain. Using this model (Dubas and Hrdina,

1978), data in Figure 5 show no marked difference in body weight gain between corresponding controls and groups of rat pups given lead for either 2, 4 or 8 weeks.

**3. Alterations in DNA synthesis** In order to assess whether lead exerts any influence on growth processes in tissues of developing rats, 1-day-old pups were given 50 μg of lead daily via gastric intubation for 3 weeks; thereafter, they received 80 ppm of metal in their drinking water for 35 days. This treatment produced no appreciable difference between corresponding controls and metal-administered animals with respect to the ability to gain weight; thus, the observed changes could be attributed to lead and not to dietary or nursing habits. The sequential alterations in the capacity of kidney, liver and lung of lead-exposed rats to incorporate labeled thymidine into DNA are illustrated in Figure 6. Lead treatment reduced the incorporation of thymidine into hepatic DNA after 2 weeks, but a return to control levels was seen at 4 weeks, followed by a 3.5-fold elevation at 8 weeks. Although metal administration failed to alter thymidine incorporation into renal DNA after 2 weeks, a reduction was noted after 4 weeks. As in the case of liver, a significant elevation in the incorporation of thymidine into kidney DNA, to 160% of control values, was observed in

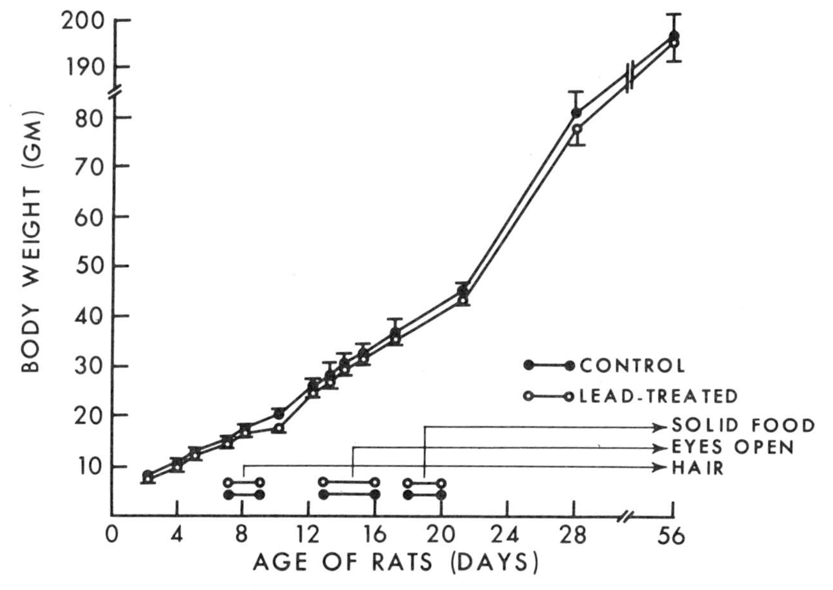


**Figure 5.** Comparison between the growth and development of control and lead-treated rats. Each point represents the mean ± SEM of at least six animals in each group. Animals were given daily oral doses of 50 μg of lead for 21 days from birth. Thereafter, for the next 35 days, groups of rats were given 80 ppm of lead in the drinking water. Animals were killed 24 hours after the cessation of treatment.

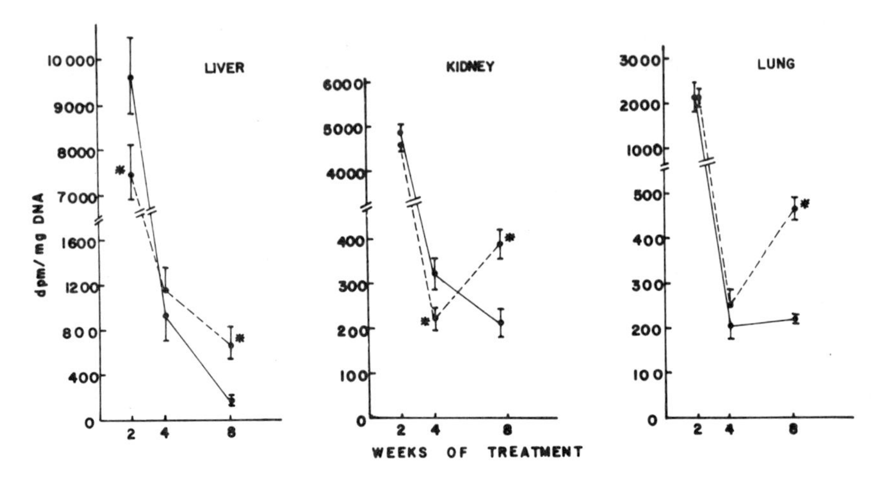


**Figure 6.** Time course of lead-inflicted changes in the incorporation of $^{14}$C-thymidine into hepatic, renal and pulmonary DNA in neonatal rats. Each point represents the mean ± SEM of at least six animals in each group. Animals were given daily oral doses of 50 $\mu$g of lead for 21 days from birth. Thereafter, for the next 35 days, groups of rats were given 80 ppm lead in the drinking water. Animals were killed 24 hours after the cessation of treatment. The incorporation of $^{14}$C-thymidine into DNA is expressed as dpm per mg DNA. Asterisks indicate statistically significant difference when compared with the values of control animals ($p < 0.05$) (Stevenson et al., (1977a).

rats treated for 8 weeks. Similarly, the incorporation of thymidine into pulmonary DNA was elevated after 8 weeks of lead treatment; however, the incorporation of thymidine into DNA remained unaltered at other time periods examined. Other heavy metals such as cobalt, nickel and zinc were found to replace magnesium ($Mg^{+2}$) in the catalysis of DNA formation by DNA polymerases of several species (Sirover and Loeb, 1976) and the quantity and fidelity DNA synthesized may be altered in the presence of metal ions other than magnesium. Since lead has been reported to produce tumors in kidney and lung (van Esch and Kroes, 1969; Kobayashi and Okamoto, 1974), the possibility exists that the lead-stimulated increase in DNA formation may reflect the synthesis of an abnormal species of DNA. In recent reports, Choie and Richter (1972a, 1972b, 1974a and 1974b) also demonstrated that in mice and rats given a single parenteral or prolonged oral dose of lead, a marked elevation resulted in the biosynthesis of renal DNA, RNA and protein. A similar augmentation in the incorporation of thymidine into mouse kidney DNA was noted by Cihak and Seifertova (1976) in mice given lead acetate by the intracardiac route. In contrast, Michaelson (1973) found that the concentration of RNA, DNA and protein remained unaltered in the brains of developing neonatal rats given lead initially via the mother and then subsequently in their diet. Treatment with silver failed to produce any significant alteration in the incorporation of thymidine into mouse renal DNA (Cihak and Seifertova, 1976). Although subacute cadmium administration resulted in a marked reduction in the capacity of kidney and liver to incorporate thymidine into DNA, significant augmentation in pulmonary DNA synthesis was found in metal-treated rats (Kacew et al., 1976a; Stoll et al., 1976). The phenomenon of heavy metal induced stimulation in the incorporation of thymidine into tissue DNA is by no means unique to lead. It has been reported in the case of kidneys obtained from folic acid-treated rats (Comber and Taylor, 1974), in livers of animals given azacytidine (Cihak and Seifertova, 1976), in renal and hepatic tissue of neonates fed chlorphentermine (Kacew et al., 1978) in parotid glands of rats given isoproterenol (Guidotti et al., 1972), as well as in lungs of animals exposed to cadmium, butylated hydroxytoluene and chlorphentermine (Kacew et al., 1976a; Witschi et al., 1976; Kacew and Narbaitz, 1977). Cell proliferation following surgical removal of one kidney or partial hepatectomy also results in enhancement of the incorporation of thymidine into renal and hepatic DNA, respectively (Hwang et al., 1974; Dicker and Shirley, 1970). Our data indicate that chronic lead administration is indeed capable of initiating renal, hepatic and pulmonary DNA synthesis; the presence of metal in the lung is not essential for the observed response. It is conceivable that the initial depression in the incorporation of thymidine into liver and kidney DNA, seen at 2 and 4 weeks, respectively, may be associated with a local reduction of blood flow to these organs.

**4. Relationship between DNA synthesis, and polyamines** Although the precise biochemical mechanism(s) underlying mammalian cell growth remains to be elucidated, evidence suggests that modulation of polyamine metabolism may be an essential prerequisite to trigger tissue development (Raina and Janne, 1975). Adminstration of testosterone or estradiol to castrated male and female rats, respectively, produced a rise in the endogenous levels of putrescine, spermidine and spermine, as well as a rise in the activities of ornithine decarboxylase and S-adenosylmethionine decarboxylase, two rate-limiting enzymes involved in polyamine synthesis, prior to the increase in accessory sex gland size and nucleic acid biosynthesis (Pegg and Williams-Ashman, 1968; Russell and Potyraj, 1972). Partial hepatectomy or unilateral nephrectomy-induced stimulation of DNA synthesis in liver and kidney also was accompanied by an augmentation in polyamine formation (Brandt et al., 1972; Russell and Snyder, 1968). A similar positive correlation between elevated incorporation of thymidine into DNA and an increase in the levels of putrescine, spermidine and spermine was found by Kacew et al. (1976a and 1977) in lungs of rat given either cadmium or chlorphentermine. In contrast, Kacew et al. (1976a; 1976b) reported a lack of any positive relationship between the effects of subacute cadmium on pancreatic and kidney DNA synthesis and polyamine metabolism. Similarly, treatment with phentermine or chlorphentermine lowered or did not affect the endogenous concentrations of renal and hepatic putrescine, spermidine and spermine, yet DNA synthesis was significantly decreased (Kacew et al., 1978). In a study of the effects of lead on neonates, Stevenson et al. (1977a) reported that, while hepatic putrescine and spermidine levels were not noticeably altered, the concentration of spermine rose after 2 weeks of lead exposure. Although liver spermine levels subsequently fell below controls after 4 weeks and were restored to control values after 2 months, it was not until the end of the 8-week treatment period that an elevation was noted in thymidine incorporation into DNA; this suggests that alterations in polyamines may precede the observed increase in DNA synthesis. In kidney, the reduction in the incorporation of thymidine into DNA noted after 4 weeks was accompanied by a fall in tissue levels of putrescine and spermidine; however, the stimulation of renal DNA formation seen at 8 weeks was associated with an elevation in the concentration of spermidine. Although the correlation between changes in renal nucleic acid synthesis and polyamine levels is not clear cut, it would appear that alterations in spermidine may play a role in DNA formation. In the lung, the concentrations of putrescine, spermidine and spermine were reduced after 4 weeks, without any concomitant depression in DNA synthesis. In contrast, pulmonary DNA synthesis was markedly stimulated after 8 weeks, at which time the polyamine levels were insignificantly different from controls, again indicating a lack of any apparent correlation between polyamine changes and nucleic acid synthesis in lead-treated rats. It is of interest that changes in polyamine biosynthesis are more closely related to the formation of RNA and protein in growing tissues (Raina and Janne, 1975). This may, in part, explain the lack of correlation between polyamine levels and DNA synthesis in liver and lung. Indeed, Fischer (1975) found that lead decreased DNA synthesis of mouse cultured fibroblast cells, yet these cells increased in size, indicating that protein formation still occurs actively and independently of DNA synthesis in the presence of lead. Thus, in the case of lead-stimulated DNA synthesis of the three tissues examined, in would seem that alterations in polyamine levels need not necessarily play a role in the observed response.

**5. Role of cyclic AMP in DNA synthesis** Since modulation in the adenylate cyclase-cyclic AMP system may be involved in the initiation of DNA synthesis in certain mammalian tissues (MacManus Whitfield, 1974), the role of this cyclic nucleotide in lead-inflicted augmentation of thymidine incorporation into hepatic, renal and pulmonary DNA was examined (Fig. 7). In three tissues examined, the endogenous levels of cyclic AMP increased approximately twofold after 4 weeks of lead treatment; the

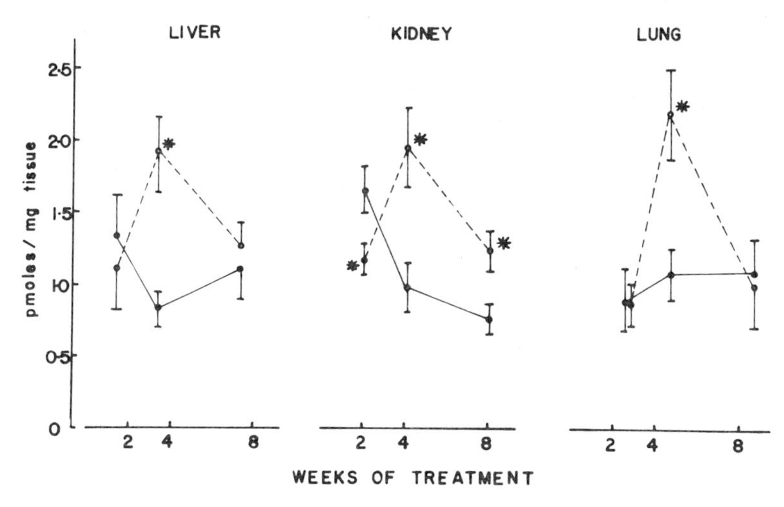


**Figure 7.** Sequential changes in hepatic, renal and pulmonary cyclic AMP levels after prolonged exposure to lead. Each point represents the mean ± SEM of at least six animals in each group. Animals were given daily oral doses of 50 μg of lead for 21 days from birth. Thereafter, for the next 35 days, groups of rats were given 80 ppm lead in the drinking water. Animals were killed 24 hours after the cessation of treatment. The endogenous concentration of cyclic AMP is expressed as pmoles per mg tissue. Asterisk indicates statistically significant difference when compared with the values of control animals ($p < 0.05$) (Stevenson et al., (1977a).

observed elevation in cyclic nucleotide content was maintained only in kidney cortex after 8 weeks. The increase in cyclic AMP levels noted in all three tissues preceded the observed enhancement in the incorporation of thymidine into DNA. In addition, the reduction of thymidine incorporation into renal DNA seen at 4 weeks was accompanied by a fall in cyclic nucleotide content in animals given lead for 2 weeks. Data thus suggest that augmentation in the incorporation of thymidine into tissue DNA is associated with alterations in cyclic AMP metabolism. The activity of pulmonary adenylate cyclase remained at control levels after 4 weeks; however, a significant fall in enzymic activity was seen after 2 or 8 weeks. It is possible that a reduction in the activity of lung cyclic nucleotide phosphodiesterase may account for the elevated concentration of cyclic AMP observed in this tissue at 4 weeks. Changes in renal and hepatic adenylate cyclase activity also preceded and accompanied the observed alterations in the incorporation of thymidine into DNA. In contrast to liver and kidney, Nathanson and Bloom (1975) demonstrated that lead produced inhibition of the activity of adenylate cyclase in rat cerebellum and suggested that alterations in cyclic AMP metabolism may be a factor in lead-induced neurologic manifestations and synaptic transmission. The observed lead-induced inhibition of rat cerebellar adenylate cyclase activity was accompanied by stimulation in the activity of phosphodiesterase. In addition, Nathanson and Bloom (1976) noted that incubation of homogenates of rat cerebral cortex, salivary gland, heart or liver with lead ions resulted in a depression of adenylate cyclase activity.

In recent years, attention also has been focused on the role of the adenylate cyclase-cyclic AMP system in the drug or surgically induced model of cell growth (MacManus and Whitfield, 1974). In regenerating rat liver following partial hepatectomy, the observed increase in DNA synthesis was preceded and accompanied by elevation in the endogenous amounts of cyclic AMP (MacManus et al., 1972). Similarly, Comber and Taylor (1974) demonstrated that the folic acid-stimulated enhancement in the incorporation of thymidine into kidney DNA was associated with a rise in tissue cyclic AMP. Further, administration of isoproterenol to rats produced an augmentation in the concentration of cyclic AMP and the activity of adenylate cyclase of the parotid gland prior to initiation of enhanced DNA synthesis (Guidotti et al., 1972; Malamud, 1969). In addition, Kacew et al. (1976a and 1977) found that the cadmium-induced elevation in the incorporation of thymidine into pulmonary and hepatic DNA was preceded and accompanied by stimulation of the adenylate cyclase-cyclic AMP system. The lead-inflicted increase in the incorporation of thymidine into renal and hepatic DNA also was associated with elevation in tissue cyclic AMP and adenylate cyclase activity. In general, Stevenson et al. (1977a) found that administration of lead enhanced the capacity of liver, kidney and lung to incorporate thymidine into DNA and produced stimulation in the adenylate cyclase-cyclic AMP system; the observed metal-inflicted responses appeared to be unrelated to alterations in polyamine metabolism.

## C. Use of Acute Treatment Lead as a Model to Study Tissue Growth

Acute treatment with lead also has been used as an experimental tool for investigating the effects of this metal on growth processes in mouse kidney. Choie and Richter (1972a, 1972b, 1974a and 1974b) demonstrated that administration of lead resulted in an elevation of the synthesis of RNA, DNA and protein in mouse and rat kidney. Similar results were obtained by Cihak and Seifertova (1976), where a single intracardiac injection of lead acetate produced an enhancement in mitotic activity and DNA synthesis of mouse kidney. In view of these findings, studies (Stevenson et al., 1977b) were undertaken to compare the influence of chronic exposure with acute lead treatment on renal, hepatic and pulmonary metabolism. A single intraperitoneal injection of lead chloride also was found to increase the incorporation of thymidine into renal DNA; this observed elevation was preceded by a rise in the incorporation of orotic acid into RNA (Figs. 8 and 9). Although acute lead administration initially enhanced the formation of

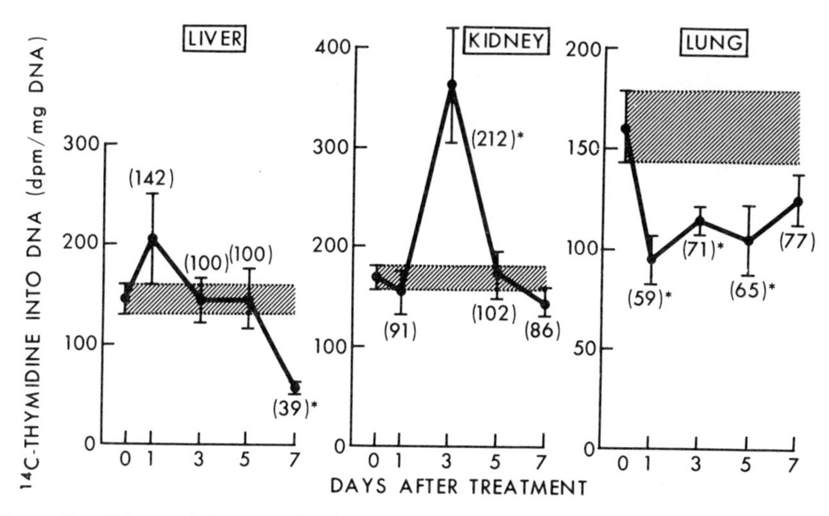


**Figure 8.** Effects of lead on the incorporation of $^{14}$C-thymidine into DNA in several tissues. Each point represents the mean ± SEM of at least five animals in each group. Rats were administered a single intraperitoneal dose of lead (10 mg/kg) and killed 1, 3, 5 or 7 days thereafter. The mean ± SEM of 24 vehicle-treated animals is shown at point zero and is represented by the area between the two dotted lines. The incorporation of $^{14}$C-thymidine into hepatic, renal and pulmonary DNA is expressed as dpm per mg DNA. Data are also given as percentages (in parentheses), with the values of control animals taken as 100%. Asterisk indicates statistically significant difference when compared with the values of control animals ($p < 0.05$). (Stevenson et al., (1977b).

RNA in rat liver and lung, the incorporation of thymidine into hepatic and pulmonary DNA was markedly reduced after metal treatment. Interestingly, addition of lead to mouse fibroblast cells *in vitro* also resulted in decreased DNA synthesis and mitotic rate, suggesting that lead exerts its blockade at the GI phase of the cell cycle (Fischer, 1975). In addition, the initial increase in the incorporation of orotic acid into hepatic and pulmonary RNA was followed by a fall in RNA synthesis on the fifth and seventh day after lead treatment, respectively. Using E. Coli RNA polymerase, calf thymus DNA and phage T4 DNA as templates, the addition of lead to the incubation medium also resulted in a decrease in overall RNA synthesis (Hoffman and Niyogi, 1977). It is possible that the observed reduction in hepatic and pulmonary DNA formation is associated with a decrease in local circulation to these organs or with a decreased capacity of the cells to incorporate thymidine into DNA.

**1. Relationship between nucleic acid formation and cyclic AMP** In order to examine the role of the cyclic AMP system in the initiation of renal nucleic acid synthesis, the influence of lead treatment on the concentration of this cyclic nucleotide was examined. Stevenson et al. (1977b) were unable to observe any correlation between lead-inflicted alterations in DNA syn-

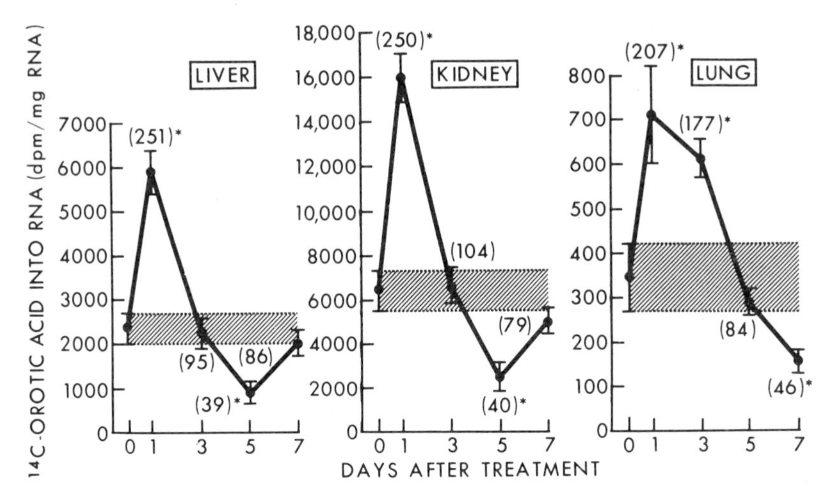


**Figure 9.** Influence of lead on incorporation of $^{14}C$-orotic acid into hepatic, renal and pulmonary RNA. Each point represents the mean ± SEM of at least five animals in each group. Rats were given a single intraperitoneal dose of lead (10 mg/kg) and killed 1, 3, 5 or 7 days thereafter. The mean ± SEM of 24 vehicle-treated animals is shown at point zero and is represented by the area between the two dotted lines. The incorporation of $^{14}C$-orotic acid into hepatic, renal and pulmonary RNA is expressed as dpm per mg RNA. Data are also given as percentages (in parentheses), with the values of control animals taken as 100%. Asterisk indicates statistically significant difference when compared with the values of control animals ($p < 0.05$). (Stevenson et al., (1977b).

thesis and the endogenous concentration of cyclic AMP in kidney as well as lung. The incorporation of thymidine into renal DNA was elevated after 3 days, but the concentration of cyclic AMP in this tissue remained depressed from the third through the seventh day of the experiment. Similarly, while pulmonary cyclic AMP levels were elevated 24 hours after lead administration, the formation of DNA in this organ was reduced throughout the treatment period. Since various investigators have reported that alterations in the adenylate cyclase-cyclic AMP system may not necessarily act as a trigger for DNA synthesis (Witschi et al., 1976; Pastan et al., 1975), it can be concluded that changes in nucleic acid formation in kidney and lung may be independent of modulation in the adenylate cyclase-cyclic AMP system. In contrast, the observed reduction in the incorporation of thymidine into hepatic DNA was preceded by a fall in the endogenous concentration of cyclic AMP in this tissue. This indicates that the effects of a single treatment with lead on hepatic cell proliferation may be associated with alterations in the cellular levels of cyclic AMP. It is of interest that this heavy metal was found to inhibit the activities of hepatic adenylate cyclase and cyclic AMP phosphodiesterase at very low concentrations ($I_{50} < 3\ \mu M$) *in vitro* (Nathanson and Bloom, 1976). In contrast, acute intravenous administration of lead failed to produce any apparent change in the endogenous concentration of cyclic AMP in rat liver *in vitro* for as long as 4 hours (Gartner, 1975). Similarly, our data (Stevenson et al., 1977b) show that hepatic and renal cyclic AMP levels were not reduced until the third day after lead administration. In contrast, the concentration of pulmonary cyclic AMP rose markedly 1 day after lead treatment. It seems evident that the varying influence of acute lead on the adenylate cyclase-cyclic AMP system of different tissues may be more complex than initially proposed by Choie and Richter (1974a and 1974b).

**2. Role of polyamines in nucleic acids synthesis** More recently, a model has been suggested which indicates that the process of cellular growth in mammals may involve stimulation of both the adenylate cyclase-cyclic AMP system and polyamine formation (Manen et al., 1978). The lead-inflicted reduction in hepatic and pulmonary nucleic acid synthesis was preceded and accompanied by a fall in the endogenous concentration of putrescine, spermidine and spermine, suggesting that a positive correlation may exist between DNA formation and polyamine levels (Table 7). In contrast, the polyamine levels remained unaltered in kidney, except for a slight yet significant elevation in spermine content which did not occur until after the lead-induced stimulation in RNA (1 day) and DNA synthesis (3 days). This indicates that the observed alterations in renal nucleic acid metabolism may be independent of changes in polyamine content. Lead treatment enhanced the formation of a specific binding protein in mouse kidney (Choie et al., 1975) and polyamine metabolism is predominately related to protein synthesis (Raina and Janne, 1975); therefore, it is possible that the observed elevation in spermine may reflect increased formation of this metal-binding protein in kidney tissue. Indeed, lead administration produced an increase in renal protein levels from the third through the seventh day of the experimental period. Other heavy metals such as mercury and cadmium also enhanced the synthesis of a specific metal-binding protein (metallothionein) in rat liver and kidney (Kimura et al., 1974; Piscator, 1964).

## II. SUMMARY AND CONCLUSIONS

The present study shows that exposure to lead disturbs glucose homeostasis and affects certain parameters involved in tissue growth. Daily administration of lead increased the process of gluconeogenesis in rat liver and kidney cortex, elevated blood glucose and serum urea and reduced the concentration of hepatic glycogen. Chronic lead treatment also produced a marked rise in liver cyclic AMP and adenylate cyclase, indicating that the observed hyperglycemia may be related to enhanced gluconeogenesis mediated through changes in the adenylate cyclase-cyclic AMP system. Discontinuation of lead administration in rats pretreated with the heavy metal failed to reverse fully the observed changes in hepatic and renal metabolism.

**Table 7.** Influence of acute lead treatment on renal, hepatic and pulmonary polyamines

| Tissue examined | Polyamine | Days after treatment | | | | |
|---|---|---|---|---|---|---|
| | | 0 | 1 | 3 | 5 | 7 |
| Kidney | Spermine | 137.0 ± 19.2 | 92.0 ± 11.8 | 160.0 ± 34.9 | 191.7 ± 1.5 | 191.1 ± 26.3 |
| | Spermidine | 18.5 ± 1.6 | 15.5 ± 1.1 | 17.9 ± 2.3 | 16.1 ± 2.4 | 17.1 ± 1.2 |
| | Putrescine | 10.8 ± 1.5 | 9.1 ± 0.7 | 9.8 ± 0.4 | 10.0 ± 1.5 | 9.3 ± 0.9 |
| Liver | Spermine | 218.4 ± 14.8 | 190.7 ± 10.4 | 181.2 ± 12.7 | 173.4 ± 5.0* | 189.7 ± 6.6 |
| | Spermidine | 46.2 ± 4.4 | 46.7 ± 2.2 | 43.8 ± 2.1 | 35.0 ± 3.3 | 29.4 ± 1.5* |
| | Putrescine | 128.1 ± 13.6 | 76.5 ± 6.9* | 75.7 ± 7.4* | 87.7 ± 11.5* | 85.1 ± 8.3* |
| Lung | Spermine | 266.0 ± 47.6 | 181.8 ± 45.3 | 215.5 ± 47.9 | 113.1 ± 19.0* | 103.5 ± 11.9* |
| | Spermidine | 6.4 ± 0.7 | 4.7 ± 1.1 | 5.0 ± 0.8 | 3.0 ± 0.2* | 2.6 ± 0.2* |
| | Putrescine | 19.5 ± 0.9 | 9.9 ± 1.2* | 10.0 ± 0.3* | 6.7 ± 0.2* | 6.3 ± 0.3* |

Means ± SEM represent at least five animals in each group. Rats were injected intraperitoneally with a single dose of lead (10 mg/kg) and groups were killed 1, 3, 5 or 7 days after treatment. The mean ± SEM of 24 vehicle-treated animals is given at time zero. Polyamines are expressed as picomoles per mg tissue.

* Significantly different from control values ($p < 0.05$)

In order to simulate a more physiologic mode of lead exposure in humans, neonatal rats were given the heavy metal either indirectly via the mothers' milk or by gastric intubation. Data demonstrate that, although lead failed to alter markedly the growth rate of newborn rats, there was resultant hyperglycemia, hypoinsulinemia, uremia and diminution of hepatic glycogen. In addition, ingestion of lead produced an elevation in the activities of the quartet of renal gluconeogenic enzymes, cyclic AMP levels and adenylate cyclase. In general, lead ingestion enhanced the incorporation of thymidine into renal, hepatic and pulmonary DNA; the observed alterations in DNA synthesis were preceded in all cases by corresponding changes in the endogenous concentration of cyclic AMP. Generally, the levels of putrescine, spermidine and spermine were unaltered or decreased in the three tissues examined. In liver and kidney, the lead-inflicted biochemical disturbances were associated with a significant retention of this heavy metal, with maximal accumulation being noted 2 weeks after exposure. In contrast, lead treatment produced a variety of metabolic alterations in rat lung in the absence of any apparent retention of the metal by this organ, suggesting that the presence of this element in pulmonary tissue need not necessarily constitute an absolute requirement for lead to inflict the observed disturbances.

The influence of a single injection of lead on nucleic acid levels and synthesis was also examined, as well as its effect on the endogenous concentrations of cyclic AMP and polyamines in liver, kidney and lung. Lead increased the incorporation of thymidine into renal DNA, but a significant reduction was seen in DNA formation by hepatic and pulmonary tissue. The incorporation of orotic acid into renal, hepatic and pulmonary RNA was initially elevated but subsequently fell below control levels. Although lead administration generally failed to alter kidney polyamine levels, the concentration of putrescine, spermidine and spermine decreased both in lung and liver. Reduction in the incorporation of thymidine into hepatic DNA was associated with a decrease in the concentration of cyclic AMP; however, lead-induced alterations in the formation of renal and pulmonary DNA appeared to to independent of changes in cyclic nucleotide levels.

The present investigation suggests that the chronic exposure of young animals to lead provides a good experimental tool for further elucidating the effects of this heavy metal on glucose homeostasis and tissue growth and development.

## ACKNOWLEDGMENTS

This work was supported by grants from the Medical Research Council of Canada. Some of the original work reviewed in this article, to which appropriate reference is made in the text, was carried out by Miss A. Stevenson in partial fulfillment of the requirements leading to the degree of Master of Science in Pharmacology. The authors thank Mr. S. Klosevych, Chief of Medical Communication Services of the University of Ottawa, and his staff for the preparation of various illustrations. Finally, we are indebted to Mrs. Diane McNeil for expert editorial assistance in the assembling of this manuscript.

## REFERENCES

Adebonojo, F.O., and Strahs, S. (1974). *Clin. Pediat. 13:*310.
Azar, A., Trochimowicz, H.J., and Maxfield, M.E. (1973). U.S. Environmental Protection Agency, In "International Symposium: Environmental Health Aspects of Lead," p. 199. CID, Luxembourg.
Barltrop, D., and Khoo, H.E. (1975). *Postgrad. Med. J. 51:*795.
Barry, P.S.I. (1975). *Brit. J. Indust. Med. 32:*119.
Bartlett, R.S., Rousseau, J.E., Frier, H.I., and Hall, R.C. Jr. (1974). *J. Nutrit. 104:*1637.
Berteloot, A., and Hugon, J.S. (1975). *Histochemistry 43:*197.
Bitensky, M.W., Russell, V., and Robertson, W. (1968). *Biochem. Biophys. Res. Commun. 31:*706.
Brandt, J.T., Pierce, D.A., and Fausto, N. (1972). *Biochim. Biophys. Acta. 279:*184.
Brown, D.R. (1975). *Toxicol. Appl. Pharmacol. 32:*628.
Carson, T.L., Van Gelder, G.A., Karas, G.C., and Buck, W.B. (1974). *Arch. Environ. Health 29:*154.
Chisolm, J.J. (1962). *J. Pediat. 60:*1.
Choie, D.D., and Richter, G.W. (1972a). *Amer. J. Pathol. 66:*265.
Choie, D.D., and Richter, G.W. (1972b). *Amer. J. Pathol. 68:*359.
Choie, D.D., and Richter, G.W. (1974a). *Lab. Invest. 30:*647.
Choie, D.D., and Richter, G.W. (1974b). *Lab. Invest. 30:* 652.
Choie, D.D., Richter, G.W., and Young, L.B. (1975). *Beitr. Path. Bd. 155:*197.
Cihak, A., and Seifertova, M. (1976). *Chem.-Biol. Interact. 13:*141.
Clary, J.J. (1975). *Toxicol. Appl. Pharmacol. 31:*55.
Comber, H.J., and Taylor, D.M. (1974). *Biochem. Soc. Transact. 2:*74.
Cornell, R.P., and Filkins, J.P. (1974). *Proc. Soc. Exp. Biol. Med. 147:*371.
Crabtree, B., Higgins, S.J., and Newsholme, E.A. (1972). *Biochem. J. 130:*391.
de la Burde, B., and Choate, M.S. (1975). *J. Pediat. 87:*638.
Dicker, S.E., and Shirley, D.G. (1970). *J. Physiol. 210:*53 P.
Dubas, T.C., and Hrdina, P.D. (1978). *J. Environ. Pathol. Toxicol. 2:*473.
Farkas, W.R. (1975). *Res. Commun. Chem. Pathol. Pharmacol. 10:*127.
Filkins, J.P. (1973). *Proc. Soc. Exp. Biol. Med. 142:*915.
Fischer, A.B. (1975). *Zbl. Bakt. Hyg., I. Abt. Orig. B 161:*26.
Friberg, L., and Vostal, J. (1972). In "Mercury and the Environment." Chemical Rubber Company Press, Cleveland, Ohio.
Gartner, S.L. (1975). *Experientia 31:*566.
Ghafghazi, T., and Mennear, J.H. (1975). *Toxicol. Appl. Pharmacol. 31:*134.
Golter, M., and Michaelson, I.A. (1975). *Science 187:*359.
Goyer, R.A. (1971a). *Curr. Top. Pathol. 55:*147.
Goyer, R.A. (1971b). *Amer. J. Pathol. 64:*167.

Gross, S.B., Pfitzer, E. A. Yeagar, D.W., and Kehoe, R.A. (1975). *Toxicol. Appl. Pharmacol. 32:*638.
Guder, W., Wiesner, W., Stukowski, B., and Wieland, O. (1971). *Hoppe-Seyler's Z. Physiol. Chem. 352:*1319.
Guidotti, A., Weiss, B., and Costa, E. (1972). *Molec. Pharmacol. 8:*521.
Hardy, H.L., Chamberlin, R.I., Maloof, C.C., Boylin, G.W., and Howell, M.C. (1971). *Clin. Pharmacol. Ther. 12:*982.
Henning, A., Seiffert, I., and Seubert, W. (1963). *Biochem. Biophys. Acta. 77:*345.
Hirsch, G.H. (1973). *Toxicol. Appl. Pharmacol. 25:*84.
Hoffman, D.J., and Niyogi, S.K. (1977). *Science 198:*513.
Hubermont, G., Buchet, J.P., Roels, H., and Lauwerys, R. (1976). *Toxicology 5:*379.
Hwang, K.M., Murphree, S.A., Shansky, C.W., and Sartorelli, A.C. (1974). *Biochim. Biophys. Acta. 366:*143.
Kacew, S., Dubas, T.C., and Stevenson, A.J. (1978). *Toxicology 10:*77.
Kacew, S., Merali, Z., and Singhal, R.L. (1976a). *Toxicol. Appl. Pharmacol. 38:*140.
Kacew, S., Merali, Z., and Singhal, R.L. (1976b). *Gen. Pharmacol. 7:*433.
Kacew, S., Merali, Z., Thakur, A.N., and Singhal, R.L. (1977). *Can. J. Physiol. Pharmacol. 55:*508.
Kacew, S., and Narbaitz, R. (1977). *Exp. Molec. Pathol. 27:*106.
Kacew, S., and Singhal, R.L. (1974). *Biochem. J. 142:*145.
Kimura, M., Otaki, N., Yoshiki, S., Suzuki, M., Horiuchi, N., and Suda, T. (1974). *Arch. Biochem. Biophys. 165:*340.
Kirschbaum, B.B., Zoltick, P.W., and Bossmann, H.B. (1973). *Res. Commun. Chem. Pathol. Pharmacol. 5:*441.
Klein, R. (1974). *Pediat. Clin. North Amer. 21:*277.
Kobayashi, N., and Okamoto, T. (1974). *J. Nat. Cancer Inst. 52:*1605.
Kostial, K., Malijkovic, T., and Jugo, S. (1974). *Arch. Toxicol. 31:*265.
MacManus, J.P., Franks, D.J., Youdale, T., and Braceland, B.M. (1972). *Biochem. Biophys. Res. Commun. 49:*1201.
MacManus, J.P., and Whitfield, J.F. (1974). *Prostaglandins* **6:**475.
Mahaffey, K.R., Goyer, R.A., and Wilson, M.H. (1974). *Arch. Environ. Health 28:*217.
Maker, H.S., Lehrer, G.M., and Silides, D.J. (1975). *Environ. Res. 10:*76.
Makman, M.H., and Sutherland, E.W. (1964). *Endocrinology 75:*127.
Malamud, D. (1969). *Biochem. Biophys. Res. Commun. 35:*754.
Manen, C.A., Costa, M., Sipes, I.G., and Russell, D.H. (1978). *Biochem. Pharmacol. 27:*219.
McLellan, J.S., vonSmolinski, A.W., Bederka, J.P., Jr., and Boulos, B.M. (1974). *Fed. Proc. 33:*288.
Merali, Z., Kacew, S., and Singhal, R.L. (1975). *Can. J. Physiol. Pharmacol. 53:*174.
Merali, Z., and Singhal, R.L. (1975). *Toxicology 4:*207.
Merali, Z., and Singhal, R.L. (1976). *Brit. J. Pharmacol. 57:*573.
Michaelson, I.A. (1973). *Toxicol. Appl. Pharmacol. 26:*539.
Michaelson, I.A., and Sauerhoff, M.W. (1974). *Toxicol. Appl. Pharmacol. 28:*88.
Moore, J.F., Goyer, R.A., and Wilson, M. (1973). *Lab. Invest. 29:*488.
Morgan, J.M., Hartley, M.W., and Miller, R.E. (1966). *Arch. Intern. Med. 118:*7.
Mouw, D.R., Wagner, J.G., Kalitis, K., Vander, A.J., and Mayor, G.H. (1978). *Environ. Res. 15:*20.

Murad, F., Chi, Y.M., Rall, T.W., and Sutherland, E.W. (1962). *J. Biol. Chem. 237:*1233.
Mylroie, A.A., Moore, L., Olyai, B., and Anderson, M. (1978). *Environ. Res. 15:*57.
Nathanson, J.A., and Bloom, F.E. (1975). *Nature 255:*419.
Nathanson, J.A., and Bloom, F.E. (1976). *Molec. Pharmacol. 12:*390.
Nomiyama, K., Sato, C., and Yamamoto, A. (1973). *Toxicol. Appl. Pharmacol. 24:*625.
Pastan, I.H., Johnson, G.S., and Anderson, W.B. (1975). *Ann. Rev. Biochem. 44:*491.
Peaslee, M.H., and Einhellig, F.A. (1977). *Experientia 33:*1206
Pegg, A.E., and Williams-Ashman, H.G. (1968). *Biochem. J. 109:*32P.
Pentschew, A., and Garro, F. (1966). *Acta Neuropathol. 6:*266.
Perkins, J.P. (1973). P. Greengard and G.A. Robison (Eds.), In "Advances in Cyclic Nucleotide Research," Vol. 3, p. 1. Raven Press, New York.
Piscator, M. (1964). *Nord. Hyg. Tidskr. 45:*76.
Pohl, S.L., Birnbaumer, L., and Rodbell, M. (1969). *Science 164:*566.
Raina, A., and Janne, J. (1975). *Med. Biol. 53:*121.
Russell, D.H., and Potyraj, J.J. (1972). *Biochem. J. 128:*1109.
Russell, D.H., and Snyder, S.H. (1968). *Proc. Nat. Acad. Sci. (U.S.A.) 60:*1420.
Sassa, S., Granick, S., and Kappas, A. (1975). *Ann. N.Y. Acad. Sci. 244:*414.
Sauerhoff, M.W., and Michaelson, I.A. (1973). *Science 182:*1022.
Scharding, N.N., and Oehme, F.W. (1973). *Clin. Toxicol. 6:*419.
Singhal, R.L., Kacew, S., Sutherland, D.J.B., and Telli, A.H. (1973). *Res. Commun. Chem. Pathol. Pharmacol. 6:*951.
Singhal, R.L., Merali, Z., Kacew, S., and Sutherland, D.J.B. (1974). *Science 183:*1094.
Sirover, M.A., and Loeb, L.A. (1976). *Biochem. Biophys. Res. Commun. 70:*812.
Stevenson, A., Merali, Z., Kacew, S., and Singhal, R.L. (1976). *Toxicology 6:*265.
Stevenson, A.J., Kacew, S., and Singhal, R.L. (1977a). *Toxicol. Appl. Pharmacol. 40:*161.
Stevenson, A.J., Kacew, S., and Singhal, R.L. (1977b). *J. Toxicol. Environ. Health 2:*1125.
Stoll, R.E., White, J.F., Miya, T.S., and Bousquet, W.F. (1976). *Toxicol. Appl. Pharmacol. 37:*61.
Sutherland, E.W., and Robison, G.A. (1969). *Diabetes 18:*797.
Vallee, B.L., and Ulmer, D.D. (1972). *Ann. Rev. Biochem. 41:*91.
Van Esch, G.J., and Kroes, R. (1969). *Brit. J. Cancer 23:*765.
Weber, G., Singhal, R.L., Stamm, N.B., Fisher, E., and Mentendick, M.A. (1964). *Adv. Enzyme Regul. 2:*1.
Weiss, I.W., Morgan, K., and Phang, J.M. (1972). *J. Biol. Chem. 247:*760.
Witschi, H., Kacew, S., Tsang, B.K., and Williamson, D. (1976). *Chem.-Biol. Interact. 12:*29.

# Lead and Heme Biosynthesis

*Michael R. Moore, Peter A. Meredith and Abraham Goldberg*

**TABLE OF CONTENTS**

## I. INTRODUCTION

The effect of lead on heme biosynthesis has been long recognized. Indeed, a great similarity has been noted between some aspects of lead poisoning and the symptomatology of acute porphyrias. Hematoporphyrinuria was noted by Stokvis (1895), Perutz (1910) and Goetzl (1911) in lead-poisoned rabbits. Laennec (1831) noted that patients who died of lead poisoning showed thinness of blood and pallor of tissues on postmortem examination. Binnendijk (cf. Stokvis, 1895) first noted the occurrence of excess porphyrin in the urine of a patient with lead intoxication and in 1927 Leibig suggested that the rise in porphyrin excretion during lead poisoning was due to the action of lead upon hemoglobin synthesis in the bone marrow. Groetpass (1932) showed that enhanced urinary porphyrins were due principally to the elevation in coproporphyrin series III isomer. More recently, Haeger (1957) found excessive urinary excretion of δ-aminolevulinic acid during lead intoxication. It is clear from these early pointers that lead has very dramatic effects upon the biosynthesis of heme.

## II. THE HEME BIOSYNTHETIC PATHWAY

The origins of heme biosynthesis lie in the citric acid cycle and in normal $C_1$ amino acid metabolism. The elucidation of the early steps of the heme biosynthetic pathway was the result of an elegant set of isotopic tracer studies carried out by Shemin (1955). In the initial stage of the pathway, succinate as succinyl CoA combines with glycine, under the aegis of the initial enzyme of the pathway, δ-aminolevulinic acid synthase (EC 2,3,1,37) and the cofactor pyridoxal phosphate to form the first of the intermediates of the pathway, δ-aminolevulinic acid. In the next stage of the biosynthetic pathway two moles of δ-aminolevulinic acid condense to form the monopyrrole, porphobilinogen, by the action of the enzyme δ-aminolevulinic acid dehydratase (porphobilinogen synthase) (EC 4,2,1,24). In the next stage of biosynthesis, four moles of porphobilinogen condense, cyclize and isomerize to form the first of the porphyrins, uroporphyrinogen, as the three isomer. This is carried out by the concerted action of two enzymes, uroporphyrinogen 1 synthase (porphobilinogen deaminase) (EC 4,3,1,8), and uroporphyrinogen cosynthase. In circumstances where there is a defi-

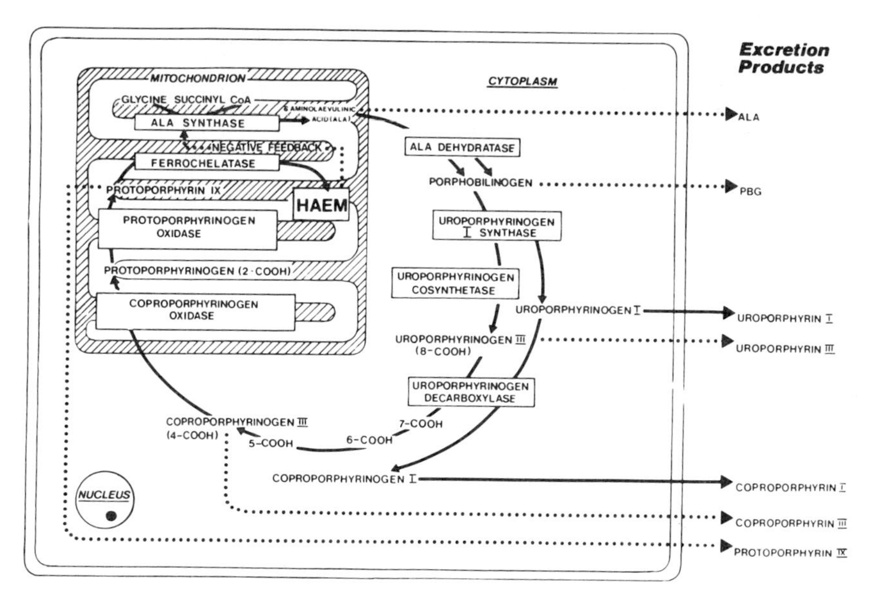


**Figure 1.** The heme biosynthetic pathway. Heme biosynthesis within the cell starts in the mitochondrion with the initial and rate-limiting enzyme ALA synthase. Synthesis then passes into the cytoplasm and returns once again to the mitochondrion for the final three stages in the pathway. Control of the pathway is by negative feedback by heme upon ALA synthase. The enzymes of the pathway are shown within the open squares and the intermediates of the pathway follow the heavy arrows to the final product, heme. Each of the intermediates may be used in the subsequent production of the following component of the pathway, excepting the series I porphyrinogens, uroporphyrinogen I and coproporphyrinogen I, which may not be utilized beyond the stage of coproporphyrinogen I and are therefore obligatorily excreted. Abbreviations: ALA, δ-aminolevulinic acid; PBG, porphobilinogen; COOH, carboxyl.

ciency of uroporphyrinogen cosynthase, series I isomer is formed. By successive decarboxylations of the octacarboxyllic porphyrin ring through the hepta, hexa and penta-carboxylic porphyrins, the tetracarboxylic coproporphyrin is formed by the reaction catalyzed by the enzyme uroporphyrinogen decarboxylase (EC 4,1,1,37). At this stage the reaction re-enters the mitochondrion, where coproporphyrinogen III is decarboxylated by coproporphyrinogen oxidase (EC 1,3,3,3) to form protoporphyrinogen IX, which is a dicarboxylic divinyl porphyrin. In the reducing milieu of the mitochondrion, protoporphyrinogen IX is oxidized to protoporphyrin by protoporphyrinogen oxidase. The final step of the pathway is the insertion of ferrous iron into protoporphyrin by ferrochelatase (EC 4,99,1,1) to form heme (Fig. 1). The one other component of this sequence that should be mentioned is the first of the biodegradative steps of heme. In this, by splitting of the heme ring with loss of iron, heme oxygenase (EC 1,14,99,3) produces bilirubin (Table 1).

**Table 1.** The enzymes of heme biosynthesis and biodegradation

| Common name(s) | Abbreviation | Systematic name | Enzyme classification nomenclature |
|---|---|---|---|
| δ-Aminolevulinate synthase | ALA.S | Succinyl-CoA: Glycine C-Succinyltransferase | 2.3.1.37 |
| Porphobilinogen synthase δ-Aminolevulinate dehydratase) | ALA.D | 5 Aminoevulinate hydro-lyase | 4.2.1.24 |
| Uroporphyrinogen 1 synthase) Porphobilinogen deaminase) | URO.S | Porphobilinogen ammonia-lyase | 4.3.1.8 |
| Uroporphyrinogen 3 cosynthetase | URO.CoS | Unclassified | |
| Uroporphyrinogen decarboxylase | URO.D | Uroporphyrinogen 3 carboxy-lyase | 4.1.1.37. |
| Coproporphyrinogen oxidase) Coproporphyrinogenase) | COPRO.O | Coproporphyrinogen: Oxygen oxidoreductase | 1.3.3.3 |
| Protoporphyrinogen oxidase | PROTO.O | Unclassified | |
| Ferrochelatase) Heme synthetase) | FERRO.C | Protoheme ferro-lyase | 4.99.1.1. |
| Heme-oxygenase | HEME.O | Heme: NADPH: Oxygen oxido-reductase | 1.14.99.3 |

### A. Delta-aminolevulinic Acid Synthase

The first step of the heme biosynthetic pathway, the formation of δ-aminolevulinic acid (ALA) from succinyl-CoA and glycine by the mitochondrial enzyme δ-aminolevulinic acid synthase represents the control point of the pathway. The work of Shemin and Rittenberg (1945) revealed the basis of the reaction, which is the condensation of glycine and succinyl-CoA by ALA synthase. The fact that pyridoxal phosphate is essential for reaction was first demonstrated by Wintrobe (1950). The enzyme-bound pyridoxal phosphate activates the methylene group of glycine preliminary to Schiff base formation. Removal of a proton from this Schiff base gives rise to a carbanion intermediate which reacts with succinyl-CoA to produce $\alpha$ amino $\beta$ ketoadipic acid, which is linked to the pyridoxal phosphate-enzyme complex. There are two possible mechanisms for the production of ALA from this complex and the reaction may proceed either way. In the first, a decarboxylation process gives an ALA-pyridoxal phosphate-enzyme complex which, on hydrolysis, yields free ALA. The alternative mechanism involves the hydrolysis of the complex to release $\alpha$ amino $\beta$ ketoadipic acid which readily decarboxylates nonenzymically to ALA (Akhtar et al., 1976).

### B. Delta-aminolevulinic Acid Dehydratase

The second enzyme of the heme biosynthetic pathway, δ-aminolevulinic acid dehydratase (ALA.D), catalyzes the condensation of two molecules of ALA to form the monopyrrole porphobilinogen (PBG). The mechanism of action of ALA dehydratase has been elucidated by Shemin (1968). The sequence is that one molecule of ALA forms a Schiff base with the enzyme. This is followed by a nucleophilic attack by the enzyme-ALA anion on the carbonyl group of a second ALA molecule. This eliminates water. Then a free amino group of the second molecule displaces the amino group of the enzyme by transamination or transaldimination to form PBG.

### C. Uroporphyrinogen 1 Synthase and Uroporphyrinogen Cosynthase

The third step of the pathway involves the formation of uroporphyrinogen III by two cytoplasmic enzymes, uroporphyrinogen 1 synthase and uroporphyrinogen III cosynthase. Uroporphyrinogen (URO) synthase polymerises four PBG molecules on the enzyme surface and forms uroporphyrinogen 1. Unlike uroporphyrinogen III, which is the precursor of heme, uroporphyrinogen 1 is an unusable intermediate of the porphyrin biosynthetic pathway and is only found in its oxidized form (uroporphyrin 1) in the excreta of humans and animals. Uroporphyrinogen III cosynthase enters into association with URO synthase and acts as a "specifier protein" of the latter, changing the mode of PBG condensation on the enzyme surface, while URO synthase catalyzes the head-to-tail condensation of the porphobilinogen molecules. The action of both URO synthase and uroporphyrinogen III cosynthase catalyses the head-to-head condensation of the porphobilinogen molecules. This results in the formation of an altered enzyme-bound dipyrrylmethane which is specifically incorporated into uroporphyrinogen III (Frydman et al., 1976; Battersby and McDonald, 1976).

### D. Uroporphyrinogen Decarboxylase

The final cytoplasmic step of heme biosynthesis involves the decarboxylation of the octa-carboxylic uroporphyrinogen to the tetra carboxylic coproporphyrinogen III. The decarboxylation catalyzed by uroporphyrinogen (URO) decarboxylase is a stepwise process, the acetic acid groups being decarboxylated in a sequential clockwise fashion starting with that on ring D and followed by those of the A, B and C rings (Jackson et al., 1976). The specificity of URO decarboxylase is for the series III isomer of uroporphyrinogen which is decarboxylated twice as rapidly as the series I isomer, but it will decarboxylate the series I isomer to produce coproporphyrinogen 1 which is excreted as coproporphyrin 1 (Fig. 2).

### E. Coproporphyrinogen Oxidase

At this stage in the pathway the reaction series re-enters the mitochondrion for the conversion of coproporphyrinogen III to protoporphyrinogen IX by the decarboxylation of two propionic side chains to vinyl groups. The enzyme coproporphyrinogen oxidase catalyzes the conversion of coproporphyrinogen III to protoporphyrinogen IX. The mechanism for this reaction is not altogether clear, but the evidence is compatible with three closely related mechanistic alternatives (Akhtar et al., 1976), whereby there is a stepwise decarboxylation of the propionyl groups to vinyl groups with a monovinyl porphyrinogen existing as an intermediate.

### F. Protoporphyrinogen Oxidase

Until recently, indirect evidence suggested that the oxidative conversion of protoporphyrinogen IX to protoporphyrin IX in the reducing milieu of the mitochondrion was controlled enzymatically (Porra and Falk, 1964). Jackson et al. (1974) conclusively demonstrated that this was true. The enzyme protoporphyrinogen oxidase has subsequently been isolated from both microorganisms (Poulson and Polglase, 1975) and mammalian tissue (Poulson, 1976).

### G. Ferrochelatase and Heme Oxygenase

The final step in this biosynthetic pathway is the insertion of ferrous iron into protoporphyrin IX by ferrochelatase to form heme. This enzyme is widely distributed in nature and as a particulate enzyme is found associated

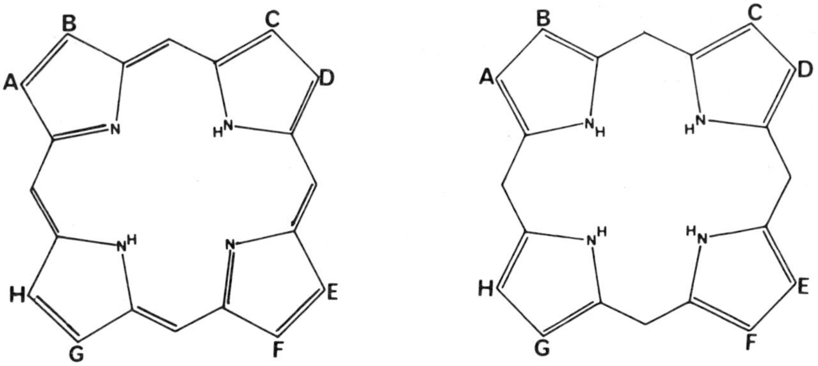

PORPHYRIN

PORPHYRINOGEN

| Position | | ACE | B D | E | F | G | H |
|---|---|---|---|---|---|---|---|
| Uroporphyrin ( ogen) | 1 | Acetyl | Propionyl | Acetyl | Propionyl | Acetyl | Propionyl |
| " | 3 | " | " | " | " | Propionyl | Acetyl |
| Coproporphyrin(ogen) | 1 | Methyl | " | Methyl | " | Methyl | Propionyl |
| " | 3 | " | " | " | " | Propionyl | Methyl |
| Protoporphyrin | 9 | " | Vinyl | " | " | " | " |

**Figure 2.** The structure of the porphyrins and porphyrinogens. By attachment of the ligands shown at the positions A to H round the porphyrin or porphyrinogen nucleus, each of the components of the biosynthetic pathway of heme is shown. Heme itself is formed by insertion of ferrous iron into the center of the protoporphyrin 9 nucleus. The porphyrin is the oxidized form of the porphyrinogen, which is the normal biosynthetic intermediate.

with the inner mitochondrial membrane in animals (Mackay et al., 1969) and bound to chromatophores in microorganisms (Neuberger and Tait, 1964). The mechanism of this enzymic reaction has not yet been clearly elucidated, but of all the factors affecting ferrochelatase activity lipids, and especially phospholipids, seem to have the greatest effect (Yoneyama et al., 1969). Heme oxygenase catalyzes the initial rate-limiting step of heme degradation by preferentially splitting the tetrapyrrole between the A and D rings, thus producing the linear tetrapyrrole bilirubin 9$\alpha$ with consequent loss of iron and of carbon monoxide. This reaction which is microsomal requires oxygen and NADPH.

## III. CONTROL OF THE HEME BIOSYNTHETIC PATHWAY

In normal circumstances, this is a highly efficient and coordinated pathway producing only sufficient amounts of the intermediates and ultimately of heme to service requirements for hemoglobin and other hemoprotein synthesis within the cell. In some circumstances, control of the pathway can be so altered by gross disturbances of the biosynthetic sequence as to produce very large quantities of redundant precursors which must then be excreted. Lead poisoning is one circumstance where such an effect can take place.

Several features combine to make the heme biosynthetic pathway easily and accurately controlled. The first important feature is that heme acts by feedback repression and inhibition upon the initial enzyme of the pathway, δ-aminolevulinic acid synthase. This step then becomes the rate-controlling step of the pathway. This control is aided by the short half-life of this enzyme; it is about 1 to 1½ hr in mammalian systems (Moore and Goldberg, 1974). Control is further aided by membrane permeability considerations, since during the sequence of the pathway the product passes from the mitochondrion to the cytoplasm and then back into the mitochondrion again. If at any stage in the pathway oxidation does take place of the porphyrinogen to the porphyrin, such porphyrin must be excreted since it cannot be further used in the biosynthetic sequence. Equally, any deficiency of the enzyme uroporphyrinogen cosynthase leading to production of series I isomer porphyrin will require that such porphyrin is also excreted since it may not be used beyond the stage of coproporphyrin I in normal biosynthetic mechanisms. It is clear, therefore, that any circumstance that can alter the relative proportions of activity of these enzymes will alter the quantity of precursor produced during the progress of the pathway.

## IV. SOURCES OF LEAD EXPOSURE

There are multiple sources of lead exposure within the environment. Most of these are summarized in Figure 3 and all derive ultimately from the lead in the earth's crust. For modern man, the greatest exposure takes place through the food that he eats and the water that he drinks, although in specific circumstances inhalation of air which is contaminated with lead or uptake of dust from industrial processes may represent important exposure vectors (Fig. 3).

## V. LOCALIZATION OF LEAD WITHIN THE BODY

The actual sites within the body in which lead is deposited are as important a feature as, and part of its effect upon, the heme biosynthetic pathway. Although during the absorptive processes lead is found in excess in the blood, following absorption it is rapidly transferred to soft tissues and ultimately ends as depot lead in bone. Within the bone lead is relatively inert, yet in excess of 95% of the whole body lead burden will be found in the bone. In these circumstances, lead is readily available to the marrow. In

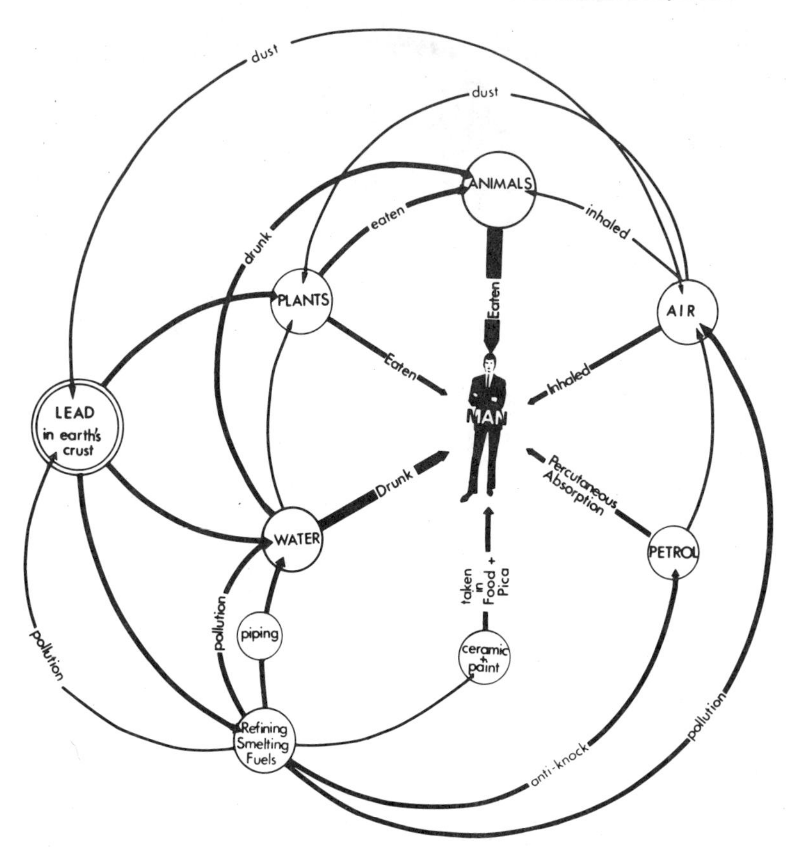


**Figure 3.** The vectors of human lead exposure. The sources from which man may be exposed to lead are shown by the arrows, with the heavier arrows indicating the greater degress of exposure. The principal source of human uptake of lead is through food and water consumption. Absorption may also take place through the air breathed, by percutaneous absorption of alkyl lead compounds in gasolines and through various sources of contamination.

consequence the processes of red cell growth and maturation and, of course, formation of hemoglobin for these cells will be inhibited. In the soft tissues, lead is found in greatest concentrations in the kidney, the principal lead excretory organ. Lower but significant concentrations of lead are also found in liver and in spleen. The distribution of lead within the body is shown in Figure 4. It may be noted that although only 10% of ingested lead is absorbed, up to 40% of inhaled lead will be taken in by the pulmonary route, providing the particle size is sufficiently small. Percutaneous absorption of lead is unlikely to play any important role in its uptake unless the lead exists in a lipid soluble form, such as the alkyl lead compounds (Fig. 4).

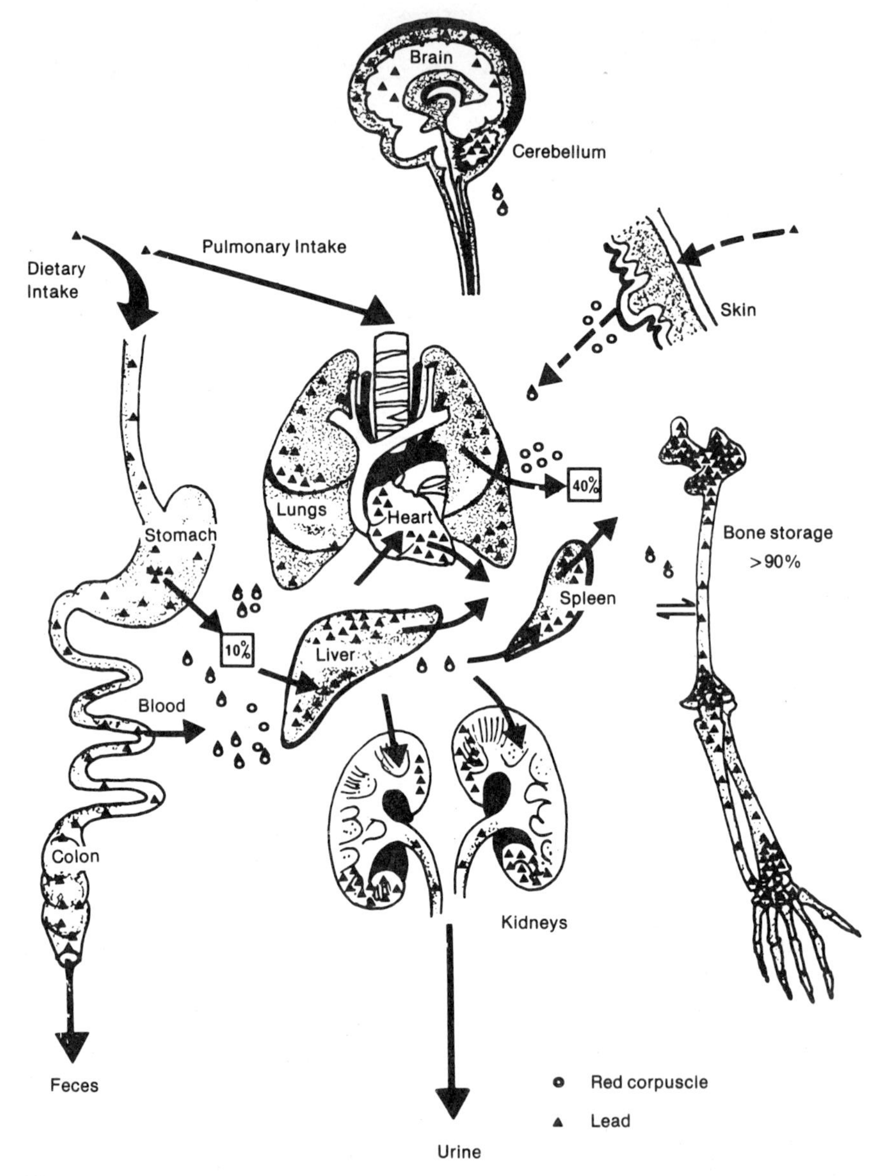


**Figure 4.** Distribution of lead within the body. The three sources of lead intake are shown here. Dietary intake will result in about 10% absorption from the gut. Pulmonary intake will result in about 40% absorption from the lungs. Percutaneous absorption will result generally in a negligible intake through the skin, excepting where lead is in the form of an alkyl lead compound which may be absorbed up to 1% of applied material. Following absorption, lead is transported by the blood to the soft tissues and to the bone. Greater than 90% of the total body lead burden will be found in bone, with relatively high concentrations in liver and kidneys. Urinary excretion accounts for the greater source of lead loss, although some may also be excreted in the feces. Red blood corpuscles shown by an open circle; Lead shown by a closed triangle.

## VI. THE EFFECTS OF LEAD ON THE ENZYMES OF THE HEME BIOSYNTHETIC PATHWAY

Lead has many diverse biochemical effects, all of which are of a deleterious nature. No evidence has been presented for an essential function for lead in metabolism. Lead has a valence shell electronic structure of $6s^2$ $6p^2$ and normally exhibits a valence of two. The vacant 6p orbitals may be filled by the formation of covalent bonds; this occurs most readily in biological systems with sulphur atoms in proteins. These biochemical effects, therefore, all probably relate ultimately to the capacity of lead to combine with specific biochemical ligands. These include sulfhydryl groups, amino groups, carboxyl groups, phenoxy groups and imidazole residues. In such circumstances, lead might be expected to alter tertiary structures of biochemical molecules and, therefore, alter or destroy their biochemical function. Consequently, lead can change enzyme activity and destroy the structure-function relationships of nucleic acids.

It is clear from many studies that lead can markedly inhibit mitochondrial respiration. In addition to the defective structure of their membranes, isolated mitochondria show impaired oxidative phosphorylation and, ultimately, impaired mitochondrial respiration.

Of the complex of enzymes that synthesise succinyl CoA from $\alpha$-keto glutarate, the dithiol enzyme lipoamide dehydrogenase is markedly inhibited by lead. This inhibition is probably representative of the effects of lead on many enzymes either containing sulfhydryl groups or activated by sulfhydryl groups.

### A. ALA Dehydratase

It is clear from consideration of excretion patterns during lead exposure that specific stages during the biosynthetic pathway are particularly vulnerable to the effects of lead. Studies have now been carried out on most enzymes of the pathway and have indicated both in animal and in human studies just which stages are most susceptible to change. The most discussed enzyme in this respect is $\delta$-aminolevulinic acid dehydratase which, during lead exposure, has clearly been shown to be markedly inhibited (Hernberg and Nikkanen, 1972) (Fig. 5). This results in excessive production and excretion of $\delta$-aminolevulinic acid. The mechanism of this inhibition remains unclear, but it is associated with alterations in the concentration of free sulfhydryl groups in biological systems. The availability of concentrations of other metals such as zinc may also contribute, since zinc has been shown to activate ALA dehydratase (Meredith and Moore, 1978).

Organic lead poisoning is rare; thus there are only a small number of studies showing its effects on the heme biosynthetic enzymes. What studies there are suggest that the effects of organic lead on these enzymes are similar to those of inorganic lead. Millar et al. (1972) showed that diethyl

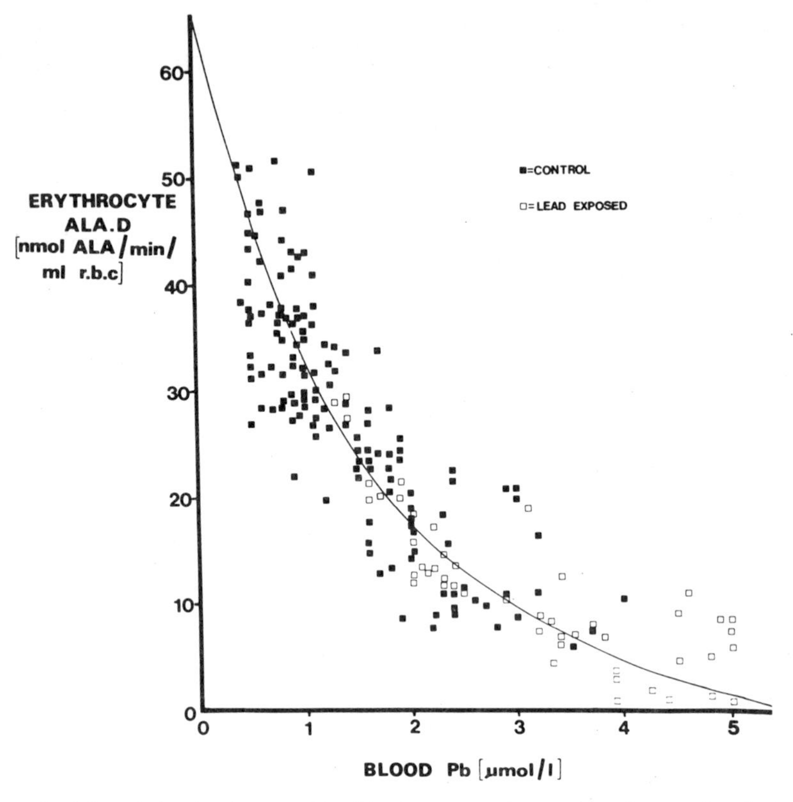


**Figure 5.** The relationship between blood lead concentrations and erythrocyte δ-aminolevulinic acid dehydratase activity. As blood lead concentrations rise in humans, so the activity of erythrocyte ALA dehydratase falls exponentially, with the greatest rate of decrease being seen at lower blood lead concentrations. For this graph, the mathematical relationship is erythrocyte ALA.D = $58.1e^{-0.62 \text{ blood lead}} + 0.06$.

lead depresses ALA.D activity while Beattie et al. (1972), in investigations on four subjects exposed to lead alkyls, demonstrated a marked depression in erythrocyte ALA.D activity and a modest elevation or erythrocyte protoporphyrin (PROTO), but no elevations in urinary ALA and COPRO.

## B. Coproporphyrinogen Oxidase and Ferrochelatase

The next stage in the pathway that is clearly inhibited by lead is at the level of coproporphyrinogen oxidation as shown by excessive urinary excretion of coproporphyrin. Recent studies have shown that lead exposure in man and animals results in highly significant depression of activity of coproporphyrinogen oxidase within the mitochondrion (Campbell et al., 1977). The last enzyme of the pathway, ferrochelatase, has also been shown to be inhibited by lead and may contribute to the sideroblastic types of anemia that develop during lead poisoning (Campbell et al., 1977) and, in addition, to the marked rise in protoporphyrin synthesis (Fig. 6).

## C. ALA Synthase and Heme Oxygenase

The activity of two enzymes in the biodegradative and biosynthetic sequence are known to be elevated during lead exposure. Recent animal experiments have demonstrated that the activity of heme oxygenase is raised by lead pretreatment (Maines and Kappas, 1976a; Goldberg et al., 1978). The net effect of diminished ferrochelatase activity and raised heme oxygenase activity is to lower the availability of free heme concentrations within biological systems by virtue of the feedback control of heme upon the activity of the initial and rate-limiting enzyme of the pathway. The result is that the activity of this enzyme δ-aminolevulinic synthase is raised during lead exposure (Maxwell and Meyer, 1976), presumably in an attempt to compensate for diminished heme production in these circumstances. The combination of increased activity of ALA synthase and decreased activity of ALA dehydratase can account in kinetic terms for the very marked increase in circulating and excreted ALA during lead exposure (Meredith et al., 1978). These changes in heme biosynthesis are detailed in Fig. 7.

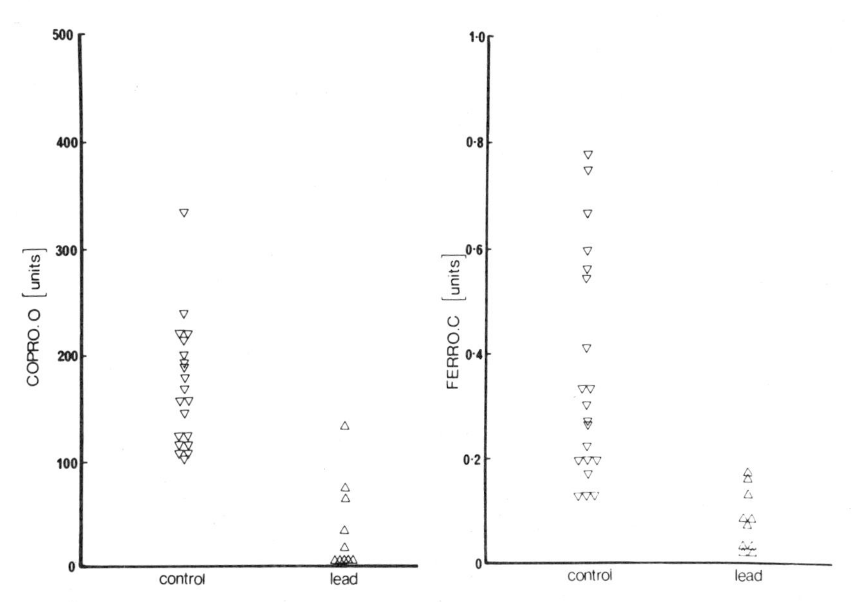


**Figure 6.** The effects of lead exposure on coproporphyrinogen oxidase and ferrochelatase. The activity of these two enzymes shows a pronounced diminution of activity after industrial lead exposure. Coproporphyrinogen oxidase, COPRO.O; units, n mol protoporphyrin utilized/g protein/hr; ferrochelatase, FERRO.C; units, percentage uptake of $^{59}$Fe into heme/hr.

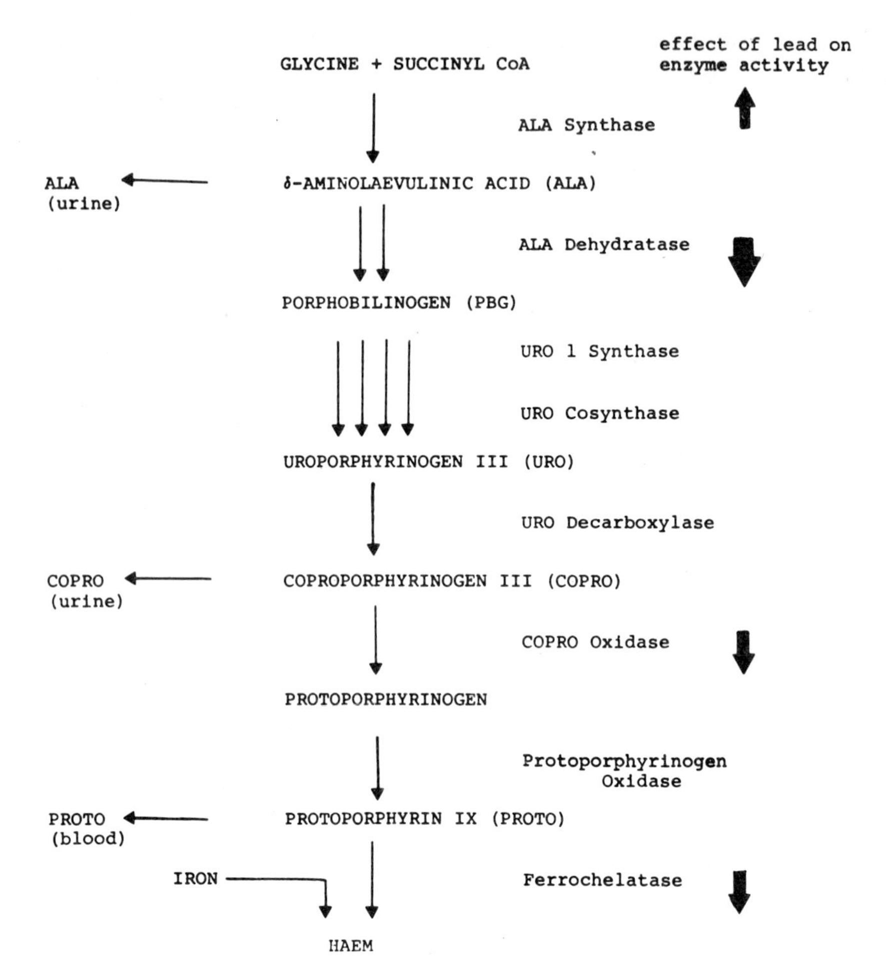


**Figure 7.** The effects of lead on heme biosynthesis. Lead inhibits three enzymes of heme biosynthesis; these are δ-aminolevulinic acid dehydratase, coproporphyrinogen oxidase and ferrochelatase. As a consequence of decreased heme production, there is a rise in the activity of the initial rate-limiting enzyme δ-aminolevulinic acid synthase. As a function of these blocks in heme biosynthesis and the increased input of ALA into the biosynthetic pathway, three intermediates of the pathway are produced in excess. The first is δ-aminolevulinic acid, which is excreted in the urine; the second is coproporphyrin, also excreted in the urine finally, protoporphyrin is produced and found in excess in the erythrocytes.

## D. Other Effects

Although these are the principal points at which lead may affect the biosynthetic pathway, other features have been noted during lead exposure. Uncommonly, porphobilinogen concentrations may rise; this has been ascribed to alterations in the activity of uroporphyrinogen I synthase (porphobilinogen deaminase). Changes in activity of uroporphyrinogen I synthase have been suggested as a secondary control point within the biosynthetic pathway (Brodie et al., 1977). In circumstances where the activity of this enzyme is raised, one would not expect elevated concentrations of porphyrin precursors. In lead exposure, however, lowered activity of ALA dehydratase is coupled with an increase in uroporphyrinogen I synthase, which will allow raised concentrations of $\delta$-aminolevulinic acid but will generally not show elevated concentrations of porphobilinogen (Brodie et al., 1977). Intriguingly, recent studies have shown that the activity of this uroporphyrinogen I synthase inhibited during *in vitro* lead exposure may be protected from the effects of lead by availability of pteridine derivatives such as pteroyl hexaglutamate (Piper and Tephly, 1974; Piper and Van Lier, 1977). Other studies have shown that at higher concentrations of lead the decarboxylation of uroporphyrinogen to coproporphyrinogen is also inhibited (Kreimer-Birnbaum and Grinstein, 1965). One enzyme which is almost certainly not affected by increased lead exposure is uroporphyrinogen cosynthase, as evidenced by the excretion of series III isomer porphyrin during lead exposure, rather than excretion of series I isomer porphyrins (Fig. 2).

## VII. CHANGES IN THE PRODUCTION AND EXCRETION OF PATHWAY INTERMEDIATES DURING LEAD EXPOSURE

### A. Delta-aminolevulinic Acid

Studies in humans of the kinetic relationships between lead exposure, the activities of $\delta$-aminolevulinic acid synthase, $\delta$-aminolevulinic acid dehydratase and blood ALA concentrations have shown a positive exponential relationship of ALA.S and blood lead (Fig. 8) and a negative exponential relationship of ALA.D and blood lead (Fig. 5). The association between blood ALA and ALA.D and their association consequently with blood lead is in the form of a power curve. Similarly, there is a power relationship between blood ALA and excretion of urinary ALA (Meredith et al., 1978).

These results have demonstrated that in lead intoxication there are large increases not only in urinary concentrations of ALA but also in circulating levels of ALA. This elevation is brought about by two factors; these are, firstly, the depression of ALA.D and, secondly, an elevation of the rate-limiting enzyme of heme biosynthesis ALA.S. Depression of ALA.D by lead is well documented, but elevation of ALA.S has only recently been shown in lead intoxication in man (Takaku et al., 1973; Campbell et al., 1977).

The relationship of raised urinary ALA and erythrocyte ALA.D shows that the consideration of such a relationship gives limited metabolic information. The absence of a straight-line relationship between blood ALA

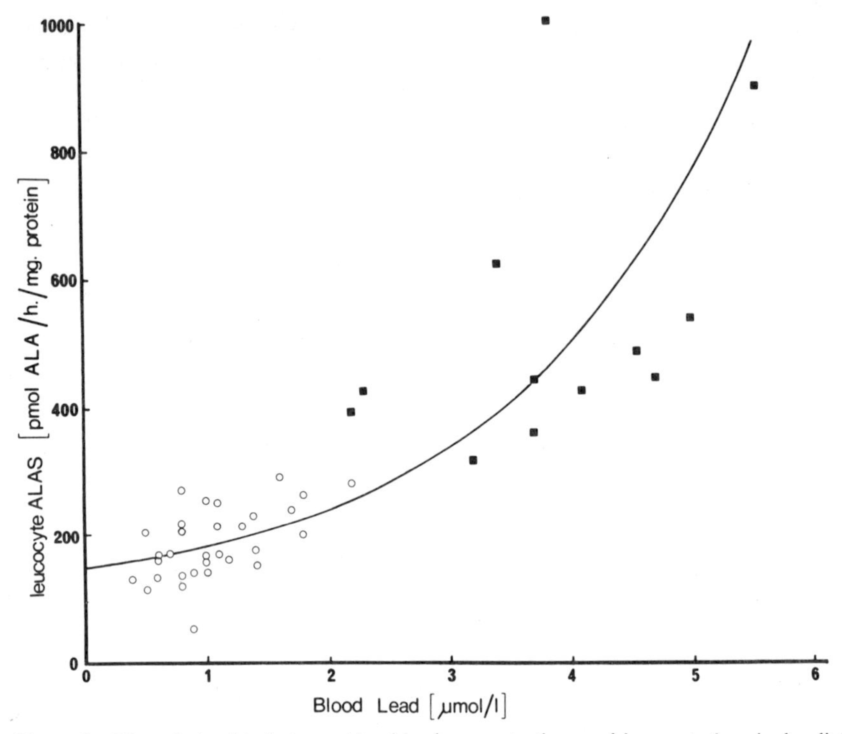


**Figure 8.** The relationship between blood lead concentrations and leucocyte δ-aminolevulinic acid synthase activity. As blood lead concentrations rise, so the activity of δ-aminolevulinic acid synthase rises exponentially. In this graph, normal subjects are shown as open circles and lead exposed workers by closed squares. The mathematical form of this equation is

$$\text{Leucocyte ALA.S} = 48.3e^{0.53\ \text{blood lead}} + 99.2.$$

and urinary ALA suggests that, at higher blood concentrations of ALA, a greater proportion is excreted into the urine. This results in a plateau of blood ALA concentration when blood lead concentration is in excess of 3 μmol/l. At this concentration, urinary ALA shows a rapid rise (Figs. 10A and 10B). Reduced renal tubular reabsorption has been considered to be a contributing factor in this rise (Druyan et al., 1965). However, Cramer and coworkers (1974) showed that ALA clearance is similar to the glomerular filtration rate, although other amino acids were excreted in excess during lead intoxication. There is then an accelerating rise in ALA excretion consistent with reduced tubular reabsorption. Thus, below blood ALA concentrations of 4 μmol/l, the relationships of ALA in urine and blood are linear. It is only when blood ALA concentrations rise above this value that there is a rapid rise in urinary ALA excretion, possibly associated with decreased tubular reabsorption. Below blood lead concentrations of 2 μmol/l, the relationships of ALA in urine and blood are linear. It is only when blood lead concentrations rise above this value that inhibition of ALA dehydratase and, consequently, heme biosynthesis is sufficiently great to depress synthesised free heme, induce ALA synthase activity and cause these accelerating rises in synthesized ALA.

The interrelationships of blood lead, leucocyte ALA.S, erythrocyte ALA.D and blood ALA are of considerable metabolic importance, provided one assumes that the concentrations of these parameters parallel those in other tissues. This is a reasonable assumption if the individual is in a steady state condition. These relationships suggest that there are "critical" tissue lead concentrations, represented by a blood lead concentration of approximately 2 μmol/l and an erythrocyte ALA.D activity of approximately 18 nmole ALA utilized per minute/ml red blood cells. Above this concentration and below this activity, the depression in heme synthesis is sufficient to raise the activity of ALA.S, the rate-limiting enzyme of heme biosynthesis, by negative feedback (Fig. 9).

## B. Porphobilinogen

It is not certain in what manner lead poisoning may influence the urinary excretion of PBG. Most studies suggest that it will be normal, although is some cases it has been reported as raised (Waldron and Stofen, 1974). The current concept of control of heme biosynthesis would suggest that the concentration should be normal in urine and plasma to correspond with raised activity of uroporphyrinogen I synthase (Brodie et al., 1977), although the activity of this enzyme has been reported to be inhibited by lead (Piper and Tephly, 1974).

## C. Coproporphyrin and Protoporphyrin

Excessive urinary porphyrin excretion has long been recognized. This is caused principally by the rise in coproporphyrin, mainly as the series III isomer. Subsequently, it has been shown that other porphyrins may be detected in increased amounts in lead intoxication. Increased concentrations of protoporphyrin may be found. So also can the concentration of uroporphyrin rise in severe cases of lead poisoning, with associated increases in intermediate porphyrins of the biosynthetic pathway (Waldron and Stofen, 1974).

A notable feature of lead poisoning is the elevation of erythrocyte porphyrin levels. Most of this is in the form of protoporphyrin, although erythrocyte coproporphyrin is also raised. The elevation of erythrocyte protoporphyrin has received considerable attention as a bioanalytical index of lead exposure. This has resulted in the publication of recommendations for the use of protoporphyrin as a primary screening test in lead intoxication (Center for Disease Control, 1975). Protoporphyrin is found both in the free state and as a zinc chelate. Both of these fluoresce under ultraviolet

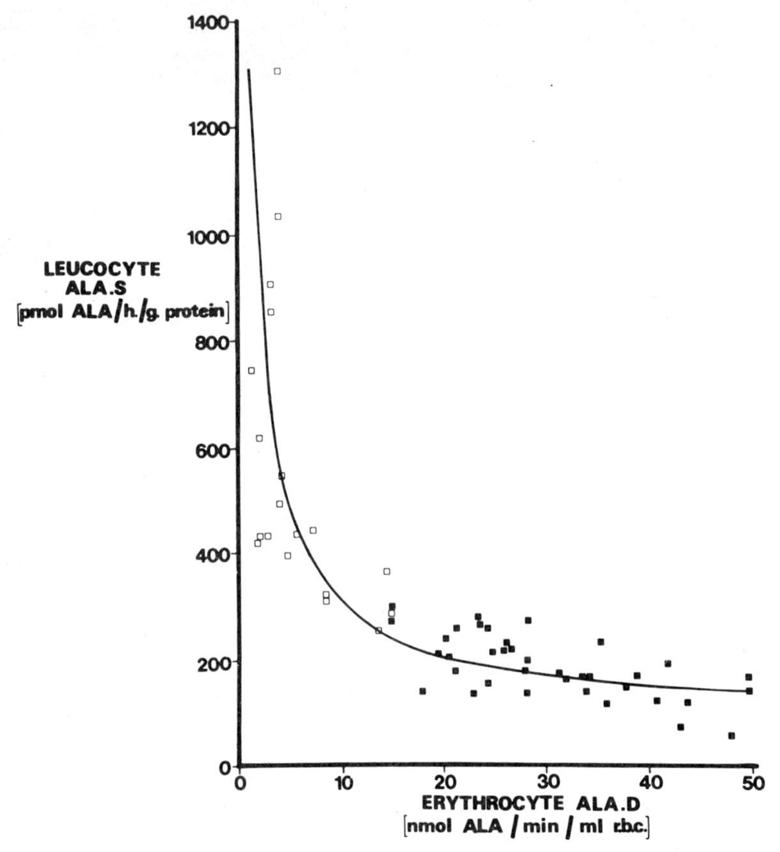


**Figure 9.** The relationship between erythrocyte δ-aminolevulinic acid dehydratase activity and leucocyte δ-aminolevulinic acid synthase activity. Corresponding to the inhibition of ALA.D by lead and the induction of ALA synthase by lead, the interrelationship of these two enzymes is in the form of a power curve. In this graph, normal subjects are shown as open squares and lead-exposed subjects as closed squares. The mathematical form of this equation is

$$\text{Erythrocyte ALA.D} = 58.1\left(\frac{\text{leucocyte ALA.S} - 99.2}{48.3}\right)^{1.17}$$

light and result in the development of fluorocytes, that is, erythrocytes which show red fluorescence under ultraviolet light (Whitaker and Vietti, 1958).

## VIII. THE DIAGNOSTIC USE OF EXCRETION AND ACTIVITY PARAMETERS

It is clear from the various features of the effect of lead upon the biosynthetic pathway that the changes in both the activity of enzymes and the production and excretion of intermediates of the pathway can be used as secondary biological indicators of lead intoxication. Although the earliest references to the biological effects of lead upon heme biosynthesis referred to increases of porphyrin in the urine, it is now certain from consideration of available evidence that such changes are certainly not the most sensitive or the most easily measured indices of exposure. For the purposes of this section, each of the different substances and enzymes and their uses will be reviewed in turn. These are the concentrations of δ-aminolevulinic acid in blood and in urine, the concentrations of coproporphyrin in urine, the concentrations of protoporphyrin in the blood, the activity of δ-aminolevulinic acid dehydratase in blood and, finally, comparison of the relative efficiency of each of these. Also included are the effects of lead on the concentration of hemoglobin in the blood, although it is certain that a very severe degree of lead intoxication must occur for changes in the concentration of hemoglobin to be found.

### A. Delta-aminolevulinic Acid Concentrations in Urine

The early studies of Haeger-Aronsen (1960) demonstrated clearly that measurement of δ-aminolevulinic acid in the urine was a useful diagnostic test for lead exposure in lead workers. The method by which ALA is measured in these circumstances has always been the colorimetric technique, whereby δ-aminolevulic acid is condensed with another compound, usually acetyl acetone, to form an Ehrlich's positive pyrrole, which may then form a pink color quantitatively with paradimethyl aminobenzaldehyde (Ehrlich's reagent). When the concentrations of urinary ALA and blood lead are compared, a linear relationship which intercepts the blood lead axis at values above the origin is seen. In the initial stages of the graph, changes in concentration are so small as the be of little use. However, when blood

**Table 2.** The concentrations of lead, porphyrins and porphyrin precursors in normal and lead-poisoned humans

| | Normal subjects | Lead-Poisoned subjects |
|---|---|---|
| *Blood lead* μmol/l | 0–1.75 | >4.0 |
| (μg/100 ml) | (0–40) | (>80) |
| *Urinary ALA* mmol/mol creatinine | 0–2.5 | >4.0 |
| (μmol/l) | 0–43 | >60 |
| *Blood ALA* μmol/l | 0.–3.5 | >5.0 |
| *Urinary coproporphyrin* | | |
| μmol/mol creatinine | 0–12.5 | >40 |
| (nmol/l) | 0–220 | >300 |
| *Erythrocyte protoporphyrin* | | |
| (μmol/l) | 0–0.75 | >1.5 |

lead concentrations exceed normal limits, *i.e.*, greater than 40 μg% (2 μM), urinary ALA concentrations show some rise and continue to do so as blood concentrations of lead rise (Fig. 10). Normal values for ALA in urine are shown in Table 2, together with the range of values in a series of industrially exposed workers.

## B. Blood Concentrations of ALA

The concentration of δ-aminolevulinic acid in the blood may also be used as an index of lead exposure, although the concentrations likely to be achieved are certainly not as great as those in acute porphyria in attack. In the normal person, blood concentrations of δ-aminolevulinic acid are small, but following industrial lead exposure they rise rapidly. The methods for measuring blood ALA are similar to those for urine, except that the blood initially must be treated with iodo-acetamide to complex interfering sulfhydryl groups, after which ALA may be extracted into acid and

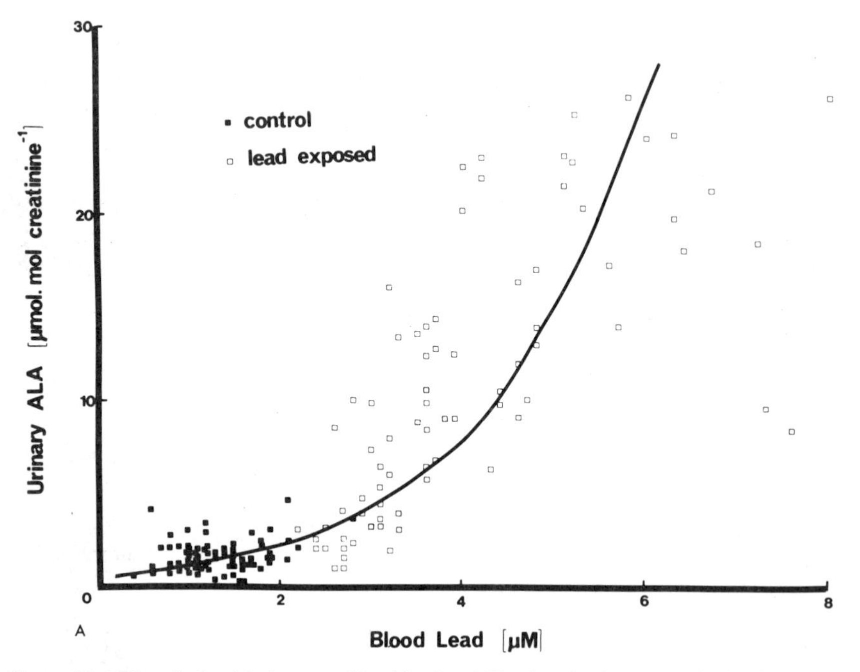


**Figure 10.** The relationship between blood lead and blood and urinary δ-aminolevulinic acid. As blood lead concentrations rise, so the concentration of the urinary excretion of δ-aminolevulinic acid rises exponentially (A). Blood ALA concentrations show a more complex relationship which is probably related to changes in the urinary excretion of ALA (B). The mathematical forms of these relationships are

$$\text{Urinary ALA} = 0.59e^{(0.63\ \text{blood lead})} + 0.41$$

$$\text{Blood ALA} = 2.2e^{(12.2e^{-0.66\ \text{blood lead}})} - 4.1e^{(-0.66\ \text{blood lead})} + 4.9$$

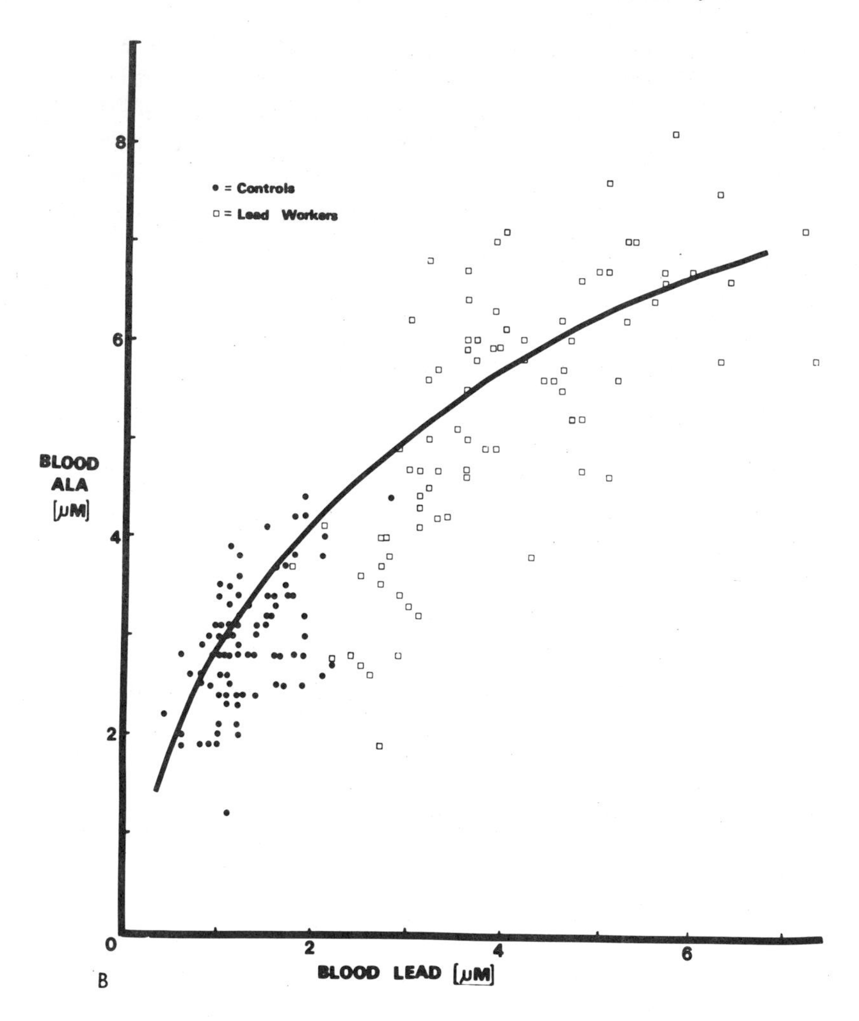


measured by methods similar to those used for urine (Haeger-Aronsen, 1960). The mathematical association between urinary ALA and blood ALA takes the form of a power curve. However in industrial exposure the relationship between blood ALA and blood lead is an exponential one; blood ALA concentration tends to plateau when the blood lead concentration is in excess of 3 μmol/l (Meredith et al., 1978) (Fig. 10B).

## C. Urinary Coproporphyrin

Urinary coproporphyrin requires more complex quantitation techniques inherently less reliable than those for δ-aminolevulinic acid; consequently, coproporphyrin in urine is generally a less satisfactory method of defining excessive lead exposure. Like δ-aminolevulinic acid, changes in the urinary coproporphyrin excretion start to take place only at greater concentrations of blood lead than those found in the general population and are therefore commonly used only in industrial lead exposure analysis. The normal values for urinary coproporphyrin and typical values found in a group of lead workers are shown in Table 2.

### D. Blood Protoporphyrin

The use of blood protoporphyrin as an index of lead exposure has received considerable attention in the past few years. Examination of this index has used classical extraction procedures (Alessio et al., 1976) and the development of microfluorimetric methods with whole blood (Kamholtz et al., 1972). Finally, the development of a portable front-face spectrofluorimeter has the advantage of rapidity of analysis and minimal preparation of material. It shows great promise in future use for examination of lead workers (Blumberg et al., 1977). Measurement of protoporphyrin in blood as an index of lead exposure is limited by the very high incidence of iron deficiency anemia, which will also raise the concentration of protoporphyrin in blood. Such objections have in part been answered by consideration of the presence and absence of the zinc complex of protoporphyrin, which is more commonly found in lead exposure of long standing. It is true to say, however, that in any disease of long standing in which protoporphyrin is raised, there are increased concentrations of this zinc chelate. The highest concentrations of protoporphyrins of all in the blood will be found in subjects with erythropoietic protoporphyria in whom the primary defect is in the activity of the enzyme ferrochelatase in the bone marrow. Normal values for blood protoporphyrin and values commonly found in lead workers are shown in Table 2. Because the association between protoporphyrin in the blood and blood lead shows a positive exponential relationship (Fig. 11), this method is of little use in examining environmental lead exposure, where a very high incidence of false positives will be found (Meredith et al., 1979).

### E. Erythrocyte Delta-aminolevulinic Acid Dehydratase

As with all enzyme measurements, the activity arrived at can vary considerably depending upon the methods used to quantitate it. This enzyme has benefited considerably from the development of a simple standardized method for the European community (Berlin and Schaller, 1974), although other methods for measurement of activity are equally acceptable. All depend ultimately upon the measurement of activity around the pH optimum of 7.4 of the rates of formation of porphobilinogen from δ-aminolevulinic acid, usually in a phosphate buffer. Substances such as

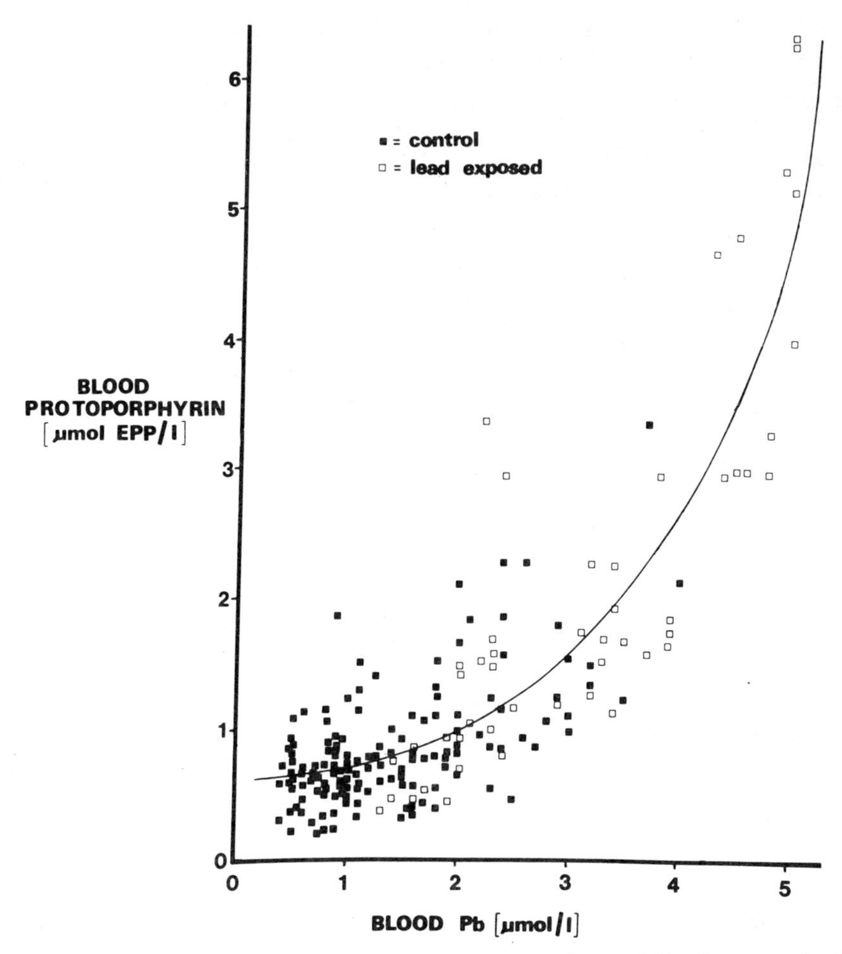


**Figure 11.** The relationship between blood lead concentrations and blood protoporphyrin concentrations. Blood protoporphyrin concentrations rise exponentially, as blood lead concentrations rise. The mathematical relationship is

$$\text{Blood protoporphyrin} = 0.14e^{(0.72\ \textit{blood lead})} + 0.43$$

alcohol (Moore et al., 1971) or zinc (Meredith and Moore, 1978) can interfere with the activity of this enzyme, but in general lead has the greatest effect of all upon the activity of this enzyme. The association between the activity of the enzyme and blood lead concentrations has been shown many times to have a negative exponential relationship (Fig. 5). This means that measurement of the activity of the enzyme is a particularly good biological index of lead exposure at environmental concentrations, *i.e.*, below 2 $\mu$M.

## F. Intercomparison of the Relative Value of Heme Biosynthetic Parameters in Assessment of Lead Exposure

It is well established that all the heme biosynthetic parameters we have considered are highly significantly correlated with blood lead levels. A high coefficient does not, however, necessarily imply a high predictive validity. Such validity can only be established by the determination of the extent to which subjects are correctly classified as to their blood lead levels on the basis of the heme biosynthetic parameter under consideration. An evaluation of this type has recently been carried out (Meredith, Moore, and Goldberg 1979) on the basis of false positive and false negative results at two blood lead levels. The two blood lead levels selected were 1.75 μmol/l and 3.0 μmol/l, the first being the now generally accepted desirable upper limit for subjects with no known lead exposure and the second being representative of a blood lead value that could be encountered in industrial lead exposure. For each of these two blood lead values, two values each were considered for urinary and blood ALA, urinary COPRO, blood protoporphyrin and erythrocyte ALA.D. These values were selected arbitrarily as being those most likely to result in high predictive validity. Predictive validity was assessed in terms of percentage of false negatives, that is, a failure to detect subjects with raised blood lead levels, and percentage of false positives, that is, a false prediction of raised blood lead levels. Results of this analysis in a group of subjects with known industrial lead exposure and subjects with no known excessive lead exposure are shown in Table 3. It is clear from this table that

**Table 3.** The predictive validity of heme biosynthetic parameters in the assessment of elevated blood lead concentrations

| | | Blood lead (μmol/l) | | | | |
|---|---|---|---|---|---|---|
| | | Environmental (>1.75) | | | Industrial (>3.0) | |
| | | False positive (%) | False negative (%) | | False positive (%) | False negative (%) |
| Blood ALA | >3.5 | 5.4 | 9.0 | >4.5 | 1.2 | 4.8 |
| (μmol/l) | >3.0 | 18.6 | 6.0 | >4.0 | 6.6 | 1.8 |
| Urinary ALA | >2.5 | 4.8 | 10.2 | >5.0 | 1.2 | 3.6 |
| (mmol/mol creatinine) | >2.0 | 11.4 | 5.4 | >4.0 | 4.2 | 3.0 |
| Urinary coproporphyrin | >12.5 | 10.2 | 7.2 | >22.5 | 3.0 | 7.2 |
| (μmol/mol creatinine) | >10.0 | 15.6 | 3.0 | >17.5 | 11.4 | 3.6 |
| Erythrocyte ALA.D | <20 | 3.7 | 7.4 | <10 | 1.6 | 4.8 |
| (nmol ALA/min/ml rbc | <25 | 6.3 | 0.5 | <12.5 | 9.0 | 0.5 |
| Blood protoporphyrin | >1.0 | 6.9 | 11.6 | >2.0 | 2.6 | 9.0 |
| (μmol EPP/1) | >0.75 | 20.6 | 4.2 | >1.5 | 9.0 | 4.2 |

only erythrocyte ALA.D offers a good index of environmental lead exposure; that is, when the number of false negative results is reduced to an acceptable level (*i.e.*, less than 5%) by reduction of the threshold value, the number of false positives remains at an acceptable level (*i.e.*, less than 10%). At the higher blood lead level it is apparent that, with the exception of urinary COPRO, all the parameters are acceptable screening tests on the basis of predictive validity.

Thus, for the screening of a broad population, the activity of erythrocyte ALA.D is a highly sensitive index of both environmental and industrial lead exposure. However, due to the relative rapidity and simplicity of blood protoporphyrin determination using the "hematofluorimeter," it is probably that the latter test is more expedient for the screening of an industrially exposed group of workers.

### G. Hemoglobin

It is only in clear cases of lead poisoning that hemoglobin levels start to fall. In general, this will only happen when blood lead levels are in excess of 4 $\mu$M (80 $\mu$g/100 ml). As shown before, ALA synthase activity is inversely related to hemoglobin and hemoprotein concentrations during lead exposure (Campbell et al., 1977; Goldberg et al., 1978); other studies have shown a similar association with blood protoporphyrin levels (Meredith et al., 1979). A rise in blood protoporphyrin levels will reflect depression of heme production more accurately than will depression of erythrocyte ALA dehydratase. This is consistent with the previous suggestion that there are critical tissue lead concentrations of about 2 $\mu$M (40 $\mu$g/100 ml). This corresponds to erythrocyte protoporphyrin levels of about 0.75 mM and erythrocyte ALA dehydratase activity of 20 nmol ALA utilized/min/ml red blood cells. At this concentration of lead, the inhibition of heme synthesis is sufficiently great to depress the synthesized free heme and therefore hemoglobin and thus to induce ALA synthase activity.

## IX. CHANGES IN DRUG METABOLISM DURING LEAD EXPOSURE

A large proportion of heme synthesized in the liver acts as the prosthetic group of the hemoprotein cytochrome P-450. This hemoprotein is the terminal oxidase of the microsomal mixed function oxidase system that is involved in the metabolism of many drugs. The possibility that drug metabolism might be altered in lead intoxication due to depressed synthesis of microsomal cytochrome P-450 has been considered in both man and animals.

## A. Animal Studies

Initial studies in animals produced contradictory results. The work of Ribeiro (1970) on the effects of arsenic, beryllium, lead and mercury on mouse liver drug-metabolizing enzymes indicated that pretreatment of mice with lead nitrate to give liver levels of $10^{-4}$ and $10^{-6}$ M failed to alter hexobarbital sleeping times or certain microsomal enzyme activities that were measured *in vitro*. When lead was added to control microsomes at concentrations between $10^{-3}$ and $10^{-5}$ M, no inhibition of hexobarbital oxidase was noted. In contrast, the results of a number of investigators (Alvares et al., 1972; Scoppa et al., 1973; Egan and Cornish, 1973; Chow and Cornish, 1978) suggest that lead has a significant effect on microsomal drug metabolism. Alvares et al. (1972) found a 40 to 50% decrease in metabolism by hydroxylation and demethylation and a significant decrease in cytochrome P-450 levels and hexobarbital sleeping times 24 hours after the intravenous injection of lead chloride. From the results of their studies, the authors considered that lead exerts its effect on drug metabolism via an inhibition of heme synthesis.

Scoppa et al. (1973) attempted to verify this hypothesis by correlating the inhibition of blood and liver ALA dehydratase by lead with drug metabolism. They were able to demonstrate an association between lowered activity of this enzyme and depression in cytochrome P-450 levels and an associated depression of *in vitro* activities of the microsomal mixed function oxidase system.

These studies did not, however, provide unequivocal evidence of the existence of a relationship between depressed heme synthesis and altered drug metabolism. Goldberg et al. (1978) provided further evidence that lead has a profound effect upon the microsomal mixed function oxidase system. An increased dose of lead to rats resulted in a progressive decrease in cytochrome P-450 content and a progressive decrease in the activities of aniline hydroxylase and aminopyrine demethylase. As had been noted previously (Maxwell and Meyer, 1976), there was an increase in the activity of the rate-limiting enzyme of heme biosynthesis, ALA synthase, associated with the impairment of the mixed function oxidase system. An inverse relationship between ALA synthase and cytochrome P-450 content was observed; this is illustrated in Figure 12. This inverse relationship provides further evidence of the control mechanism whereby a decrease in the regulatory pool of heme, as assessed by the microsomal content of the hemoprotein cytochrome P-450, results in an increase in the activity of ALA synthase by a negative feedback mechanism.

From *in vitro* studies (Scoppa et al., 1973; Goldberg et al., 1978), it is evident that lead has no direct marked effect on the microsomal mixed function oxidase system. It is apparent that the depletion of cytochrome P-450

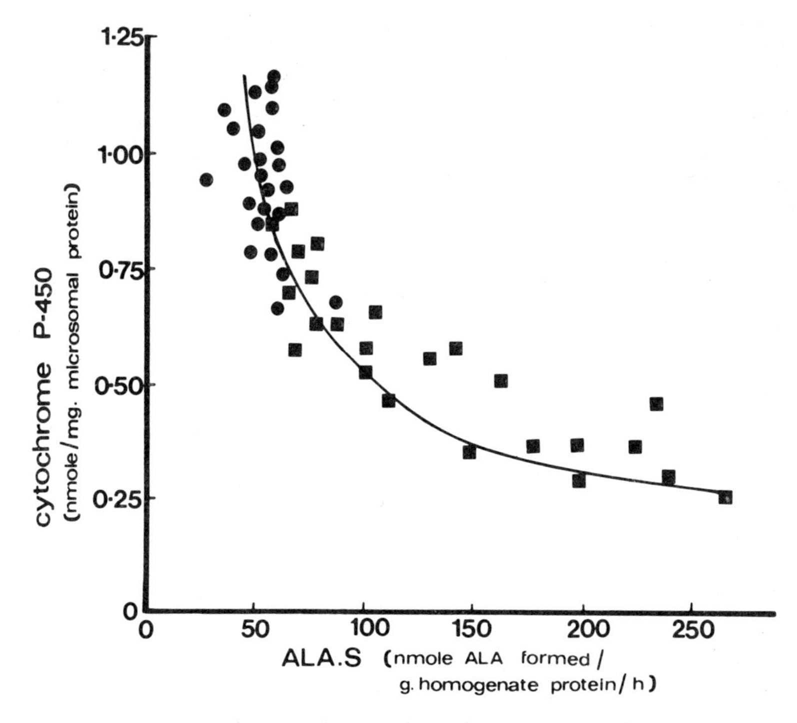


**Figure 12.** The changes in hepatic microsomal cytochrome P-450 and hepatic ALA synthase in lead-treated rats. In this graph, the lead-treated animals are shown as closed squares and the untreated animals are shown as closed circles. As lead concentration rises, so the activity of ALA synthase rises; concurrently, there is an exponential decrease in the concentration of hepatic microsomal cytochrome P-450. The mathematical form of this relationship is

$$\text{Cytochrome P-450} = \frac{41.4}{\text{ALA.S}} + 0.098$$

in lead intoxication in animals can be mediated through two possible mechanisms; the first is associated with a reduction in the activity of the intermediate enzymes of heme biosynthesis ALA dehydratase, COPRO oxidase and ferrochelatase synthesis (Goldberg et al., 1978). The second possible mechanism is associated with an increase in heme degradation due to increased activity of the heme degradative enzyme heme oxygenase (Maines and Kappas, 1976b; Goldberg et al., 1978). This latter mechanism conflicts with a previous model suggesting that ALA synthase and heme oxygenase are interrelated with respect to their regulation by heme Bissell and Hammaker (1976). They further predicted that these enzyme activities vary reciprocally. This theory has experimental support from the studies of Schacter (1975). From this theory, one would expect depletion of the regulatory heme pool, which would activate ALA.S but depress heme oxygenase. Experimental evidence does not support this theory. Lead pretreatment of rats results in inhibition of hepatic heme synthesis and depletion of the regulatory heme pool. Thus there is not only an associated increase in the activity of ALA synthase, but also a rise in the activity of heme oxygenase.

These findings are, however, consistent with the theory postulated by Maines and Kappas (1976c) that the mode of action of metals in inducing heme oxygenase is based on a repressor component in the regulatory mechanism of heme oxygenase. They suggested that this repressor has an −SH active constituent, the oxidation-reduction capacity of which is necessary for controlling heme oxygenase production. If the oxidation-reduction cycle of the −SH groups of this component are blocked, as would occur after treatment with lead, which has a high affinity for −SH groups, the regulatory function of this cellular constituent on heme oxygenase would be lost. The system would then function without regressive regulation, leading to an exaggerated synthesis of the enzyme.

Thus, the impairment of the mixed function oxidases demonstrated in lead treated animals has clearly been shown to be mediated through a decrease in heme biosynthesis and an increase in heme degradation. As yet, it is not possible to conclude which of these two mechanisms is the more important in the reduction of microsomal P-450 content; both mechanisms are possible and they are not mutually exclusive.

### B. Human Studies

The attention focused on the effects of lead on *in vitro* drug metabolism in animals has resulted in a number of investigations into the effects of lead exposure in man on *in vivo* drug metabolism rates. A preliminary study was carried out by Alvares et al. (1973) in two children who showed clinical and biochemical manifestations of acute lead poisoning. As a measure of *in vivo* drug metabolism rates, phenazone (antipyrine) half-lives were determined both before and after chelation therapy. The determination of phenazone half-life in man offers an index of drug metabolizing capacity because an oral dose is completely absorbed and is metabolized exclusively by the hepatic cytochrome P-450 dependent microsomal mixed function oxidase system. The phenazone half-lives in the two subjects were found to be significantly longer than in a group of normal children, while chelating therapy led to a reduction in half-life in both subjects. The same authors (Alvares et al., 1976) were, however, unable to demonstrate a statistically significant change in the rate of elimination of phenazone from the plasma of eight industrially exposed lead workers following chelation therapy. They did note that in seven of the eight subjects there was a reduction in phenazone half-life following chelation therapy.

Subsequent studies have demonstrated that in acute lead intoxication in man there is a depression in the rate of drug metabolism (Meredith et al.,

1977; Fischbein et al., 1977). In the first of these two studies (Meredith et al., 1977), plasma phenazone elimination rates were determined in a group of 10 demolition workers admitted to hospital with clinical and biochemical symptoms of excessive lead exposure before, immediately after and again at least 12 weeks after the cessation of chelation therapy. At the same time, erythrocyte ALA dehydratase activity, blood lead levels and hemoglobin levels were determined. Chelation therapy was found to be associated with a highly significant ($p < 0.005$; paired *t* analysis) decrease in phenazone half-life and an increase in clearance both immediately after therapy and again when measured at least 12 weeks after chelation therapy (Fig. 13). The changes in phenazone elimination rates following chelation therapy were also associated with decreased blood lead levels and increases in both hemoglobin levels and erythrocyte ALA dehydratase activity. The increase in phenazone elimination rates in the subjects following chelation therapy was significantly correlated with the fall in blood lead levels ($p < 0.005$).

In a similar study on five subjects with biochemical manifestations of chronic lead intoxication, Fischbein et al. (1977) noted a slight but statistically significant decrease in phenazone half-life following chelation therapy. However, the rate of elimination of phenylbutazone was unaffected by chelation therapy.

It is apparent that the acute administration of lead to rats is associated with decreases in cytochrome P-450 content of hepatic microsomes. The increase in *in vivo* drug metabolism rates measured following chelation therapy of lead-intoxicated human subjects (Meredith et al., 1977; Fischbein et al., 1977) is probably due to depressed levels of the hepatic microsomal hemoprotein cytochrome P-450.

## X. THE ANEMIA OF LEAD POISONING

Three mechanisms are involved in the development of anemia during lead poisoning. Firstly, there is depression of heme biosynthesis; secondly, there is disturbance of globin synthesis; and, finally, there is hemolysis of red blood cells. The development of such anemia will occur at two times, during red blood cell formation and during red blood cell destruction. Since lead is preferentially deposited into bone (Fig. 4), it is clear that during red cell formation in the bone marrow the concentrations of lead in the marrow will be very much greater than that in the blood (Albaharry, 1972). In these circumstances, the developing cells within the bone marrow will be particularly affected by lead. Therefore one would expect defective erythropoiesis (Landow et al., 1973). Berk et al. (1970) suggested that erythropoiesis is completely effective during lead intoxication in a human adult with lead poisoning. However, Dagg et al. (1965), in a study of two patients with lead

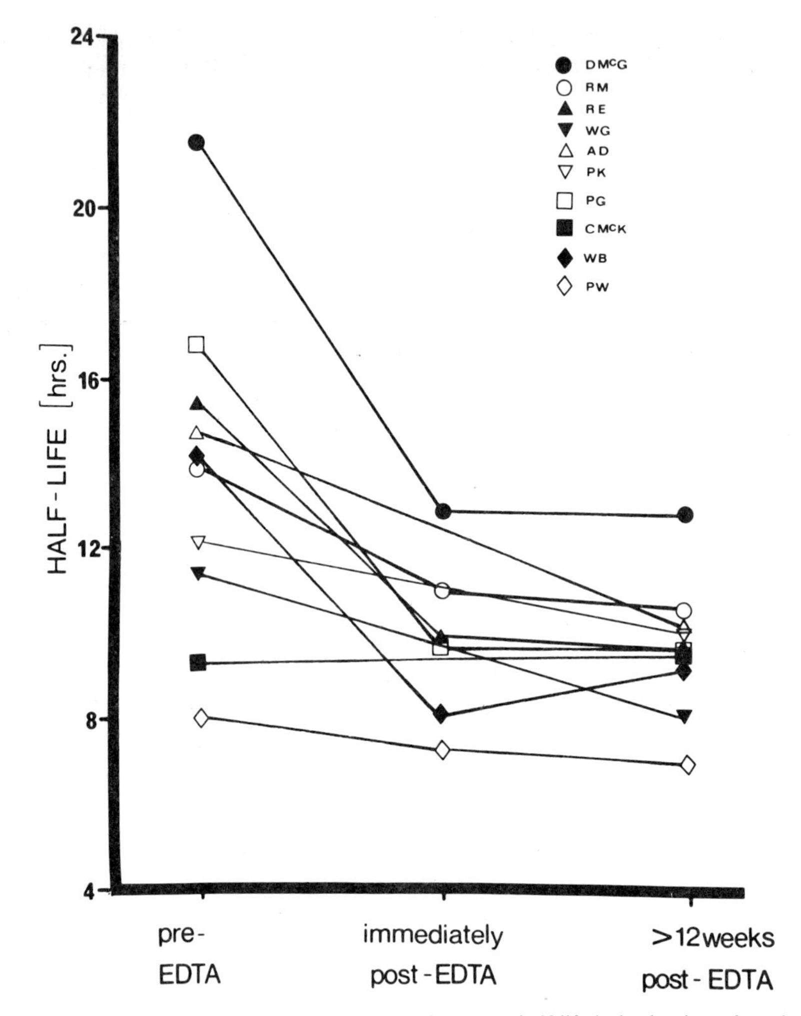


**Figure 13.** The effects of chelation therapy on phenazone half-life in lead-poisoned workers. The rate of phenazone metabolism was examined in 10 lead workers at the time of lead poisoning, immediately after ethylene diamino tetra-acetic acid (EDTA) chelation therapy and more than 12 weeks after chelation therapy. In all cases, the rate of phenazone metabolism rose following chelation of lead from the systems of these lead-poisoned subjects, as evidenced by a decrease in the phenazone half-life. This could not be attributed to existing concentrations of EDTA.

poisoning before and after treatment with penicillamine, showed that the anemia of lead poisoning is due partly to hemolysis and partly to inhibition of iron utilization.

Lead affects protein synthesis; specifically, it will inhibit the synthesis of globin. It interferes with amino acid incorporation and, when reticulocytes are incubated with lead, one finds disaggregation of polyribosomes. *In vitro* experiments have shown that, in addition to this decrease in globin synthesis, the rate of production of alpha and beta chains become out of phase, with suppression of alpha chain production being more marked than that of the beta chains. *In vivo*, there appears to be some compensation for this to restore the balance of synthesis; the significance of this finding is in some doubt (White and Harvey, 1972). The final effect that lead may have on the red cell is its effect on the red cell membrane. The normal life span of the red cell is about 120 days. In hemolytic anemia, the mean life span of these cells measured by radioisotope techniques is reduced (Griggs and Harris, 1958; Waldron, 1966). This shortening of the red cell life span is probably due to direct toxic effect upon the cell membrane.

Lead interferes with the $Na^+$-$K^+$ dependent ATPase activity in the red cell (Hasan et al., 1967); as a consequence, red cells incubated with high concentrations of lead leak large quantities of $K^+$. This loss is irreversible and is thought to be caused by interference with this energy-requiring pump. At the same time, there is little change in the input of $Na^+$ into the cell; in consequence there is a rise in $Na^+$ concentration in lead poisoned erythrocytes, associated with potassium-mediated shrinkage of the cell. This $K^+$ leakage almost certainly leads to an increased mechanical fragility of the red blood cell and consequent lysis. The importance of anemia in lead poisoning is readily seen in the incidence of the various features of lead poisoning shown in Figure 14, where about 90% of lead-poisoned workers were found to have some degree of anemia. As the degree of lead poisoning increases, the hemoglobin concentration in the blood decreases. At the same time, there are rises in the activity of $\delta$-aminolevulinic acid synthase, presumably in an attempt to compensate for the decreased concentrations of free heme within the cell. The type of anemia found in these workers is usually a normocytic, normochromic anemia. In affected children, it may become macrocytic and hypochromic anemia.

## XI. A COMPARISON OF LEAD POISONING AND ACUTE HEPATIC PORPHYRIA

### A. General Considerations

Although acute porphyria and lead poisoning are conditions of different etiology, one being genetically determined and the second acquired, it is of interest to compare the similarities of their clinical features and of the excretion patterns of porphyrin precursors and porphyrins. These similarities often reflect similarities in the abnormalities of enzyme activity and porphyrin metabolism (Dagg et al., 1965; Gibson and Goldberg, 1970; Campbell et al., 1977; Doss and Tiepermann, 1977). The features of acute

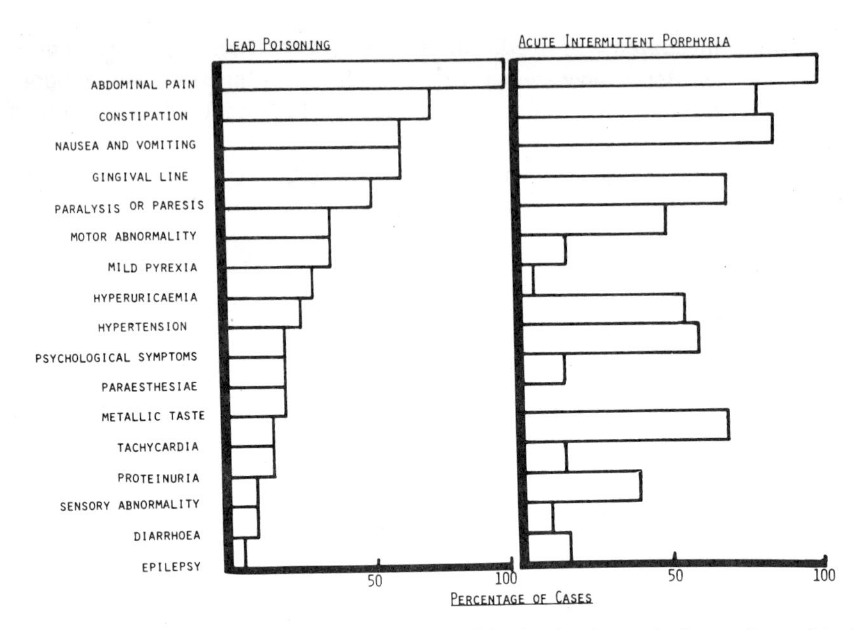


**Figure 14.** The incidence of the typical features of lead poisoning and of acute intermittent porphyria. The incidence of the effect of lead poisoning in 10 industrially exposed subjects, all of whom had blood lead concentrations exceeding 4 μmol/l, is demonstrated in this histogram, together with the similar incidence of features of acute intermittent porphyria in 50 subjects with this disease. Some features are seen only in lead poisoning, such as hyperuricemia, metallic taste and, in some cases where poor dental hygience co-exists with lead poisoning, the gingival blue line.

porphyria, which are abdominal pain, constipation, vomiting, paralysis or paresis, are similar to those found in lead intoxication and, indeed, the incidence of these in both conditions is also similar (Fig. 14). In lead poisoning, the paralysis or paresis is entirely of the motor neuron type, while in acute intermittent porphyria 10% of the subjects show upper motor neuron involvement and organic brain disorder (Goldberg, 1968).

In terms of excretion of either porphyrins or their precursors, the overall pattern found in lead poisoning is in no way identical to any of the porphyrias, although some similarities exist. In all acute porphyrias there is excessive excretion of δ-aminolevulinic acid; the same is found in lead poisoning. In terms of porphyrin excretion, none of the porphyrias produce a pattern similar to lead poisoning; exceptions are the urinary excretion of coproporphyrin, which is found in hereditary coproporphyria, and the excessive production of protoporphyrin seen in variegate porphyria and erythropoietic protoporphyria. A comparison of the patterns of change of enzyme activity in lead poisoning and in acute intermittent porphyria and hereditary coproporphyria shows the reason for these differences (Fig. 15).

In acute intermittent porphyria, there is depression of porphobilinogen deaminase activity leading to increased activity of ALA synthase, while in hereditary coproporphyria there is depression of coproporphyrinogen oxidase activity with subsequent increases in the activity of ALA synthase. In lead poisoning, three enzymes are clearly inhibited (ALA dehydratase, coproporphyrinogen oxidase and ferrochelatase) with consequent elevation in δ-aminolevulinic acid synthase.

There is thus one feature common to each of these porphyrias and lead poisoning; that is a rise in the activity of ALA synthase. The only other feature of two of the acute porphyrias that might be found in lead poisoning is a porphyrin-mediated solar photosensitization of the skin. In variegate porphyria, increased protoporphyrin leads to photosensitivity; in a proportion of cases with hereditary coproporphyria, a photosensitive skin eruption may be found, although this is never the case in acute intermittent porphyria. Photosensitization in lead poisoning is uncommon, but it has

| | ALA.S | ALA.D | URO.S | URO.D | COPRO.O | FERRO.C. |
|---|---|---|---|---|---|---|
| LEAD POISONING | ↑ | ↓ | ↑ | 0 | ↓ | ↓ |
| ACUTE INTERMITTENT PORPHYRIA | ↑ | 0 | ↓ | ↓ | 0 | 0 |
| HEREDITARY COPROPORPHYRIA | ↑ | 0 | 0 | 0 | ↓ | 0 |

LEGEND TO FIGURE

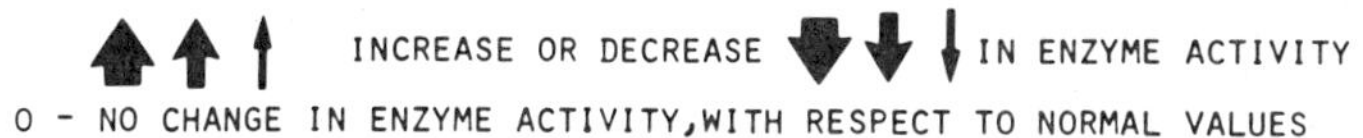


0 - NO CHANGE IN ENZYME ACTIVITY, WITH RESPECT TO NORMAL VALUES

ALA.S - δAMINOLAEVULINIC ACID SYNTHASE
ALA.D - δAMINOLAEVULINIC ACID DEHYDRATASE
URO.S - UROPORPHYRINOGEN 1 SYNTHASE
URO.D - UROPORPHYRINOGEN DECARBOXYLASE
COPRO.O - COPROPORPHYRINOGEN OXIDASE
FERRO.C - FERROCHELATASE

**Figure 15.** A comparison of the changes in enzyme activity in lead poisoning, acute intermittent porphyria and hereditary coproporphyria. Although there are many similarities in the excretion of intermediates of the heme biosynthetic pathway in lead poisoning and these two acute porphyrias, the only truly common point to all is the rise in activity of δ-aminolevulinic acid synthase. There are also similarities in depression of coproporphyrinogen oxidase in both lead poisoning and hereditary coproporphyria.

been reported (Allan et al., 1975). Various reasons have been advanced for the difference in photosensitizing lesions in lead poisoning and the cutaneous porphyrias, such as the presence of zinc chelates of protoporphyrin in lead poisoning (Lamola et al., 1975) and altered plasma binding of porphyrins (Piomelli et al., 1975). It is possible that depression of enzyme activity at an earlier stage in heme biosynthesis in lead poisoning than in hereditary coproporphyria means that lesser quantities of substrate are being metabolized by the depressed coproporphyrinogen oxidase. In consequence, lesser quantities of coproporphyrin are able to build up in the circulation. The defect may also lie in oxidation-reduction factors, since within the body the porphyrin normally exists as the reduced porphyrinogen which is only photoactive when oxidized to the porphyrin form (Fig. 2). This view is supported by the feature of altered terminal oxidation during lead exposure.

The liver is the main site of disturbance of porphyrin metabolism in acute porphyria. When ferrochelatase activity in the bone marrow is depressed by lead, however, hepatic cells show a compensatory increase in the activity of this enzyme (Otrzonsek, 1967). This may explain some of the clinical differences between lead poisoning and acute porphyria, in which the principal effects of acute porphyria are in the liver, whereas those of lead poisoning would appear to be in the bone marrow.

## B. The Influence of delta-aminolevulinic Acid

The role of the heme biosynthetic precursor ALA within the body has been the subject of several investigations, owing to its involvement in the pathogenesis of lead poisoning, acute porphyria and hereditary tyrosinanemia (Becker and Kramer, 1977). Attacks of acute porphyria are characterized, among other signs, by ascending paralysis, which can progress to quadriplegia and to respiratory failure. During these attacks, the blood and urinary levels of ALA reach very high values. In lead intoxication, blood ALA concentrations may double or more (Meredith et al., 1978) and hepatic concentrations of lead in these circumstances have been found to be increased fourfold (Suketa et al., 1975). In normal circumstances, blood ALA concentrations are in the region of 3 μmol/l (Meredith et al., 1978). During attacks of acute porphyria, the concentration of ALA in plasma has been reported to be as high as 24 μg/ml (200 μmol/l), at which time the ALA concentration in cerebrospinal fluid was 2.8 μg/ml (23 μmol/l) (Sweeney et al., 1970). These concentrations are commonly much lower (Moore and Meredith, 1976; Percy and Shanley, 1977).

In laboratory animals, high levels of ALA in the body are associated with various behavioral changes, including some impairment of voluntary movement. This effect has been observed after direct administration of

ALA (Moore and Meredith, 1976; Shanley et al., 1975; Yuwiler et al., 1970). ALA has also been shown to have various pharmacological activities (McGillion et al., 1975; Loots et al., 1975; Becker et al., 1976) and indeed to be taken up by various tissues including the brain (Becker et al., 1974; Moore and Meredith, 1976; Shanley et al., 1975). Some workers, however, have suggested that there are few behavioral effects from ALA in animals (Shanley et al., 1975) or in man (Doss, 1976) or toxic effects in mice (Kennedy et al., 1976) or in rats (Pierach et al., 1977).

In other experiments, effects of $\delta$-aminolevulinic acid (ALA) on social behavior were examined by ethological analysis of encounters between male mice. Thirty to forty minutes after receiving an injection of ALA, mice explored the cage and scanned less frequently than saline-injected controls and showed longer periods of immobility (Cutler et al., 1979b). Blockage of neuromuscular function appears to be the predominant effect of ALA in those preparations where the muscle becomes unresponsive to nerve stimulation. A separate toxic action, possibly affecting membrane transport or muscle metabolism, appears to predominate when inhibition of muscular contraction precedes blockage of the action potential, at which time the muscle becomes temporarily unresponsive to direct electrical stimulation.

Some of the known actions of ALA that may be responsible for its effect on nerve-muscle function include an inhibition of membrane sodium transport (Eales et al., 1971) and an inhibition of $Na^+$-$K^+$ dependent ATPase (Becker et al., 1971) with associated increase in capillary permeability (McGillion et al., 1975). Becker et al. (1975) found that ALA significantly reduced the resting membrane potential of frog sartorius muscle; this is an effect which would be expected from an inhibition of $Na^+$-$K^+$ dependent ATPase. However, Feldman et al. (1968), using the rat phrenic nerve hemidiaphragm preparation, obtained results suggestive of hyperpolarization. They found that ALA in amounts that could be expected in the acute phase of porphyria produced a presynaptic inhibition of the potassium-augmented miniature endplate potential frequency and appeared to reduce membrane excitability. It is possible that ALA may inhibit the transport of sodium in both directions across the cell membrane by blocking the channels through which sodium is transported into the cell during propagation of an impulse, in addition to its effects on ATPase. Some interference with ion transport across the cell membrane was indicated in the experiments of Cutler et al. (1979a) by the presence of polyphasic action potentials that occurred in frog nerve-muscle preparations following application of ALA.

Inhibition of $Na^+$-$K^+$ dependent ATPase may also affect neuromuscular function. Many different experimental conditions known to inhibit this enzyme deplete acetylcholine concentrations on presynaptic sites (Vizi,

1977). Such an effect would provide an explanation for the ALA-induced spontaneous contractions of muscle and for the initial increase in muscle contraction prior to blockage.

A further study of the mode of action of ALA is likely to provide some interesting information that may shed light on the role of ALA in lead poisoning. Such studies do not exclude the possibility of effects from the metabolites of ALA, the monopyrroles such as porphobilinogen, which have also been shown to cause presynaptic neuromuscular inhibition (Feldman et al., 1971).

These various studies have clearly demonstrated that the interaction of lead with two enzymes of heme biosynthesis results in marked increases in both urinary and blood levels of ALA. This passes through the blood-brain barrier and can cause marked changes in spontaneous activity of rodents. The possibility therefore exists that ALA can act either directly or indirectly on neuromuscular function and cause some or all of the neurological and behavioral effects of lead toxicity.

## XII. SUMMARY AND CONCLUSIONS

The influence of lead upon heme biosynthesis and upon the initial stages of heme degradation has been reviewed. The principal effects of lead are upon the biosynthetic enzymes. The enzymes inhibited are δ-aminolevulinic acid dehydratase, coproporphyrinogen oxidase and ferrochelatase. The greatest diminution of activity is in the first of these, δ-aminolevulinic acid dehydratase. As a consequence of depressed synthesis of heme, the activity of the initial and rate controlling enzyme of the pathway, δ-aminolevulinic acid synthase, is increased, as is the activity of heme-oxygenase, the first enzyme in the biodegradative sequence.

As a consequence of these changes in enzyme activity there is an increase in the production and excretion of the intermediates of the biosynthetic pathway, δ-aminolevulinic acid, coproporphyrin and protoporphyrin. The changes in these intermediates have been used together with δ-aminolevulinic acid dehydratase as secondary indices of lead exposure.Since synthesis of heme is depressed, so also is the synthesis of hemoproteins such as cytochrome P-450 and hemoglobin. Depression of cytochrome P-450 by lead will cause alterations in hepatic drug metabolism in both animals and man.

Anemia will also develop during lead poisoning due to three factors. These are depressed heme synthesis, altered globin synthesis and red cell fragility resulting in hemolysis.

Finally, because lead poisoning is so similar in presentation to the acute porphyrias, these are compared and contrasted. It is considered possible that the associated neurological manifestations of these diseases occur in part through the overproduction of the porphyrin precursor δ-aminolevulinic acid, which is shown to be neuroactive.

## REFERENCES


Akhtar, M., Abboud, M.M., Barnard, G., Jordan, P., and Zaman, Z. (1976). *Phil. Trans. Roy. Soc., London B., 273:*117.

Albaharry, C. (1972). *Am. J. Med., 52:*367.

Alessio, L., Bertazzi, P.A., and Monelli, O., Foa, V. (1976). *Int. Archiv. Occup. Environ. Hlth., 37:*89.

Allan, B.R., Moore, M.R., and Hunter, J.A.A. (1975). *Brit. J. Derm. 92:*715.

Alvares, A.P., Fischbein, A., Sassa, S., Anderson, K.E., and Kappas, A. (1978). *Clin. Pharmac. Ther., 19:*183.

Alvares, A.P., Kapelner, S., Sassa, S., and Kappas, A. (1973). *Clin. Pharmac. Ther., 17:*179.

Alvares, A.P., Leigh, S., Cohn, J., and Kappas, A. (1972). *J. Exptl. Med., 135:*1406.

Battersby, A.R., and McDonald, E. (1976). *Phil. Trans. Roy. Soc., London B. 273:*161.

Beattie, A.D., Moore, M.R., and Goldberg, A. (1972). *Lancet ii:*12.

Becker, D., and Kramer, S. (1977). *Medicine, 56:*411.

Becker, D.M., Goldstuck, N., and Kramer, S. (1975). *S. Afr. Med. J., 49:*1790.

Becker, D.M., Kramer, S., and Viljoen, J.D. (1974). *J. Neurochem. 23:*1019.

Becker, D.M., Viljoen, J.D., and Kramer, S. (1971). *Biochem. Biophys. Acta, 225:*26.

Becker, D.M., Viljoen, D., and Kramer, S. (1976) M. Doss (Ed.), In "Porphyrins in Human Diseases," p. 163. Karger, Basel.

Berk, P.D., Tschudy, D.P., Shepley, L.A., Waggoner, J.G., and Berlin, N.I. (1970). *Am. J. Med., 48:*137.

Berlin, A., and Schaller, K.H. (1974). *Z. Klin. Chem. Klin. Biochem. 12:*389.

Bissell, M.D., and Hammaker, L.E. (1976). *Arch. Biochem. Biophys. 176:*103.

Blumberg, W.E., Eisinger, J., Lamola, A.A., and Zuckerman, D.M. (1977). *Clin. Chem. 23:*270.

Brodie, M.J., Moore, M.R., Thompson, G.G., Campbell, B.C., and Goldberg, A. (1977). *Biochem. Soc. Trans. 5:*1466.

Campbell, B.C., Brodie, M.J., Thompson, G.G., Meredith, P.A., Moore, M.R., and Goldberg, A. (1977). *Clin. Sci. Mol. Med., 53:*335.

Center for Disease Control. (1975). *J. Pediat. 87:*824.

Chow, C.P., and Cornish, H.H. (1978). *Toxicol. Appl. Pharmacol. 43:*219.

Cramer, K., Goyer, R.A., Jagenburg, R., and Wilson, M.H. (1974). *Brit. J. Ind. Med., 31:*113.

Cutler, M.G., Dick, J.M., and Moore, M.R. (1979b). *Life Sci. 23:*2233.

Cutler, M.G., Moore, M.R., and Ewart, F. (1979a). *Psychopharmacol. 61:*131.

Dagg, J.H., Goldberg, A., Lochhead, A., and Smith, J.A. (1965). *Quart. J. Med.* 134:163.

Doss, M. (1976) In "Supplement to Proceedings of International Porphyrin Meeting on Porphyrins in Human Diseases," M. Doss and P. Nawrocki (Eds.), p. 49. Falk, Freiburg im Briesgau.

Doss, M., and Tiepermann, R.V. (1977). In "Clinical Chemistry and Chemical Toxicology of Metals," S.S. Brown (Ed.), p. 183. Elsevier/North Holland, New York.

Druyan, R., Haeger-Aronsen, B., Von Studnitz, W., and Waldenstrom, J. (1965). *Blood, 26:*2.

Eales, L., Douglas, R., and Isaacson, L.C. (1971). *Experientia, 27:*276.

Egan, G.F., and Cornish, H.H. (1973). *Toxicol. Appl. Pharmacol. 25:*467.

Feldman, D.S., Levere, R.D., and Lieberman, J.S. (1968). *J. Clin. Invest., 47:*33a.

Feldman, D.S., Levere, R.D., Lieberman, J.S., Cardinal, R.A., and Watson, C.J. (1971). *Proc. Nat. Acad. Sci., 68:*383.

Fischbein, A., Alvares, A.P., Anderson, K.E., Sassa, S. and Kappas, A. (1977). *J. Toxicol. Environ. Hlth., 3:*431.

Frydman, B., Frydman, R.B., Valasinas, A., Levy, E.S., and Feinstein, G. (1976). *Phil. Trans. Roy. Soc. London B., 273:*137.

Gibson, S.L.M., and Goldberg, A. (1970). *Clin. Sci. 38:*63.

Goetzl, A. (1911). *Wien. Klin. Wschr. 24:*1727.

Goldberg, A. (1968). *Seminars in Hematology, 5:*424.

Goldberg, A., Meredith, P.A., Miller, S., Moore, M.R., and Thompson, G.G. (1978). *Br. J. Pharmac. 62:*529.

Griggs, R.C., and Harris, J.W. (1958). *Clin. Res. 6:*188.

Grotepass, W. (1932). *Z. Physiol. Chem. 205:*193.

Haeger, B. (1957). *Scand. J. Clin. Lab. Invest. 9:*211.

Haeger-Aronsen, B. (1960). *Scand. J. Clin. Lab. Invest.* 12:suppl. 47.

Hasan, J., Hernberg, S., Metsala, P., and Vihko, V. (1967). *Arch. Environ. Hlth., 14:*309.

Hernberg, S., and Nikkanen, J. (1972). *Pracov, Lek. 24:*77.

Jackson, A.H., Games, D.E., Couch, P., Jackson, J.R., Belcher, R.B., and Smith, S.G. (1974). *Enzyme, 17:*81.

Jackson, A.H., Sancovich, H.A., Ferramola, A.M., Evans, N., Games, D.E., Matlin, S.H., Elder, G.H., and Smith, S.G. (1976). *Phil. Trans. Roy. Soc. London B. 273:*191.

Kamholtz, L.P., Thatcher, L.G., Blodgett, F.M., and Good, T.A. (1972). *Pediatrics, 50:*625.

Kennedy, G.L., Arnold, D.W., and Calandra, J.C. (1976). *Fd. Cosmet. Toxicol. 14:*45.

Kreimer-Birnbaum, M., and Grinstein, M. (1965). *Biochim. Biophys. Acta, 111:*110.

Laennec, R.T.H. (1831). In "Traite de L'auscultation Mediate," 4th ed., p. 79. Chaude, Paris.

Lamola, A.A., Piomelli, S., Poh-Fitzpatrick, M.B., Yamane, T., and Harber, L.C. (1975). *J. Clin. Invest. 56:*1528.

Landow, S.A., Schooley, J.C., and Arroyo, F.L. (1973). *Clin. Res. 21:*559.

Liebig, N.S. (1927). *Archiv. fur Exp. Path. Pharm. 125:*16.

Loots, J.M., Becker, D.M., Meyer, B.J., Goldstuck, N., and Kramer, S. (1975). *J. Neural. Transmission, 36:*71.

McGillion, F.B., Moore, M.R., and Goldberg, A. (1975). *Clin. Exp. Pharm. Physiol. 2:*365.

McKay, R., Druyan, R., Getz, G.S., and Rabinowitz, M. (1969). *Biochem. J. 114:*455.

Maines, M.D., and Kappas, A. (1976a). *Proc. Nat. Acad. Sci. 73,12:*4428.

Maines, M.D., and Kappas A. (1976b). M. Doss (Ed.) In "Porphyrias in Human Diseases," P. 43. Karger, Basel.

Maines, M.D., and Kappas, A. (1976c). *Ann. Clin. Res. 8:*39.

Maxwell, J.D., and Meyer, U.A. (1976). *Europ. J. Clin. Invest. 6:*373.

Meredith, P.A., and Moore, M.R. (1978). *Biochem. Soc. Trans. 6:*760.

Meredith, P.A., and Moore, M.R. (1979). *Biochem. Soc. Trans. 7:*39.

Meredith, P.A., Campbell, B.C., Moore, M.R., and Goldberg, A. (1977). *Europ. J. Clin. Pharmac. 12:*235.
Meredith, P.A., Moore, M.R., Campbell, B.C., Thompson, G.G., and Goldberg, A. (1978). *Toxicology, 9:*1.
Meredith, P.A., Moore, M.R., and Goldberg, A. (1979). *Clin. Sci. 56:*61.
Millar, J.A., Thompson, G.G., Goldberg, A., Barry, P.S.I., and Lowe, E.M. (1972). *Brit. J. Indust. Med. 29:*317.
Moore, M.R., Beattie, A.D., Thompson, G.G., and Goldberg, A. (1971). *Clin. Sci. 40:*81.
Moore, M.R., and Goldberg, A. (1974). A Jacobs and M. Worwood (Eds.) In "Iron in Biochemistry and Medicine," p. 118. Academic Press, New York.
Moore, M.R., and Meredith, P.A. (1976). D. Hemphill (Ed.) In "Proceedings of 10th Annual Conference on Trace Substances in Environmental Health, p. 363. University of Missouri, Columbia.
Moore, M.R. and Meredith, P.A. (1979). *Biochem. Soc. Trans. 7:*37.
Neuberger, A., and Tait, G.H. (1964). *Biochem. J. 90:*607.
Otrzonsek, N. (1967). *Int. Archiv. Gewerbepath. Gewerbehyg. 24:*66.
Percy, V.A., and Shanley, B.C. (1977). *S.A. Med. J. 52:*219.
Perutz, A. (1910). *Wein. Klin. Wschr. 23:*4.
Pierach, C.A., Guidon, L., Petryka, Z.J. Baur, H.R., and Watson, C.J. (1977). *Experientia, 33:*873.
Piomelli, S., Lamola, A.A., Poh-Fitzpatrick, M.B., Seaman, C. and Harber, L.C. (1975). *J. Clin. Invest. 56:*1519.
Piper, W.N., and Tephly, T.R. (1974). *Fed. Proc. 33:*588.
Piper, W.N., and Van Lier, R.B.L. (1977). *Mol. Pharmacol. 13:*1126.
Porra, R.J., and Falk, J.E. (1964). *Biochem. J. 90:*69.
Poulson, R. (1976). *J. Biol. Chem. 251:*3730.
Poulson, R., and Polglase, W.J. (1975). *J. Biol. Chem. 250:*4.
Ribeiro, H.A. (1970). *Western Pharmacol. Soc. Proc. 13:*13.
Schacter, B.A. (1975). C.A. Goresky and M.M. Fisher (Eds.), In "Jaundice," p. 80. Plenum, New York.
Scoppa, P., Roumengous, M., and Penning, W. (1973). *Experientia, 29:*970.
Shanley, B.C., Neethling, A.C., Percy, V.A., and Carstens, M. (1975). *S.A. Med. J. 49:*576.
Shemin, D. (1955). G.W. Wolstenholme (Ed.) In "Ciba foundation Symposium on Porphyrin Biosynthesis and Metabolism," p. 4. Churchill, London.
Shemin, D. (1968). *Symp. Biochem. Soc. 28:*75.
Shemin, D., and Rittenberg, D. (1945). *J. Biol. Chem. 159:*567.
Stokvis, B.J. (1895). *Z. klin. Med. 28:*1.
Suketa, Y., Aoki, M., and Yamamoto, T. (1975). *J. Toxicol. Environ. Hlth. 1:*127.
Sweeney, V.P., Pathak, M.A., and Asbury, A.K. (1970). *Brain, 93:*369.
Takaku, F., Aoki, Y., and Urata, G. (1973). *Jap. J. Clin. Hemat. 14:*1303.
Vizi, E.S. (1977). *J. Physiol. London 267:*161.
Waldron, H.A. (1966). *Brit. J. Indust. Med. 23:*83.
Waldron, H.A., and Stofen, D. (1974). "Subclinical Lead Poisoning," p. 80. Academic Press, New York.
Whitaker, J.A., and Vietti, T.J. (1958). *Pediatrics, 24:*734.
White, J.M., and Harvey, D.R. (1972). *Nature, 236:*71.
Wintrobe, M.M. (1950). *Harvey Lectures, 45:*87.
Yoneyama, Y., Sawada, H., Takeshita, M., and Sugita, Y. (1969). *Lipids, 4:*321.
Yuwiler, A., Wetterberg, L., and Geller, E. (1970). *Biochem. Pharmac. 19:*189.

# Lead and Energy Metabolism

*Richard J. Bull*

**TABLE OF CONTENTS**



## I. INTRODUCTION

Conclusive evidence does not exist to show that lead (Pb) toxicity is directly dependent upon an effect on energy or intermediary metabolism. On the other hand, no physiologic effect of Pb has been described in biochemical and molecular terms sufficiently well to refute involvement with cellular energetics. A variety of mitochondrial functions have been demonstrated to be sensitive to Pb *in vitro* and mitochondrial abnormalities are common findings with Pb toxicity *in vivo*. More limited evidence suggests that at least some of these effects occur at low enough exposure levels *in vivo* to be of physiologic significance.

Investigations into the effects of Pb on energy metabolism have not been comprehensive, in only a few cases have they been systematic. Although there is a substantial amount of evidence that Pb affects a number of steps in intermediary metabolism, electron and ion transport, there has been little effort to demonstrate how such effects may be involved in the systemic toxicity of Pb. The purpose of this monograph is not only to describe biologic effects of Pb, but also to provide a framework for future experimentation. Consequently, the present chapter will attempt to place effects of Pb on energy metabolism into the context of normal physiologic function. In many cases, this results in the stating of hypotheses in the absence of experimentally established ties between energy metabolism and a toxicologic effect of Pb. Accordingly, it is necessary to distinguish experimental fact from hypothesis.

Energy metabolism occupies a central position in all biologic processes. Biologic systems have evolved many complex mechanisms for integrating the controlled breakdown of carbon to hydrogen bonds as a means of deriving energy to perform physiologic functions. Few would argue against the idea that substantial interference with one or more steps in cellular energetics can have immediate and disastrous effects on the usual functional outputs of a cell, tissue or organ. It is now becoming apparent that mitochondria play much more subtle roles in cellular function than simply capturing the energy of the carbon to hydrogen bond. It is less clear how Pb interference with these mitochondrial properties may result in subtle toxicologic effects.

Toxicologic effects of Pb in mammalian systems are most often described in terms of physiologic function. As a result, there is a tendency to think that such effects are specific for the function of specific tissues or organs. It will become evident that different aspects of energy metabolism probably play greater or lesser roles in the function of specific tissues. Consequently, target organ specificity of observed effects can arise from activity of a cellular site common to a variety of tissues, but with more serious or, at least, more noticeable consequences in the tissue most dependent upon the affected subsystem.

One of the better understood ways in which the toxic effects of a chemical occur relates to the pervasive role mitochondria play in the regulation of intermediary metabolism. The function of all systems is directly dependent upon provision of energy in the form of adenosine triphosphate (ATP). In others, function may also be dependent upon generation of key intermediates. The ability of mitochondria to divert the flow of carbon and influence certain cellular functions by controlling intracellular concentrations of ions and other regulatory factors only recently has been recognized. Although mitochondria from different tissues share many properties, they do display intertissue and even intratissue heterogeneties with respect to these latter properties.

## II. BASIC ELEMENTS OF ENERGY METABOLISM

The literature available concerning cellular energetics is perhaps the most diverse and extensive in the biologic sciences; here only the features essential to providing a context for Pb effects on energy metabolism will be discussed, and even these will be described with some oversimplification. A variety of recent reviews deal with different aspects of mitochondrial function in some detail (Vignais, 1976; Papa, 1976; Packer and Gomez-Puyou, 1976; Hanstein, 1976; Baltscheffsky and Baltscheffsky, 1974; Green, 1974; Moore, 1972; Palmer and Hall, 1972; Slater, 1971; Lehninger, 1971; Beattie, 1971; van Dam and Meyer, 1971; Rabinowitz and Swift, 1970).

### A. Mitochondrial Structure

Quantitatively, the major capacity for conservation of the energy derived from oxidation of physiological substrates in aerobic cells is found in mitochondria. These subcellular organelles are bounded by an inner and an outer membrane, separated by intermembrane space. The outer membrane and the intermembrane space contain rotenone-insensitive NADH-cytochrome *c* reductase, nucleoside diphosphokinase, the fatty acid elongating system, adenylate kinase and monoamine oxidase (Palmer and Hall, 1972). Charged and uncharged molecules with a molecular weight of up to 10,000 appear to freely cross the outer membrane. The inner membrane, however, is very selective in its permeability characteristics. Essentially, all traffic of charged species across this membrane takes place by virtue of carrier mechanisms (Lehninger, 1971; Williamson, 1976). Included in the need for transport are the anionic substrates of tricarboxylic acid cycle (TCA) enzymes which are isolated from the cytosol in the mitochondrial matrix by the inner membrane (Chappel, 1968). Respiratory chain components are associated with the infoldings of the inner membrane, referred to as cristae, and are also effectively isolated from the cell sap (Palmer and Hall 1972).

It is of some interest that mitochondria contain nucleic acids and a capacity for protein synthesis which is independent of nuclear DNA (Rabinowitz and Swift, 1970; Beattie, 1971). It is now clear that mitochondria do not contain even a substantial portion of the compliment of deoxy ribonucleic acid (DNA) required for their own duplication (Mahler, 1976). However, the few polypeptides that are synthesized in mitochondria do appear to be essential to the formation of mitochondria fully competent in oxidative phosphorylation (Davidian and Penwell, 1971; Oerter and Bass, 1973; Tzagoloff, 1971; Mahler, 1976).

### B. Electron Transport

The function of the mitochondrion as a specialized cellular organelle centers around its capability for electron transport and the coupling of this transport to phosphorylation of adenosine diphosphate (ADP). Although features of this energy conserving system have been known for many years and a wealth of information exists concerning its characteristics, neither the electron transport chain, nor the exact mechanism of coupling of electron transport to phosphorylation of ADP can yet be specified.

The electron transport chain is most simply described in terms of a transfer of reducing equivalents from NADH or reduced flavoprotein to molecular oxygen as follows.

$$\text{NADH} \xrightarrow{} fp_1 \xrightarrow{\text{Site I}} \underset{\underset{fp_2}{\uparrow}}{\text{UQ}} \rightarrow \text{cyt. b} \xrightarrow{\text{Site II}} \text{cyt. } c_1 \rightarrow \text{cyt. c} \rightarrow \text{cyt. a} \xrightarrow{\text{Site III}} \text{cyt. } a_3 \rightarrow O_2$$

where $fp_1$ and $fp_2$ represent different flavoproteins, UQ represents ubiquinone and cyt. designates the respective cytochromes. Sites I, II and III are the approximate locations where electron transport is coupled with phosphorylation of ADP. This scheme neglects the as yet incompletely understood nonheme iron components in the $fp_1$ and UQ regions (Chance, 1972; Estrada-O and Gomez-Lojero, 1976). Similarly, energized forms of cytochromes *b* and $a$-$a_3$ thought by some to be involved in oxidative phosphorylation are not depicted (Wilson and Erecinska, 1975; Wilson et al., 1974). It is important to keep this basic sequence in mind since chemicals differentially affect sites in the electron transport chain and in oxidative phosphorylation; Pb is no exception. Establishing the region of such interactions is important in mechanistic terms and may have implications in the expression of toxicity.

There are two basic hypotheses, and some variations on each theme, which attempt to explain coupling of electron transport to phosphorylation of ADP. The chemical hypothesis, suggested by analogy to substrate-level phosphorylation in the glycolytic pathway (Slater 1953), was the first proposed. Clearly separable from this is the chemiosmotic hypothesis (Mitchell, 1961). The latter hypothesis suggests that electrochemical gradients (largely, pH gradients) generated from electron transport serve as the energy source for oxidative phosphorylation. The major shortcoming of the chemical hypothesis is the failure to demonstrate a high energy intermediate. Similarly, lack of conclusive proof of the assumption that transfer of reducing equivalents along the electron transport chain is coupled with a translocation of protons that results in a membrane potential is the major weakness of the chemiosmotic hypothesis (Slater, 1971).

A major variant of the chemical hypothesis involves high energy conformational states of the cytochromes (Slater, 1971; Chance, 1972; Wilson et al., 1974; Green, 1974). Dissipation of the energy present in these states is suggested to be directly coupled to phosphorylation or transport phenomena. Models of this type of coupling are based on the observation of a red shift in the cytochrome *b* absorbance spectra accompanied by shifts in half-reduction potentials depending upon the (ATP)/(ADP)($P_i$) ratio ($P_i$ equals $PO_4^{\equiv}$) (Wilson and Erecinska, 1975; Chance et al., 1970). Thermodynamic data support a similar energized form for cytochrome $a+a_3$ (Wilson et al., 1974).

## C. Intermediary Metabolism

Substrates and intermediates of cellular energy conserving systems are diverse. Traditionally, the metabolic pathways involved in energy metabolism are considered to be glycolysis, the hexose monophosphate shunt, $\beta$-oxidation of fatty acids and the transaminases; which all feed into the tricarboxylic acid cycle (TCA cycle). These pathways can be viewed as capable of directing either carbohydrates, lipids or amino acids towards complete oxidation to $CO_2$ in the TCA cycle; alternatively, there may be interconversion of carbon skeletons between these substrate classes. The individual steps of these metabolic pathways are familiar to virtually all students of biology and can be found in elementary texts of biochemistry. Consequently, these details will not be dwelt on here.

Less generally recognized are some specialized relationships between energy metabolism and cellular function in particular tissues. In some instances, this involves intermediary metabolism; in others, regulation of the cellular environment occurs. Subsequent sections will attempt to deal with specific examples of this type of phenomena.

## D. Ion Transport

Mitchondria are now well known for their ability to accumulate and release ions. As indicated earlier, virtually all charged molecular species are excluded from simple diffusion across the inner mitochondrial membrane. Consequently, specialized carriers are required to move chemicals which exist as anions or cations at physiological pH in and out of mitochondria.

Anion carriers exist for the transport of inorganic phosphate ($P_i$), $OH^-$, dicarboxylic acids, tricarboxylic acids, $\alpha$-ketoglutarate, pyruvate, glutamate, glutamate-aspartate and the adenine nucleotides (Williamson, 1976). These carriers are essential to carrying out normal processes of energy conservation in mitochondria. They are also involved in a variety of biosynthetic functions of mitochondria, such as gluconeogenesis and ammonia detoxification (Williamson, 1976). Some anions are simply excluded from mitochondria because of the lack of a carrier mechanism. These include $Cl^-$, $Br^-$ and $SO_4^{=}$ (Chappel, 1968).

The ability of mitochondria to concentrate calcium was observed by Cleland and Slater in 1953. However, the importance of this finding was not immediately recognized. This was partially a result of the nature of the early experiments, which involved relatively high concentrations (>1 mM) of calcium (Vasington and Murphy, 1962). Such conditions produced badly damaged mitochondria and appeared as an essentially irreversible process of unknown physiological significance. However, it became apparent that electron transport is used to accumulate $Ca^{++}$ in preference to phosphorylation of ADP (Rossi and Lehninger, 1964). Secondly, accumulation of calcium under limited loading conditions (<500 $\mu$M) was more consistent with free calcium concentrations encountered within the cell (Rasmussen, 1971). Under these conditions, uptake of calcium was reversible and produced no discernible functional damage to the mitochondria. All mitochondria which have been studied display the capability to transport $Ca^{++}$ by a process displaying virtually identical affinities for the ion and with the same stoichiometry with electron transport (2 $Ca^{++}$/ATP equivalent) (Carafoli and Lehninger, 1971).

A number of other monovalent and divalent cations are taken up by isolated mitochondria. Monovalent ions, particularly $K^+$, are accumulated at a rate proportional to the rate of respiration (Brierley et al., 1971). These ions may also be exchanged for $H^+$ across the mitochondrial membrane in an energy-independent manner (Brierley, 1976). In this case, $Na^+$ appears more effectively exchanged than $K^+$. Although it is not clear if an exchange reaction is involved, $Na^+$ also induces the release of $Ca^{++}$ from the mitochondria of particular tissues (Crompton et al., 1978).

Divalent cations such as $Mn^{++}$, $Mg^{++}$, $Fe^{++}$ and $Zn^{++}$ also appear to be accumulated by mitochondria by energy-dependent processes (Gunter et al., 1978; Brierley and Knight, 1967; Flatmark and Romslo, 1975). These appear to involve a mechanism analogous to but distinctly different from that of the $Ca^{++}$ carrier.

## III. INTEGRATION OF ENERGY METABOLISM WITH CELLULAR FUNCTION

### A. Control of Energy Metabolism

The rate at which substrate is oxidized and the rate of electron transport in mitochondria can be directly related to the cellular energy demand. Such close regulation of the utilization of substrate is extremely important to the economy of the cell. Uncontrolled utilization of substrate would compromise the ability to divert carbon for biosynthetic and maintenance functions of the cell. Consequently, a complex web of control sites exist within pathways of intermediary metabolism and electron transport that allow the cell to divert the flow of carbon consistent with the needs of the moment.

A number of metabolic states can be produced in isolated mitochondria which in large part describe the possibilities for *in vivo* control of energy metabolism. The conventional states of mitochondria are described in Table 1. State 1 represents mitochondria as they are isolated from tissues. Since levels of substrate, $P_i$, and ADP are all limited, the (ATP)/(ADP) ($P_i$) ratio remains high and the rate of respiration low. If excess $P_i$ and ADP are added, substrate rapidly becomes depleted and limits respiration. Consequently the (ATP)/(ADP)($P_i$) ratio remains low in State 2. Addition of substrate puts the system into State 3, where respiration proceeds rapdily and ATP is synthesized to increase the (ATP)/(ADP)($P_i$) ratio. As $P_i$ and ADP are depleted in the synthesis of ATP, respiration is slowed in State 4. If oxygen is removed from the system, electron transport ceases and the (ATP)/(ADP)($P_i$) ratio falls in State 5. State 6 represents an energized-inhibited state of mitochondria which can be produced by the addition of low concentrations of $Ca^{++}$ ($<50\ \mu M$) to mitochondria in the absence of permanent anions, for example, $P_i$ (Chance and Schoener, 1966). This state is related to the energized forms of the cytochromes that are postulated to act as intermediates in the conformational hypothesis of coupled phosphorylation dealt with in a previous section.

The State 4 → State 3 → State 4 transitions are viewed as the most basic means of assessing mitochondrial function. These transitions measure the ability of mitochondria to respond to an impressed energy demand and the integrity of metabolic control over the utilization of substrate. In addition to the respiratory rates measured under the two conditions, two other parameters are commonly applied. The first is the respiratory control ratio (RCR), which is defined as the State 3 rate of respiration divided by the State 4 rate. The second is the ADP/O ratio which represents the number of ATP molecules formed from ADP per oxygen consumed. The RCR is a measure of metabolic control and the ADP/O ratio is a measure of the tightness of coupling between electron transport and phosphorylation of ADP. Since there are three phosphorylation sites between NADH and $O_2$ a ratio of 3.0 is accepted as the theoretical maximum. Oxidation of reduced flavoproteins, such as the succinic dehydrogenase coenzyme, bypasses site I; therefore, a maximum ADP/O ratio with substrates such as succinate is 2.0.

The basic control mechanism of energy metabolism is the extra-mitochondrial phosphorylation state which is indicated by the ratio of (ATP)/(ADP)($P_i$) (Owen and Wilson, 1974). Dependence upon this ratio has been demonstrated over a wide range of ATP, ADP and $P_i$ concentrations in cell suspensions, as well as in isolated mitochondria (Erecinska et

**Table 1.** Metabolic states of mitochondria*

| State | $O_2$ | Substrate | $P_i$ | Acceptor | $(ATP)/(ADP)(P_i)$ | Cation | Respiration |
|---|---|---|---|---|---|---|---|
| 1 | Excess | Endogenous | Endogenous | Endogenous | High | — | Slow |
| 2 | Excess | Near 0 | Excess | Excess | Low | — | Very slow |
| 3 | Excess | Excess | Excess | Excess | Medium | — | Fast |
| 4 | Excess | Excess | Excess | Excess | Low | — | Slow |
| 5 | None | Excess | Excess | Excess | Low | — | None |
| 6 | Excess | Excess | None | Present | Very high | $Ca^{++}$, $Mn^{++}$ | Slow |

* Chance and Hollinger, 1963; Chance and Schoener, 1966

al., 1977; Erecinska et al., 1978). Put into context of the normal functioning of the cell, this states that activities of the cell which utilize ATP directly control the rate at which energy metabolism operates. An increase in the functional activity of a cell, therefore, increases the rates of substrate oxidation and electron transport.

An alternate proposal made by Atkinson (1971) suggested that energy metabolism is governed by the cellular "energy charge." Energy charge is defined as follows:

$$\frac{\text{ATP} + 1/2\ \text{ADP}}{\text{ATP} + \text{ADP} + \text{AMP}}$$

This ratio, however, appears to be valid only under conditions of constant intracellular $P_i$ concentrations (Erecinska et al., 1977). Nevertheless, the formulation of the energy change concept offers a broader insight into problems of metabolic control.

Energy charge as a regulatory scheme attempted to conceptualize the integration of ATP-regenerating and ATP-utilizing sequences within the cell (Atkinson, 1971). It is based on the observation that the energy charge of tissues under normal conditions *in vivo* was always maintained within a very narrow range (0.8 to 0.95). Consequently, a regulatory range for ATP-utilizing versus ATP-regenerating would be based on a scheme such as that depicted in Figure 1. Regulation of ATP-utilizing systems (*e.g.*, biosynthetic

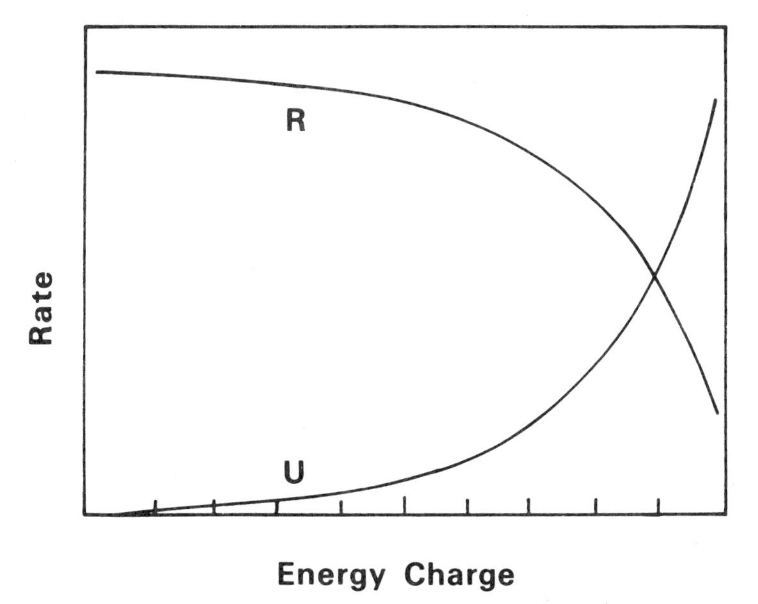


**Figure 1.** Common regulatory region of ATP-regenerating (R) and ATP-utilizing (U) systems under the assumptions of the energy charge concept (Atkinson, 1968).

pathways) is dependent both upon energy charge and the level of product. If product is low and energy charge is adequate, a high rate of synthesis would result. If product is high, energy charge may be utilized for alternate pathways.

One of the predictions of the energy charge concept is that energy charge potentials below 0.6 are out of the regulatory range and incompatible with a steady state (Ridge, 1972). In certain tissues, a system exists that can stabilize the energy charge under conditions of overwhelming energy demand. The purine nucleotide cycle is one such system; it was first described in skeletal muscle (Lowenstein, 1972). In the cycle, AMP is deaminated to inosine monophosphate (IMP) by adenylate deaminase. Adenylosuccinate synthetase catalyzes the formation of adenylosuccinate from IMP at the expense of one molecule of GTP. Adenylosuccinase in turn regenerates AMP through the removal of fumurate from adenylosuccinate. The rate of deamination of AMP in liver increases sharply with decreasing energy charge (Chapman and Atkinsin, 1973) with the net result of a contraction of the total adenine nucleotide pool. The adenine nucleotide pool has been shown to contract both *in vivo* and *in vitro* in a number of tissues in time of moderate to servere metabolic stress (Kleihues et al., 1974; Bull and O'Neill, 1975; Holmsen and Robkin, 1977). The significance of this response to metabolic stress may lie in an ability to briefly shut down systems dependent on ATP as a substrate to allow the cell to recoup its losses by ATP-regenerating systems which operate maximally at very low concentrations of ADP (Holmsen and Robkin, 1977). The net result may then be a short term sacrifice of cellular function to allow re-establishment of metabolic competency. Conversely, if the crisis is severe enough to result in permanent functional deficits, adenine nucleotide levels will be restored. Consequently, operation of protective mechanisms such as the purine nucleotide cycle can mask the role of cellular energetics in the development of functional damage. For example, prolonged ischemia (1 hr) produces substantial losses of the adenylate energy charge in the brain and permanent central nervous system damage. However, both the energy charge and the ATP content of the brain can be substantially recovered with several hours of recirculation (Kleihaus et al., 1974).

The dynamic and closely integrated relationship of function of a cell or tissue and energy metabolism has been demonstrated in a wide variety of systems. Increased work in the isolated heart is associated with increased metabolism, but the integration of control is so fine that adenine nucleotide concentrations do not vary significantly over a fairly wide range of workloads (Neeley et al., 1972). Similarly, rates of energy metabolism can be shown to change proportionately to isometric tension in vascular smooth muscle (Peterson and Paul, 1974), light flashes induce metabolic transients in the retina (Lolley, 1969), nerve stimulation produces increased metabolism in isolated sympathetic ganglia (Horowicz and Larrabee, 1962) and peripheral stimulation of nerves produces increased metabolic activity in the receptive areas in the brain (Kennedy et al., 1975). More speculative hypotheses consider the oscillations that can be generated through pathways involved in cellular energetics as the basis for circadian rhythms (Hess and Boiteux, 1971). These are but a few of many possible examples indicating a tight coupling between function and metabolism.

## B. Intermediary Metabolism

The common function of glycolysis, $\beta$-oxidation of fatty acids and TCA cycle oxidations in all tissues is either direct synthesis of ATP or the provision of reducing equivalents for synthesis of ATP through oxidative phosphorylation. These intermediates, in turn, serve as the immediate substrates of electron transport. The operation of these pathways is controlled, as is electron transport, to a large extent by the (ATP)/(ADP)($P_i$) ratio.

It is important to note, however, that there are other modifiers of intermediary metabolism in addition to the phosphate potential. Key steps in the glycolytic pathway and the TCA cycle are modified by a variety of effectors. These include end products, extracellular factors and cations, in addition to simple availability of substrate. It is beyond the scope of this review to detail the regulation of intermediary metabolism in depth. However, the complexity of the situation can be illustrated by a brief consideration of the regulatory inputs on the pyruvate dehydrogenase complex.

The level of active pyruvate dehydrogenase complex in a cell is determined by the activities of two other enzymes, pyruvate dehydrogenase kinase and pyruvate dehydrogenase phosphatase (Chiang and Sacktor, 1975). Phosphorylation of the pyruvate dehydrogenase by the Mg-ATP dependent kinase inactivates whereas the phosphatase activates the enzyme (Reed et al., 1973). In turn, the activities of these two enzymes are controlled by a variety of effectors. The kinase reaction is inhibited by ADP, CoA, NAD, pyruvate, pyrophosphate, $Ca^{++}$ and $Mg^{++}$ whereas the phosphatase reaction is activated by $Ca^{++}$ and $Mg^{++}$ (Cooper et al., 1975; Reed et al., 1975). These effectors would act to activate pyruvate dehydrogenase and increase the channeling of pyruvate into TCA cycle oxidations. On the other hand, the kinase is activated by ATP and the phosphatase is inhibited by Acetyl CoA and NADH (Cooper et al., 1975; Reed et al., 1973). These effectors, therefore, decrease the activity of pyruvate dehydrogenase and decrease the flow of pyruvate into the TCA cycle. In addition, there is evidence of direct end product inhibition of pyruvate dehydrogenase by acetyl CoA and NADH. These modifications are thought to play a role in decreasing the utilization of pyruvate when fatty acids are being utilized as the chief carbon source for TCA cycle oxidations (Reed et al., 1973). It is obvious that the result of certain of these effectors would definitely parallel the effects of the (ATP)/(ADP)($P_i$) ratio on electron transport. However, the effects of the cations add an additional dimension to the regulation of the enzyme that may be related to extracellular influences upon metabolism (Rasmussen, 1971).

There are a number of pathways of intermediary metabolism which appear peculiarly involved in the function of particular tissues which deserve mention. The close association of such pathways to energy metabolism is such that impairment of energy metabolism could have a rather direct effect on these functional pathways, giving rise to the appearance of target organ specific effects.

Maintenance of blood glucose concentration by the liver is perhaps the best established pathway. The function of certain organs, most notably the brain, is critically dependent upon the maintenance of a stable level of blood glucose concentration (McIlwain and Bachelard, 1971). Regulation of the storage of export of glucose by the liver is controlled by a complex of endocrine factors. Briefly, these control mechanisms direct that glucose is stored in the liver as glycogen when present in excess (*e.g.*, following a meal). Under the reverse circumstance, this glycogen can be exported to maintain the blood glucose concentration. As glycogen is depleted, glucose is also generated by gluconeogenesis from fatty acids and amino acids derived from lipid and protein catabolism. This is accomplished through the conversion to TCA cycle and/or glycolytic intermediates and effective reversal of glycolytic pathway.

The means by which tissues dispose of ammonia derived from catabolism of amino acids as source of energy represents a second such specialized pathway. Ammonia is excreted as urea, which is formed in the liver via the Krebs-Hensliet pathway (urea cycle). Other tissues, such as the brain, do not possess this pathway and must detoxify ammonia produced by oxidative deamination (Hawkins et al., 1973; McIlwain and Bachelard, 1971). Detoxification of ammonia in the brain depends upon the addition of ammonia to glutamate to form glutamine by an energy-dependent process (McIlwain and Bachelard, 1971). Glutamine diffuses from the brain and is carried to the liver via the blood, where glutaminase makes the ammonia available for urea synthesis. It is well established that removal of ammonia from brain is intrinsically dependent upon provision of glutamate from nonamino acid sources by transamination of $\alpha$-ketoglutarate. In isolated cerebral tissues, only glucose serves effectively as a carbon source for this purpose (Weil-Mahlerbe, 1974).

It is significant that the capacities of cells for intermediary metabolism varies widely between tissues. Specialized adaptions of intermediary metabolism of particular tissues have long been recognized (e.g., the virtually complete dependence of the adult brain on glucose as the substrate for energy metabolism). In recent years it has also become apparent that intermediary metabolism is heterogeneous within tissues. This is particularly well established with amino acid metabolism. At least two distinct TCA cycles can be demonstrated in brain *in vivo* by the paradoxical appearance of glutamine-glutamate specific activity ratios greater than 1.0 following the administration of certain radioactively labeled substrates such as acetate (Van den Berg, 1972). Such a ratio is incompatible with a substrate/product relationship in a single compartment system. This has led to the postulate of a large glutamate pool in the brain with low metabolic turnover and a small glutamate pool with a rapid turnover (Cheng, 1972). Glutamine is synthesized rapidly from glutamate in the small pool and slowly, if at all, in the large pool. Other, similar relationships suggest the existence of at least one additional compartment in the brain (Cheng, 1972). Evidence of metabolic compartmentation is also available for other tissues, such as heart, liver and adipose tissue (Denton, 1972; Garfinkel et al., 1972) and appears to relate more to subcellular compartments. In brain, separable TCA cycles are generally ascribed to different cellular types within the tissue, with a possible third compartment represented by synaptic endings (Balazset al., 1972; Cheng, 1972).

In the brain, the arrangement of distinct TCA cycles is of particular interest because of the close association of TCA cycle activity to the synthesis of certain putative neurotransmitters. Of these, acetylcholine is best established as a neurotransmitter (Whittaker, 1968), but strong evidence also exists to implicate the amino acids glutamate and $\delta$-aminobutyric acid (GABA). A special adaptation of intermediary metabolism in the brain is reflected by the much higher concentrations of these amino acids relative to other organs (McIlwain and Bachelard, 1971). The dependence of synthesis and breakdown of these two amino acids on TCA cycle activity has already been pointed out.

It has been known for some time that the acetate used for acetylcholine synthesis in the brain is efficiently derived only from glucose or pyruvate *in vivo* or *in vitro* (Browning and Schulman, 1968; Sollenberg and Sorbo, 1970; McIlwain and Bachelard, 1971; Tucek and Cheng, 1974). Exogenous acetate is poorly incorporated into acetylcholine. More recent data has shown that relatively slight impairment of energy metabolism results in a proportionate decrease in the incorporation of labeled choline into acetylcholine (Gibson et al., 1975 and 1978; Blass, 1976). The level of inhibition utilized was insufficient to produce alterations in brain adenine nucleotides; consequently, it was concluded that synthesis of acetylcholine is directly dependent on TCA cycle activity. This is consistent with the finding of neurologic deficits that precede the depletion of energy-rich phosphates in hypoglycemia (Ferrendelli and Chang, 1973) and hypoxia (Seisjo and Swetnow, 1970a and 1970b). Other workers have maintained that high affinity choline uptake is rate limiting to acetylcholine synthesis. However, this transport system has yet to be clearly separated kinetically from choline acetylase (Simon et al., 1976).

### C. Ion Transport

The function of certain tissues, notably electrically excitable tissues, depends greatly upon the distribution of ions within the intracellular compartment and in relationship to the extracellular fluids. In these tissues, the ability of mitochondria to transport and release calcium, and perhaps other ions, can play an important role in cellular function. This section examines some of the evidence indicating that modification of these mitochondrial properties may have implications for certain target tissues.

Lehninger et al. (1978) have recently described one basis on which mitochondrial calcium transport can serve as a coordinator of cellular energetics. Essentially, the redox state of mitochondrial pyridine nucleotides appears to determine whether calcium is taken up or released by mitochondria. A relatively reduced level promotes uptake and retention of calcium, whereas an oxidized state results in release of calcium independent of the rate of electron transport. In other words, energy-depleted states would promote the release of calcium from mitochondria; in energy repleted states, calcium would accumulate. Due to the near equilibrium condition that exists between the (ATP)/(ADP)($P_i$) ratio and the (NADH)/($NAD^+$) ratio and the rapidity with which calcium may be taken up or released, it is an extremely attractive means of integrating a variety of cellular functions. However, in anoxic or ischemic states, $Ca^{++}$ would presumably be accumulated in mitochondria, thus disrupting the "normal" relationship between the phosphate potential and levels of reduced pyridine nucleotides. Consequently, some of the pathologic sequelae of such states may result from inappropriate distributions of cellular $Ca^{++}$.

Efflux of calcium from mitochondria of the adrenal cortex, parotid gland, myocardium and skeletal muscle is dependent upon $Na^+$ (Compton et al., 1978). On the other hand, mitochondria from liver, kidney, lung, uterus and ileum do not display this property. The former tissues are electrically excitable, utilizing local currents of $Na^+$ and $K^+$ to generate propagated action potentials. Consequently, sodium-induced efflux of calcium from mitochondria has the potential of coupling a variety of metabolic events with the functional state of the cells. Free calcium in the cytosol of these neural systems is a known prerequisite for secretory activity (Whitaker, 1968) and, in the case of muscle, triggers contraction (Krebs et al., 1973). Free cytosolic calcium is also closely linked with a variety intercellular and intracellular regulatory phenomena (Rasmussen, 1971). These properties of mitochondria have led to postulates that mitochondria may play a regulatory role in such activities as myocardial contractility through their capability of controlling intracellular calcium concentrations

(Haugaard et al., 1969). Such interpretations are consistent with the effects of cardiac glycoside-induced increased contractility of the myocardium, an effect presumably mediated by their well known effects on the $Na^+ - K^+$ stimulated ATPase, resulting in the elevation of intracellular free $Ca^{++}$ (Akera and Brody, 1978).

Brain mitochondria have a very high capability for calcium transport, exceeding heart mitochondria by at least 20-fold in the initial rate at which they will accumulate calcium (Nichols, 1978). This may be taken to indicate a very high sensitivity of brain mitochondria to changes in cystolic calcium concentration. Elevated cytosolic calcium concentrations are incompatable with excitability of the neuronal membrane (Tasaki, 1968). Consequently, mitochondria may play some role in regulating mechanisms involved in nerve transmission.

A second area where the calcium transport property of mitochondria are suggested to perform a physiologic function is in mineralization processes (Lehninger, 1977). Calcium accumulated in mitochondria is deposited as amorphous tricalcium phosphate and is apparently stabilized as such, rather than being converted to crystalline hydroxapatite. This property may be important in preventing the pathologic calcification of soft tissues. Formation of hydroxyapatite in biologic systems is a nearly irreversible process. Granules of tricalcium phosphate have been observed in mitochondria of osteoclasts, osteocytes and chondrocytes of healing bone. On this basis, it has been suggested that mitochondrial calcium accumulation might serve to localize calcium phosphate at sufficiently high concentrations to be utilized for normal physiologic processes such as bone mineralization (Lehninger, 1977; Becker, 1976).

## IV. EFFECTS OF LEAD ON ISOLATED MITOCHONDRIA

### A. Mitochondrial Accumulation of Lead

Several *in vitro* studies have clearly demonstrated the ability of isolated mitochondria to accumulate $Pb^{++}$ from incubation media. Both active and passive processes appear to be involved (Scott et al., 1971). Passive binding of Pb to mitochondria is biphasic on Scatchard plots, indicating two populations of sites which are saturated at approximately 50 and 140 $\mu$moles Pb/mg of mitochondrial protein for high and low affinity sites, respectively. The extent of Pb binding is affected by the composition of the incubation media, being maximal in the presence of the nonpermanent anions $NO_3^-$ and $Cl_3^-$, reduced considerably in acetate and substantially absent in the presence of $P_i$. In the latter case, the precipitation of insoluble lead phosphate salts is probably responsible.

Active transport of Pb into mitochondria appears to be closely related to calcium transport described in previous sections. The addition of Pb to brain or heart mitochondria in state 4 results in increased respiration (Scott et al., 1971; Holtzman et al., 1977). This stimulation of respiration requires the presence of a permeant anion such as $P_i$ or acetate. As shown in Figure 2, an additional uptake of Pb from incubation media can be demonstrated with the addition of oxidizable substrate (TMPD and ascorbate), using murexide as an indicator of free Pb concentration (Scott et al., 1971). This binding is reversed by the uncoupler m-chlorocarbonyl-cyanide phenylhydrazone (CCP) and inhibited by lanthanum ($La^{+++}$). Additionally, Pb produces an energy-dependent alkalization of the intramitochondrial space in a manner analgous to that produced by $Ca^{++}$, which is also competitively reduced by $Ca^{++}$ and $La^{+++}$.

The accumulation of Pb in isolated mitochondria is also demonstrable by the formation of electron-dense granules (Walton, 1973). Similar, but less electron-dense, granules could be demonstrated by incubating mitochondria in 4 mM $CaCl_2$ ("massive loading conditions"). The formation of such granules is dependent upon energy and is decreased substantially by inhibitors of electron transport. Unlike calcium granules, Pb granules could not be dispersed by incubation with 0.5 mM dinitrophenol (DNP). It should be noted that these experiments were conducted under substantially different conditions than those used by Scott et al. (1971), which showed reversible uptake of Pb. The principle difference was the inclusion of 4 mM phosphate in the medium and the high

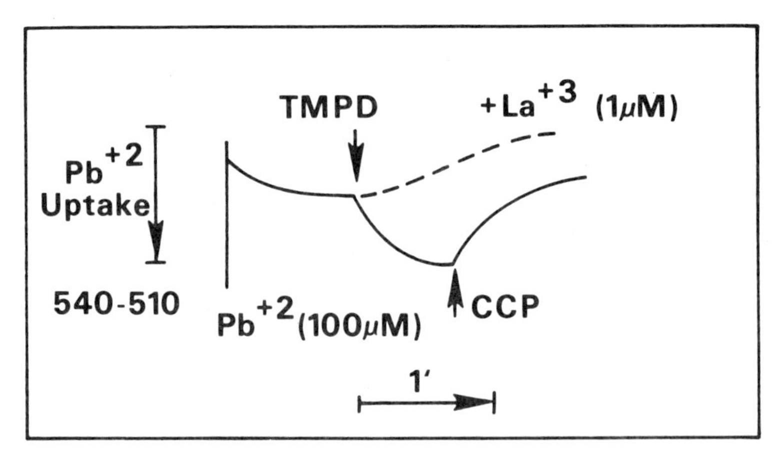


**Figure 2.** Respiration dependent uptake of $Pb^{++}$ in isolated heart mitochondria. Spectrophotometric trace of 540–510 nm reflects a decrease in the concentration of a colored Pb-murexide complex. Uptake initiated by addition of TMPD and ascorbate as substrates for respiration (Scott et al. 1971).

concentrations of Pb and $Ca^{++}$ utilized by Walton (1973). This allows for the formation of insoluble calcium and lead phosphates within the mitochondria.

## B. Effects of Pb on Respiration of Isolated Mitochondria

If proper consideration is given to the precipitation of Pb in phosphate containing media, mitochondrial respiration can be shown to be quite sensitive to Pb (Cardona et al., 1971; Scott et al., 1971; Parr and Harris, 1975; Holtzman et al., 1977; Loeppe and Miller, 1970). As an example, succinate supported respiration is inhibited 50% in intact heart mitochondria or submitochondrial particles by 10 $\mu$M Pb acetate in the absence of $P_i$ (Fig. 3), but only 10 to 15% in the presence of 100 $\mu$M Pb acetate in the presence of 10 mM $K_2HPO_4$ (Cardona et al., 1971). Similar results have been reported in brain mitochondria (Holtzman et al., 1977). Summaries of the literature which addresses inhibition of respiration in the presence and absence of $P_i$ are given in Tables 2 and 3, respectively.

As indicated earlier, the initial addition of Pb to mitochondria in the presence of a permeant anion such as $P_i$ results in enhanced respiratory rates (Scott et al., 1971; Koeppe and Miller, 1970; Holtzman et al., 1977). As Pb concentrations are increased above 0.25 mM, this increase becomes transient and is replaced by a progressive inhibition of respiration when NADH-linked substrates are utilized (Holtzman et al., 1977). This does not occur with succinate or $\alpha$-glycerophosphate. However, respiration with both NAD-linked substrates and succinate lose their sensitivity to ADP at concentrations of Pb of 0.5 mM and above. This activation of respiration closely resembles the effect of valinomycin, in that higher rates of respiration are induced in the presence of $K^+$ than are induced by $Na^+$ or tetramethylammonium ion (Scott et al., 1975). Additionally, respiration is greatly enhanced in the presence of permeant anions such as $P_i$ or acetate. Increased respiration by Pb is also partially related to the transport of Pb into mitochondria. However, to explain a maintained rate as is observed with succinate in brain mitochondria or NADH-linked substrates in heart mitochondria, a continual uptake and release of Pb would be necessary (Holtzman, 1977). There is no direct evidence to suggest cycling of Pb in and out of mitochondria.

Given that Pb at low concentrations inhibits respiration in isolated mitochondria, the question is, by what mechanism? A classical means of addressing this question is to utilize substrates that enter the cytochrome chain at different levels. As indicated earlier, NADH enters prior to site I; the reduced flavoprotein of succinic dehydrogenase enters between site I and site II. A combination of TMPD-ascorbate is capable of nonenzymatic reduction of cytochrome *c* and allows assessment of the electron transport through site III.

**Table 2.** Effect of Pb on substrate-dependent respiration *in vitro*, absence of phosphate

| Substrate | Preparation | Concentration ($\mu M$) | Effect | Reference |
|---|---|---|---|---|
| Pyruvate + malate | Heart IM* | 2.3 | −35% | Parr and Harris, 1976 |
| Pyruvate + malate | Heart IM | 10 | −50% | Cardona et al., 1971 |
| Succinate | Heart IM | 2.5 | −50% | Scott et al., 1971 |
| Glutamate + malate | Heart SM† | 100 | — | Scott et al., 1971 |
| $\beta$-Hydroxybutyrate | Heart SM | 2500 | — | Scott et al., 1971 |
| Ascorbate + TMPD | Heart SM | 300 | none | Scott et al., 1971 |
| NADH | Heart SM | 80 | −50% | Scott et al., 1971 |
| Glutamate + malate | Brain IM | 10–50 | — | Holtzman et al., 1977a |
| Succinate | Brain IM | 10–50 | — | Holtzman et al., 1977a |
| Succinate | Corn IM | 12.5 | −80% | Koeppe and Miller, 1970 |
| NADH | Corn IM | 50 | +174–640% | Koeppe and Miller, 1970 |

* IM, intact mitochondria

† SM, submitochondrial particles

**Table 3.** Effect of Pb on substrate dependent respiration *in vitro*, presence of phosphate

| Substrate | Respiration | Concentration ($\mu$M) | Effect | Reference |
|---|---|---|---|---|
| Pyruvate + malate | Heart IM* | 70 | −50% | Parr and Harris, 1975 |
| Pyruvate + malate + ATP | Heart IM | 7 | −50% | Parr and Harris, 1975 |
| Succinate | Heart IM | 100 | −10 to 15% | Cardona et al., 1971 |
| Succinate | Brain IM | 250–5000 | + | Holtzman et al., 1977 |
| Succinate | Brain IM | 7500 | − | Holtzman et al., 1977 |
| $\alpha$-Glycerophosphate | Brain IM | 150 | + | Holtzman et al., 1977 |
| $\alpha$-Glycerophosphate | Brain IM | 2500–7500 | − | Holtzman et al., 1977 |
| Glutamate + malate | Brain IM | 250–5000 | + | Holtzman et al., 1977 |
| Glutamate + malate + 10 min | Brain IM | 50 | − | Holtzman et al., 1977 |
| Pyruvate | Brain IM | 250–5000 | + | Holtzman et al., 1977 |
| Pyruvate + 10 min | Brain IM | 50 | − | Holtzman et al., 1977 |
| $\alpha$-Ketoglutarate | Brain IM | 250–5000 | + | Holtzman et al., 1977 |
| $\alpha$-Ketoglutarate + 10 min | Brain IM | 50 | − | Holtzman et al., 1977 |
| Malate | Brain IM | 250–5000 | + | Holtzman et al., 1977 |
| Glutamate | Brain FT† | 1000 | − | Holtzman et al., 1977 |
| NADH | Brain FT | 10000 | No effect | Holtzman et al., 1977 |

* IM, intact mitochondria

† FT, frozen and thawed mitochondria

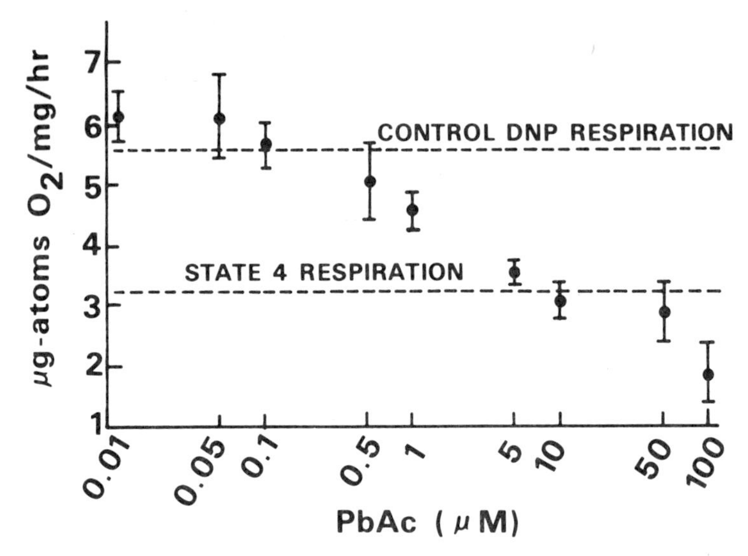


**Figure 3.** Pb-induced inhibition of dinitrophenol (DNP)-stimulated respiration in isolated heart mitochondria oxidizing succinate (10 mM) (Cardona et al., 1971).

Scott et al. (1971) examined the effect of Pb on respiration supported by NADH, succinate and TMPD-ascorbate in heart submitochondrial particles. Submitochondrial particles are used for this purpose to avoid limitations imposed by substrate transport. As can be seen in Figure 4, succinate oxidation is the most sensitive to Pb, whereas TMPD-ascorbate appears completely insensitive to Pb. NADH oxidation was inhibited by Pb, but at approximately eightfold the concentrations required for succinate oxidation. From these data it is apparent that inhibitory effects of Pb on electron transport occur prior to cytochrome *c*. Spectrophotometric evidence, showing oxidation of cytochrome *b* upon addition of Pb to heart mitochondria metabolizing succinate, indicates the sensitive site is prior to cytochrome *b* (Scott et al., 1971). These data indicate that the most sensitive site to Pb involves TCA cycle enzymes or is confined to the nonheme protein and ubiquinone segments of the electron transport chain. However, the inhibition of succinate oxidation at lower concentrations of Pb suggests that effects on matrix enzymes may be considerably more sensitive to Pb than the effect on electron transport.

Inhibition of respiration by Pb is substrate specific. On the surface, the pattern of Pb inhibition of respiration seems to display a substantial specificity for the organ from which mitochondria are derived. Although such a conclusion might appear to be reasonable from the literature, it cannot be clearly stated because conditions under which mitochondria from different tissues have been examined vary widely. In intact heart mitochondria incubated without $P_i$; succinate-supported respiration was very sensitive to Pb, while glutamate and malate oxidation was relatively insensitive (Scott et al., 1971). Brain mitochondria incubated in the presence of $P_i$ display a greater sensitivity of glutamate and malate oxidation to inhibitory effects of Pb than for succinate respiration (Holtzman et al., 1977a). Approximately equivalent concentrations of Pb were required to inhibit respiration supported by the two substrate systems in brain mitochondria incubated in the absence of $P_i$. Although not an *in vitro* experiment, reduced pyruvate oxidation was observed without reduced succinate oxidation in mitochondria obtained from kidneys of Pb-treated rats (Krall et al., 1974). The results of these studies taken together suggest the possibility of tissue-specific effects of Pb on cellular energetics that may serve as basis for apparent organ-specific toxicity produced by Pb. This is an area of Pb toxicology that requires much more systematic study.

### C. Pb Effects on Intermediary Metabolism

*In vitro* studies of Pb effects upon intermediary metabolism generally have been confined to mitochondrial enzymes. Relatively few specific enzymatic activities or substrate transport systems have been examined. However, the

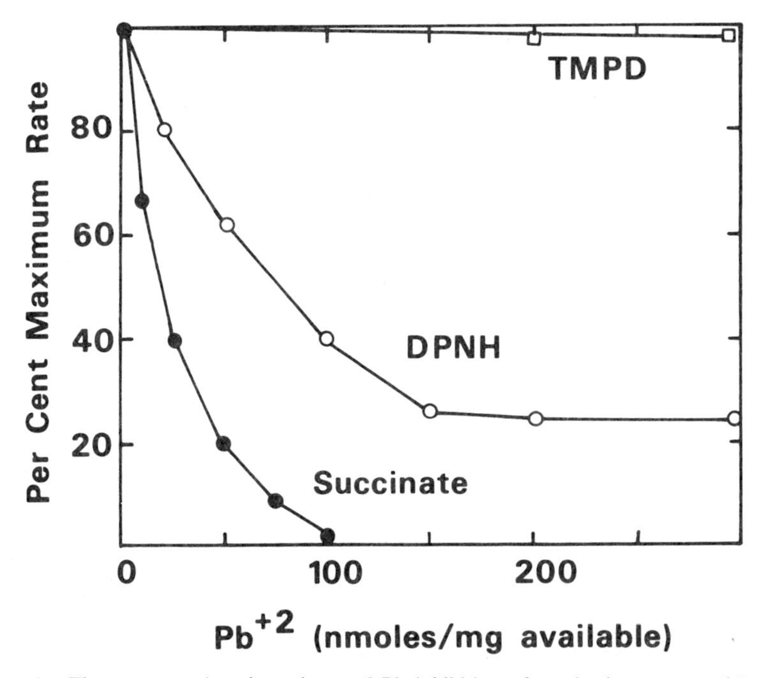


**Figure 4.** The concentration dependence of Pb inhibition of respiration supported by succinate, DPNH and TMPD-ascorbate in heart submitochondrial particles (Scott et al., 1971).

literature reviewed in the previous section suggests that inhibition of mitochondrial enzymes is probably related to inhibitory effects of Pb on respiration. Table 4 indicates the range of enzyme activities and substrate carriers affected by Pb in isolated mitochondria.

Scott et al. (1971) observed that Pb inhibited pyruvate oxidation at very low concentrations ($I_{50} = 2.5\ \mu M$) in isolated heart mitochondria. The site of this inhibition is not clear, although it appears likely to involve the pyruvate dehydrogenase complex. Several enzymatic activities are involved with this system. The thiamine pyrophosphate-dependent pyruvate dehydrogenase, dihidroxypolyl transacetylase and dihydrolipoyl dehydrogenase are sequentially involved in the generation of acetyl CoA from pyruvate (Reed et al., 1973). Within this complex, at least two systems have been demonstrated to be sensitive to Pb. Heart dihydrolipoyl dehydrogenase, a dithiol enzyme, is inhibited by 50% at a Pb concentration of 6.5 $\mu$M (Ulmer and Vallee,1969). Pyruvate transport may also be involved in that uptake of pyruvate by mitochondria is inhibited to the same degree by concentrations of Pb as low as 5 $\mu$M (Parr and Harris, 1976; Krall et al., 1971).

Succinate dehydrogenase is a second enzyme reported to be sensitive to Pb in isolated mitochondria. Cardona et al. (1971) indicated that succinic dehydrogenase in intact heart mitochondria was inhibited by Pb at concentrations of 10 $\mu$M. In heart submitochondrial particles, the sensitivity of this enzyme is apparently less, being 50% inhibited by 80 $\mu$M of Pb (Scott et al., 1971; Brierley, 1977). This difference may be attributed to accumulation of Pb in intact mitochondria or perhaps is related to the inhibition of succinate transport observed at 5 $\mu$M Pb in intact mitochondria (Scott et al., 1971; Krall et al., 1971). Succinate dehydrogenase in isolated kidney mitochondria appears to be less sensitive to Pb. Up to 2 mM of Pb produced no change in the activity of the enzyme (Iannaccone et al., 1974). Also in contrast to work in heart mitochondria (Scott et al., 1977), kidney succinate oxidase activity was reduced by 0.5 mM of Pb. Again, it is tempting to suggest that these differences in sensitivity to Pb indicate some tissue specific properties of these mitochondria. This is strengthened somewhat by the observation of reduced pyruvate oxidation without reduced succinate oxidation in mitochondria obtained from kidneys of Pb treated rats (Krall et al., 1974). However, the conditions of the Iannaccone et al. (1974) study differ so much from studies using heart mitochondria that such a conclusion is untenable. Basically, this study utilized a preincubation with Pb *in vitro* at 37°C, followed by centrifugation and washing. Such a procedure could allow for selective loss of enzyme as a result of Pb-induced damage to mitochondrial membranes, as well as loss of accumulated Pb.

Iannaccone et al. (1974) also report substantial inhibition of NADH cytochrome *c* reductase, glutamate dehydrogenase and cytochrome oxidase with Pb concentrations of 0.5 mM in kidney mitochondria. Although dif-

**Table 4.** Enzymes associated with cellular energetics that are inhibited by Pb *in vitro*

| Enzyme | Concentration ($\mu$M) | Effect | Reference |
|---|---|---|---|
| Adenyl cyclase | 2.4 | −50% | Nathanson and Bloom, 1976 |
| Succinic dehydrogenase (heart) | 80 | −50% | Scott et al., 1971 |
| Glycine cleavage enzyme | 2000 | −50% | Suketa et al., 1976 |
| Dihydrolipoyl dehydrogenase | 6.5 | −50% | Ulmer and Vallee, 1969 |
| Glutamate dehydrogenase | 500 | −65% | Iannaccone et al., 1974 |
| Cytochrome oxidase | 500 | −21% | Iannaccone et al., 1974 |
| Succinate oxidase | 500 | −43% | Iannaccone et al., 1974 |
| NADH cytochrome-c reductase | 500 | −82% | Iannaccone et al., 1974 |
| Succinic dehydrogenase (kidney) | 500 | None | Iannaccone et al., 1974 |

ficult to evaluate because of the concentrations used, as well as the problems stated above, these results differ qualitatively from those observed in heart mitochondria. Glutamate or TMPD-ascorbate supported respiration were relatively insensitive to Pb in heart mitochondria (Scott et al., 1971). These measures should reflect activities similar to those measured in kidney mitochondria. The data of Iannaccone et al. (1974) does qualitatively resemble the relative sensitivities of succinate-versus glutamate-supported respiration in brain mitochondria in the presence of $P_i$ (Holtzman et al., 1977).

An enzyme, which is not strictly one of intermediary metabolism but is involved in the regulation thereof, that is quite sensitive to Pb *in vitro* is adenyl cyclase (Nathanson and Bloom 1976). Both hormone-dependent and basal activities were inhibited at concentrations of $<3\ \mu M$ of Pb. The myocardial enzyme was most sensitive to Pb, being inhibited 50% at a concentration of 1 $\mu M$.

Unfortunately, very few definite conclusions can be formed concerning the effect of Pb on intermediary metabolism in isolated mitochondria. There seems to be little doubt that pyruvate oxidation is sensitive to $\mu M$ concentrations of Pb in mitochondria from all tissues examined. In the heart, succinate oxidation also appears quite sensitive. It is possible that effects on enzyme activities reported on mitochondria from other tissues may reflect tissue-specific characteristics of mitochondria. However, this cannot be taken as established and requires further experimentation.

## D. Effects of Pb on Ion Transport

In recent years, the property of mitochondria most studied *in vitro* in relationship to Pb effects has been ion transport. Most thoroughly investigated has been the transport of calcium, although it is clear that other ion transport systems are also affected. Studies which have addressed these points are listed in Table 5.

**Table 5.** Effects of Pb on ion accumulation by isolated mitochondria

| Ion | Preparation | Uptake | Efflux | Reference |
|---|---|---|---|---|
| $Ca^{++}$ | Heart | – | ? | Scott et al., 1971; Parr and Harris, 1976. |
| $K^+$ | Heart | + | + | Scott et al., 1971 |
| $NH_4^+$ | Heart | + | ? | Scott et al., 1971 |
| $Li^+$ | Heart | + | ? | Scott et al., 1971 |
| $Na^+$ | Heart | + | ? | Scott et al., 1971 |
| Acetate | Heart | + | ? | Scott et al., 1971 |
| $Cl^-$ | Heart | + | ? | Scott et al., 1971 |
| Pyruvate | Heart | – | – | Krall et al., 1971 |
| Succinate | Heart | – | – | Krall et al., 1971; Scott et al., 1971 |
| $Ca^+$ | Brain | – | ? | Goldstein, 1977 |

Pioneering studies by Scott et al. (1971) documented that Pb promoted osmotic swelling of mitochondria in media containing $K^+$, $NH_4^+$, $Li^+$ or $Na^+$ and a permeant anion. This is essentially a passive transfer and does not require an energy source. Heart mitochondria incubated in $K^+$ acetate, however, will accumulate additional $K^+$ by a respiration-dependent mechanism which produces rapid swelling of mitochondria. This process is enhanced by Pb as well. Addition of Pb also increases loss of accumulated $K^+$ from mitochondria in exchange for $H^+$, which is also dependent upon respiration and blocked by uncouplers of oxidative phosphorylation. These effects on active $K^+$ exchange can be induced by as little as 10 $\mu$M Pb. It is suggested that these changes may be related to swelling of mitochondria in *in vivo* studies described in later sections.

Interactions of Pb with calcium transport have been described in mitochondria obtained from heart and brain. Scott et al. (1971) were the first to demonstrate Pb transport into heart mitochondria by an energy-dependent system that had characteristics similar to the calcium transporter. Pb and $Ca^{++}$ transport in this system was demonstrated to be mutually competitive. However, the kinetics of the interaction were not described further. In brain, the uptake of Pb also appears to be energy dependent (Goldstein, 1977; Holtzman et al., 1977). Although calcium uptake in brain mitochondria appears to be competitively inhibited by Pb (Goldstein, 1977), mutual antagonism has yet to be demonstrated.

A more complete kinetic analysis of the Pb and Ca interaction in heart mitochondria has been reported by Parr and Harris (1976). Two distinct phases of Pb inhibition of calcium transport were observed in the absence of $P_i$ (Fig. 5). The more sensitive process was inhibited with a $K_i$ of 0.4 $\mu$M and had the properties of a competitive inhibition. Inhibition of the second phase was virtually complete at 0.8 $\mu$M Pb. If ATP was added to the incubation media, the distinction between the phases was lost, leaving a monotonic inhibition with a $K_i$, of 1.7 $\mu$M (Fig. 6). The basis for this effect of ATP has not been established.

The effect of Pb on uptake of calcium in brain mitochondria has not been as thoroughly investigated. In general, inhibition of calcium uptake (in the presence of ATP) appears to be competitive (Fig. 7). At a constant $Ca^{++}$ concentration of 10 $\mu$M (Fig. 8), $Ca^{++}$ uptake is inhibited 50% by Pb concentrations of 5 $\mu$M (Goldstein, 1977).

As indicated earlier, Pb does inhibit uptake of certain anions. Most notable are the substrates, pyruvate and succinate (Scott et al., 1971; Krall et al., 1974). Problems of substrate transport have been dealt with in a previous section. However, it is apparent that Pb does promote the transport of other anions. This can be of significance when an anion normally excluded from mitochondria apparently penetrates the mitochondrial membrane (*e.g.*, $Cl^-$). Pb does induce swelling in isolated heart mitochondria in a KCl

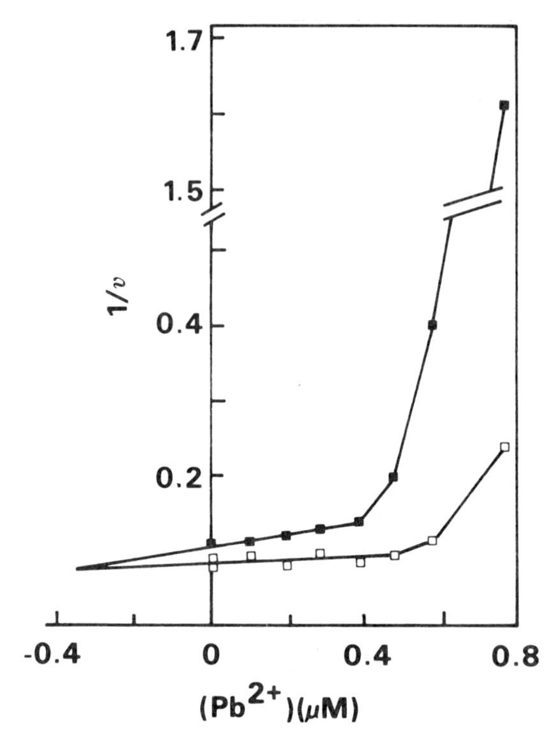


**Figure 5.** Dixon plot of the effect of $Pb^{++}$ on the reciprocal of $Ca^{++}$- removal rate of rat heart mitochondria. Incubation carried out in the absence of $P_i$ at concentrations of 109, (□), or 42, (■), μM $Ca^{++}$ (Parr and Harris, 1976).

media, which can be taken as an indication of damage to the mitochondrial membrane (Scott et al., 1971). Although less sensitive than effects of Pb on $Ca^{++}$ and $K^+$ transport, this damage may be related to observations of swollen mitochondria in tissues of Pb-treated animals.

## V. EFFECT OF Pb ON ENERGY METABOLISM OF ISOLATED TISSUES

A fairly diverse literature exists which examines the effects of Pb on the function of isolated tissues or cells. However, little consideration has been given to the possibility that such functional changes may result indirectly from a Pb effect on cellular energetics. These studies will not be reviewed here, since attempts to relate the phenomena observed would be entirely speculative. However, one has to conclude that in the absence of specific data to rule out an effect on energy metabolism, other mechanistic interpretations of data are equally speculative.

Distribution of Pb in isolated tissue or cell preparations has not been well investigated. However, some distribution to mitochondria is suggested by the development of electron-dense granules in the mitochondria of kidney cells (Walton and Buckley, 1977) and isolated brain capillaries (Goldstein et al. 1977). In the former tissues, these granules were observed in the absence of $Ca^{++}$, thereby avoiding confusion with calcium granules, which may be observed under certain conditions in isolated mitochondria (Walton, 1973). However, the uptake of $Ca^{++}$ by isolated brain capillaries from the incubation media was greatly enhanced by the Pb treatment (Goldstein et al., 1977). In both studies, cells were exposed to quite high levels of Pb, 100 $\mu$M and 400 to 1000 $\mu$M. Consequently, the absence of such granules *in vivo* does not argue against the validity of such a distribution where blood Pb levels would not be expected to exceed 5 to 10 $\mu$M even with severe lead poisoning.

Bull et al. (1975) investigated the *in vitro* effects of Pb on $K^+$-stimulated metabolism in cerebral cortical slices. Pb (67 $\mu$M) was observed to

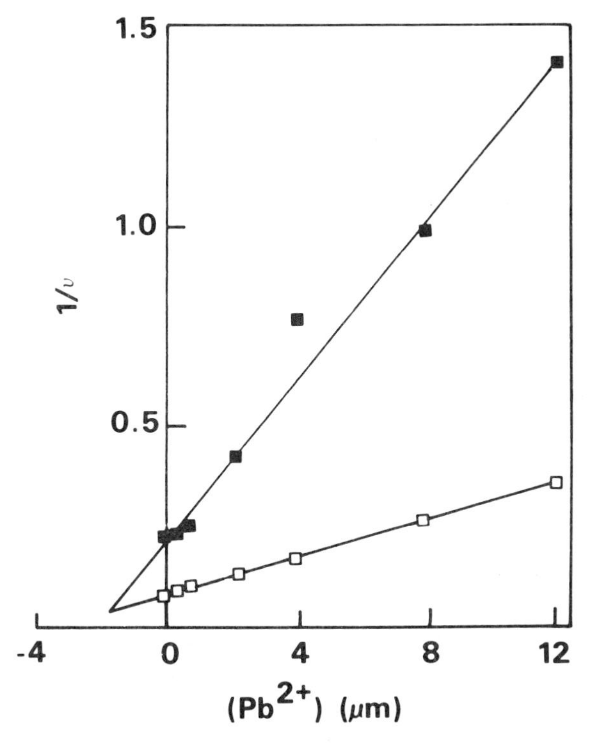


**Figure 6.** Dixon plot of the effect of $Pb^{++}$ on the reciprocal of $Ca^{++}$ removal rate of rat heart mitochondria in the presence of 1.2 $\mu$mol ATP/mg protein. Concentrations of $Ca^{++}$ were 102, (□), and 34, (■), $\mu$M $Ca^{++}$. Other conditions same as Figure 5 (Parr and Harris, 1976).

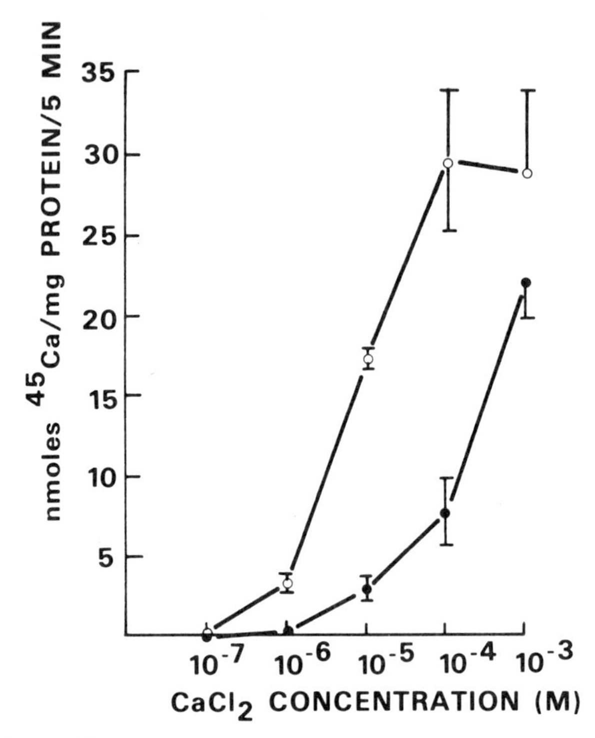


**Figure 7.** Effect of $Pb^{++}$ on the uptake of $Ca^{++}$ by isolated brain mitochondria. Pb was present at a concentration of 10 $\mu$M, (●), and is compared to control, (○), in the absence of $Pb^{++}$. $Ca^{++}$ bound in the absence of ATP was subtracted from the data (Goldstein, 1977).

decrease the immediate respiratory response to $K^+$ by 50%. This inhibition was associated with a decreased oxidation of tissue NAD(P)H* following $K^+$. (Changes in tissue pyridine nucleotides with stimulation were followed spectrophotometrically. NADH and NADPH cannot be distinguished from one another by these methods. Consequently, NAD(P)H is used to designate this uncertainty.) A small decrease was noted in the rate of aerobic glycolytic rate both before and after the addition of $K^+$. However, no significant changes were observed in the nonstimulated rate of oxygen consumption.

These data point up a well known feature of isolated brain tissue. Stimulated rates of energy metabolism almost universally show a higher sensitivity to toxic agents than resting rates (McIlwain and Bachelard, 1971). This in some cases can be attributed to a specific block to excitation processes, as has been demonstrated through the use of saxitoxin (Bull and Trevor, 1972) or tetrodotoxin (McIlwain, 1967). Alternatively, such changes may be the result of direct interference with energy metabolism, simply because such manipulations cause the tissue to respire at rates more comparable to those observed *in vivo* (McIlwain and Bachelard, 1971). The changes induced by such stimulation are rationalized on the basis that brain tissue *in vitro* exhibits little or no spontaneous electrical activity, in sharp contrast to the brain *in vivo*.

It is of interest that inhibition of $K^+$-stimulated respiration in brain slices by Pb (Bull et al. 1973) was substrate specific. Data described in the previous paragraph were obtained with glucose as substrate. With pyruvate at a concentration of 10 mM as substrate, no inhibition of K-stimulated respiration was observed, although a very small inhibition of $K^+$-induced NAD(P)H oxidation was observed. On the other hand, the respiratory response to $K^+$ with 10 mM of lactate as substrate was inhibited. The primary difference between oxidation of pyruvate from that of glucose or lactate is the obligatory donation of a cytosolic reducing equivalent (from the glycolytic pathway or lactic acid dehydrogenase, respectively). This suggests that some step in the aerobic oxidation of this reducing equivalent is responsible for impaired respiration with glucose or lactate. However, this reducing equivalent

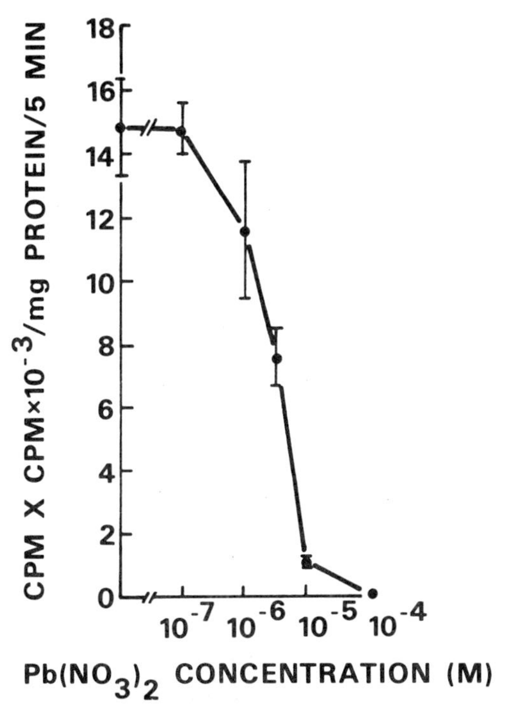


**Figure 8.** Concentration dependence of $Pb^{++}$-inhibition of ATP-dependent calcium uptake in isolated brain mitochondria (Goldstein, 1977).

apparently does not accumulate in the cytosol itself, because aerobic lactate output by the tissue was not increased with glucose as substrate. As discussed later, these results may be related to $Ca^{++}$ influence on the intramitochondrial compartmentation of pyridine nucleotides (Vinogradov et al. 1972).

Inhibited $K^+$-stimulated respiration in isolated brain tissue by Pb is probably best interpreted on the basis of inhibited energy metabolism. The demonstration of a substrate-specific inhibition of NAD(P)H oxidation is the principle direct evidence of this conclusion (Bull et al. 1975). It is difficult to account in a simple manner for such an effect by an indirect mechanism.

Apparently, no other tissue has been investigated *in vitro* with examination of parameters specifically related to energy metabolism. However, three studies deserve mention, since they provide evidence of disturbances in the control of cellular $Ca^{++}$ concentrations. While not specifically linked to cellular energetics experimentally, the possible relationships are quite evident from previous sections of this paper.

Uptake of $Ca^{++}$ by slices of kidney cortex is inhibited 37% by 200 $\mu$M of Pb (Kapoor and van Rossum, 1977). Mitochondrial $Ca^{++}$ accumulation was reduced to a greater extent than the remainder of the tissue (55% reduction). There appeared to be no effect of Pb on the extrusion of $Ca^{++}$ from renal cortex slices.

In a study aimed at the mechanism of Pb-induced contraction of arterial smooth muscle, Picinini et al. (1977) found that such tissues accumulate $Ca^{++}$ in the presence of $Pb^{++}$. In control tissues, $Ca^{++}$ concentrations were substantially independent of the media concentration of $Ca^{++}$. At relatively high concentrations of Pb, tissue concentrations of $Ca^{++}$ varied with the media $Ca^{++}$ concentrations. Compartmental analysis of calcium efflux from $^{45}Ca$-preloaded tissues indicated that turnover of intracellular $Ca^{++}$ was sharply inhibited by Pb. Half-times of efflux from this compartment were increased from 34.5 to 53.8 min by 1.2 mM of Pb. This change in tissue calcium status is associated with increased activation of $K^+$-induced contracture by the addition of increasing concentrations of $Ca^{++}$. These data suggest that intracellular free $Ca^{++}$ concentrations are increased to a greater extent by the addition of extracellular calcium in the presence of Pb. Such results can be related to inhibition of either the active extrusion of $Ca^{++}$ from the cell or the active uptake of this ion by mitochondria on the sarcoplasmic reticulum. Energy metabolism can, of course, be involved in all three of these mechanisms. However, no further evidence exists to substantiate this point or to differentiate between the three above possibilities.

Isolated brain capillaries in the presence of 100 $\mu$M of Pb avidly accumulate $Ca^{++}$ from the incubation media (Goldstein et al. 1977). On the basis of a lack of a Pb effect on $^{86}Rb$ uptake by the capillaries (as an analogue of $K^+$) and an inhibition of ATP-dependent calcium binding to isolated capillary membranes at $10^{-5}$ M Pb, the authors conclude that the effect was due to inhibited extrusion of $Ca^{++}$ from the cell. Further studies demonstrated an equivalent inhibition of calcium uptake by isolated brain mitochondria (Goldstein, 1977). It therefore remains to be established whether one of these mechanisms or both is responsible for the accumulation of $Ca^{++}$ induced by Pb.

## VI. *IN VIVO* EVIDENCE OF Pb EFFECTS ON ENERGY METABOLISM—ADULT ANIMALS

A variety of observations have been made to support the idea that changes in cellular energetics may be involved in certain toxicologic effects of Pb. The evidence available has been obtained using widely different experimental procedures and often using only a single parameter. Attempts to relate the effects of Pb on energy metabolism with a specific functional deficit in a tissue or organ are relatively rare.

### A. Subcellular Distribution of Pb *In Vivo*

A limited amount of experimental work has been directed toward determining the subcellular distribution of Pb *in vivo*. Two approaches have been utilized; one involves subcellular fraction of tissues following administration of a Pb isotope and another uses autoradiographic localization of the Pb isotope in the tissue. The former technique has the advantage of precise quantitation, but is of questionable significance because of the energy-dependent distribution of Pb implied in the foregoing discussions. Procedures used in deriving subcellular fractions are likely to destroy any energy-dependent gradients of Pb.

It is perhaps not surprising that little indication of Pb to mitochondrial fractions can be found on a milligram of protein basis using subcellular fractionation techniques (Barltrop et al., 1971). The exception to a number of tissues (kidney, liver and heart) appears to be spleen, which does seem to concentrate Pb preferentially in mitochondria. However, this may be secondary to deposition of Pb in an insoluble form in the mitochondrial matrix. It is of interest to note that, even with the deficiencies of the methodology used, Pb is at least equally distributed to mitochondria on this basis. Further, it has been shown that the distribution of Pb within liver mitochondria under similar conditions is largely associated with the so-called heavy mitochondrial fraction (Sabbioni and Marafante, 1976). This fraction consists of the inner membrane of mitochondria and the matrix containing respiratory enzymes. Consequently, there seems to be adequate evidence to indicate Pb does gain relatively free access to TCA cycle enzymes and the respiratory chain *in vivo*. That such distributions may be even more preferential to mitochondria *in vivo* is suggested by the autoradiographic observations of Murakami and Hirosawa (1973), who reported that more than half of the $^{210}Pb$ grains were associated with mitochondria in kidney epithelial cells. It is apparent that more extensive data is needed to establish this point for kidney and that such observations should be extended to other tissues.

The distribution of Pb in the tracer studies just described must be clearly differentiated from experiments using very high Pb concentrations in the diet. Exposure to Pb in the diet at concentration of 0.1% or more results in the formation of nuclear inclusion bodies in the kidney (Beaver, 1961). These bodies contain very large amounts of Pb which may be poorly exchangeable; this results in a considerable distortion in the distribution of Pb to the nucleus (Goyer, 1971).

### B. Ultrastructural Evidence of Damage to Mitochondria *In Vivo*

The most common evidence of mitochondrial involvement in Pb toxicity involves structural alterations in mitochondria of tissues taken from Pb-treated animals. Unfortunately, most of this evidence has been incidental to the main thrust of most studies; almost none of it is quantitative. This creates substantial problems in interpretation, because mitochondrial changes are one of the most common artifacts of poor tissue fixation.

The most systematic effort to identify ultrastructural alterations in mitochondria with Pb toxicity has been done in the kidney. Goyer (1968) demonstrated that mitochondria of the proximal renal tubule are swollen, assuming an oval or rounded shape in rats receiving 1% Pb acetate in their diet for up to 20 weeks. Similar observations had been made earlier by Beaver (1961). The swelling of mitochondria was related to aminoaciduria observed in the same animals (Sun et al., 1966; Goyer, 1968). Mitochondria isolated from the kidneys of Pb-treated rats display depressed respiratory control and ADP/O ratios (Goyer et al., 1968; Goyer, 1971). It is of interest that observations of swollen mitochondria have been made in kidney biopsies of Pb-exposed humans (Wedeen et al., 1975; Biagini et al., 1977). Consequently, this observation is an important link in relating animal experiments to the human experience.

Similar mitochondrial alterations have been observed in other tissues of Pb-exposed animals. Using intravenous injections of approximately 15 mg of Pb acetate/kg to rats, Hoffman et al. (1972) observed mitochondria, swollen, some to the point of complete loss of cristae, in Kupffer cells of the liver, among other pathologic changes. Liver parenchymal cells showed similar but less severe abnormalities. Asokan (1974) found similar changes in the myocardium of rats with blood levels of Pb averaging 112 $\mu$g/dl. Moore et al. (1975) observed swelling and loss of cristae structure in heart mitochondria of rats with blood Pb concentrations of 42 to 68 $\mu$g/dl. Press (1977) reported mitochondrial swelling as well as other structural changes in the cerebellum of neonatal rats treated with Pb.

## C. Evidence of Pb Interference with Energy Metabolism *In Vivo*

Demonstrating an effect on *in vivo* energetics is a difficult task. Certain conclusions can be made on the basis of measuring *in vivo* levels of metabolic intermediates. A second alternative is to use radiolabeled substrates and the measurement of incorporation into metabolic intermediates or end products such as $CO_2$. Additionally, tissues, subcellular fractions or enzymes may be removed from treated animals and subjected to analysis *in vitro*. A summary of particular enzyme activities, and respiratory systems which have been altered by Pb treatment *in vivo* is provided in Table 6.

Klausa and Rehner (1977) examined the levels of pyruvate, citrate, isocitrate and $\alpha$-ketoglutarate in kidneys of Pb exposed rats. Pyruvate and citrate concentrations were elevated in a dose-dependent manner over a range of 0.27 to 0.82% Pb in the diet. Concentrations of isocitrate and $\alpha$-ketoglutarate were not significantly changed from control. There were no significant changes in activity of phosphofructokinase, pyruvate kinase, isocitrate dehydrogenase or malate dehydrogenase. Citrate synthetase activity was increased by approximately 24%, while pyruvate and citrate levels were elevated 60% and 70%, respectively. A straight forward interpretation of these findings is not possible. The elevation of citrate concentrations does not necessarily refute the possibility that inhibition at the pyruvate dehydrogenase complex (which was not measured) was involved. Pyruvate is a regulator of its own oxidation (Cooper et al., 1975). Consequently, elevation of tissue pyruvate may compensate for an inhibition of Pb allowing the maintenance of citrate concentration. Actual increase in citrate concentrations may be accounted for by the small activation of citrate synthetase. Interpretation of this data would be greatly facilitated by measuring the rate of pyruvate oxidation to $CO_2$. In other experiments utilizing kidney slices obtained from rats exposed to 2 to 4% Pb acetate in the diet, pyruvate oxidation was not significantly inhibited as measured by the decarboxylation of ($^{14}C$)-pyruvate in adult rats. However, this same treatment did result in significant inhibition in 30-day-old rats (Hirsch, 1973).

In similarly treated rats, rat kidney mitochondria gave evidence of lower ADP/O ratios and depressed respiratory control ratios (Goyer and Krall, 1969; Krall et al., 1971; Goyer, 1971). State 4 rates of respiration were elevated and could be further increased by the addition of a uncoupler. These data do not support the idea that pyruvate oxidation was rate limiting. Rather, they suggest that pyruvate (as well as citrate) levels in the tissue were elevated as a result of feedback activation of pathways producing pyruvate for TCA cycle oxidations. Although the pyruvate dehydrogenase step is sensitive to Pb *in vitro* (Ulmer and Vallee, 1969), uncoupling of oxidative phosphorylation also would be expected to elevate tissue concentrations of pyruvate and citrate.

**Table 6.** Effect of Pb *in vivo* on enzyme activities or substrate-dependent respiration of isolated mitochondria

| Enzyme or substrate | Dose | Route | Effect | Reference |
|---|---|---|---|---|
| Kidney | | | | |
| Glutamate dehydrogenase | 0.1% Pb acetate | Diet | – | Iannaccone et al., 1976 |
| Malate dehydrogenase | 0.1% Pb acetate | Diet | – | Iannaccone et al., 1976 |
| Glucose-6-phosphate dehydrogenase | 0.1% Pb acetate | Diet | + | Iannaccone et al., 1976 |
| Dt diaphorase | 0.1% Pb acetate | Diet | + | Iannaccone et al., 1976 |
| Pyruvate carboxylase | 2% Pb acetate to dams-pups 1–21 days followed by 20–80 mg Pb/1 to 15 days of age | Diet | + | Stevenson et al., 1976 |
| Phosphoenolpyruvate carboxy kinase | " | Diet | + | Stevenson et al., 1976 |
| Fructose-1,6-diphosphatase | " | Diet | + | Stevenson et al., 1976 |
| Glucose-6-phosphatase | 10 mg/Kg per day | ip | + | Stevenson et al., 1976 |
| Phosphofructokinase | 0.5 to 1.5% Pb acetate 40 weeks | Diet | None | Klausa and Rehner, 1976 |
| Pyruvate kinase | " | Diet | None | Klausa and Rehner, 1976 |
| Citrate synthetase | " | Diet | + | Klausa and Rehner, 1976 |
| Isocitrate dehydrogenase | " | Diet | None | Klausa and Rehner, 1976 |
| Malate dehydrogenase | " | Diet | None | Klausa and Rehner, 1976 |
| Liver | | | | |
| $CO_2$ fixation | 5 and 10 mg/Kg | iv | – | Amatruda et al., 1977 |
| Brain | | | | |
| Glutamate-malate respiration | 4% Pb acetate to dams-pups 14–28 days of age | Diet | –RCR* | Holtzman et al., 1976 |
| Glutamate-malate respiration | 1% Pb acetate to dams-pups 0–28 days of age | Diet | –RCR | Holtzman et al., 1978b |
| Succinate respiration | " | Diet | – | Holtzman et al., 1978b |

* RCR, respiratory control ratio

Amatruda et al. (1977) have demonstrated the ability of intravenously administered Pb acetate to sharply decrease carbon dioxide fixation in hepatic mitochondria. This effect could not, however, be associated with inhibition of pyruvate carboxylase, a major means of fixing $CO_2$ in mitochondria. The authors concluded that the effect was due primarily to a dose-dependent depletion of mitochondrial ATP concentrations. (ATP is a cosubstrate of the pyruvate carboxylase reaction.)

As detailed in other sections of this monograph, heme synthesis is known to be sensitive to Pb. Interference with heme synthesis or Pb stimulation of heme oxygenase (Maines and Kappas, 1976) has the potential of decreasing tissue content of cytochromes. In either case, decreased levels of tissue cytochromes would be expected at some level to have impact on cellular energetics. Decreased levels of cytochrome $a+a_3$ have been observed in kidney mitochondria of rats fed a diet of 1% Pb acetate (Rhyne and Goyer, 1971). Whether inhibited heme synthesis or increased heme breakdown is responsible for this observation remains to be established.

Other enzyme activities of the kidney are modified by Pb treatment. Rats fed 0.1% Pb acetate displayed decreased glutamate dehydrogenase and malate dehydrogenase and increased levels of Dt diaphorase and glucose-6-phosphate dehydrogenase activities (Iannaccone et al., 1976). Decreases in glutamate dehydrogenase and malate dehydrogenase may be related to the demonstrated sensitivity of NADH-dependent respiration in certain isolated mitochondrial preparations (Holtzman et al., 1977). Increased glucose-6-phosphate dehydrogenase and Dt diaphorase activities were suggested by the authors as an adaptation to impaired pyruvate oxidation. However, increases in the former enzyme also occur in response to various types of cellular damage which increase the demand for NADPH used in biosynthetic reactions. Consequently, these changes are not clearly related to impaired cellular energetics.

Bull et al. (1975) have shown that $K^+$-stimulated respiration is inhibited in cerebral cortical slices taken from rats treated with intraperitoneal injections of Pb. As was the case with studies *in vitro*, the inhibition was accompanied by spectrophotometric evidence of impaired NAD(P)H oxidation and was observed when glucose, but not when pyruvate, served as substrate. Inhibition of the metabolic responses to $K^+$ is depicted in Figure 9. Inhibited oxidation of NAD(P)H was observed with total doses of Pb as low as 12 mg/kg administered over a 2-week period. This treatment resulted in blood and brain Pb concentrations of 72 $\mu$g/100 ml and 0.41 $\mu$g/g, respectively. This indicates that the energy metabolism of brain tissue is sensitive to Pb at levels not considered pathologic in the adult.

An increased turnover of intracellular $Ca^{++}$ accompanies the respiratory response of cerebral cortical tissues to $K^+$ (Bull, 1977). Tissues

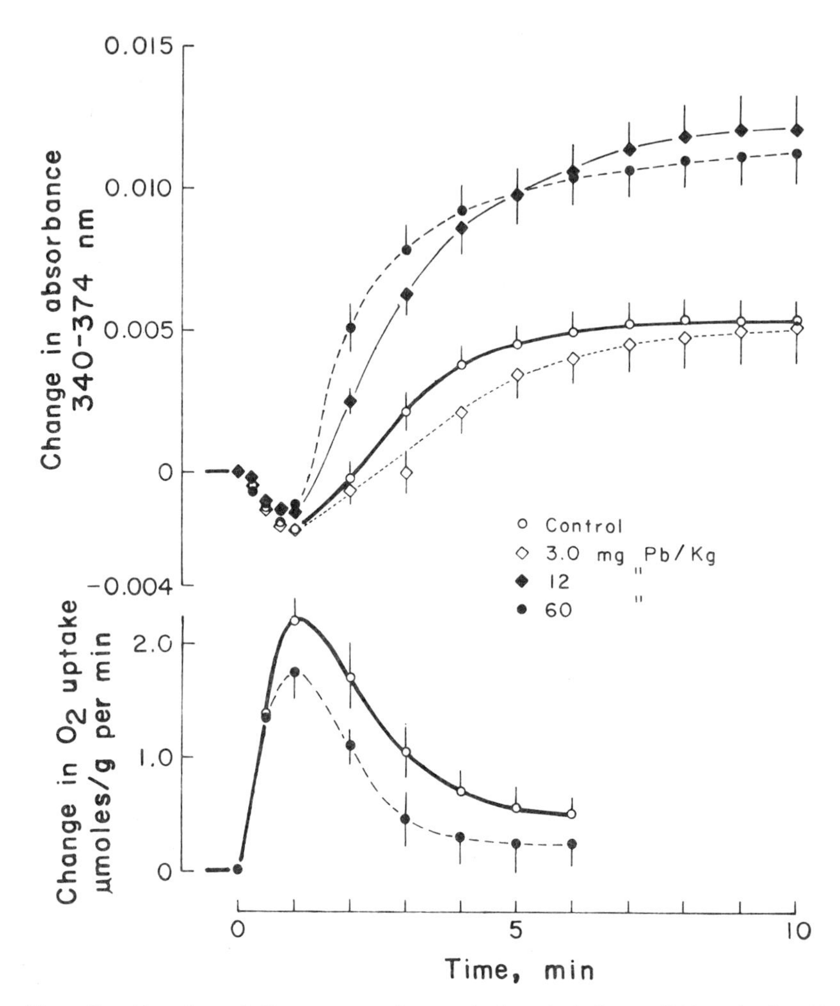


**Figure 9.** Altered metabolic responses of rat cerebral cortical slices to $K^+$ in animals subjected to 6 intraperitoneal injections over a 2 week period to total doses indicated. KCl concentration was increased from 3 to 30 mM at zero time. Upper trace is averaged responses of tissue NAD(P)H response. Downward deflection denotes oxidation. Lower trace denotes increases in tissue oxygen consumption induced by $K^+$ (Bull et al., 1975).

obtained from animals receiving a total of 60 mg Pb/kg (10 mg/kg × 6 over a 2-week period) displayed a small, and insignificant, decrease in the rate at which $^{45}Ca$ was lost from the tissue in the nonstimulated state (Fig. 10). However, the rapid efflux of $Ca^{++}$ which accompanies the respiratory response to $K^+$ was significantly inhibited in tissues taken from Pb-treated animals. Note should be taken of the fact that Pb treatment changed the kinetics of $^{45}Ca$ efflux, rather than the total amount released. Tissues obtained from control animals displayed a very rapid increase in $Ca^{++}$ efflux, which returned to baseline levels monotonically. $K^+$-induced efflux of $^{45}Ca$ from tissues of Pb-treated animals displayed a similar but reduced initial peak release of $Ca^{++}$, which returned to baseline in a distinctly biphasic manner. These data may indicate a particular cellular compartment of $Ca^{++}$ with a turnover markedly inhibited by Pb. In normal tissues, turnover of this $Ca^{++}$ pool is presumably rapid enough to be masked by the initial peak of $^{45}Ca$ efflux. The identify of this $Ca^{++}$ pool has not been established. However, its close association with altered metabolic responses to $K^+$ suggests that it could be of mitochondrial origin. Previous work has documented a $K^+$ and $Ca^{++}$ interaction as being involved in the respiratory responses of brain slices to $K^+$ (Bull and Cummins, 1973).

A hypothesis that Pb interfers with turnover of mitochondrial $Ca^{++}$ in brain is consistent with observations of increased concentrations of calcium in the brain of Pb-treated guinea pigs (Bouldin et al., 1975; Thomas et al., 1971). Such accumulations of calcium on tissues of Pb-treated animals is

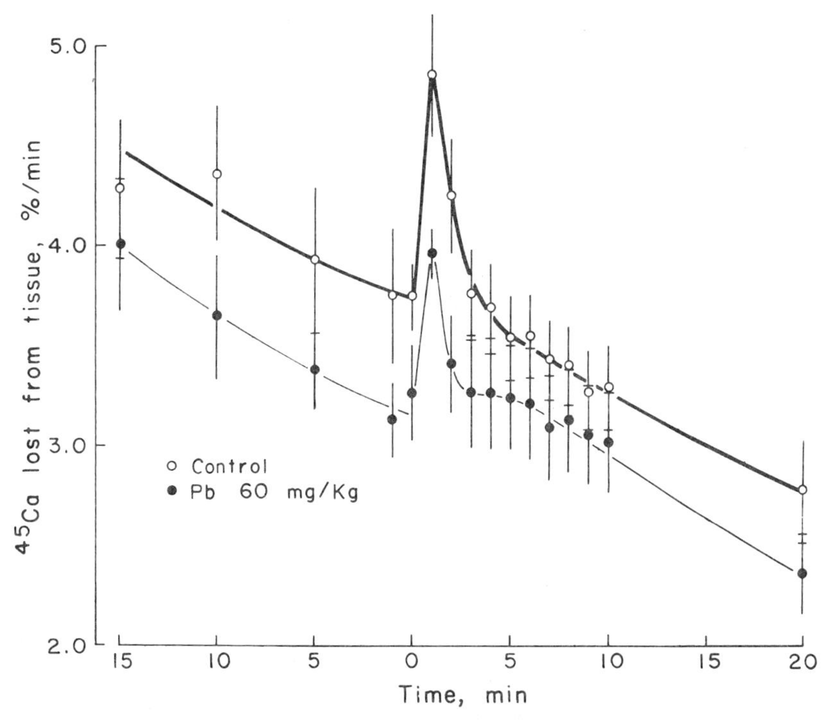


**Figure 10.** Effect of Pb treatment of rats (6 doses of 10 mg/Kg ip over a two week period) on the $K^+$-induced efflux of $^{45}Ca$ from cerebral cortical slices taken from treated and control animals. KCl concentration of the incubation media was increased from 3 to 30 mM at zero time (Bull, 1977).

not restricted to the brain. Recent data by Kapoor and van Rossum (personal communication) and Yamamoto et al. (1974) have demonstrated such accumulations in kidney and liver, respectively. However, direct evidence that this is secondary to interference with mitochondrial function has yet to be obtained.

## VII. *IN VIVO* EVIDENCE OF Pb EFFECTS ON ENERGY METABOLISM IN THE DEVELOPING ANIMAL

Recent work in Pb toxicology has tended to focus on the developing animal. Some of the difficulties encountered in applying the Pentschew and Garro (1966) model of lead encephalopathy are discussed in other chapters of this monograph. Basically, the difficulty is that female rats and mice exposed to Pb in food and water at concentrations at or above approximately 1000 ppm during the lactation period display decreased food and water consumption (Bornschien et al., 1977; Marker et al., 1975). The offspring of such animals are severely retarded in growth. This suggests that many of the effects observed with such exposures arise from decreased milk production and the ensuing undernutrition in pups. That the retarded growth is not a direct effect of Pb is demonstrated very clearly by the normal growth of pups directly intubated with doses of Pb many times those obtained in the mother's milk (Golter and Michaelson, 1975). Many of the ultrastructural behavioral, neurochemical and neurologic changes which have been associated with this model of Pb exposure have been shown previously to result from undernutrition alone. In most studies using exposure of this magnitude, there was no consideration of undernutrition. Therefore, it is difficult to attribute the effects observed simply to Pb. Consequently, this presentation will distinguish clearly between studies in which significant growth retardation was observed or would be expected and those in which this possibility has been excluded. However flawed the literature may be in this area, certain studies indicate that cellular energetics were closely associated with the final outcome.

Only three studies have addressed questions of cellular energetics and intermediary metabolism using high doses of Pb to the dam during the lactation period. Two studies addressed brain biochemistry, and the third investigated biochemical changes in the kidney. Patel et al. (1974a and 1974b) examined the central nervous system (CNS) metabolism of $^{14}$C-glucose in pups 7 to 24 days of age. Utilization of glucose (Himwich, 1970) and incorporation of $^{14}$C-glucose into amino acids of brain increased sharply between 10 to 30 days of postnatal age in the rat. Pb treatment significantly inhibited incorporation of $^{14}$C-glucose into cerebral and cerebellar amino acids at 7, 14, and 24 days of age. Although the level of $^{14}$C-glucose incorporation was constantly increasing with age, the specific activities of the amino acids were reduced by approximately the same proportion at each age (glutamate to between 49 to 56% of control, aspartate, 55 to 71% and glutamine, 23 to 34%). Substrates utilized by brain for cellular energetics change substantially during this developmental period (Land and Clark, 1975). However, the possibility of the depressed glucose incorporation into amino acids as the result of the use of alternate substrates was ruled out by the demonstration of a similar decrease in the incorporation of $^{14}C$-acetate into the amino acids (Patel, 1974b). Additionally, glutamate and aspartate concentrations in brain were significantly depressed in 19-day-old animals treated similarly. This is very strong evidence that Pb treatments of this magnitude severely limit TCA cycle oxidations. A question remains as to whether direct inhibition by Pb was involved or whether a severe delay in maturation of intermediary metabolism explained the results. The disproportionate decrease of incorporation $^{14}C$-glucose into glutamine suggests, in addition to other things, that the brain's ability to detoxify ammonia may be severely limited in the Pb-treated animal.

A developmental delay appears to be least partially involved in the observed changes in the intermediary metabolism of glucose in the brain of young rats. This is suggested by the smaller ratio of glutamine to glutamate specific activity in 19-day-old Pb-treated animals than in controls when $^{14}C$-acetate is used as the labeled substrate. The paradoxical glutamine/glutamate specific activity ratio was explained earlier as a measure of metabolic compartmentation in the brain. Since this ratio increases with age, it may be taken as an index of maturation of metabolic compartments; a depressed ratio is an indication of retarded development (Balazs et al., 1972). This interpretation is consistent with delays in morphologic development of animals subjected to similar Pb treatments (Krigman et al., 1974).

The second major study implicating effects on cellular energetics in the developing animal with high Pb exposures was that of Holtzman and Hsu (1976). Although exposure to Pb was as 4% Pb carbonate in the dam's diet, scheduling of the exposure was unique in this study. The Pb diet was not introduced until the 14th postnatal day, whereas other studies have introduced the Pb exposure on the day of birth. This level of Pb was maintained in the diet until 28 days of age, at which time pups from half the litters had developed encephalopathy. Of particular interest in the study of Holtzman and Hsu (1976) was the 40% increase in state 4 respiratory rate of cerebellar mitochondria within 1 day of treatment (Fig. 11). While still apparent after 7 days of treatment (21 days of postnatal age), the effect was lost at 14 days of treatment (28 days of age). However, at this point there is also observed a substantial inhibition of state 3 (ADP-dependent) respiration that was not apparent at the earlier times. These data strongly suggest that a very early effect of Pb treatment is an uncoupling of oxidative phorylation. Although there was a significant weight loss with a single day

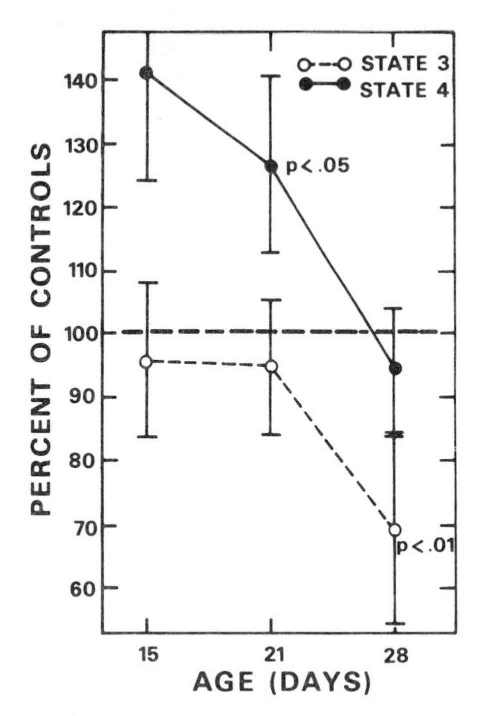


**Figure 11.** State 3 and State 4 respiratory rates of cerebellar mitochondria of rat pups whose dams were fed 4% Pb carbonate from 2 weeks postpartum relative to rates observed in control animals (Holtzman and Hsu, 1976).

of treatment, it is difficult to rationalize such an immediate effect of undernutrition on oxidative phosphorylation.

Stevenson et al (1976) treated female rats with 2% Pb acetate in the drinking water from 1 day following the birth of pups until weaning at 21 days of age. At 21 days of age, the pups were weaned to drinking water containing 20 to 80 ppm Pb. Although not specifically addressed in the paper, it is hard to imagine that these animals were not substantially retarded in their growth. The principle findings were an elevated blood glucose concentration and increased levels of gluconeogenic enzymes in the kidneys of Pb-treated animals at 56 days of age. This latter effect was associated with substantial increases in cyclic-3′,5′-AMP in the renal cortex. Since the response of adult animals to intraperitoneal doses of Pb presented only a partial duplication of this effect, it is difficult to determine whether the observed effects were due to the extremely large Pb exposure during the lactation period or to the more modest post-weaning exposure. It actually appears that there may be some interaction between the two levels of exposure, since most parameters displayed a dose-response relationship to the postweaning exposure. It is notable that a *decrease* in blood glucose was observed with somewhat greater exposures during the 0 to 24 days of age period in the Patel et al. (1974b) study. In terms of cellular energetics, one can only conclude that utilization of glucose is inhibited. The authors (Stevenson et al. 1976) suggest that the results relate to depressed pancreatic function on the basis of a depressed insulinogenic index (serum insulin concentration/blood glucose concentration). Other endocrine factors (*e.g.*, glucagon) were not examined, so this conclusion is not entirely warranted. An alternate possibility is that increased glucose concentrations represent an adaptation to decreased ability of the brain to utilize glucose (Patel et al., 1974a and 1974b). The increased renal concentrations of cyclic-3′, 5′-AMP may be evidence of such a hormonal adjustment.

Experiments which have investigated the role of cellular energetics in the toxicity to the developing animal at doses of Pb not complicated by undernutrition are much more limited. Two approaches have been taken experimentally to avoid undernutrition. The first lowers Pb levels in the diet or drinking water to a point that signs of undernutrition are not observed and the second is to directly intubate the neonatal animal. The former

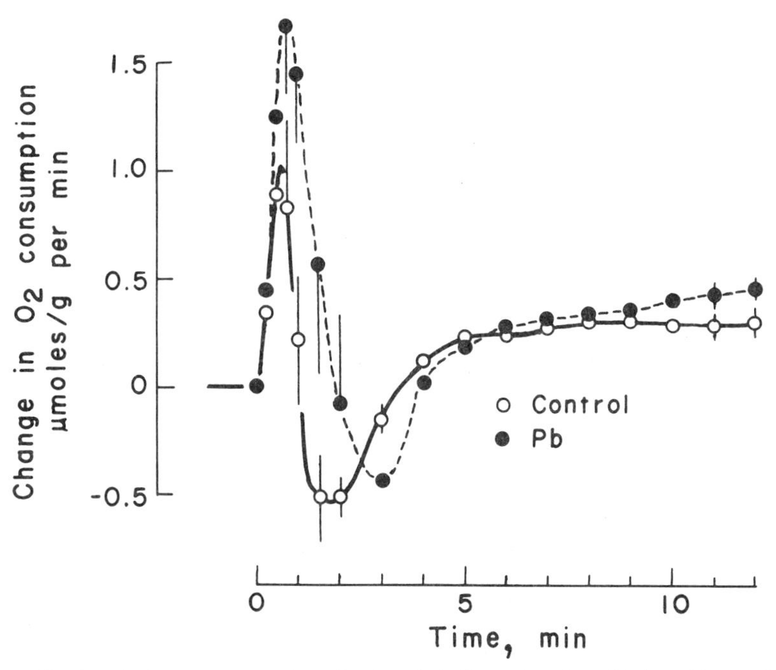


**Figure 12.** Respiratory responses of cerebral cortical slices taken from 15-day-old control and Pb-treated animals. Pb treatment is described in the text. Respiratory response was initiated by increasing the KCl concentration 3 to 30 mM in the incubation media at zero time (McCauley et al., in 1979).

approach is limited by the blood and brain Pb concentrations that can be achieved and the second, by the stressful mode of administration. To fully appreciate the range of effects of Pb, both methods will be necessary, since the maximum concentration of Pb that can be administered via the drinking water appears to be 400 mg/l of drinking water with a standard laboratory chow (Purina 5001). This leads to an average blood Pb concentration of about 50 to 60 μg/dl in rat pups at weaning.

Bull et al. (1979) and McCauley et al. (1979) utilized a protocol for treating female rats from 14 days prior to breeding through weaning of pups. The maximum dose used was 200 mg Pb/l of drinking water, which led to blood Pb concentrations of 36 μg/g wet weight in pups at 21 days of age. Metabolic responses of cerebral cortical slices taken from Pb-exposed pups at 15 days of age to $K^+$ were significantly altered. The $K^+$-induced respiratory response of cerebral cortical slices involves three distinct phases; these are an initial burst of oxygen consumption, interrupted by a sharp inhibitory period, followed by a second net increase in respiration (Fig. 12). While the basis of the three phases of this response is not completely understood, it was of interest that the Pb-induced change in the respiratory response only involved an enhancement of the initial respiratory burst (Fig. 12). This suggests that Pb is affecting energetics of a specific metabolic compartment at 15 days of age.

Glucose uptake by cerebral cortical slices of 15-day-old Pb-treated pups was also significantly greater than in tissues from control pups (Table 6). This increased uptake was specifically associated with $K^+$-stimulation, since glucose uptake in normal media $K^+$ concentrations was unaltered. lactic acid output by the tissues was not significantly altered in the presence or absence of elevated $K^+$ concentrations.

These data suggest an uncoupling of energy metabolism in the sense that increased respiration and utilization of substrate are associated with a

**Table 7.** The effect of pre- and postnatal treatment with Pb on the normal increase in cerebral cytochrome $c + c_1$ concentrations between 10 to 15 days of age in the rat pup and blood Pb concentrations at weaning (For further details, see Bull et al., 1978.)

| | Number | Blood Pb, 21 days of age | OD, 550 nm, 10–15 days |
|---|---|---|---|
| Control | 25 | 8.1 ± 2.0* | 100.0 ± 8.5† |
| 5 mg Pb/l | 12 | 11.7 ± 2.3 | 89.9 ± 10.0 |
| 30 mg Pb/l | 10 | 21.3 ± 1.9 | 82.4 ± 10.9 |
| 200 mg Pb/l | 20 | 35.7 ± 3.4 | 70.8 ± 8.1 |

* μg Pb/100 g ± SEM

† Percent of control change ± SEM

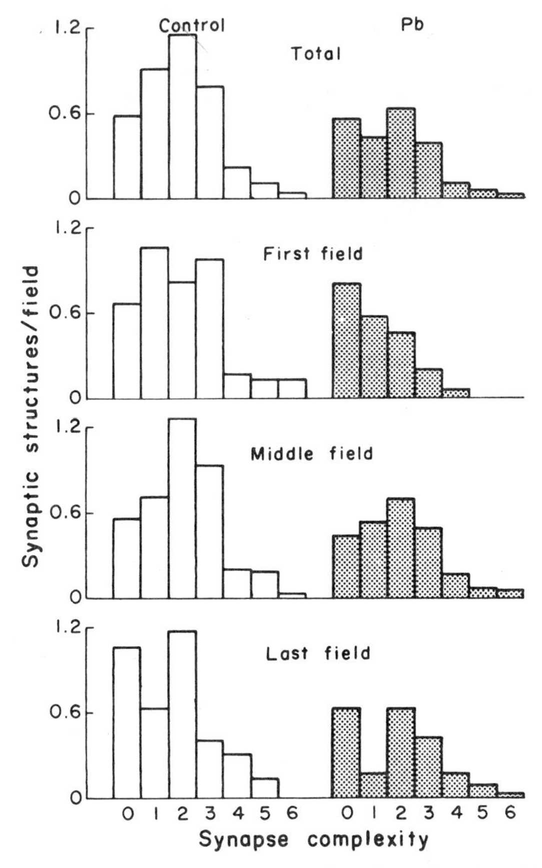


**Figure 13.** Relative synaptic counts in the cerebral cortices of 15-day-old control and Pb-treated rats (200 mg Pb/1 of drinking water). Synaptic complexity refers to the average number of presynaptic dense projections associated with ethanolphosphotungstic acid stained synaptic figures in not less than 6 fields from 7 nonlittermate animals. First, middle and last field approximate layers I, II, and III of the parietal cortex (McCauley et al., 1979).

standard stimulus (an increase of 3 to 30 mM in media $K^+$ concentration). This could arise from an actual uncoupling of oxidative phosphorylation or from inefficient operation of some energy-utilizing system (*e.g.*, transport of $K^+$ through a membrane which is structurally altered). However, observations of Pb effects on isolated brain mitochondria subjected to either *in vitro* (Holtzman et al., 1977) or *in vivo* (Holtzman and Hsu, 1976) treatment with Pb are consistent with an uncoupling of oxidative phosphorylation. An interesting aspect to these studies was that during the same time period there were signs of delayed brain development. In the developing rat there is a sharp increase in the capacity of the brain for oxidative metabolism (Himwich, 1970; Bull et al., 1979). Between 10 and 21 days of age, the normal increases in cerebral cytochrome concentrations was seen to be delayed by approximately 2 days. Table 7 documents the degree to which the rate of cytochrome $c + c_1$ increases were delayed by Pb treatment between the ages of 10 to 15 days of age in rat pups. The rates of cytochrome $c + c_1$ accumulation were depressed in a dose-dependent manner over a range of Pb concentrations from 5 to 200 mg/l of drinking water administered as described above. Blood Pb concentrations of litter mates taken at 21 days of age correlated significantly ($p < 0.01$) with the depressed rates of cytochrome $c + c_1$ accumulation. The delay could not be associated with any significant accumulations of brain $\delta$-aminolevulinic acid. On this basis, the delay could not be associated with a specific inhibition of heme synthesis (McCauley et al., 1978). During the period in which cerebral cytochrome concentrations are increasing, the cerebral cortex is undergoing substantial morphologic development and the animal's behavior is changing. For example, the number of synapses in the surface layers of the cerebral cortex increase markedly (Aghajanian and Bloom, 1967). Synaptic counts taken in the parietal cortex of Pb-treated animals using ethanol-phosphotungstic acid staining were also depressed at 15 days of age (Fig. 13). A decreased synaptic complexity in Pb-treated rats was consistent with an interpretation of a delay in synaptic development. Finally, it was documented that initial exploratory and locomotor activity of the Pb-treated pups was significantly depressed from that of control animals during the period of 15 to 21 days of age (Crofton et al. 1979). In summary, it appears that uncoupling of energy metabolism observed in cortical tissues of Pb-treated animals may be causally related to delays in cerebral cortical development.

## VIII. SUMMARY AND CONCLUSIONS

Experimental work in the past decade has described a number of rather specific interactions of Pb with enzymes of intermediary metabolism and oxidative phosphorylation. Evidence is beginning to accumulate that associates these phenomenologic observations with the *in vivo* toxicology of Pb. The evidence that modifications of cellular energetics may be responsible for toxicologic effects of Pb in intact animals can be summarized as follows.

1. Isolated mitochondria avidly take up Pb by an active process, presumably via the $Ca^{++}$ transporter.
2. A variety of mitochondrial functions have been shown to be very sensitive to Pb *in vitro*. These effects include uncoupling of oxidative phosphorylation, inhibited substrate oxidation and modification of the permeability of the mitochondrial membrane and ion transport, particularly $Ca^{++}$ transport.

3. The energy metabolism of isolated tissues treated with Pb is demonstrably inhibited; intracellular calcium compartments display altered kinetics. Decreased turnover of intracellular calcium stores appears to be a common denominator of the effect of Pb in kidney, liver, brain and smooth muscle.
4. Tracer studies of Pb distribution *in vivo* suggest relatively free access, if not preferential distribution of systemic Pb to the mitochondrial compartment of various tissues.
5. Observations of altered structure and uncoupled oxidative phosphorylation have occurred in mitochondria; uncoupled energy metabolism, inhibited respiration and altered kinetics of cellular $Ca^{++}$ stores have been documented in tissues obtained from animals exposed to Pb *in vivo*.
6. In certain instances, specific toxic effects of Pb (for example, the aminoaciduria of high Pb exposures or delayed brain development at low Pb exposures) are at least circumstantially linked with uncoupled energy metabolism. In the case of delayed brain development, effects on cellular energetics occurred at systemic Pb concentrations low enough to satisfy virtually any criteria of dose dependency for a physiologic effect.

Beyond these general points, there are questions that remain to be addressed. For example, how are effects of Pb on cellular energetics linked to functional deficits of a particular tissue? With respect to effects of Pb on brain energy metabolism, the close relationship of TCA cycle oxidations with the synthesis of neurotransmitter substances are as acetylcholine, $\gamma$-aminobutyrate and glutamate, provide one possible linkage. Acetylcholine synthesis has been shown to be substantially reduced in the brains of mice exposed to high concentrations of Pb (5000 mg/l) in drinking water (Carroll et al., 1977; Shih and Hanin, 1977).

A wide range of possibilities for linkages between effects of Pb on energy metabolism and altered functional activity arises from the pervasive role $Ca^{++}$ plays in regulating and even effecting cellular function. The finding that the levels of mitochondrial pyridine nucleotides reveals a means by which cellular energetics may directly modulate functional activity. There is evidence that suggests that Pb may uncouple oxidative phosphorylation *in vivo*. This would be expected to decrease the level of pyridine nucleotide reduction in mitochondria and inappropriately release mitochondrial $Ca^{++}$ stores. If direct evidence of interference of Pb with this system can be obtained *in vivo*, it could serve as a common basis for such diverse effects of Pb as promotion of pathologic calcification of tissues and altered neurotransmitter metabolism.

Perhaps the most challenging area of Pb toxicology involves its effects on the developing animal. Parameters used in investigating developmental toxicity, particularly functional parameters, must be changed to fit the developmental period. Data presently available strongly suggests that an effect on cellular energetics occurs at a low enough exposure and at a reasonable time to account for a delay in synaptogenesis. If this association is causal, how is it mediated? Does Pb effect developmental events which occur in an earlier or later period? Preliminary data indicate an effect on one metabolic compartment at 15 days of age and another at 21 days of age (Bull, unpublished). Does this herald a delay in glial development and myelination, a later developmental event? When 200 mg Pb/l of drinking water is administered to dams, delays in metabolic development are observed and may persist as late as 50 days of age, even if Pb exposure is terminated at 21 days of age. Is this a further indication that glial development and myelination may be inhibited? Under these same conditions, synaptic development appears to recover. Would continued exposure or exposure to higher levels of Pb render a synaptic deficit permanent? What are the peculiar properties of brain mitochondria early in life that could account for the sensitivity of this developmental period? Holtzman et al. (1977) were unable to document significant differences of effects of Pb on immature and mature brain mitochondria. Does this mean that the developing nervous system is simply more sensitive to interference with energetics? More recent data suggests that mature brain mitochondria have a much reduced capability for taking up Pb (Holtzman, personal communication). Perhaps this provides answers for some of these questions. However, it is apparent that the surface has hardly been scratched concerning the role of cellular energetics in the developmental toxicity of Pb. It is apparent that a great many questions remain to be answered before the underlying mechanisms of Pb toxicity can be adequately explained. The present work suggests that if a common underlying mechanism is involved in the vague and diverse functional impairments produced by Pb, cellular energetics is at the very least an attractive candidate.

## REFERENCES

Aghajanian, G.K., and Bloom, F.E. (1967). *Brain Res. 6:*716.

Akera, T., and Brody, T.M. (1978). *Pharmacol. Rev. 29:*187.

Amatruda, J.M., Staton, A.J., and Kiesow, L.A. (1977). *Biochem. J. 166:*75.

Atkinson, D.E. (1968). *Biochemistry 7:*4030.

Balazs, R., Patel, A.J., and Richter, D. (1972). R. Balazs and J.E. Cremer (Eds.), In "Metabolic Compartmentation in the Brain," p. 167. John Wiley and Sons, New York.

Baltscheffsky, H., and Baltscheffsky, M. (1974). *Ann. Rev. Biochem. 43:*871.

Beattie, D.S. (1971). *Sub.-Cell. Biochem. 1:*1.

Beaver, D.L. (1961). *Am. J. Pathol. 39:*195.

Becker, G.L. (1977). *J. Oral. Pathol. 6:*307.

Biagini, G., Misciattelli, M.E., Baccarani, M.C., Vangelista, A., Raffi, G.B., and Candarella, R. (1977). *Lav. Um. 29:*179.
Bouldin, T.W., Mushak, P., O'Tuama, L.A., and Krigman, M.R. (1975). *Environ. Health Persp. 12:*81.
Brierley, G.P., Jurkowitz, M., Scott, K.M., and Merola, A.J. (1971). *Arch. Biochem. Biophys. 147:*545.
Brierley, G.P. (1976). L. Packer and A. Gomez-Puyou (Eds.), In "Mitochondria: Biogenesis, and Membrane Structure," p. 3. Academic Press, New York.
Brierley, G.P. (1977). S.D. Lee (Ed.), In "Biochemical Effects of Environmental Pollutants," p. 397. Ann Arbor Science, Ann Arbor, Mich.
Brierley, G.P., and Knight, V.A. (1967). *Biochemistry 6:*3892.
Browning, E.T., and Schulman, M.P. (1968). *J. Neurochem. 15:*1391.
Bull, R.J. (1977). S.D. Lee (Ed.), In "Biochemical Effects of Environmental Pollutants, p. 425. Ann Arbor Science, Ann Arbor, Mich.
Bull, R.J., and Cummins, J.T. (1973). *J. Neurochem. 21:*923.
Bull, R.J., Lutkenhoff, S.D., McCarty, G.E., and Miller, R.G. (1979). *Neuropharmacology 18:*83.
Bull, R.J., and O'Neill, J.J. (1975). *Psychopharmacol. Comm. 1:*109.
Bull, R.J., Stanaszek, P.M., O'Neill, J.J., and Lutkenhoff, S.D. (1975). *Environ. Health Perspect. 12:*89.
Carafoli, E., and Lehninger, A.L. (1971). *Biochem. J. 122:*681.
Cardona, E., Lessler, M.O., and Brierley, G.P. (1971). *Proc. So. Exp. Biol. Med. 136:*300.
Carroll, P.T., Silbergeld, E.K., and Goldberg, A.M. (1977). *Biochem. Pharmacol. 26:*397.
Chapman, A.G., and Atkinson, D.E. (1973). *J. Biol. Chem. 248:*8309.
Chance, B. (1972). *FEBS Letters, 23:*3.
Chance, B., and Hollunger, G. (1963). *J. Biol. Chem. 278:*418.
Chance, B., and Schoener, B. (1966). *J. Biol. Chem. 241:*4577.
Chance, B., Wilson, D.F., Dutton, P.L., and Erecinska, M. (1970). *Proc. Nat. Acad. Sci. (USA). 66:*1175.
Chappell, J.B. (1968). *Brit. Med. Bull. 24:*150.
Cheng, S.C. (1972). R. Balazs and J.E. Cremer (Eds.), In "Metabolic Compartmentation in the Brain," p. 105. John Wiley and Sons, New York.
Cheng, S.-C., and Nakamura, R. (1972). *Brain Res. 38:*355.
Chiang, P.K., and Sacktor, B. (1975). *J. Biol. Chem. 250:*3399.
Cooper, R.H., Randle, P.J., and Denton, R.M. (1975). *Nature 257:*808.
Crompton, M., Moser, R., Ludi, H., and Carafoli, E. (1978). *Eur. J. Biochem. 82:*25.
Dawson, E.B., Cravy, W.D., Clark, R.R., and McGanity, W.J. (1969). *Am. J. Obst. Gynec. 103:*253.
Davidian, N. McC., and Penniall, R. (1971). *Biochem. Biophys. Res. Comm. 44:*15.
Denton, R.M. (1972). R. Balazs and J.E. Cremer (Eds.), In "Metabolic Compartmentation in the Brain," p. 339. John Wiley and Sons, New York.
Erecinska, M., Stubbs, M., Miyata, Y., Ditre, C.M., and Wilson, D.F. (1977). *Biochim. Biophys. Acta 462:*20.
Erecinska, M., Wilson, D.F., and Nishika, K. (1978). *Am. J. Physiol: Cell Physiol. 3*(2):C82.
Estrada-O, S., and Gomez-Lojero, C. (1976). L. Packer and A. Gomez-Puyou (Eds.), In "Mitochondria, Bioenergetics, Biogenesis and Membrane Structure," p. 167. Academic Press, New York.

Flatmark, T., and Romslo, I. (1975). *J. Biol. Chem. 250:*6433.
Gabbiani, G., Tuchweber, B., and Perrault, G. (1970). *Calc. Tiss. Res. 6:*20.
Garfinkel, D., Ochs, M.J., and Dzubow, L. (1974). *Fed. Proc. 33:*176.
Garfinkel, D., Ochs, M.J., and Garfinkel, L. (1972). R. Balazs and J.E. Cremer (Eds.), In "Metabolic Compartmentation in the Brain," p. 353. John Wiley and Sons, New York.
Gibson, G.E., and Blass, J.P. (1976). *J. Neurochem. 27:*37.
Gibson, G.E., Blass, J.P., and Jendon, D.J. (1978). *J. Neurochem. 30:*71.
Gibson, G.E., Jope, R., and Blass, J.P. (1975). *Biochem. J. 148:*17.
Goldstein, G.W., Asbury, A.K., and Diamond, I. (1974). *Arch. Neurol. 31:*382.
Goldstein, G.W. Wolinsky, J.S., and Csejtey, J. (1977). *Ann. Neurol. 1:*235.
Golter, M., and Michaelson, I.A. (1975). *Science 187:*359.
Goyer, R.A. (1968). *Lab. Invest. 19:*71.
Goyer, R.A. (1971). *Am. J. Pathol. 64:*167.
Goyer, R.A., and Krall, R. (1969). *J. Cell. Biol. 41:*393.
Goyer, R.A., Krall, A., and Kimball, J.P. (1968). *Lab. Invest. 19:*78.
Green, D.E. (1974). *Biochem. Biophys. Acta 346:*27.
Gunter, T.E., Gunter, K.K., Puskin, J.S., and Russell, P.R. (1978). *Biochemistry 17:*339.
Hanstein, W.G. (1976). *Biochim. Biophys. Acta 456:*129.
Haugaard, N., Haugaard, E.S., Lee, N.H., and Horn, R.S. (1969). *Fed. Proc. 28:*1657.
Hawkins, R.A., Miller, A.L., Nielsen, R.C., and Veech, R.L. (1973). *Biochem. J. 134:*1001.
Hess, B., and Boiteux, A. (1971). *Ann. Rev. Biochem. 40:*237.
Himwich, H.E. (1970). H.A. Himwich (Ed.), In "Developmental Neurobiology," p. 22. Charles C Thomas, Springfield, Ill.
Hirsch, G.H. (1973). *Toxicol. Appl. Pharmacol. 25:*84.
Hoffman, E.O., Trejo, R.A., DiLuzio, N.R., and Lamberty, J. (1972). *Exp. Mol. Pathol. 17:*159.
Holmsen, H., and Robkin, L. (1977). *J. Biol. Chem. 252:*1752.
Holtzman, D., and Hsu, J.S. (1976). *Pediat. Res. 10:*70.
Holtzman, D., Hsu, J.S., and Mortell, P. (1977). *Neurochem. Res. 3:*195.
Horowicz, P., and Larrabee, M.G. (1962). *J. Neurochem. 9:*1.
Iannaccone, A., Boscolo, P., and Bambardieri, G. (1976). *Life Sciences 19:*427.
Kapoor, S.C., and van Rossum, G.D.V. (1977). *Pharmcologist 19:*180.
Kennedy, C., Des Rosiers, M.H., Jehle, J.W., Reivich, M., Sharpe, F., and Sokoloff, L. (1975). *Science 187:*850.
Klausa, A., and Rehner, G. (1977). *Nutr. Metab. 21:*205.
Kleihaus, P., Kobayashi, K., and Hossman, K.-A. (1974). *J. Neurochem. 23:*417.
Koeppe, D.E., and Miller, R.J. (1970). *Science 167:*1376.
Krall, A.R., Meng, T.J., Harmon, S.J., and Dougherty, W.J. (1971). *Fed. Proc. 30:*1285 (Abs. #1357).
Krigman, M.R., Druse, M.J., Taylor, F.D., Wilson, M.H., Newell, L.R., and Hogan, E.L. (1974). *J. Neuropath. Exp. Neurol. 33:*671.
Krigman, M.R., and Hogan, E.L. (1974). *Environ. Health Perspect. 7:*187.
Land, J.M., and Clark, J.B. (1975). F.A. Hommes and C.J. VanDenBerg (Eds.), In "Normal and Pathological Development of Energy Metabolism," p. 155. Academic Press, New York.
Lehninger, A.L. (1971). *Adv. Cytopharmacol. 1:*199.
Lehninger, A.L. (1977). *Horizons Biochem. Biophys. 4:*1.

Lehninger, A.L., Vercesi, A., and Bababunmi, E.A. (1978). *Proc. Natl. Acad. Sci.* (*USA*) *75:*1690.
Lolley, R.N. (1969). *J. Neurochem. 16:*1469.
Lowenstein, J.M. (1972). *Physiol. Rev. 52:*382.
Mahler, H.R. (1976). L. Packer and A. Gomez-Puyou (Eds.). In "Mitochondria: Bioenergetics, Biogenesis, and Membrane Structure," p. 213. Academic Press, New York.
Maines, M.D., and Kappas, A. *Biochem. J. 154:*125.
McCauley, P.T., Bull, R.J., and Lutkenhoff, S.D. (1979). *Neuropharmacology. 18:*93.
McIlwain, H. (1967). *Biochem. Pharmacol. 16:*1389.
McIlwain, H., and Bachelard, H.S. (1971). "Biochemistry and the Central Nervous System," pp. 76 and 196. Churchill Livingstone, Edinburgh and London.
Mergner, W.J., Smith, M.W., Sahaphong, S., and Trump, B.F. (1977). *Virchows. Arch. B. Cell. Path. 26:*1.
Moore, C.L. (1972). P. Mortell (Ed.), In "Metabolic Pathways," 3rd Ed., Vol. VI. "Metabolic Transport," p. 573. Academic Press, New York.
Mitchell, P. (1961). *Nature* (*Lond.*) *191:*144.
Nathanson, J.A., and Bloom, F.E. (1976). *Mol. Pharmacol. 12:*390.
Neeley, J.R., Denton, R.M., England, P.J., and Randle, P.J. (1972). *Biochem. J. 128:*147.
Nicholls, D.G. (1978). *Biochem. J. 170:*511.
Oerter, D., and Bass, R. (1972). *Naunyn-Schmiedeberg's Arch. Pharmacol. 272:*239.
Packer, L., and Gomez-Puyou, A. (1976). "Mitochondria: Bioenergetics, Biogenesis and Membrane Structure." Academic Press, New York.
Palmer, J.M., and Hall, D.O. (1972). *Prog. Biophys. Mol. Biol. 24:*125.
Papa, S. (1976). *Biochem. Biophys. ACTA 456:*39.
Parr, D.R., and Harris, E.J. (1975). *FEBS Letters 59:*92.
Parr, D.R., and Harris, E.J. (1976). *Biochem. J. 158:*289.
Patel, A.J., Michaelson, I.A., Cremer, J.E., and Balazs, R. (1974a). *J. Neurochem. 22:*581.
Patel, A.J., Michaelson, I.A., Cremer, J.E., and Balazs, R. (1974b). *J. Neurochem. 22:*591.
Pentschew, A., and Garro, F. (1966). *Acta Neuropathologia 6:*268.
Peterson, J.W., and Paul, R.J. (1974). *Biochem. Biophys. Acta 357:*167.
Piccinini, F., Favalli, L., and Chiari, M.C. (1977). *Toxicology 8:*43.
Press, M.F. (1977). *Acta Neuropathologia* (*Berl.*) *40:*259.
Rabinowitz, M., and Swift, H. (1970). *Physiol. Rev. 50:*376.
Reed, L.J., Pettit, F.H., Roche, T.E., and Butterworth, P.J. (1973). F. Hiujing and E.Y.C. Lee (Eds.), In "Protein Phosphorylation in Control Mechanisms," p. 83. Academic Press, New York.
Rhyne, B.C., and Goyer, R.A. (1971). *Exp. Mol. Pathol. 14:*386.
Rossi, C.S., and Lehninger, A.L. (1964). *J. Biol. Chem. 239:*3971.
Sabbioni, E., and Marafante, E. (1976). *Chem.-Biol. Interactions* 15, 1.
Scott, K.M., Hwang, K.M., Jurkowitz, M., and Brierley, G.P. (1971). *Arch. Biochem. Biophys. 147:*557.
Secchi, G.C., Alessio, L., and Cirla, A. (1970). *Clin. Chem. Acta 27:*467.
Shih, T.-M., and Hanin, I. (1977). *Fed. Proc. 36:*3733.
Siesjo, B.K., and Zwetnow, N.N. (1970a). *Acta Physiol. Scand. 79:*114.
Siesjo, B.K., and Zwetnow, N.N. (1970b). *Acta Neurol. Scand. 46:*187.
Simon, J.R., Atweh, S., and Kuhar, M.J. (1976). *J. Neurochem. 26:*909.

Slater, E.C. (1953). *Nature*, (*Lond.*) *172:*975.
Slater, E.C. (1971). *Quarterly Rev. Biophys. 4:*35.
Sollenberg, J., and Sorbo, B. (1970). *J. Neurochem. 17:*201.
Stevenson, A., Merali, Z., Kacew, S., and Singhal, R.L. (1976). *Toxicology 6:*265.
Sun, C.N., Goyer, R.A., Mellies, M. and Yin, M.W. (1966). *Arch. Path.* (Chicago) *82:*156.
Tasaki, I. (1968). "Nerve Excitation," p. 116. Charles C Thomas, Springfield, Illinois.
Thomas, J.A., Dallenback, F.D., and Thomas, M. (1971). *Virchows Arch. Abt. A. Path. Anat. 352:*61.
Tucek, S., and Cheng, S.-C. (1974). *J. Neurochem. 22:*893.
Tzagoloff, A. (1971). *J. Biol. Chem. 246:*3050.
Ulmer, D.D., and Vallee, B.L. (1969). D.D. Hemphill (Ed.), In "Trace Substances in Environmental Health II," p. 7. University of Missouri Press, Columbia, Mo.
Van Dam, K., and Meyer, A.J. (1971). *Ann. Rev. Biochem. 40:*115.
Van den Berg, C.J. (1972). R. Balazs and J.E. Cremer (Eds.), In "Metabolic Compartmentation in the Brain," p. 137. John Wiley and Sons, New York.
Vignais, P.V. (1976). *Biochim. Biophys. Acta 456:*1.
Vinogradov, A., Scarpa, A., and Chance, B. (1972). *Arch. Biochem. Biophys. 152:*646.
Vasington, F.D., and Murphy, J.V. (1962). *J. Biol. Chem. 237:*2670.
Walton, J.R. (1973). *Nature 243:*100.
Walton, J., and Buckley, I.K. (1977). *Exp. Mol. Pathol. 27:*167.
Wedeen, R.P., and Maesaka, J.K., Weiner, B., Lipat, G.A., Lyons, M.M., Vitale, L.F., and Joselow, M.M. (1975). *Amer. J. Med. 59:*630.
Weil-Malherbe, H. (1974). *Mol. Cell. Biochem. 4:*31.
Whittaker, V.P. (1968). *Neurosci. Res. Prog. Bull. 6*(*Suppl.*) 27.
Williamson, J.R. (1976). L. Packer and A. Gomez-Puyou (Eds.), In "Mitochondria: Bioenergetics, Biogenesis, and Membrane Structure, p. 79. Academic Press, New York.
Wilson, D.F., and Erecinska, M. (1975). *Arch. Biochem. Biophys. 167:*116.
Wilson, D.F., Erecinska, M., and Dutton, P.L. (1974). *Ann. Rev. Biophys. Bioengin. 3:*203.
Yamamoto, T., Yamaguchi, M., and Suketa, Y. (1974). *Toxicol. Appl. Pharmacol. 27:*204.

# Effects of Lead on Mammalian Reproduction

*John U. Bell and John A. Thomas*

**TABLE OF CONTENTS**



## I. INTRODUCTION

It has been suggested that lead poisoning was partly responsible for the decline of the ancient Roman Empire (*cf* Gilfillan, 1965). Henderson (1824), upon examining recipes of the Romans and the concentration of this heavy metal as a constituent of their cooking ware, speculated that lead poisoning was quite common. Further, Rosenblatt (1906) analyzed ancient bones for their lead content and recorded extraordinarily high levels during, but not before or after, the height of the Roman civilization. Lead contamination was prevalent in ancient Rome and lead was reportedly present in high concentrations even in wine. Perhaps for that reason, Roman wives were not permitted to drink wine because of an inference that those who did bore fewer children. Some evidence also exists indicating that Greek wines produced sterility and miscarriages (Athenaeus of Naucrates, *cf* Gilfillan, 1965).

In the late 1800's and early 1900's, women in pottery and white lead factories recognized lead as an abortifacient. Paul (1860) reported that lead had a profound effect upon fertility and upon the viability of offspring. There was a higher incidence of sterility, miscarriages and stillbirths in women employed in British pottery factories than in those not employed in such environments (*cf* Hamilton and Hardy, 1974). As early as 1881, lead was reported to be a teratogen, as evidenced by an increased incidence of fetal macrocephaly (Rennert, 1881). Later studies by Chyzzer (1908) and by Oliver (1911) further strengthened the relationship between lead exposure and macrocephaly.

Lead-related abortions in women employed in the printing industry have also been reported (*cf* Hamilton and Hardy, 1974; Bourret and Mehl, 1966). A recent review by Rom (1976) has discussed the historical aspects of lead toxicity, particularly in the female.

Not only was plumbism observed in the female, but evidence began to indicate that male workers employed in lead-related industries had a high incidence of sterile marriages (*cf* Hamilton and Hardy, 1974). Deneufbourg (1905) reported that the incidence of abortion and stillbirth was higher for the wives of husbands who were employed in the lead industry than for those females whose spouses were working in non-lead related industries. It is therefore apparent that lead is capable of interfering with both female and male reproductive systems. The purpose of the present review is to relate more recent findings about the toxic effects of lead on mammalian reproduction.

## II. LEAD AND MALE REPRODUCTIVE CAPACITY

Most of the literature dealing with the toxic effects of lead upon the reproductive system has been concerned with the female, since this heavy metal is known to be a teratogen in a number of species. Nevertheless, there is evidence to indicate that lead can exert deleterious actions upon male reproductive organs (Table 1).

**Table 1.** Effects of lead on the male reproductive system

| Species | Effect | Reference |
|---|---|---|
| Rat | Infertility | Puhac et al. (1963) |
| Rat | Germinal epithelium damage | Timm and Schulz (1966) |
| Rat | Oligospermia and testicular degeneration | Golubovich et al. (1968) |
| Rat | Decreased sperm motility and prostatic hyperplasia | Hilderbrand et al. (1973) |
| Mouse | Infertility | Varma et al. (1974) |
| Mouse | Increase in abnormal sperm | Eyden et al. (1978) |
| Human | Teratospermia, hypospermia and asthenospermia | Lancranjan et al. (1975) |

### A. Spermatotoxicity

Early studies by de Quatrefages (1850; *cf* Mann, 1964) described in great detail the marked toxicity of lead upon spermatozoa. Lead salts are among the oldest known spermicidal agents. Puhac et al. (1963) added lead nitrate to the rations of male rats, using several dosage levels. After 30 days of lead exposure, the rats were paired with untreated females. The lead-treated males did not sire any pups until 45 days after the cessation of lead exposure, suggesting that lead can produce infertility in male rats. Lead acetate can cause sperm to become less motile (Hilderbrand et al., 1973); Timm and Schulz (1966) have reported the presence of lead in various loci in the testes in normal rats. The lead was situated in testicular tubules and along the edge of the lumen or at the tails of the spermatozoa. Following the parenteral administration of lead, as well as in chronic lead poisoning, lead deposits were visible in the lumina of the seminiferous tubules (Timm and Schulz, 1966).

When the male offspring ($F_1$ generation) of lead-exposed female rats were mated with control, nonexposed females, the resulting offspring had reduced birth weight (Stowe and Goyer, 1971). This reduction was interpreted by the authors as being caused by defects in the spermatozoa which fertilized the normal ova. Eyden et al. (1978) administered lead acetate to male mice via their food; although a dietary lead level of 1% had no effect during the first 4 weeks of exposure, there was a significant increase in the sperm abnormality count by 8 weeks. Unfortunately, exposure to that level of lead caused a severe weight loss in the animals, raising the possibility that the increased incidence of abnormal spermatozoa was secondary to systemic metabolic disturbances rather than to a direct effect on the testes. The administration of lead subacetate to male Swiss mice produced infertility, as evidenced by reduced pregnancy rates (Varma et al., 1974). The incidence of pregnancy was 52.7% in control animals, but only 27.6% in the group containing the lead-exposed males. The authors suggested that the reduced fertility was the result of a genetically related mutagenic change in the spermatozoa.

Lancranjan et al. (1975) studied the reproductive capacity of 150 men occupationally exposed to lead. The workers were divided into four groups, including lead-poisoned workers (with average blood lead levels of 74 $\mu$g/100 ml) and those showing moderate (blood lead levels of 53 $\mu$g/100 ml), slight (blood lead levels of 41 $\mu$g/100 ml) or physiologic (blood lead levels of 23 $\mu$g/per 100 ml) absorption of lead. The fertility of the lead-poisoned and moderate lead absorption groups was decreased; the reduction correlated with an increased frequency of asthenospermia, hypospermia or teratospermia. Spermatozoan damage in the form of oligospermia has also been reported in the rat following exposure to lead subacetate (Golubovich et al., 1968).

## B. Testes, and Sex Accessory Organs

Tipton and Cook (1963) estimated that all the heavy metals could be detected in the human testes for the population of the United States and judged the average testicular concentration of lead to be 12 ppm. Golubovich et al. (1968) reported that lead acetate could cause testicular degeneration in rats; the ribonucleic acid content of lead-toxic rat testes was significantly lower than in normal rat testes, suggesting that reduced ribosomal activity and impaired protein synthesis could be associated with lead toxicity. Studies by Hilderbrand et al. (1973), however, indicated that lead acetate (5 or 100 μg/day for 30 days po) had no effect upon testicular weights in mature rats. Der et al. (1976) also failed to observe any reduction in testicular weight in rats treated with lead. Interestingly enough, Der et al. (1976) found that small amounts of lead could prevent cadmium-induced testicular atrophy.

Rats treated with lead failed to reveal any significant reduction in the weights of the epididymus (Der et al., 1976). As in the testes, however, the cadmium-induced loss of epididymal weight was offset by the presence of small amounts of lead.

A 30-day treatment period with lead acetate had no effect upon rat seminal vesicle weights (Hilderbrand et al., 1973). Similar findings were reported by Der et al. (1976). However, cadmium-induced weight losses in the seminal vesicles could be antagonized by the presence of small amounts of lead (Der et al., 1976).

Studies by Hilderbrand et al. (1973) reported that lead acetate treatment caused significant increases in the weights of the rat prostate gland. These organs also revealed a distinct histologic hyperplasia. Later studies

**Table 2.** Human maternal and fetal blood levels at delivery

| Reference | Maternal blood lead (μg/100 ml) | Fetal blood lead (μg/100 ml) |
|---|---|---|
| Barltrop (1968) | 13.9 | 10.8 |
| Harris and Holley (1972) | 13.2 | 12.3 |
| Haas et al. (1972) | 16.9 | 14.9 |
| Gershanik et al. (1974) | 10.5 | 9.4 |
| Fahim et al. (1976) | 13.1 | 4.3 |
| Clark (1977) | 14.7 | 11.8 |
| Zetterlund et al. (1977) | 7.9 | 6.6 |
| Lauwerys et al. (1978) | 10.2 | 8.4 |
| Hubermont et al. (1978) | 10.6 | 8.8 |
| Buchet et al. (1978) | 10.0 | 8.1 |

(Der et al., 1976) using a different strain of rat failed to reveal any significant weight changes in the prostate glands of animals treated with lead. Cadmium-induced decreases in rat prostrate gland weights can be antagonized by small amounts of lead (Der et al., 1976). Khare et al. (1978) injected 1 mg of lead acetate directly into the prostate of adult male rats. After 60 days, there was no change in the weights of testes, epididymis or seminal vesicles; however, there was a significant reduction in prostate weight. In addition, the prostate displayed edema of stroma and had excessive periprostatic fibrosis.

### C. Other Endocrine Related Effects

Studies by Lancranjan et al. (1975) failed to reveal any lead-induced changes in androgens in workmen exposed to lead. These investigators reported that total urinary neutral 17 ketosteroid levels were normal in workmen exposed to lead. The finding that total urinary gonadotropin levels were unaffected by lead excluded hypothalamic-adenohypophysial factors as being responsible for the altered spermatogenesis. This germinal insufficiency was due to a direct toxic action of lead upon the gonads (Lancranjan et al., 1975). Since lead can inhibit hepatic detoxification by modifying microsomal enzyme systems (Hilderbrand et al., 1973), it is quite possible that steroid metabolism might be affected. There are not only sex differences in hepatic microsomal enzyme systems (Conney, 1967), but it has been demonstrated that lead accumulates to a greater extent in females than in males (Klevay, 1972).

## III. SOURCES OF LEAD EXPOSURE IN THE EMBRYO, FETUS AND NEONATE

### A. Transfer of Lead via Seminal Fluid

Although there is considerable evidence that lead can influence spermatogenesis, there have been very few studies where the seminal lead content has actually been determined. Monkiewicz et al. (1975) found that there was an increase in seminal lead content in bulls exposed to lead environmentally. They also reported that there was an inverse relationship between survival of spermatozoa and seminal lead content.

From a summary of the literature, there appears to be only one paper which reports on lead content of human semen. In the study by Plechaty et al. (1977), the seminal lead content was measured in healthy men who were not occupationally exposed to lead. The average seminal lead content in 21 volunteers was 5.9 $\mu$g/100 ml, while the average blood lead content in the group was 13.1 $\mu$g/100 ml. In each case, the seminal lead content was equal to or less than the blood lead content. The authors found no correlation between the levels of blood lead and seminal lead or between the levels of seminal lead and sperm counts.

These findings necessitate a re-evaluation of studies in which effects on the conceptus were observed following lead exposure only to the male. While lead-induced gametotoxicity may indeed by the cause of embryo- or fetotoxicity, the study of Plechaty et al. (1977) suggests that there may be a passage of lead from the male via the semen which could influence the conceptus directly.

## B. Transfer of Lead via the Placenta

Although it has been recognized for a number of years that lead can pass from mother to fetus, the mechanisms involved and the kinetic parameters are still unknown. A large portion of the human data in this area merely correlates maternal blood levels of lead with fetal (cord blood) levels at birth. Table 2 illustrates very clearly that the human placenta offers little barrier to the passage of lead. Perhaps even more important is the observation that there are significant levels of lead in both maternal and fetal blood, even when there has been no occupational exposure to the metal. As Scanlon (1972) has pointed out, we do not know what effect these "normal" fetal levels of lead might have on development. Biochemical effects of low level lead have been demonstrated by Hernberg and Nikkanen (1970), who found a decreased activity of delta-aminolevulinic acid dehydratase in erythrocytes at blood levels between 5 and 15 μg/100 ml. This was also confirmed by Kuhnert et al. (1976), who found that blood lead levels of 16 to 19 μg/100 ml were associated with inhibition of delta-aminolevulinic acid dehydratase in both maternal and fetal erythrocytes.

There is some controversy as to whether the accumulation of lead occurs during embryonic development or is restricted until later in gestation. Barltrop (1969) demonstrated that placental transfer of lead began as early as the 12th week of gestation and that the lead content of the fetus increased throughout pregnancy. He suggested that the total amount of lead transferred during pregnancy was less than 300 μg and that the distribution of this lead throughout the fetal tissues resembled the pattern found in adults, with the highest levels being found in bone and liver. Casey and Robinson (1978) reported that fetal (22 to 43 weeks of gestation) tissue levels of lead were highest in bone (approximately 2.2 μgPb/g tissue dry weight), liver (0.9 μg/g tissue dry weight), and heart (0.9 μg/g tissue dry weight). Chambe et al. (1972) measured lead levels in embryonic and fetal tissue obtained following legal abortion. They found that lead could be detected in two-thirds of the subjects aborted during the first trimester, with levels ranging from 0.38 to 2.0 μg/g. Therefore, there seems to be little doubt that the embryo/fetus is exposed to lead very early in development.

In a number of studies, the lead content of human placental tissue has been reported (Table 3). Although there is considerable variation from

**Table 3.** Lead content of human placental tissue

| Reference | Country | Lead content (μg/100 g) |
|---|---|---|
| Horiuchi et al. (1959) | Japan | 57 |
| Collucci et al. (1973) | USA | 96 |
| Einbrodt et al. (1973) | Germany | 56 |
| Thieme et al. (1974) | Germany | 40 |
| Baglan et al. (1974) | USA | 30 |
| Fahim et al. (1976) | USA | 6 |
| Wibberley et al. (1977) | Britain | 93 |
| Karp and Robertson (1977) | USA | 29 |
| Hubermont et al. (1978) | Belgium | 10 |

study to study, as well as within individual studies, the presence of lead in the placenta raises the possibility that placental metabolic function may be affected.

In a very interesting study, Fahim et al. (1976) determined lead levels in umbilical cords, placentae, placental membranes and maternal and fetal blood collected from 253 women who delivered in a lead-mining region and compared them with samples obtained from 249 women who delivered in a non-mining region. They further subdivided their deliveries into term, term with premature membrane rupture and preterm. Their data indicated that the pattern of lead distribution in term subjects was not different when the mining and non-mining regions were compared. Although there were no significant differences in the lead content of placentae and umbilical cord tissue, it was found that the lead content of fetal membranes was three to six-fold higher in term with premature membrane rupture and preterm deliveries than in normal term deliveries, irrespective of the region.

The authors suggested that the apparent accumulation of lead in the fetal membranes may decrease the collagen content, leading to membrane rupture. A word of caution must be added at this point. Although the lead levels in maternal blood (30 μg/100 g or less) and fetal blood (17 μg/100 g or less) obtained from term with premature membrane rupture and preterm deliveries were significantly higher than levels found in term deliveries, they were still well within what are considered to be "normal" blood lead levels.

There have been relatively few experimental studies on the placental permeability of lead in lower mammals. The first report demonstrating placental transfer of lead in a mammal was that of Baumann (1933), using radioactive lead and whole body autoradiography. She showed that lead crossed the placenta rapidly and localized mainly in developing fetal bone. Work by Morris et al. (1938) and Dalldorf and Williams (1945) in the rat, and Calvery et al. (1938) in the dog, confirmed that lead could traverse the mammalian placenta. There was then very little research in this area until the early 1970's, when McClain and Becker (1970) reported briefly on the placental transfer of $^{210}Pb(NO_3)_2$ in the rat and the mouse. The lead solution was infused iv at a rate of 0.5 mg/min/kg late in gestation. Six hours after a single administration of 50 mg/kg, the animals were sacrificed and lead levels were determined. Even though the maternal dose of $Pb(NO_3)_2$ had been the same in the mice and rats, the rat fetus contained a far higher level of lead than did the mouse fetus (0.3 $\mu$g/g versus 0.02 $\mu$g/g). In these experiments, the placenta also appeared to impede the maternal-fetal passage of lead to a certain extent.

These same authors examined in greater detail the transfer of lead nitrate in rats (McClain and Becker, 1975). After the iv administration of 50 mg/kg of $^{210}Pb(NO_3)_2$ on Day 17 of gestation, maternal blood levels were found to decline exponentially over a 24-hr observation period. The decline in placental lead levels paralleled the fall in maternal blood. Whole fetus lead levels increased continually during the 24-hr observation period to a level of 8 $\mu$g/g. Placental lead levels at that time were 13 $\mu$g/g. During the continuous infusion of $^{210}Pb(NO_3)_2$, maternal blood lead levels increased in a linear fashion, reaching 425 $\mu$g/g (plasma, 310 $\mu$g/ml) by 64 min. Whole fetus lead concentrations at that time were 2.4 $\mu$g/g, confirming that the placenta does retard the passage of lead from the maternal blood to the fetus despite the existence of a favorable concentration gradient.

Placental transfer of lead following acute administration at various states of pregnancy (Days 11 to 20) was evaluated by Green and Gruener (1974). They administered 5 mg/kg of lead acetate ip and 20 pCi/kg of $^{210}Pb$ on different days of gestation and then measured radioactivity in the whole fetus. It was reported that from 0.7 to 3.4% of the total administered radioactivity was transferred to the fetus and, interestingly, 23 to 38% of fetal activity was found in the fetal head.

The only available experimental data on the placental transfer of lead during the period of embryonic differentiation is in a study by Carpenter et al. (1973) using the pregnant golden hamster. This model was chosen because the major stages of organogenesis occur during a 24-hr period between Days 8 and 9 of gestation (Ferm, 1976). Pregnant hamsters were injected iv with $^{210}Pb(NO_3)_2$ at doses equivalent to 2.4 and 6.1 $\mu$gPb/kg on Day 7 or 8 of gestation. Gestation sacs were removed as early as 15 min after injection and subjected to autoradiography. Detectable radioactivity was present in embryonic tissue as quickly as 15 min following injection, with peak levels occurring between 1 and 4 hr after injection. This study demonstrated that placental transfer of lead can occur rapidly even at relatively low maternal levels and that the yolk sac placenta is the primary transfer site in the hamster. The transferred lead is widely distributed throughout the embryo.

The influence of chronic lead exposure on placental transfer was investigated by Singh et al. (1976). They administered $Pb(NO_3)_2$ in the diet for 45 days at a dose of 200 mg/kg body weight/day to adult male and female rats. These animals then mated and lead levels were determined in the resulting offspring within 30 min of their birth. Although their sample size was small, the authors were able to show that tissue lead levels were extremely high (liver, 740 μg/100g; kidney, 2811 μg/100g; heart, 1103 μg/100g; brain, 910 μg/100g; and blood 92 μg/100 ml). At 24 hr after birth, tissue levels had dropped dramatically (liver, 166 μg/100g; kidney, 245 μg/100g; heart, 205 μg/100g; brain, 88 μg/100g; and blood, 48 μg/100 ml). The authors suggested that fetal lead was loosely bound within the tissues and possibly maintains an equilibrium with the maternal body burden of lead.

A report by Buchet et al. (1977) suggests that lead stored in the female can be mobilized during pregnancy and transferred to the fetus. In their study, utilizing female rats, group 1 received 1 ppm of lead nitrate in their drinking water for 150 days prior to mating, during pregnancy and for 21 days following delivery. Group 2 received 1 ppm of lead nitrate in their drinking water for 150 days prior to mating, but it was withdrawn during pregnancy and after delivery. Group 3 received 1 ppm of lead nitrate in their drinking water for 150 days; the animals were then kept free of lead exposure for 50 days before mating, during pregnancy and following delivery. The authors discovered that the lead content of blood, liver, kidney, heart and brain of the newborns was significantly higher in group 3 than in the other groups, despite the fact that lead exposure of the mothers had been terminated 50 days prior to mating. Maternal blood lead levels were also higher in group 3 than in groups 1 or 2. The authors suggested that the observation could be due to a reduction in osteocytic or osteoclastic resorption of bone in groups 1 and 2, whereas the 50-day lead-free period in group 3 might have been sufficient to restore osteoclastic activity.

In conclusion, there seems to be little doubt that lead is capable of crossing the human placenta; however, more work is needed to clarify the mechanisms involved. Despite the obvious limitations, placental transfer of lead must be more extensively studied in common laboratory animal models.

### C. Transfer of Lead via the Milk

There is very little information available concerning the transfer of lead to the human neonate via maternal milk. This is particularly true when one attempts to find data from mothers who would be considered higher risks, such as those living near lead-emitting industries. A survey of the literature provides some normal human milk lead levels (Table 4). What is significant about these studies is the relatively low lead content detected (i.e., with the exception of the studies of Tracy and McPheat (1943) and Noirfalise et al. (1967), mean lead concentrations were less than 0.06 μg/ml). Obviously, information must be obtained from high risk mothers before any real con-

**Table 4.** Lead content of human milk

| Reference | Country | Lead concentration ($\mu$g/ml) |
|---|---|---|
| Kehoe et al. (1933) | USA | 0.056 |
| Tracy and McPheat (1943) | Scotland | 0.163 |
| Noirfalise et al. (1967) | Belgium | 0.277 |
| Murthy and Rhea (1971) | USA | 0.012 |
| Lamm et al. (1973) | USA | 0.020 |
| Dillon et al. (1974) | USA | 0.026 |

clusions can be made concerning milk transfer as a source of neonatal lead toxicity in humans.

There are a number of studies which examine the milk transfer of lead in laboratory animals. Millar et al. (1970) fed maternal rats a diet containing 4% lead. On examination of the suckling rats, they found a significant increase in the lead content of liver, kidney, spleen, blood and brain. In addition, they found that there was an 80 to 90% reduction in the activity of delta-aminolevulinic acid dehydratase in both brain and blood, as well as a 60% reduction in the liver.

Green and Gruener (1974) administered lead acetate ($^{210}Pb$) to maternal rats immediately following birth and then determined the total radioactivity in the suckling rats at various stages during the next 9 days. Although the authors did not state the dose of lead acetate they administered and their numbers were small, their data confirm that milk transfer of lead does occur.

In an interesting study, Brown (1975) administered lead carbonate to nursing rats by gavage for 1 to 10 or 11 to 20 days following parturition and then measured blood lead levels in the suckling rats. The dose administered to the mothers (35 $\mu$gPb/kg/day) was low enough so that no significant effect on organ weights was produced. When the suckling rats were exposed to lead via the milk from 1 to 10 days after parturition, there was a significant increase in blood lead in the 11-day-old pup. However, administration of lead from 11 to 20 days after parturition (directly to the dam; indirectly to the suckling rats) did not produce any increase in blood lead in the 21-day-old pup. Apparently lead can be transferred via maternal milk, but the results of this study suggest that there are age-related differences in uptake, distribution and/or retention of lead.

Roels et al. (1977) administered 0.1, 10 and 100 ppm $Pb(NO_3)_2$ to lactating rats in drinking water from the day of delivery up to Day 21 following delivery. Both mothers and their newborns were sacrificed on Day 21. The authors reported that in the suckling rats, blood lead was significantly increased in the groups given 10 and 100 ppm Pb. Liver, kidney and brain lead content was significantly increased in the group given 100 ppm and kidney lead content was increased in the group given 10 ppm. The accumulation of Pb in kidney tissue of the 100 ppm group was associated with a marked increase in free tissue porphyrin concentration and a slight decrease in delta-aminolevulinic acid dehydratase activity. The authors stress that the effects observed in the group given 100 ppm of Pb were associated with blood lead levels of only 15 $\mu$g/100 ml whole blood, well within the range of blood lead found in human newborns in Belgium. They concluded that lead is easily transferred from mother to suckling rats and that the "no effect" level of lead in drinking water during lactation is approximately 1 ppm.

Momcilovic (1978) studied the transfer of $^{203}Pb\ Cl_2$ from mother to newborn following a single ip injection to Day 15 of lactation. After 48 hr, almost 20% of the dose administered to the mother was found in the litter.

In conclusion, despite a shortage of human data, it appears that maternal milk might be a source of lead for the neonate, particularly when levels in the mother are elevated.

## IV. LEAD TOXICITY IN THE MAMMALIAN EMBRYO AND FETUS

There has been a good deal of speculation that lead can cause teratogenesis in the human embryo. Early reports, around the turn of the century, linked a form of macrocephaly in children to lead exposure by pregnant women who were employed in the pottery glazing industry (Rennert, 1881; Chyzzer, 1908; Oliver, 1911). At the present time, however, there is no sound evidence implicating lead as a human teratogen (Scanlon, 1972) and there is at least one report describing an apparent case of intrauterine plumbism with no detectable congenital malformations (Palmisano et al., 1969).

This is not the case, however, with laboratory animals. Although a number of studies have reported that lead is not teratogenic in rabbits (Jessup, 1967), cows (Shupe et al., 1967) or sheep (James et al., 1966), there is convincing evidence that lead-induced malformations can occur in rats, mice and hamsters.

### A. Rat

McClain and Becker (1970) administered a single iv dose of $Pb(NO_3)_2$ to pregnant rats at various stages of gestation. They observed a variety of teratogenic effects, including hydronephrosis and nonossification of cervical centri following doses of 35 mg/kg and 50 mg/kg, respectively, administered on Day 10 of gestation.

In a more detailed report (McClain and Becker, 1975), these same authors found that iv administration of $Pb(NO_3)_2$ produced external congenital malformations when administered on gestational Days 8 or 9. The malformations observed ranged from absent or shortened tails to complete absence of the posterior portion of the body. Urorectal malformations and the absence of external genitalia were also noted, as were soft tissue anomalies of the urogenital system and axial skeletal defects. The authors reported a resorption rate of 15 and 20% on Days 8 and 9, respectively; however, the rate increased dramatically between Days 10 and 15 to values approaching 100% before dropping to 8% on Day 16. They concluded that the rat embryo was most sensitive to lead-induced malformations on Days 8 and 9 of gestation, but was most sensitive to the lethal effects of lead between gestational Days 10 and 15.

Kennedy et al. (1975) administered lead acetate to pregnant rats by oral intubation at dosage levels of 0, 7.14, 71.4 and 714 mg/kg body weight on Days 6 to 16 of gestation. As a consequence of maternal toxicity, the highest dose had to be discontinued after 3 days. Animals were killed on Day 21 and the uterine horns were examined for implantation sites, resorption sites and viable fetuses, as well as for malformations. Although fetal growth was retarded at the highest dose level, there were no specific skeletal abnormalities that could be associated with exposure to lead. The authors concluded that lead was not teratogenic even at doses which produced maternal toxicity.

These findings were confirmed by Kimmel et al. (1976), who administered lead acetate to female rats in their drinking water from weaning through mating and gestation. Although the highest level of lead (250 ppm) caused a slight but nonsignificant increase in resorptions, no significant teratogenic effects were detected over the dose range tested (0, 0.5, 5, 50 and 250 ppm).

The contradictory findings of these studies might be a reflection of differences in the strain of rat used or, more importantly, in the route of lead administration. This latter point was discussed by McClain and Becker (1975), who found that the specific effects they observed following iv administration of lead did not occur when lead was administered orally.

### B. Mouse

As is the case with rats, there are a number of conflicting reports concerning the teratogenic potential of lead in mice. When McClain and Becker (1970) administered an intravenous injection of lead nitrate (50 mg/kg) to pregnant mice at various stages of gestation, they found that cleft palate occurred in 33% of the fetuses exposed on Day 12 and nonossification of cervical ventri occurred in 27% of the fetuses exposed on Day 10. When lead was administered orally or by the intraperitoneal route, few teratogenic changes were observed.

The administration of lead acetate to pregnant mice by gavage at levels up to 714 mg/kg was not associated with any teratogenic response (Kennedy et al., 1975). There were slight delays in ossification in fetuses recovered from females who were given 714 mg/kg, although no specific skeletal anomalies were detected. The authors concluded that no teratogenic response was induced in mice following in utero exposure to maximally tolerated doses of lead acetate. McLellan et al. (1974) also reported that no teratogenic effects were noted following administration of lead chloride (360 nM/g) to pregnant mice on Gestational Day 9. In addition to generally incomplete and/or delayed ossification, they demonstrated that the body weights of the lead-exposed fetuses were only 61% of the control fetal weights.

In the first of a series of very interesting studies, Jacquet et al. (1975) exposed female mice to dietary lead levels of 0, 0.125, 0.250 and 0.500% from Day 1 of pregnancy and sacrificed them on Days 16 to 18 of gestation. Each of the treatment groups contained 50 mice that had displayed evidence of mating (i.e., the presence of a vaginal plug). Although the authors found that lead caused growth retardation of the embryos and an increase in the number of postimplantation deaths, no gross malformations were observed in the lead-treated embryos. Although they apparently did not fully appreciate it at the time, they also observed that exposure to lead caused a dose-dependent decrease in the number of pregnancies. They defined pregnancy as the presence of at least one implant when the females were sacrificed on Days 16 to 18 of gestation. The incidence of pregnancies was 26 out of 50 in the 0% lead (control) group, 28 out of 50 in the 0.125% lead group, 11 out of 50 in the 0.250% lead group and only 8 out of 50 in the 0.500% lead group.

In a second study (Jacquet et al., 1976), they again allowed mice to mate and then exposed them to a diet containing 0, 0.125, 0.250, 0.500 and 1% lead. After only 48 hr, the animals were sacrificed and the embryos were classified into four categories depending on their stage of development (*i.e.*, nondivided embryos, two-cell embryos, four-cell embryos and eight-cell embryos). In the control group, 52% of the embryos had not divided, whereas most of the rest (48%) had reached the eight-cell stage. Dietary exposure to 0.125, 0.250 and 0.500% lead produced an approximately equal number of embryos in the four and eight-cell stages, although the proportions of embryos in the one and two-cell stage remained the same. At 1% dietary lead levels, the number of embryos undergoing no divisions had increased. Although they cited a number of possible causes, the authors speculated that lead caused a delay in first cell division in part of their embryonic population.

The whole picture began to fit together in a third paper by Jacquet (1977). He found that despite the interference of lead with early cell division, all the embryos reached the blastocyst stage by Day 5 (although lead-treated blastocysts did appear to be smaller than control blastocysts). By Day 7, no embryos from lead-treated mothers had progressed beyond the blastocyst stage and, more importantly, they remained unattached to the uterus. Morphologically, there was a failure to differentiate trophoblastic giant cells or to evoke a deciduous reaction of the uterus after lead treatment, even though this normally occurs on Day 5. A number of embryos showed signs of degeneration and there appeared to be some regression of the corpora lutea.

As a partial explanation, Jacquet et al. (1977) noted that lead treatment of pregnant mice caused a reduction in the maternal plasma progesterone levels, which normally peak around Day 5.5 to 6, without altering estradiol levels, which normally peak on Day 4. The estrogen peak apparently sensitizes the uterus to permit transformation of the embryo from morula to blastocyst, whereas the progesterone apparently is required for activation of the embryo, induction of the deciduous response and continuation of pregnancy. Although the estrogen-dependent phase appeared to proceed normally, as evidenced by blastocyst formation, the lack of progesterone during the critical phase of implantation could interfere with decidual formation and the differentiation of trophoblastic giant cells. These observations combined with the observation that the corpora lutea appeared to be degenerating suggested that the implantation problems may be secondary to lead effects in the mother.

Wide and Nilsson (1977) also reported that lead treatment during the "attachment period" in mice prevented any sign of decidual reaction. The invasion of the trophoblast and the formation of the primitive streak were also sensitive to the dosage of $PbCl_2$ employed in their study (1 mg/animal iv).

### C. Golden Hamster

The sensitivity of the golden hamster embryo to lead has been detailed in a series of papers published between 1967 and 1977 (Ferm, 1967 and 1976; Ferm and Ferm, 1971; Carpenter and Ferm, 1974 and 1977). In this animal, exposure of the embryo to inorganic lead salts on the morning of the eight gestational day can result in severe malformations of the sacral-tail region.

In a very interesting report, Carpenter and Ferm (1977) detailed the time sequence of the teratogenic effect. They observed that, within 30 hr of lead administration on the morning of Day 8, there was edema of the developing tail region, which developed into dorsal blisters and hematomas on gestational Days 9 and 10. This local accumulation of fluid apparently tends to distort the caudal neural tube and nearby dorsal tissues, which in turn disturbs the normal morphogenetic interaction between the tissues, leading to extensive agenesis or dysgenesis. Embryonic mortality increased between Days 11 and 13 (reaching 70%); the embryos that survived displayed various degrees of caudal malformation, including stunted, deformed or missing tails and abnormalities of the lower lumbar and sacral spinal cord and vertebrae. Although the authors suggested a number of possibilities, the actual mechanism involved in the initiation of the teratogenic change remains unresolved.

In conclusion, although lead exposure during *in utero* development has been shown to retard normal growth, there is also evidence in the literature that exposure of the embryo to the heavy metal can produce congenital malformations. Although this has been demonstrated for a number of animal species (particularly the golden hamster), there is still not sufficient evidence to implicate lead as a teratogen in humans.

## V. SUMMARY AND CONCLUSIONS

Although it has been known for a long time that lead can have profound effects on reproduction in laboratory animals, its full impact on human reproduction remains unresolved. Obviously, placental transfer of lead does occur in the human, as indicated by measurable cord blood levels in pregnancies free from occupational exposure to the heavy metal. Also, the finding of significant levels of lead in normal human semen raises the possibility of embryonic lead exposure via the father at the time of conception. There have been suggestions that lead is teratogenic in the human embryo; some fairly substantive evidence has been presented that the metal can induce malformations in various animal species. In order to come to any conclusions concerning the effects of lead on the human embryo, a good deal of epidemiologic data must be compiled. In this respect, it is not enough just to study those individuals with occupational exposure to high levels of lead. Long term exposure to low (subclinical) levels of lead must also be studied, particularly during *in utero* development.

## REFERENCES

Baglan, R.J., Brill, A.B., Schulert, A., Wilson, D., Larsen, K., Dyer, N., Mansour, M., Schaffner, L., Hoffmann, L., and Davies, J. (1974). *Envir. Res. 8:*64.

Barltrop, D. (1968). In "Mineral Metabolism in Paediatrics," p. 135. Blackwell Scientific Publications, Oxford.

Barltrop, D. (1969). *Pediat. 49:*145.

Baumann, A. (1933). *Arch. F. Gynaköl. 153:*584.

Bourret, J., and Mehl, J. (1966). *Arch. Mal. Prof. 27:*1.

Brown, D.R. (1975). *Toxicol. Appl. Pharmacol. 32:*628.

Buchet, J.P., Lauwerys, R., Roels, H., and Hubermont, G. (1977). *Int. Arch. Occup. Envir. Hlth. 40:*33.

Buchet, J.P., Roels, H., Hubermont, G., and Lauwerys, R. (1978). *Envir. Res. 15:*494.

Calvery, H.O., Laug, E.P., and Morris, H.J. (1938). *J. Pharmacol. Exp. Therap. 64:*364.

Carpenter, S.J., and Férm, V.H. (1974). *Anat Rec. 178:*323.

Carpenter, S.J., and Ferm, V.H. (1977). *Lab. Invest. 37:*369.

Carpenter, S.J., Ferm, V.H., and Gale, T.F. (1973). *Experient. 29:*311.
Casey, C.E., and Robinson, M.F. (1978). *Br. J. Nutr. 39:*639.
Chambe, S., Swinyard, C.A., and Nishimura, H. (1972). *Teratol. 5:*253.
Chyzzer, A. (1908). *Chir. Presse* (*Budapest*) *44:*906.
Clark, A.R.L. (1977). *Postgrad. Med. J. 53:*674.
Collucci, A.V., Hammer, D.I., Williams, M.E., Hinners, T.A., Pinkerton, C., Kent, J.L., and Love, G.J. (1973). *Arch. Envir. Hlth. 27:*151.
Conney, A.H. (1967). *Pharmacol. Rev. 19:* 317.
Dalldorf, G., and Williams, R.R. (1945). *Science 102:*668.
Deneufbourg, H. (1905), *These de Paris.*
Der, R., Fahim, Z, Yousef, M., and Fahim, M. (1976). *Res. Comm. Chem. Path. Pharmacol. 14:*689.
Dillon, H.K., Wilson, D.J., and Schaffner, W. (1974). *Am. J. Dis. Child. 128:*491.
Einbrodt, H.J., Schiereck, F.W., and Kinny, J. (1973). *Arch. F. Gynaköl. 213:*303.
Eyden, B.P., Maisin, J.R., and Mattelin, G. (1978). *Bull. Envir. Contam. Toxicol. 19:*266.
Fahim, M.S., Fahim, Z., and Hall, D.G. (1976). *Res. Comm. Chem. Path. Pharmacol. 13:*309.
Ferm, V.H. (1967). *Lab. Anim. Care 17:*452.
Ferm, V.H. (1976). *Curr. Top. Pathol. 62:*145.
Ferm, V.H., and Ferm, D.W. (1971) *Life Sci. 10:*35.
Gershanik, J.J., Brooks, G.G., and Little, J.A. (1974). *Am. J. Obstet. Gynecol. 119:*508.
Gilfillan, S.C. (1965). *J. Occup. Med. 7:*53.
Golubovich, E., Aukimenko, M.M., and Chirkova, E.M. (1968). *Toksik. Khim. Veshchestv 10:*64.
Green, M., and Gruener, N. (1974). *Res. Comm. Chem. Path. Pharmacol. 8:*735.
Haas, T., Wieck, A.G., Schaller, K.H., Mache, K., and Valentin, H. (1972). *Zbl. Bakt. Hyg. I. Abt. Orig. B. 155:*341.
Hamilton, A., and Hardy, M.L. (1974). In "Industrial Toxicology," p. 119. Publishing Sciences Group, Acton, Maine.
Harris, P., and Holley, M.R. (1972). *Pediat. 49:*606.
Henderson, A. (1824). "A History of Ancient and Modern Wines," p. 339.
Hernberg, S., and Nikkanen, J. (1970). *Lancet 1:*63.
Hilderbrand, D.C., Der, R., Griffin, W.T., and Fahim, M.S. (1973). *Am. J. Obstet. Gynecol. 115:*1058.
Horiuchi, K., Horiguchi, S., and Suekane, M. (1959). *Osaka City Med. J. 5:*41.
Hubermont, G., Buchet, J.P., Roels, H., and Lauwerys, R. (1978). *Int. Arch. Occup. Envir. Hlth. 41:*117.
Jacquet, P. (1977). *Arch. Pathol. Lab. Med. 101:*641.
Jacquet, P., Gerber, G.B., Leonard, A., and Maes, J. (1977) *Experient. 33:*1375.
Jacquet, P., Leonard, A., and Gerber, G.B. (1975). *Experient. 31:*1312.
Jacquet, P., Leonard, A., and Gerber, G.B. (1976). *Toxicol. 6:*129.
James, L.F., Lazar, V.A., and Binns, W. (1966). *Am. J. Vet. Res. 27:*132.
Jessup, D.C. (1967). U.S. National Technical Information Service Report PB-201, p. 139.
Karp, W.B., and Robertson, A.F. (1977). *Envir. Res. 13:*470.
Kehoe, R.A., Thamann, F., and Cholak, J. (1933). *J. Indust. Hyg. 15:*301.
Kennedy, G.L., Arnold, D.W., and Calandra, J.C. (1975). *Fd. Cosmet. Toxicol. 13:*629.
Khare, N., Der, R., Ross, G., and Fahim, M. (1978). *Res. Comm. Chem. Path. Pharmacol. 20:*351.

Kimmel, C.A., Grant, L.D., and Sloan, C.S. (1976). *Teratol. 13:*27A.
Klevay, L.M. (1972). *Fed. Proc. 31:*734.
Kuhnert, P.M., Kuhnert, B.R., and Erhard, P. (1976). D.D. Hemphill (Ed.), In "Trace Substances in Environmental Health," Vol. 10, p. 373. University of Missouri Press, Columbia, Mo.
Lamm, S., Cole, B., Glynn, K., and Ullmann, W. (1973). *N. Engl. J. Med. 289:*574.
Lancranjan, I., Popescu, H.I., Gavanescu, O., Klepsch, I., and Servanescu, M. (1975). *Arch. Envir. 30:*396.
Lauwerys, R., Buchet, J.P., Roels, H., and Hubermont, G. (1978). *Env. Res. 15:*278.
Mann, T. (1964). In "The Biochemistry of Semen and of the Male Reproductive Tract." John Wiley and Sons, New York.
McClain, R.M., and Becker, B.A. (1970). *Pharmacol. 29:*347.
McClain, R.M., and Becker, B.A. (1975). *Toxicol. Appl. Pharmacol. 31:*72.
McLellen, J.S., VanSmolinski, A.W., Bederka, J.P., and Boulos, B.M. (1974). *Fed. Proc. 33:*288.
Millar, J.A., Battistini, V., Cumming, R.L.C., Carswell, F., and Goldberg, A. (1970). *Lancet 2:*695.
Momcilovic, B. (1978). *Arch. Envir. Hlth. 33:*115.
Monkiewicz, J., Jaczewski, S., and Dynarowicz, I. (1975). *Med. Weter. 31:*684.
Morris, H.P., Lang, E.P., Morris, H.J., and Grant, R.L. (1938). *J. Pharmacol. Exp. Therap. 64:*420.
Murthy, G.K., and Rhea, U.S. (1971). *J. Dairy Sci. 54:*1001.
Noirfalise, A., Hensghcm, C., and Legros, J. (1967). *Arch. Belg. Med. Soc. 25:*73.
Oliver, T. (1911). *Brit. Med. J. 1:*1096.
Palmisano, P.A., Sneed, R.C., and Cassady, G. (1969). *Pediat. 75:*869.
Paul, C. (1860). *Arch. Gen. Med. 5:*513.
Plechaty, M.M., Noll, B., and Sunderman, F.W. (1977). *Ann. Clin. Lab. Sci. 7:*515.
Puhac, I., Hrgovic, N., Stankovic, M., and Popvic, S. (1963). *Acta Verterin. 13:*3.
Rennert, O. (1881). *Arch. F. Gynköl. 16:*109.
Roels, H., Lauwerys, R., Buchet, J.P., and Hubermont, G. (1977). *Toxicol. 8:*107.
Rom, W.N. (1976). *Mt. Sin. J. Med. 43:*542.
Rosenblatt, M. (1906). *Beitrage Zur Kenntnis Der chronischen Bleivergiftuny.* p. 107.
Scanlon, J. (1972). *Clin. Pediat. 11:*135.
Shupe, J.L., Binns, W., James, L.F., and Keeler, R.F. (1967). *J. Am. Vet. Med. Assoc. 151:*198.
Singh, N.P., Thind, I.S., Vitale, L.F., and Pawlow, M. (1976). *J. Lab. Clin. Med. 87:*273.
Stowe, H.D., and Goyer, R.A. (1971). *Fertil. Steril. 22:*755.
Thieme, R., Schramel, P., Klose, B.J., and Waidl, E. (1974). *Geburtshilfe Frauenheilkunde* 34:36.
Timm, F., and Schulz, G. (1966). *Histochem. 7:*15.
Tipton, I.H., and Cook, M.J. (1963). *Health Phys. 9:*103.
Tracy, A., and McPheat, J. (1943). *Biochem. 37:*683.
Varma, M.M., Joshi, S.R., and Adeyemi, A.O. (1974). *Experient. 30:*486.
Wibberley, D.G., Khera, A.K., Edwards, J.H., and Rushton, D.I. (1977). *J. Med. Genet. 53:*674.
Wide, M., and Nilsson, D. (1977). *Teratol. 16:*273.
Zetterlund, B., Winberg, J., Lundgren, G., and Johansson, G. (1977). *Acta Paediat. Scand. 66:*169.

# Effects of Lead on the Kidney

*David D. Choie and Goetz W. Richter*

TABLE OF CONTENTS



## I. INTRODUCTION

It has been known for many years that functional and structural abnormalities of the kidneys accompany both acute and chronic lead poisoning. These pathologic changes affect primarily (in the acute phase, almost exclusively) the proximal convoluted tubules. They are similar in man and in experimental animals such as rats, mice and dogs. Modern developments in renal physiology, electron microscopy, cell fractionation and bio- and histochemistry have made it possible to analyze renal lesions associated with lead poisoning in detail. Nevertheless, detailed investigations have barely begun. We hope that this chapter serves to indicate these beginnings and to point out major problems that have to be explored and solved.

## II. RELATIVE LEAD CONTENT OF KIDNEYS

Once absorbed, lead (Pb) is carried to various soft tissues through the blood. Although about 90% of lead in the blood is bound to erythrocytes, it reaches soft tissues in maximum concentration within 1 hr. At this time, about 50% of administered lead is found in three tissues, the kidney, the liver and the blood (Morgan et al., 1977; Castellino and Aloj, 1964). Approximately 20% of total lead absorbed becomes localized in the kidneys (Morgan et al., 1977). This represents the highest concentration of lead in any organ in the early stages of acute lead intoxication. In time, the lead in soft tissues is remobilized and is either excreted in the urine or redistributed to other tissues, mostly to bones. At first the rate of mobilization of renal lead is rapid; then it gradually decreases. The half-life of renal lead is estimated to be 1 to 2 days (Conrad and Barton, 1978). One week after a single dose, about 2% remains in the kidneys. There are few data on the subcellular distribution of lead. One study (Barltrop et al., 1971) showed that, 48 hr after a single ip injection of $^{203}Pb$ into rats, nuclei, mitochondria, microsomes and lysosomes of the kidneys contained 13%, 12%, 23% and 1.2% of total lead in the kidney, respectively. The largest fraction was found in the cytosol (36%).

Since lead has a long biological half-life, a body burden is built up which is cumulative with age. In man, over 90% of the total body burden is in bones (Barry, 1975; Schroeder and Tipton, 1968). In the kidneys, the mean concentration of lead is about 50 times greater in the cortex than in the medulla (Barry, 1975).

In a study of the lead content of 129 human cadavers (aged between 0 and 99 years), Barry (1975) found that, apart from bones, the major depository of lead, the liver, aorta and kidney, in that order, were richest in lead. For example, the mean lead content of renal cortex from male adults in England, with no known occupational exposure to lead, ranged from 0.15 to 1.85 ppm (wet weight), with a mean of 0.78 ± 0.38 (SD) ppm. In males with known occupational exposure the range was 0.33 to 2.20 ppm, with a mean of 0.66 ± 0.56 (SD) ppm. There was considerable variation with age in both groups. More specific information about the deposition of lead in the kidney in lead poisoning has come from experimental studies.

## III. EXCRETION OF LEAD THROUGH KIDNEYS

The principal routes of excretion of absorbed lead are through the kidneys and the intestine. During the first 24 hr after iv administration of lead ~25% is excreted in urine and about 10% in feces (Conrad and Barton, 1978; Morgan et al., 1977). Renal excretion of lead involves glomerular filtration and transtubular transport (Vostal and Heller, 1968); however, recent evidence also indicates that a large fraction of lead in glomerular ultrafiltrate is reabsorbed by the tubules (Vander et al., 1977). There is some evidence that renal excretion of lead has a maximum limit above which retention of lead increases (Goyer, 1971).

In kinetic terms, total excretion of lead follows a biphasic pattern. Immediately after a single dose, the rate of excretion is high; about 50% of absorbed lead is lost during the first week. In the second phase, the rate of excretion is much slower, with a biologic $t_{1/2}$ of about 6 months. This biphasic excretion is apparently independent of the dose administered (Castellino and Aloj, 1964). About two-thirds of the total amount excreted passes through the kidneys and nearly all of the remainder passes through the gut. A small fraction of lead is excreted by other routes, such as sweat and sebum.

## IV. EARLY EFFECTS OF LEAD ON KIDNEYS

Lead nephropathy is conveniently considered in two phases. The early phase may last from days to months and is associated with acute or repeated acute clinical manifestations of lead poisoning, referred to in this chapter as "early effects." Chronic lead nephropathy occurs after years of exposure to lead and is associated with severe changes in renal structure, failure of renal function and sometimes hypertension and saturnine gout.

### A. Structural Changes Seen by Light Microscopy

Whereas a number of heavy metals, such as uranium, mercury and cadmium, can cause massive tubular necrosis, with sloughing of tubular epithelium (Cuppage and Tate, 1967), acute intoxication with lead produces subtle changes in renal structure, mainly in the epithelial cells of the proximal tubules. One of the earliest morphologic changes in response to moderate doses of lead is enlargement of nuclei in the proximal tubular cells. In rats and other experimental animals, conspicuous variation in nuclear size may be detectable within a few days after oral administration of lead. The nucleoli become prominent, a phenomenon that indicates augmented nuclear activity. There may also be an increase in mitotic figures in the proximal tubular epithelium. Frank tubular necrosis is absent, even after a relatively large dose of lead (Choie and Richter, 1972a). Continuous treatment of rats with 1% lead acetate in the diet for at least a month usually produces the most characteristic morphologic feature of lead intoxication at the level of light microscopy, *i.e.*, formation of intranuclear inclusion bodies in the proximal tubular epithelial cells (Fig. 1). Such "inclusions" were discovered by Blackman (1936) in the kidneys of children that had died of lead poisoning. Since then, similar intranuclear inclusions in proximal tubular cells have been found in lead-poisoned rats, mice, dogs, rabbits, swine, fowl and other species (Goyer and Rhyne, 1973). They have also been noted,

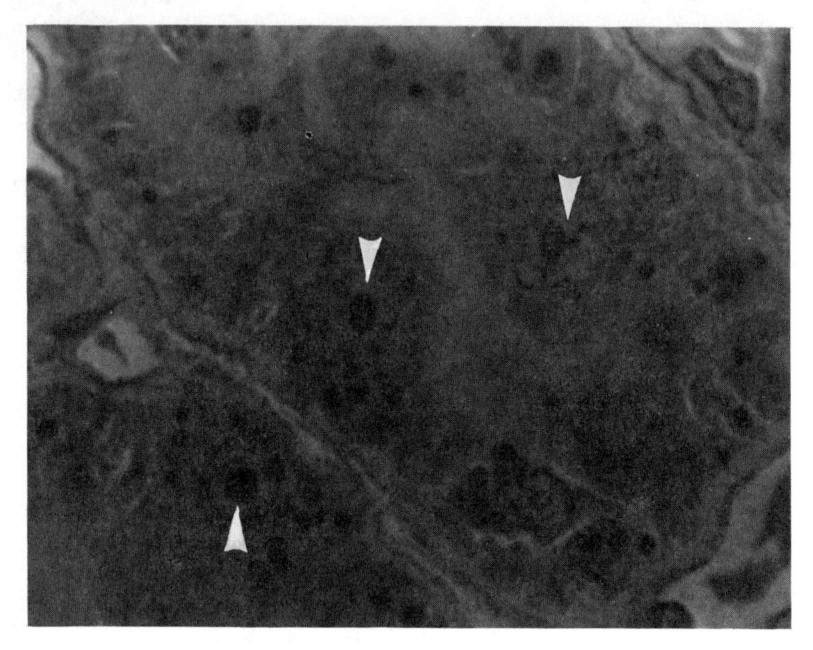

**Figure 1.** Light micrograph of lead protein inclusion bodies (arrows) in three nuclei in proximal convoluted tubule of rat given 12 intraperitoneal injections of lead subacetate over a period of 92 days. Tissue fixed in 10% neutral formalin. Iron-hematoxylin stain. Magnification, ×1700.

although much less frequently, in other sorts of cells outside the kidneys, *e.g.*, in astrocytes (Goyer and Rhyne, 1973) and osteoclasts (Hsu et al., 1973; van Mullem and Stadhouders, 1974). These inclusions are distinct from nucleoli, although, like the latter, they are roughly spherical or oval and vary in size. Their staining characteristics are somewhat variable and have been reviewed elsewhere in detail (Landing and Nakai, 1959; Richter et al., 1968). In sections stained with hematoxylin and eosin, they tend to be eosinophilic. They can be acid-fast when stained with carbolfuchsin. They stain with mercuric bromphenol blue, but not with fast green after a wash with trichloroacetic acid. Sometimes the inclusions are closely surrounded by Feulgen-positive material, but they do not themselves contain such material (Richter et al., 1968). As exposure to lead continues, the number of proximal tubular cells containing intranuclear inclusion bodies increases. It can be inferred from serial observations that up to a point the intranuclear inclusion bodies enlarge as lead poisoning continues. Two or more bodies may be found in one nucleus. After several months of exposure to lead, hyperplasia of proximal tubular epithelium may be observed. The epithelium lining some of the tubules becomes crowded, heaped up or otherwise irregularly disposed; many cells have intensely basophilic cytoplasm. If administration of lead to rats is continued for 6 months or longer, some epithelial cells in the hyperplastic regions become so atypical as to suggest premalignant change (Zollinger, 1953, Choie and Richter, 1972b).

As seen by light microscopy, the acute changes in the cytoplasm of proximal tubular epithelial cells are nonspecific and unimpressive. In rats, for example, vacuolation or swelling of the cytoplasm may be present. Glomeruli, distal and collecting tubules appear uninvolved.

Available data indicate that histologic changes in the early phase of lead intoxication in man and experimental animals are similar (Goyer, 1971).

### B. Ultrastructural Changes in the Early Phase

The ultrastructural changes in acute experimental lead nephropathy comprise nonspecific and specific lesions. Both are characteristically situated in the proximal portions of the renal tubules. Nonspecific changes in proximal tubular epithelial cells include dilation of endoplasmic reticulum (ER), particularly of the smooth portion (SER), blebbing of the outer nuclear membrane, enlargement of autophagosomes and conspicuous changes in the mitochondria. In the early phase of experimental lead nephropathy, enlargement of mitochondria associated with reduction and shortening of cristae is common and develops within four weeks of dietary exposure to lead. As lead nephropathy progresses for months, nearly all mitochondria in the proximal tubular cells develop structural lesions, and the total number of mitochondria per cell becomes reduced (Goyer, 1971). Autoradiographic evidence indicates that lead actually penetrates mitochondria (Murakami and Hirosawa, 1973). The mitochondrial lesions are believed to be the principal basis for renal tubular dysfunction in lead poisoning. Goyer et al. (1968) have postulated that the mitochondrial lesions are responsible for impairment of energy metabolism in renal tubular cells, manifested by diminished utilization of $O_2$, partial uncoupling of oxidative phosphorylation and reduced production of adenosine triphosphate (ATP) (Table 1).

**Table 1.*** Respiratory and phosphorylative abilities of mitochondria isolated from kidneys of control rats and rats fed 1% lead for 10 weeks†

| | | $\mu$Atoms $O_2$/g protein/min | | | |
|---|---|---|---|---|---|
| Mitochondria | Substrate | State IV | State III | Respiratory control | ADP:O |
| Control | Pyruvate | 43.2 | 130.3 | 3.3 | 2.7 |
| Lead | Pyruvate | 46.5 | 101.8 | 2.2 | 2.1 |
| Control | Succinate | 76.9 | 178.4 | 2.4 | 1.8 |
| Lead | Succinate | 62.4 | 140.0 | 2.4‡ | 1.8‡ |

* From Goyer and Rhyne (1973)

† Modified from Goyer and Krall (1969)

‡ No significant difference between lead and control mitochondria

Since tubular reabsorption of amino acids from glomerular filtrate is by active transport requiring ATP, amino acids are excreted in the urine if this requirement is not met.

The characteristic specific ultrastructural lesions in acute and chronic lead nephropathy are the intranuclear inclusions in the epithelial cells of the proximal tubules. A typical inclusion is composed of an electron-dense core and a loose mesh of fibrils that extend to the periphery of the inclusion (Fig. 2). Sometimes a ring of intermediate density can be seen between central core and outer fibrillary structures. The protruding fibrils at the periphery are about 120 Å in thickness and vary in length up to about 3000 Å (Richter et al., 1968).

Although it had long been thought that intranuclear inclusions appear only after a latent period of 4 to 7 weeks of exposure to lead (Goyer, 1971), recent experimental data clearly show that, at least in experimental animals, they arise soon after uptake of lead. By electron microscopy, Choie and Richter (1972c) demonstrated that well developed intranuclear inclusions were present in the tubular epithelial cell 24 hr after a single ip injection of lead acetate (50 $\mu$g/g weight) in rats. This made it seem likely that the inclusions had begun to develop many hours earlier. In further experiments with rats and mice, intranuclear inclusions were detected as early as 6 hr after an intracardiac injection of lead (30 $\mu$g/g weight) (Choie and Richter, 1975). At a dose of 5 to 10 $\mu$g of lead/g weight, the inclusions appeared after 8 hr in mouse kidneys. Figure 3 shows an example 4 days after injection of lead.

These findings indicate that the formation of intranuclear inclusion bodies is one of the earliest manifestations of lead poisoning. The formation of intranuclear inclusions appears to be triggered at a minimum effective concentration of lead in renal tissues, but, on the basis of the findings of Choie et al. (1975), this minimum concentration may be lower than the 10 to 20 $\mu$g lead per g of wet kidney weight estimated by Goyer and Rhyne (1973). This renal concentration would be attained rapidly if lead is injected intravascularly or intraperitoneally, and more slowly if lead is given in the diet, perhaps over weeks. The earliest intranuclear inclusion bodies of lead nephropathy are too small to be visible by light microscopy, but they consist of a characteristic fibrillar mesh and central dense, amorphous material. They have been found only in the proximal tubular epithelial cells (Choie and Richter, 1972c; Choie et al., 1975; Richter, 1976).

In mice and rats, these rapidly formed intranuclear inclusions are present in conjunction with similar fibrillar structures in the cytoplasm (Choie and Richter, 1972c; Richter, 1976) which are generally smaller in size and have less well developed central cores (Fig. 3). The cytoplasmic structures are usually present in regions that are rich in free polyribosomes. Experiments with rats indicated that within 2 days after a single intra-

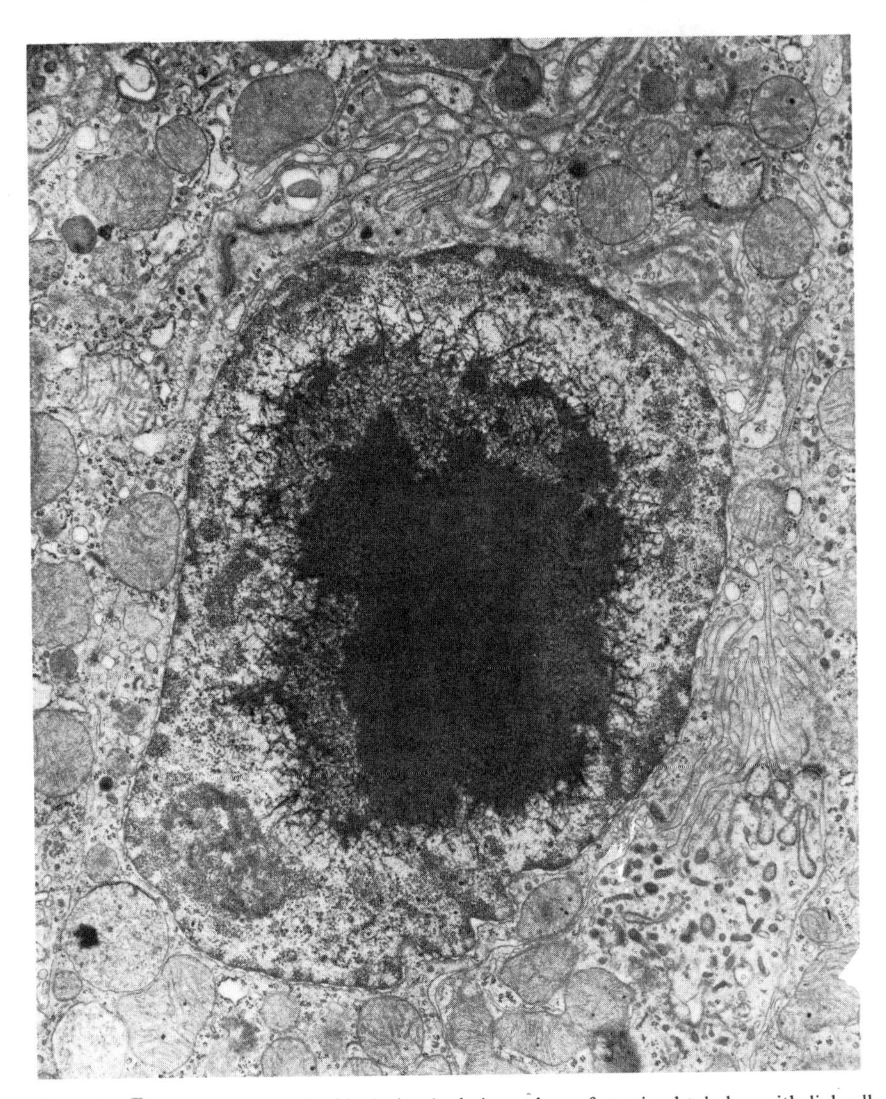

**Figure 2.** Electron micrograph of inclusion body in nucleus of proximal tubular epithelial cell of rat given 14 intraperitoneal injections of lead subacetate over a period of 135 days. Fixed in phosphate-buffered 1% $OsO_4$ solution. Fibrils extend from dense core to periphery of the inclusion body. A nucleolus, which appears normal, is at lower left. Stained with uranyl acetate and lead citrate. Magnification ×16,000. (Richter et al., 1968.)

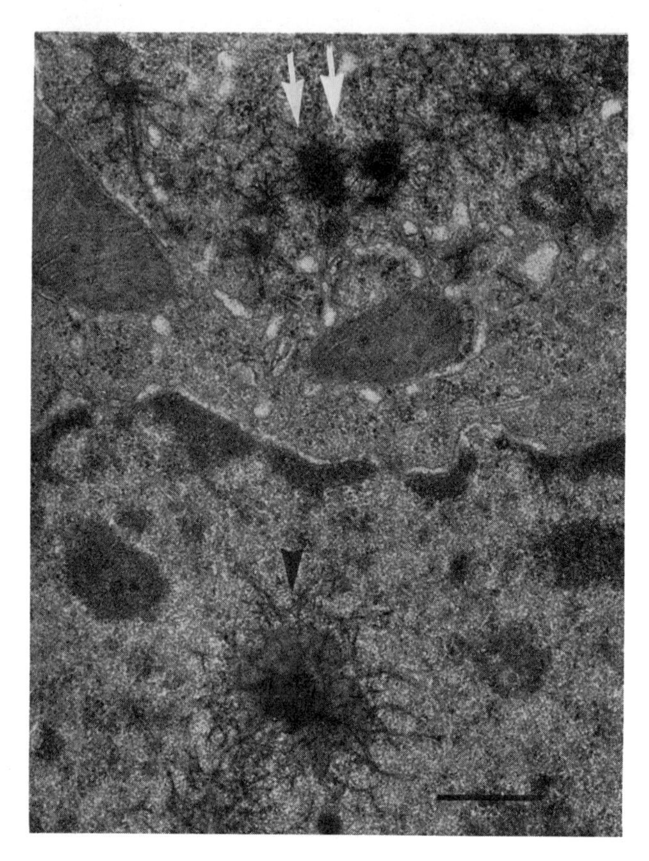

**Figure 3.** Electron micrograph of part of proximal tubular epithelial cell from a rat 4 days after a single intraperitoneal dose of lead acetate (0.15 mg Pb/g body weight). Cytoplasmic structures (white arrows) resemble the intranuclear inclusion (black arrowhead). Stained with uranyl acetate and lead citrate. Scale: 0.5 μm. (Choie and Richter, 1972c.)

cardiac injection of lead (*e.g.*, 10 μg/g) into rats, cytoplasmic fibrillar bodies are more frequent than intranuclear inclusions in the proximal tubular epithelial cells (Richter, 1976). Beyond the fourth day after injection of lead, the cytoplasmic fibrillar bodies disappeared, apparently by lysosomal autophagy; 10 days after the injection of lead, intranuclear inclusions were still present, but cytoplasmic fibrillar bodies were rarely encountered.

Another feature of the earliest phase of lead nephropathy is the presence of clusters of ferritin in the immediate vicinity of cytoplasmic fibrillar bodies (Richter, 1976 and 1979). In rats, these clusters were present 2 to 3 days after an injection of lead. They tended to disappear rapidly after the fourth day, together with the cytoplasmic fibrillar bodies.

Other lesions frequently encountered in renal tubular epithelial cells, proximal as well as distal, in experimental acute lead nephropathy are laminated membranous structures ("myelin figures") and autophagy of mitochondria by lysosomal derivatives (Richter et al., 1968; Goyer, 1971).

## C. Composition, Function and Origin of Intranuclear Inclusions

Although analyses of the intranuclear inclusions have, almost without exception, been performed on material derived from chronically lead-poisoned kidneys, the results are likely to bear a close relationship to the early inclusion bodies as well and are therefore summarized here.

Histochemically, intranuclear inclusions do not contain deoxy ribonucleic acid (DNA), as evidenced by negative Feulgen reactions (Richter et al., 1968). They probably contain acidic proteins because they stain with mercuric bromophenol blue and not with fast green after trichloroacetic acid; they also contain lead (Richter et al., 1968). They can be degraded by treatment with pronase, but not by DNase or RNase (Richter et al., 1968). These results indicate that the inclusions contain proteins but not DNA or ribonucleic acid (RNA). The presence of lead has been demonstrated by autoradiography (Dallenbach, 1965), electron probe X-ray microanalysis (Carroll et al., 1970) and histochemistry (Grzybek, 1972).

Intranuclear inclusion bodies can be isolated from kidney homogenate by differential centrifugation in sucrose solutions (Goyer et al., 1970a). Chemical analysis of such isolated inclusions has confirmed that they contain protein as well as lead, probably in the form of a lead-protein complex (Goyer et al., 1970a; Moore et al., 1973). The inclusion bodies are insoluble in aqueous solutions or organic solvents but can be dispersed by denaturing agents such as 6M urea and 1% sodium dodecyl sulfate (SDS) (Moore et al., 1973). Treatment of the isolated inclusion bodies with a chelating agent disodium ethylenediaminetetraacetate (EDTA) removes lead and fragments the bodies. This suggests that lead is ionically bound to the protein in the inclusion, perhaps at sites of sulfhydryl groups. Data on the amino acid composition indicate that the protein(s) in inclusion bodies is acidic, with a relatively large content of aspartic and glutamic acids (Moore et al., 1973).

The biologic significance of the intranuclear inclusion body in lead nephropathy is not clear. It may be the end product of a detoxification process (Goyer et al., 1970b). It is certain that mature inclusions are rich in lead. It has been estimated that 80 to 90% of renal lead is concentrated within cell nuclei and that at least 50% of intranuclear lead in proximal tubular cells is contained within inclusion bodies (Goyer, 1971). Preparations of relatively old inclusion bodies have been found to contain approximately 50 $\mu g$ of lead per mg of protein, which is a 60 to 100 times greater concentration of lead than that in the kidney as a whole (Goyer et al., 1970a). On the basis of these findings, Goyer and Rhyne (1973) postulated that intranuclear inclusions serve to protect cells from the toxic effects of lead on cell metabolism by binding lead. By analogy, the same would apply to the cytoplasmic fibrillar bodies, although their lead content has not been determined.

The origin of the intranuclear inclusion in lead nephropathy has been a subject of considerable interest. For example, it has been postulated that the protein matrix of inclusions results from degradation and restructuring of a pre-existing intranuclear protein (Richter et al., 1968). Thus, it seemed possible that lead enters the nucleus and binds to soluble protein molecules. Thereupon, the lead-protein complexes form fibrils, perhaps through a process of polymerization. In this connection, electron microscopy suggested that the fibrils of the inclusion are composed of subunits (Richter et al., 1968). Alternatively, lead may bind to a pre-existing soluble cytoplasmic protein, be transported with the protein into the nucleus and there form an inclusion body. In this case, the cytoplasmic fibrillar bodies seen in the very early stage of lead nephropathy would be products of the same process as the intranuclear bodies, with polymerization (formation of fibrils) taking place in the cytoplasm as well as in the nucleus. Another hypothesis is that the protein moiety of the intranuclear inclusion is derived from carrier proteins involved in the transport of lead in blood plasma (Goyer et al., 1970a).

A few years ago, evidence for the origin of inclusion body protein was obtained which indicates that *de novo* synthesis of protein is required (Choie et al., 1975). A single intracardiac injection of lead acetate (5 to 10 $\mu$g/g body weight) was given to mice to induce intranuclear inclusions. Eight hours later there were typical small fibrillar structures in the cytoplasm as well as the nucleus of many proximal tubular cells. When these mice were treated with cycloheximide, an inhibitor of protein synthesis, immediately after injection of lead, neither intranuclear nor cytoplasmic fibrillar bodies were formed (Choie et al., 1975). This inhibitory effect of cycloheximide suggests that *de novo* synthesis of protein is required for the biogenesis of the fibrillar bodies in cytoplasm and nucleus. Since it has been shown that lead in appropriate dosage can stimulate synthesis of at least some proteins in mouse kidney (Choie and Richter, 1974b), the inference that *de novo* protein synthesis is essential to the formation of intranuclear inclusion bodies is reasonable, but it is not clear how this synthesis is related to the formation of the inclusion bodies.

## V. Pathologic Changes in the Kidney after Chronic Intoxication with Lead

The pathologic features of chronic lead nephropathy have remained controversial. At present, there is little clinical documentation of the pathogenesis of chronic lead nephropathy. Early in this century, it was thought that the most common feature was interstitial nephritis leading to contracted kidneys (Morgan et al., 1966). Although this has not always been substantiated by later observations, significant support for this idea has come from a series of studies on chronic lead poisoning in Queensland, Australia. There, extensive use of lead-based paint in houses resulted in widespread lead poisoning among children. Between 1890 and 1930, there was a high incidence of chronic renal failure and a great increase in mortality from chronic nephritis and hypertension (Emmerson, 1973). In a series of follow-up studies, Henderson and Inglis (1957) found that the bones of patients dying with granular contracted kidneys tended to have a high lead content. This suggested that the increased mortality from chronic renal failure in Queensland was related to excess exposure to lead in childhood.

In a more recent study, Inglis et al. (1978) analyzed kidneys of patients known to have had chronic exposure to lead on the basis of 1) significant increase of lead content of skull bone (>4.1 mg/100 g moist bone) or 2) significant increase in the urinary excretion of lead (>0.6 mg/4 days) after a standard infusion of CaEDTA. In patient falling into either of these categories, the kidneys were symmetrically contacted (individual kidneys weighing less than 112 g, and in some cases, less than 30 g). In severe cases the cortical surfaces of such kidneys were granular and the cortices were very thin. Sometimes chalky deposits, suggesting urates, were found in the renal medullary substance. Microscopically, the granularity was seen to be due to alternating nodules of dilated and atrophic tubules. A large proportion of glomeruli had disappeared. The remaining glomeruli and their related tubules were greatly enlarged. There was occasional periglomerular and interstitial fibrosis. Although the genesis of the primary renal lesion is unknown, it has been claimed on the strength of histologic evidence from a few cases that lead can cause destruction of glomeruli (Inglis et al., 1978).

In the US a study of chronic lead nephropathy was done in patients known to have suffered lead intoxication (Morgan et al., 1966). These patients had been chronic drinkers of "moonshine whiskey," which is often contaminated by lead. Renal biopsies showed that the kidneys had intranuclear lead inclusions, obsolescent glomeruli, tubular degeneration, various vascular lesions and interstitial fibrosis. Inflammatory infiltrate was minimal and urine cultures were negative. Similar histologic findings were reported by Wedeen et al. (1975), who studied kidney biopsies from lead workers. In both studies, the kidneys of patients with excessive exposure to lead appeared to be near normal in size. Absence of contracted kidneys in these patients suggested that lead intoxication in the Australian children had been more severe and prolonged (Emmerson, 1973). Although none of the clinical studies conclusively prove that prolonged exposure to lead causes chronic nephropathy, clinical signs and histologic evidence indicate that lead may be an etiologic factor.

Because of limitations and uncertainties inherent in clinical histories, sampling and controls, studies with experimental animals are essential if we are to understand the pathogenesis of chronic lead nephropathy. As an experimental model, Goyer (1971) induced chronic lead nephropathy in rats by continuous feeding of 1% lead acetate for long periods. In the first 20 weeks, the main histologic changes were "cloudy swelling" and intranuclear inclusion bodies in the proximal tubular cells. By electron microscopy, mitochondrial swelling and other nonspecific cytoplasmic changes were seen. In the next stage (>20 weeks, <52 weeks) the incidence of tubular epithelial cells with intranuclear inclusions became maximal. As determined by Choie and Richter (1972b) in another study, about 40% of all proximal tubular cells contained recognizable nuclear inclusions in rats that had been treated with lead for 6 months. In both studies there was a progressive increase in interstitial fibrous tissue and the epithelium of some tubules was hyperplastic, while in other tubules it was atrophic. In the rats studies by Choie and Richter (1972b), some hyperplastic tubular cells appeared atypical and many of these did not contain intranuclear inclusions. In Goyer's study (1971), the kidneys showed interstitial scarring and fibrous and some sclerotic glomeruli after a year or more of lead poisoning. Eventually, renal neoplasms developed.

Although total weight gain of rats is retarded by chronic lead poisoning, kidney weight is greater in leaded rats than in controls (Goyer, 1971). This contrasts with the contracted kidneys described by Emmerson (1973) for chronic lead nephropathy in man. It has still to be established that in man the early pathologic changes in the kidneys progress to chronic lead nephropathy.

## VI. PATHOPHYSIOLOGIC CORRELATES

Lead impairs renal function in successive phases. Early damage to proximal tubular cells causes tubular dysfunction, affecting mainly tubular reabsorption (Goyer, 1971; Vander et al., 1977). The tubular injury, however, is not permanent; normal function can be restored by proper treatment (Goyer, 1971). If exposure to lead is prolonged, there may be progressive tubular damage (Goyer, 1971). Whether chronic lead nephropathy in man is regularly associated with diminished glomerular filtration, secondary to extensive scarring, as might be expected, has not been determined.

The effects of lead on renal tubular cell functions are related to its affinity for ligands such as sulfydryls, phosphates, hydroxyls, carboxyls, amino groups and others. Thus, lead inhibits most enzymes with a functional sulfhydryl group (Nordberg, 1976). Some enzymes, *i.e.*, δ-aminolevulinic acid dehydrase (ALAD), are more sensitive than others. Lead can form complexes with the phosphate groups of nucleotides, thereby affecting conformation of nucleic acids. It also catalyzes a nonenzymatic hydrolysis of nucleoside triphosphates (Vallee and Ulmer, 1972). It binds to membranes and may thus alter membrane permeability and interrupt substrate transport through membranes. Lead also inhibits sodium and potassium-dependent adenosine triphosphate (Nechay and Saunders, 1978). Reduction of $Na^+$ and $K^+$ ATPase activity by lead would disrupt cellular cation transport. Lead may interfere with iron metabolism by blocking the heme ferrochelatase activity that mediates the incorporation of Fe into the tetrapyrrole ring of heme, perhaps causing accumulation of ferruginous micelles in mitochondria (Bessis et al., 1957; Chisolm, 1964; Trump et al., 1978). Exposure to lead does not necessarily result in an inhibition of all such metabolic activities. At certain concentrations, lead can enhance the catalytic activity of some enzymes, although catalytic specificities may also change. For instance, treatment with lead increases gluconeogenic enzyme activities of pyruvate carboxylase, phosphoenolpyruvate carboxylase, fructose 1,6 diphosphatase and glucose 6-phosphatase in rat kidney (Stevenson et al., 1976). It has also been reported that lead increases activity of lactate dehydrogenase, glutamic dehydrogenase and glucose- 6-phosphate-dehydrogenase and that in low concentrations ($10^{-7}$ M) lead can stimulate heme synthesis in rabbits (Vallee and Ulmer, 1972).

Acute renal dysfunction due to lead is clinically manifested as an excessive excretion of amino acids, glucose and phosphates (in other words, as diminished reabsorption in the proximal tubules). The aminoaciduria of lead poisoning is thought to be a consequence of diminished available energy in the proximal tubular epithelial cells. Since lead causes aberrations in mitochondria, with reduction in respiratory rates and oxidative phosphorylation, there is a decrease in production of the ATP necessary for active transport of glomerular ultrafiltrate across the tubular cell membrane. In the presence of insufficient ATP in the proximal tubular cell, there is incomplete absorption of amino acids. Although a dose-response relationship for this effect of lead has not been established, a renal concentration of lead above 50 $\mu g/g$ wet weight was found to be associated with increased urinary excretion of amino acids in experimental animals (Goyer, 1971). Nearly all amino acids are excreted above normal levels in severe lead poisoning.

## VII. SATURNINE GOUT AND LEAD NEPHROPATHY

The incidence of gout is much higher in chronic lead nephropathy than in other types of chronic renal disease. For this reason it has been assumed for a long time that chronic lead nephropathy leads to a form of gout called saturnine gout (Goyer, 1971). There are few clinical data on the distinction between saturnine and idiopathic gout. Saturnine gout usually occurs rather early in adult life, without familial predisposition. It is associated with anemia and, although it often affects the lower limbs, arthritic attacks are less frequent than in idiopathic gout. A history of lead poisoning can usually be obtained (Morgan et al., 1966; Goyer, 1971). About half of the patients examined for chronic lead nephropathy in Queensland, Australia, were suffering from gouty arthritis (Inglis et al., 1978). In the US the frequency of joint disease is particularly high among patients known to have consumed illicit whiskey for years (Morgan et al., 1966; Ball and Sorenson, 1969).

Patients with chronic lead nephropathy usually have hyperuricemia. The excess urate is believed to be the underlying cause of the joint disease. The high plasma urate concentration and frequent gouty arthritis in chronic nephropathy are generally attributed to low renal clearance of urate (Emmerson et al., 1971). Renal excretion of urate is mediated by a number of factors, including renal blood flow, glomerular filtration rate, tubular reabsorption and secretion, urine flow rate, metabolites, hormones and drugs (Emmerson and Ravenscroft, 1975). It has been proposed that, in chronic lead nephropathy, the reduced renal excretion of urate is due to a defect in active tubular secretion of urate. However, recent evidence suggests that low clearance of urate is a result of excessive tubular reabsorption (Emmerson et al., 1971). The hyperuricemia appears to be the consequence of reduced renal excretion of urate.

Another important factor in saturnine gout is that lead can disrupt purine metabolism. Farkas et al. (1978) reported that lead inhibited guanine aminohydrolase, which catalyzes the hydrolytic deamination of guanine to xanthine. They showed that an iv injection of lead acetate into a pig resulted in a 4.5-fold increase in the concentration of guanine in the urine and formation of crystalline accretions on the epiphyseal plate of one of the femoral head after 48 hr. The accretions appeared to be guanine.

## VIII. RENAL HYPERTENSION

Although it has been thought that arteriolar sclerosis and even arteriosclerosis are primary lesions in chronic lead nephropathy (Cantarow and Trumper, 1944), recent evidence indicates that arteriolar changes in the kidneys are not a constant feature of lead poisoning and that they are a late manifestation (Inglis et al., 1978). In early studies, the incidence of hypertension appeared to be increased among workers exposed to lead; proliferation of cells in the intima of various arterioles was thought to be an early lesion of chronic lead poisoning (Cantarow and Trumper, 1944). In the Queensland experience, the incidence of hypertension was increased some 15 years after acute lead poisoning in childhood (Inglis et al., 1978), yet no cause-effect relationship between lead nephropathy and hypertension was established. While some evidence is suggestive, it is still insufficient. For example, Lilis et al. (1977) reported that, in a group of 158 lead workers, 17% of those with 3 to 10 years of exposure to lead had hypertension, while among those with more than 10 years of exposure, the incidence was 60%. Of the 26 cases with hypertension, 16 had both elevated systolic and diastolic pressures, 7 had elevated systolic pressure only, and 3 had elevated diastolic pressure only. About half had some signs of lead intoxication, including central nervous system (CNS) symptoms, gastrointestinal (GI) symptoms, muscle and joint pains and elevated blood lead levels. There were indications of impaired renal function in 18% of the workers, as evidenced by increased blood urea nitrogen (>21 mg/100 ml) levels.

Clinical observations indicate that hypertension is rarely seen during the early phase of lead intoxication. Thus it seems that the hypertension would have to be secondary to renal damage in chronic lead poisoning. Certain histologic changes in renal vessels have been invoked to suggest possible relations of chronic lead nephropathy to hypertension. Morgan et al. (1966) reported that some patients suffering from chronic lead poisoning had microscopic lesions in small arteries and arterioles. These included medial thickening, intimal proliferation and, occasionally, hyalinization of the arterial walls. These lesions were most conspicuous in the hypertensive patients. There was also fibrosis of the adventitia of small arteries. Wedeen et al. (1975) found lesions of renal arterioles in lead workers, *i.e.*, endothelial proliferation obliterating the arteriolar lumen. In the Queensland patients with contracted kidneys after childhood exposure to lead, the afferent arterioles of glomeruli showed subintimal hyalin, as in benign hypertension. Occasionally, fibrinoid necrosis of afferent arterioles of glomeruli was seen, suggesting the changes of malignant hypertension (Inglis et al., 1978). These observations indicate a possible relationship between chronic lead nephropathy and renal hypertension. However, all of the observed changes may well have been due to other factors.

Some attempts have been made to produce hypertension in experimental animals by long continued treatment with lead. However, the results have not been conclusive (Goyer, 1971).

## IX. LEAD AND CELL PROLIFERATION

The adult mammalian kidney normally maintains a low level of cell renewal activity. The renal tubular cells, however, can be stimulated to proliferate. A classic example is the compensatory hyperplasia in the kidney that remains after unilateral nephrectomy (Johnson and Vera Roman, 1966). A number of chemicals are known to be mitogenic in the kidneys of experimental animals, *e.g.*, folic acid (Threlfall et al., 1966) and isoproterenol (Malamud and Malt, 1971). Until recently it was generally thought that lead has an inhibitory effect on cell proliferation because of its many toxic effects. Indeed, in relatively high concentrations, lead is toxic enough to cells *in vitro* to cause severe depression of cell replication (Fischer, 1975; Walton and Buckley, 1977). But in a dose range that is toxicologically more relevant, lead can markedly stimulate cell replication, at least in renal tubules. Thus, contrary to common assumptions, lead has turned out to be a potent mitogenic agent in the renal tubular epithelium.

It was first reported by Choie and Richter (1972a) that a single dose of lead can greatly stimulate proliferation of renal tubular epithelium in rats. An ip injection of lead (40 $\mu g/g$ weight) produced an approximately 40-fold increase in cell proliferation of the proximal tubular epithelium, as determined by autoradiography. By comparison, unilateral nephrectomy produced only an eightfold increase in cell proliferation in the remaining kidney. When unilateral nephrectomy was combined with an injection of lead, the peak cell proliferation in the remaining kidney was about 65 times above that in controls. Thus, unilateral nephrectomy and a dose of lead had a synergistic effect on tubular cell proliferation. In mice, an intracardiac injection of lead acetate (5 $\mu g/g$ weight) also stimulated uptake of $^3$H-thymidine in renal tubular epithelium greatly, with a maximum of 45 times above that in controls after 33 hr (Choie and Richter, 1974a) (Fig. 4). Similar findings have been reported by Cihak and Seifertova (1976). Cell proliferation is increased in all parts of renal tubules but is most pronounced in the proximal tubular epithelium (Choie and Richter, 1974a). Increased incorporation of $^3$H-thymidine into nuclear DNA is followed by a wave of mitoses in the tubular epithelium.

Histologic and cell kinetic studies have made it clear that the acute lead-induced cell proliferation in the renal tubules is not a regenerative response. Thus, there was no tubular necrosis after treatment with lead in the doses employed. In true acute tubular necrosis (for instance, after a dose of $HgCl_2$), increased cell proliferation follows sloughing of renal tubular epithelial cells (Cuppage and Tate, 1967). The increased incorporation of $^3$H-thymidine by tubular cells after administration of lead is also not a result of circadian variation, nor is it due to repair synthesis of DNA damaged by lead. Circadian variation can be ruled out because the level of cell renewal is normally very low in rat and mouse kidneys throughout the 24-hr cycle. In repair synthesis, incorporation of $^3$H-thymidine occurs without subsequent cell division. In cell replication, on the other hand, an increase in labeling activity is followed by a surge of mitoses. In the rodents treated with lead, there was a wave of mitotic activity about 6 hr after peak labeling activity. Hence, lead induced cell proliferation, rather than repair synthesis of DNA.

In both rat and mouse kidneys, the temporal changes associated with cell proliferation after a single dose of lead are remarkably discrete. The initial increase in DNA replication occurs about 24 hr after administration of lead and is followed by a narrow peak activity after 30 to 35 hr (Fig. 4). By 48 hr, proliferative activity is substantially reduced. There is a wave of mitotic activity 6 hr after the peak labeling activity in the renal tubular

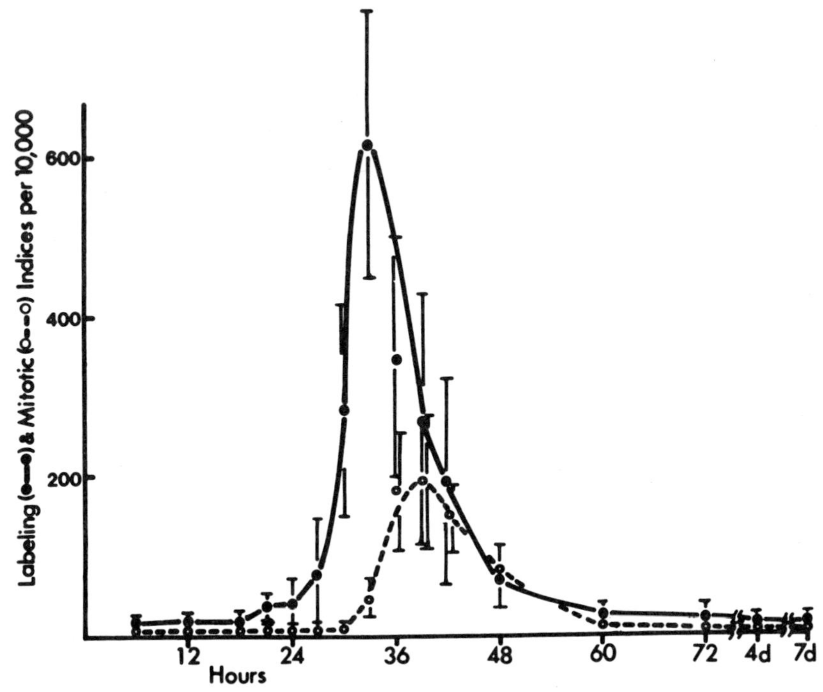


**Figure 4.** Lead-induced cell proliferation in mouse kidney. Mice were given an intracardiac injection of lead acetate (5 μg lead/g body weight) at 0 hr. Each mouse was pulse-labeled with $^3$H-thymidine for 1 hr before sacrifice. Mean labeling (filled circles with solid lines) and mitotic (open circles with broken lines) indices per 10,000 cells were determined for proximal tubular epithelial cells from autoradiographs. Each point represents mean ± SD for 5 mice. Mean labeling and mitotic indices for controls were 14 and 3 per 10,000 cells, respectively (not shown). (Choie and Richter, 1974a.)

epithelium. It is not clear why the lead-induced proliferative activity is only of brief duration. One possibility is that after the initial transport into renal tubular cells, the intracellular concentration of lead is effectively reduced by compartmentalization and incorporation into fibrillar protein. Lead is also excreted in the urine, and too little may be available to stimulate DNA replication by the second day. If the induction of renal cell proliferation depends partly on the intracellular concentration of soluble lead, one may expect a second wave of proliferative activity if a second dose of lead is given at 48 hr. This was borne out by experiment. When first, second and third doses of lead were given to rats at 48-hr intervals, first, second and third waves of proliferation of renal tubular cells were elicited (Choie and Richter, 1973). However, the second and third waves were smaller than the first (Fig. 5).

The nature of the stimulation of renal tubular cell proliferation by lead is unknown. DNA replication in mammalian cells requires a complex series

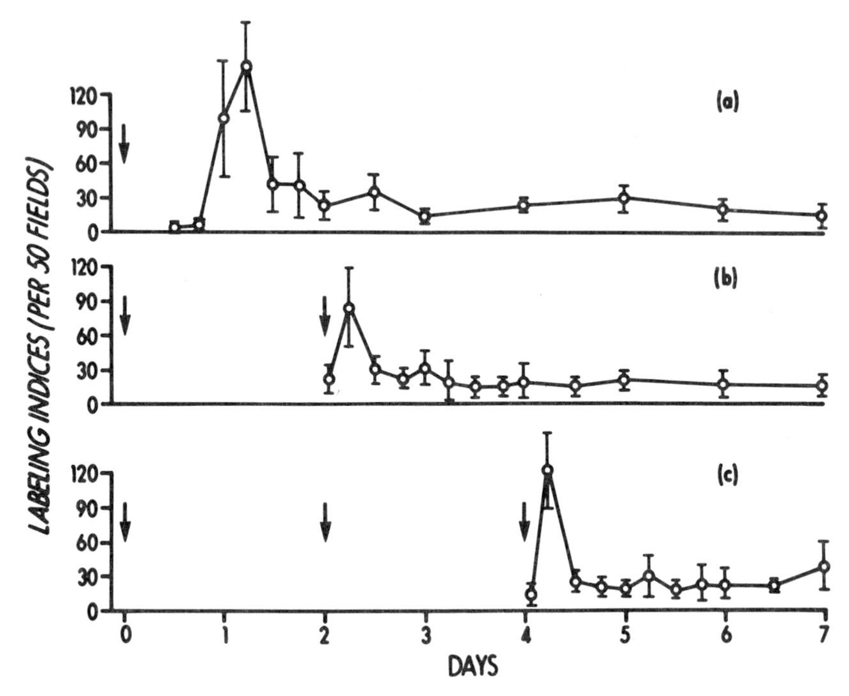


**Figure 5.** Labeling indices of nuclei in epithelial cells of proximal and distal convoluted tubules in rat kidneys after one (a), two (b) and three (c) intraperitoneal injections of lead acetate (0.05 mg Pb/g body weight). Cells were pulse-labeled with $^{3}$H-thymidine for 1 hr. Each point represents the mean ± SD of labeled cells per 50 microscopic fields (3600 ± 550 cells) for three rats. Injections of lead are indicated by arrows. Mean labeling index of nuclei in controls, given sodium acetate (0.06 mg/g body weight), was 4 per 50 microscopic fields. (Choie and Richter, 1973.)

of biochemical events before the cells can proceed from the $G_1$ to the S phase of the cell cycle (Baserga, 1968). During the prereplicative period, there is an increase in synthesis of RNA and protein. The new RNA and proteins include m-RNA and enzymes necessary for DNA replication. Synthesis of RNA and protein in response to lead has been investigated in relation to DNA synthesis (Choie and Richter, 1974a and 1974b). Incorporation of $^{3}$H-uridine and $^{14}$C-leucine into RNA and protein, respectively, was substantially greater in mice treated with lead than in controls. New synthesis of RNA and protein began within 3 hr after injection of lead. Maximal synthesis of protein, 50% above controls, occurred after 24 hr and coincided with the onset of DNA synthesis. When RNA synthesis was inhibited by treatment with actinomycin D, DNA synthesis did not occur. Similarly, when protein synthesis was suppressed by cycloheximide, DNA synthesis was inhibited. These findings showed that new synthesis of RNA and protein are prerequisites for the DNA synthesis induced by lead in renal tubular cells; they imply that lead may affect the cellular control of gene expression by stimulating RNA synthesis. However, the nature of the newly synthesized RNA has not yet been determined.

Since a single dose of lead stimulates an immediate increase in cell proliferation in rat and mouse kidneys, one may ask whether the proliferative activity continues under a prolonged treatment with lead. In one experiment a group of rats was injected with lead acetate ip at weekly intervals for 6 months. Autoradiography of the kidneys after pulse labeling with $^3$H-thymidine showed that the labeling activity in the proximal tubular epithelium was on the average 15 times greater in rats treated with lead than in controls (Choie and Richter, 1972b) (Fig. 6). The mitotic index of tubular epithelium was also increased. About one-half of the proximal tubular epithelial cells contained characteristic intranuclear inclusion bodies; the nuclear DNA in some of the cells with inclusions was labeled (Table 2). Although the presence of intranuclear inclusions did not prevent DNA synthesis, the labeling index of cells with intranuclear inclusions was much lower than that of other tubular epithelial cells. These results indicate that lead-induced proliferative activity in the tubular epithelium persists during prolonged treatment. The level of renal cell proliferation, however, was less intense after treatment for 6 months than after a single dose of lead. Histologically, many tubular epithelial cells were enlarged and irregular in shape and had prominent basophilic nucleoli. Although signs of acute tubular necrosis were absent, there were some desquamated cells in the lumens of renal tubules. Foci of hyperplasia of proximal tubular cells, with hyperchromatic, enlarged or oddly shaped nuclei were also common.

## X. CARCINOGENICITY OF LEAD

Inorganic and organic compounds of lead have been shown to induce tumors in experimental animals (Zollinger, 1953; Boyland et al., 1962; Van Esch et al., 1962; Epstein and Mantel, 1968), as have compounds of other metals, including Be, Cd, Co, Cr, Fe, Ni, Ti and Zn (Sunderman, 1978). No metal has been proven to cause cancer in man, but As, Cd, Cr and Ni are suspected of being carcinogenic in human beings. Malignant tumors of the kidney are relatively uncommon in man, representing about 1% of all malignancies exclusive of skin cancers. In experimental animals, spontaneous tumors of the kidneys are rare (Hamilton, 1975).

Zollinger (1953) first demonstrated that inorganic lead can induce epithelial tumors in the kidneys of rats. He found that 19 out of 29 surviving rats developed renal tumors after oft-repeated subcutaneous injections of a 2% suspension of lead phosphate. The total dose of lead per rat was 120 to 680 mg over a period of 10 months or more. Since then, Boyland et al. (1962) have induced renal tumors in male rats by administering 1% lead

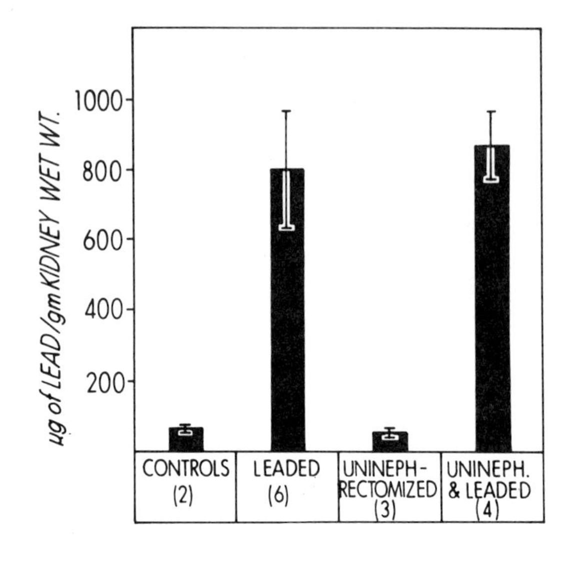


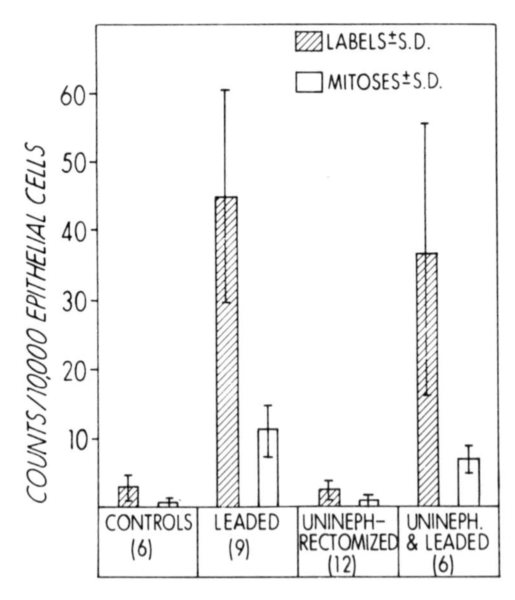


**Figure 6.** Accumulation of lead and increase in cell proliferation in kidneys of rats given lead (1 to 7 mg/week) intraperitoneally for 6 months. A, Mean lead content of kidneys (μg Pb/g wet weight ± SD). B, Proliferative activities in epithelia of proximal tubules of kidneys in four experimental groups. Counts of labeled cells ± SD (hatched columns) and of mitoses ± SD (open columns) per 10,000 cells. (Choie and Richter, 1972.)

**Table 2.** Mean labeling indices of four experimental groups*

| Experimental group | Number rats | Number kidneys | Labeled cells without inclusions† | SE | Labeled cells with inclusions† | SE | Cells with inclusions† | SE |
|---|---|---|---|---|---|---|---|---|
| Controls | 612 | 12 | 3.0 ± 1.8 | 0.7 | 0 | — | 0 | — |
| Leaded | 9 | 17‡ | 42.1 ± 13.8 | 4.6 | 2.7 ± 2.2 | 0.7 | 4074 ± 273 | 91 |
| Uninephrectomized | 12 | 12 | 2.6 ± 1.5 | 0.4 | 0 | — | 0 | — |
| Uninephrectomized and leaded | 6 | 6 | 34.5 ± 20.0 | 8.1 | 2.2 ± 1.3 | 0.5 | 4216 ± 370 | 151 |

* From Choie and Richter (1972b); all animals were labeled with $^{3}$H-thymidine for 1 hr

† The counts are per 10,000 epithelial cells of the proximal tubules in the kidneys ± SD

‡ Both kidneys were counted in all rats but one

acetate in the diet for 11 months or longer. In this case, the incidence of renal tumors was 15 out of 16 surviving animals. Similarly, Van Esch et al. (1962) produced renal tumors in rats after treatment with 0.1% or 1% basic lead acetate in the diet. Mice and hamsters were also tested for carcinogenic response. Chronic treatment of 0.5% basic lead acetate in the diet induced tumors in mouse kidneys, but failed to do so in one attempt in hamsters (Van Esch and Kroes, 1969).

Lead-induced renal tumors are bilateral and often multiple. They are either adenomas or adenocarcinomas of the renal cortex. The adenomas may be solid, papillary, tubular or mixed. The incidence of renal tumors after treatment with lead is not significantly different in male and female rats. The incidence of renal tumors induced by lead depends on the concentration of lead in the kidney and the duration of exposure. Experimental data indicate that, up to a tolerance limit, a larger dose of lead induces a greater incidence of renal tumors in a shorter period of time, at least in rats (Van Esch et al., 1962).

Perhaps there is a threshold level of carcinogenicity for compounds of lead in experimental animals, but such a level has not been determined. A rough estimate of the effective minimum carcinogenic amount of lead, based on available data, is around 100 mg per rat absorbed over a period of about 10 months. This is the amount of lead actually absorbed, and thus, the total dose given in the diet should be much greater, because of fractional absorption from the GI tract. Physiologic and dietary factors that modify the metabolism of lead must also influence the rate of absorption and hence the carcinogenic effect of lead.

At one time there was a question whether porphyrins rather than lead itself are the actual renal carcinogens, since lead poisoning causes excretion of a large amount of coproporphyrin in the urine. However, production of increased urinary porphyrin by feeding Sedormid has failed to induce any renal tumor in rats over a period of 1 year (Boyland et al., 1962).

Hormonal effects on tumorigenesis induced by lead in the kidney have been investigated to a limited extent. Testosterone and xanthopterin, which influence renal growth, have little effect on tumor induction by lead (Roe et al., 1965). It has not been determined whether progesterone or corticosteroids influence tumor induction by lead in kidneys. Another factor to consider is the effect of unilateral nephrectomy on tumor induction by lead in the remaining kidney. There is an indication that unilateral nephrectomy accelerates the appearance of renal carcinomas induced by diethylstilbestrol (DES) in hamsters (Hamilton, 1975).

A number of studies indicates that lead can act synergistically with other carcinogens. Lead and chromate are weak carcinogens when applied individually to experimental animals. If lead and chromate are administered simultaneously, however, the carcinogenic effects in the kidneys and in other organs in rats are potentiated (Furst et al., 1976). Likewise, a combination of 1% lead acetate and 0.06% 2-acetylaminofluorene (2-AAF) in the diet greatly increases the incidence of renal tumors in rats (Shakerin et al., 1965). Weekly intratracheal administration of lead oxide (PbO) to Syrian golden hamsters (1 mg/animal) for 10 weeks has been reported to produce alveolar metaplasia and adenomatous proliferation in the lungs, but no neoplasms. Administration of the carcinogen benzo[a]pyrene (BP) alone induces little change in lungs. When PbO and BP are given together, adenomas and adenocarcinomas are induced (Kobayashi and Okamoto, 1974). Apparently, lead acts synergistically with BP as a cocarcinogen in pulmonary tumorigenesis in hamsters.

Thus, in experimental animals, inorganic lead can induce tumors in organs other than the kidneys. There are reports that prolonged treatment with lead produces mammary adenomas in female and testicular and prostatic tumors in male rats (Van Esch et al., 1962; Hamilton, 1975). Also in rats, prolonged administration of 1% lead acetate in the diet results in an increased incidence of gliomas and cancers of the liver and bladder, in addition to renal tumors (Oyasu et al., 1970). When tumors are present in two or more organs in the same animal, metastatic spread from one primary site must always be considered. Histologic evidence, however, indicates that, in rats treated with lead, primary tumors do arise in different organs.

The carcinogenic potential of organic lead has not been investigated as much as that of inorganic lead. One report states that four weekly sc injections of tetraethyl lead (TEL) into newborn mice induced lymphomas (Epstein and Mantel, 1968). A significant increase in the incidence of lymphomas occurred 36 weeks after the injections were stopped. At a dose of 0.1 to 0.4 mg of TEL per injection, lymphomas developed in 12% of females, but in only 4% of male mice. No renal tumors were noted in these mice.

Perhaps the carcinogenic effect of lead is related to the interaction of lead with DNA. Although it was once thought that lead might inhibit mitosis of renal tubular cells and thereby produce nuclear abnormalities leading to neoplastic transformation (Van Esch et al., 1962), it is now clear that lead stimulates proliferation of the renal tubular epithelium in rats and mice (Choie and Richter, 1972a and 1974a). The mitogenic effect of lead in the kidneys is characteristic and occurs over a wide range of dose levels (Choie and Richter, 1974a). The stimulatory action of lead on renal cell proliferation may have preneoplastic significance. Many chemical carcinogens, such as 2-acetylaminofluorene (2-AAF) and dimethylnitrosamine (DMN), are known to stimulate cell proliferation prior to the development of neoplasia in target organs (Farber, 1973).

A consistent feature of the effects of lead in kidneys is atypical hyperplasia of tubular epithelium. After 6 months of repetitive treatment with lead acetate, the nuclei of the epithelial cells in the proximal tubules are often enlarged and irregular in shape. The hyperplastic foci are easily distinguished from the surrounding tissues by the hyperbasophilic cytoplasm of the cells and by the increase in number (crowding) of epithelial cells (Zollinger, 1953; Choie and Richter, 1972b). It seems likely, therefore, that the renal tumors which appear at later stages arise in hyperplastic nodules.

At present, there is no evidence that lead is carcinogenic in man. It should be pointed out, however, that improved health standards have significantly reduced the levels of occupational exposure to lead in the US over the last 50 years and that concentrations that are carcinogenic in rodents are now unlikely to occur in man. A recent survey of a group of 7032 men who worked 1 year or more in the lead industry in the US revealed that deaths from kidney tumors were no more frequent than in the general population (Cooper and Gaffey, 1975). However, deaths from all other neoplasms were in slight excess among lead workers. It is possible that man is not as susceptible to the carcinogenic effect of lead as are rodents. It should be noted that lead nitrate is teratogenic in hamsters (Carpenter and Ferm, 1977) and lead acetate can cause chromosomal aberrations in monkey lymphocytes (Deknudt et al., 1977a). Among a group of lead workers in Europe, chromosomal fragmentation and other abnormalities in lymphocytes were significantly more frequent than in a control group (Deknudt et al., 1977b).

It may be concluded that inorganic lead is carcinogenic in the kidneys of rats and mice. Organic lead is suspected of being carcinogenic in experimental animals, but further data are needed to establish this carcinogenicity. There is at present no conclusive evidence that lead is carcinogenic in human kidneys.

## XI. SUMMARY AND CONCLUSIONS

The effects of lead on the kidney are a function of dose and duration of exposure. In general, the renal toxicity of lead is closely related to its selective accumulation in the kidneys. Lead can produce acute and chronic nephropathies. In acute lead nephropathy, lesions are situated mainly in the proximal tubular epithelium and they involve, *inter alia*, changes in membrane functions and in energy metabolism, structural aberrations of mitochondria, and the appearance of pathognomonic morphological features in nuclei and cytoplasm of proximal tubular epithelial cells. Clinically, severe acute lead nephropathy is manifested by aminoaciduria, glycosuria, and hypophosphatemia, which result from depressed tubular reabsorption. Prolongation of the acute phase leads to chronic lead neophropathy, which has not been well characterized. It is often associated with gout and hypertension. Lead also influences tubular cell proliferation in the kidney.

At least in rats and mice, it can stimulate proliferation of renal tubular epithelium in the absence of tubular necrosis. The mitogenic effect of lead may be related to its carcinogenicity in experimental animals.

## REFERENCES


Ball, G.V., and Sorensen, L.B. (1969). *New Engl. J. Med. 280:*1199.
Barltrop, D., Barret, A.J., and Dingle, J.T. (1971). *J. Lab. Clin. Med. 77:*705.
Barry, P.S.I. (1975). *Brit. J. Ind. Med. 32:*119.
Baserga, R. (1968). *Cell Tissue Kinet. 1:*167.
Bessis, M., and Breton-Gorius, J. (1957). *Rév. d'Hémmatol. 12:*43.
Blackman, S.S. (1936). *Bull. Johns Hopkins Hosp. 58:*384.
Boyland, E., Dukes, C.E., Grover, P.L., and Mitchley, B.C.V. (1962). *Brit. J. Can. 16:*283.
Cantarow, A., and Trumper, C. (1944). In "Lead Poisoning," Williams & Wilkins Co., Baltimore, Md. p. 72.
Carpenter, S.J., and Ferm, V.H. (1977). *Lab. Invest. 37:*339.
Carroll, K.G., Spinelli, F.R., and Goyer, R.A. (1970). *Nature 227:*1056.
Castellino, N., and Aloj, S. (1964). *Brit. J. Ind. Med. 21:*308.
Chisolm, J.J. (1964). *J. Pediat. 64:*174.
Choie, D.D., and Richter, G.W. (1972a). *Am. J. Pathol. 66:*265.
Choie, D.D., and Richter, G.W. (1972b). *Am. J. Pathol. 68:*359.
Choie, D.D., and Richter, G.W. (1972c). *Science 177:*1194.
Choie, D.D., and Richter, G.W. (1973). *Proc. Soc. Exp. Biol. Med. 142:*446.
Choie, D.D., and Richter, G.W. (1974a). *Lab. Invest. 30:*647.
Choie, D.D., and Richter, G.W. (1974b). *Lab. Invest. 30:*652.
Choie, D.D., Richter, G.W., and Young, L.B. (1975). *Beitr. Path. 155:*197.
Cihak, A., and Seifertova, M. (1976). *Chem. Biol. Interact. 13:*141.
Conrad, M.E., and Barton, J.C. (1978). *Gastroenterology 74:*731.
Cooper, W.C., and Gaffey, W.R. (1975). *J. Occup. Med. 17:*100.
Cuppage, F.E., and Tate, A. (1967). *Am. J. Pathol. 51:*405.
Dallenbach, F.D. (1965). *Verh. Deut. Ges. Pathol. 49:*179.
Deknudt, G.H., Colle, A., and Gerber, G.B. (1977a). *Mut. Res. 45:*77.
Deknudt, G.H., Manuel, Y., and Gerber, G.B. (1977b). *J. Tox. Environ. Hlth. 3:*885.
Emmerson, B.T. (1973). *Kidney International 4:*1.
Emmerson, B.T., Mirosch W., and Douglas, J.B. (1971). *Aust. N. Z. J. Med. 4:*353.
Emmerson, B.T., and Ravenschroft, P.J. (1975). *Nephron. 14:*62.
Epstein, S.S., and Mantel, N. (1968). *Experientia 24:*580.
Farber, E. (1973). H. Busch (Ed.), In "Methods in Cancer Research," Vol. 7, p. 345. Academic Press, New York.
Farkas, W.R., Stanawitz, T., and Schneider, M. (1978). *Science 199:*786.
Fischer, A.B. (1975). *Zbl. Bakt. Hyg. B. 161:*26.
Furst, A., Schlauder, M., and Sasmore, D.P. (1976). *Cancer Res. 36:*1779.
Goyer, R.A. (1971). H.W. Altmann (Ed.), In "Current Topics in Pathology," Vol. 55, p. 147. Springer Verlag, New York.
Goyer, R.A., and Krall, A. (1969). *Fed. Proc. 28:*619.
Goyer, R.A., and Rhyne, B.C. (1973). G.W. Richter and M.A. Epstein (Eds.), In "International Review of Experimental Pathology," Vol. 12, p. 1. Academic Press, New York.

Goyer, R.A., Krall, A., and Kimball, J.P. (1968). *Lab. Invest. 19:*78.
Goyer, R.A., May P., Cates, M.M., and Krigman, M.R. (1970a). *Lab. Invest. 22:*245.
Goyer, R.A., Moore, J.F., Rhyne, B., and Krigman, M.R. (1970b). *Arch. Env. Health 20:*705.
Grzybek, H. (1972). *Mikroskopie 28:*292.
Hamilton, J.M. (1975). G. Klein and S. Weinhouse (Eds.), In "Advances in Cancer Research," Vol. 22, p. 1. Academic Press, New York.
Henderson, D.A., and Inglis, J.A. (1957). *Aust. Ann. Med. 6:*145.
Hsu, F.S., Krook, L., Shively, J.N., Duncan, J.R., and Pond, W.G. (1973). *Science 181:*447.
Inglis, J.A., Henderson, D.A., and Emmerson, B.T. (1978). *J. Pathol. 124:*65.
Johnson, H.A., and Vera Roman, J.M. (1966). *Am. J. Pathol. 49:*1.
Kobayashi, N., and Okamoto, T. (1974). *J. Natl. Can. Inst. 52:*1605.
Landing, B.H., and Nakai, H. (1959). *Am. J. Clin. Pathol. 31:*499.
Lilis, R., Fischbein, A., Eisinger, J., Blumberg, W.E., Diamond, S., Anderson, H.A., Rom, W., Rice, C., Sarkozi, L., Kon, S., and Selikoff, I.J. (1977). *Env. Res. 14:*255.
Malamud, D., and Malt, R.A. (1971). *Lab. Invest. 24:*140.
Moore, J.F., Goyer, R.A., and Wilson, M. (1973). *Lab. Invest. 29:*488.
Morgan, A., Holmes, A., and Evans, J.C. (1977). *Brit. J. Ind. Med. 34:*37.
Morgan, J.M., Hartley, M.W., and Miller, R.E. (1966). *Arch. Intern. Med. 118:*17.
Murakami, M., and Hirosawa, K. (1973). *Nature 245:*153.
Nechay, B.R., and Saunders, J.P. (1978). *J. Tox. Environ. Health 4:*147.
Nordberg, G.F. (1976). G.F. Nordberg (Ed.), In "Effects and Dose-Response Relationships of Toxic Metals," p. 15. Elsevier Scientific Publ. Co., New York.
Oyasu, R., Battifora, H.A., Clasen, R.A., McDonald, J.H., and Hass, G. (1970). *Cancer Res. 30:*1248.
Richter, G.W. (1976). *Am. J. Pathol. 83:*135.
Richter, G.W., Velasquez, M.J. & Shedd, R. (1979). *Am. J. Pathol. 94:*483.
Richter, G.W., Kress, Y., and Cornwall, C.C. (1968). *Am. J. Pathol. 53:*189.
Roe, F.J.C., Boyland, E., Dukes, C.E., and Mitchley, B.C.V. (1965). *Brit. J. Cancer 19:*860.
Schroeder, H.A., and Tipton, I.H. (1968). *Arch. Environ. Health. 17:*965.
Shakerin, M., Paloucek, J., Oyasu, R., and Hass, G.M. (1965). *Fed. Proc. Fed. Amer. Soc. Exp. Biol. 24:*684.
Stevenson, A., Merali, Z., Kacew, S., and Singhal, R.L. (1976). *Toxicology 6:*265.
Sunderman, F.W. (1978). *Fed. Proc. 37:*40.
Threlfall, G., Taylor, D.M., and Buck, A.T. (1966). *Lab. Invest. 15:*1477.
Trump, B.F., Berezesky, I.K., Jiji, R.M., Mergner, W.J., and Bulger, R.E. (1978). *Lab. Invest. 39:*375.
Vallee, B.L., and Ulmer, D.D. (1972). *Annu. Rev. Biochem. 41:*91.
Van Esch, G.J., and Kroes, R. (1969). *Brit. J. Cancer 23:*765.
Van Esch, G.J., Van Genderen, H., and Vink, H.H. (1962). *Brit. J. Cancer 16:*289.
Van Mullem, P.J., and Stadhouders, A.M. (1974). *Virchows Arch. B. 15:*345.
Vander, A.J., Taylor, D.L., Kalitis, K., Mouw, D.R., and Victery, W. (1977). *Amer. J. Physiol. 233:*F532.
Vostal, J., and Heller, J. (1968). *Environ. Res. 2:*1.
Walton, J., and Buckley, I.K. (1977). *Exp. Mol. Pathol. 27:*167.
Wedeen, R.P., Maesaka, J.K., Weiner, B., Lipat, G.A., Lyons, M.M., Vitale, L.F., and Joselow, M.M. (1975). *Amer. J. Med. 59:*630.
Zollinger, H.U. (1953). *Virch. Arch. Pathol. Anat. Physiol. 323:*694.

# Chronic Effects of Lead in Nonhuman Primates

*Robert F. Willes, D. C. Rice and J. F. Truelove*

TABLE OF CONTENTS



## I. INTRODUCTION

Studies concerning the toxicity of lead (Pb) in nonhuman primates are conducted basically for two reasons; first, to acquire data on the effects of lead on animals, thereby providing guidelines on the most sensitive and relevant toxicologic parameters to investigate in lead-exposed humans, and, second, to establish the relationship between lead dose and toxic effects under controlled experimental conditions.

There is a paucity of published data on the effects of long-term exposure of nonhuman primates to lead, especially at the lower levels of exposure that could occur from food, water and air. There are several reports of acute lead poisoning in nonhuman primates which demonstrate that at least the acute signs of lead toxicity are similar to those observed in humans and other animal species (Ferraro and Hernandez, 1932a; 1932b; Fisher, 1954; Hopkins, 1970; Houser and Frank, 1970; Hausman et al., 1961; Goode et al., 1973; Clasen et al., 1974; Hopkins and Dayan, 1974; Zook et al., 1976). Young nonhuman primates appear to be more sensitive to lead intoxication than adults or juveniles (Allen et al., 1974; Zook et al., 1976). Similarly, greater sensitivity of the young to lead intoxication has been observed in other animal species (Forbes and Reina, 1972) and humans (Lin-Fu, 1972; Waldron, 1975).

This chapter summarizes the current knowledge of the effects of lead on nonhuman primates, concentrating particularly on the effects of long-term exposure to inorganic lead. There are no reported data, to the authors' knowledge, on the effects of long-term exposure of nonhuman primates to organolead compounds, although tetraethyl lead appears more toxic than tetramethyl lead under conditions of acute exposure (Heywood et al., 1978).

A study of the effects of long-term lead exposure on infant monkeys (*Macaca fascicularis*) is being conducted at the Health Protection Branch (HPB), Ottawa, Canada, by the authors of this chapter. Since the HPB study will be referred to repeatedly throughout this chapter, an outline of the experiment is presented in Table 1. In the main body of the study, infant monkeys were dosed orally with lead acetate at doses of 700, 140, 70, 35 and 0 μg Pb/kg body wt given 5 days per week (Monday to Friday), resulting in daily doses equivalent to 500, 100, 50, 25 and 0 μg Pb/kg body wt/day. Dosing was initiated at 1 day of age. All infants were reared from birth in an infant primate nursery (Willes et al., 1977a).

**Table 1.** Design of infant monkey (*Macaca fascicularis*) long term lead toxicity study (HPB study)*

| Lead dose (μg Pb/kg body wt/day) | Number of infants | Age dosing initiated (days) | Mean age at weaning ± SEM (days) |
|---|---|---|---|
| 2000 | 3 | 108 ± 18 | 467 ± 18 |
| 500 | 4 | 0 | 206 ± 3 |
| 100 | 6 | 0 | 204 ± 2 |
| 50 | 8 | 0 | 204 ± 2 |
| 25 | 8 | 0 | 204 ± 2 |
| 0 | 12 | — | 204 ± 2 |

* Study conducted by R. Willes, D. Rice and J. Truelove, Health Protection Branch, Health and Welfare Canada, Ottawa, Canada.

In addition to the main body of the study, a group of three monkeys was dosed with 2000 μg Pb/kg body wt/day beginning at about 100 days of age. This group of monkeys was used for exploratory work to assess a variety of parameters which may be affected by lead exposure.

All infant monkeys were fed a milk substitute (hereafter referred to as infant formula) from birth (Willes et al., 1977a). Several cubes of adult primate diet were presented to the infants on an *ad libitum* basis beginning at 40 to 60 days of age. The 2000 μg Pb/kg body wt/day infants were taken off infant formula (weaned) at 360 days of dosing. All other infants were weaned at approximately 200 days of treatment. A commercial primate diet was fed *ad libitum* after weaning.

## II. Lead Absorption in Nonhuman Primates

The increased susceptibility of young children to the toxic effects of lead is, at least in part, due to their increased capacity to absorb lead from the gastrointestinal tract (GIT) as compared to adults (Alexander, 1974; Chisolm, et al., 1974). Lead retention is also greater in infant monkeys (65 to 70% lead retention) than adults (3% lead retention), as assessed by measurements of lead retention over a 96-hr period using $^{210}Pb(NO_3)_2$ (Table 2; Willes et al., 1977b). Fecal excretion of lead was proportionately higher in adults, indicating that the increased retention of lead by infant monkeys was probably due to increased lead absorption (Willes et al., 1977b).

A variety of dietary factors are known to influence lead absorption from the GIT. Dietary levels of calcium, phosphates, iron, fat, protein, vitamin D and vitamin E are known to affect lead absorption in rodents (Barltrop and Khoo, 1975; Stephens and Waldron, 1975). Also, the type of diet consumed (*ie.*, milk diet versus solid foods) alters lead absorption by rodents (Kello and Kostial, 1973; Stephens and Waldron, 1975; Kostial et al., 1978). In light of these observations, the question arises whether dietary factors were responsible for the increased lead absorption observed in infant monkeys.

Infant monkeys reared on commercial infant formula in a primate nursery have higher intakes of fat, protein and iron than adult monkeys on a standard primate diet (Willes et al., 1977b). Barltrop and Khoo (1975) demonstrated increased lead absorption in rats on iron-deficient diets. However, since the infant monkeys had higher iron intake than the adults, it is doubtful if differences in iron intake accounted for the higher lead absorption observed in the infant monkeys. Higher fat and protein intake may be responsible for the greater lead absorption in infant monkeys, although other constituents in the diet may also be involved.

The influence of type of diet on lead absorption was also evident in infant monkeys. Infant monkeys dosed chronically with lead acetate show a decline in blood lead levels after weaning from infant formula onto an adult

**Table 2.** Lead retention by infant and adult monkeys 96 hr after oral dosing with $^{210}Pb(NO_3)_2$*

| | Age | | |
|---|---|---|---|
| | 10 days | 150 Days | Adults |
| Percentage of oral lead dose retained† | 64.5 ± 2.5 | 69.8 ± 2.5 | 3.2 ± 3‡ |

* Willes et al., 1977a.

† Mean ± SEM

‡ Percentage retention significantly different ($p < 0.01$) from all other age groups.

primate diet (Figs. 1A and 1B). This phenomenon does not appear to be age dependent, since the decrease in blood lead was observed whether infants were weaned after 200 days or 360 days of lead treatment (Figs. 1A and 1B). In these experiments, the daily dose of lead per kg body wt was constant throughout the study for each group of infants; consequently, a decrease in lead absorption would result in a concomitant decrease in blood lead concentration. The phenomenon of increased absorption of metals from the gastrointestinal tract when consuming milk has also been observed with other heavy metals such as cadmium, mercury, manganese, copper (Kostial et al., 1978) and methylmercury (authors' unpublished observations).

The mechanism for this change in absorption after weaning has not been elucidated. Kello and Kostial (1973) and Kostial et al. (1978) have proposed that certain factors present in milk may facilitate lead absorption. Alternatively, factors in the adult diets may act to reduce lead absorption. To further elucidate these effects of milk versus solid diets on lead absorption, blood lead levels of infant monkeys (HPB study) dosed orally with 2000 $\mu$g Pb/kg body wt/day were monitored while the diet fed was changed from an adult primate diet to a milk substitute diet (infant formula). The addition of milk to an adult primate diet produced only a small transient change in blood lead values (Fig. 1B). However, a milk diet alone resulted in a substantial increase in blood lead levels (Fig. 1B). These data suggest that substances may be present in the adult primate diet which act to reduce lead absorption.

Regardless of the mechanisms involved, these data demonstrating increased lead absorption while consuming milk diets have important implications in the assessment of acceptable levels of lead intake by young infants for whom milk is the major dietary component.

## III. TISSUE DISTRIBUTION OF LEAD IN NONHUMAN PRIMATES

The tissue distribution of a single oral dose of lead differs considerably between infant and adult monkeys. The levels of $^{210}$Pb in various bones and teeth have been shown to be significantly higher in infant monkeys than in adults 96 hr after receiving the same dose of $^{210}Pb(NO_3)_2$ (Table 3; Willes et al., 1977b). Presumably, this is due to the more rapid deposition of new bone in young infants and the substitution of lead for calcium in bone (Goyer and Mahaffey, 1972).

Interestingly, even though infant monkeys absorbed about 14 times as much of the orally administered $^{210}Pb$ as adults, their blood lead levels were

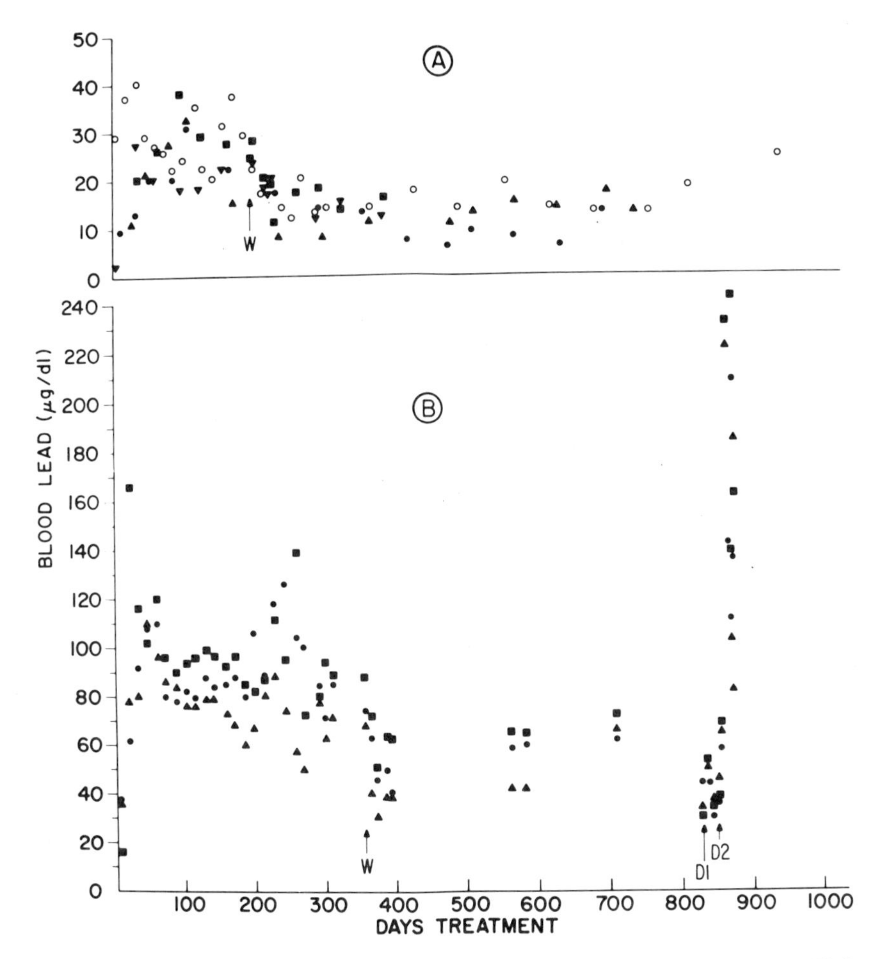


**Figure 1.** A, Blood lead concentrations of five monkeys dosed from birth with 100 μg Pb/kg body wt/day. The diet fed to these monkeys was changed from milk (S.M.A., Wyeth Ltd., Toronto, Canada) plus primate diet to primate diet (Teklab Primate Diet, Winfield, Ind.) only at 200 days of treatment (marked W). B, Blood lead concentrations of three monkeys dosed from about 100 days of age with 2000 μg Pb/kg body wt/day. The diet fed these monkeys was changed from milk plus primate diet to primate diet only at 360 days of treatment (marked W), then to primate diet plus milk at 830 days treatment (marked D1), then to milk only at 851 days treatment (marked D2).

not significantly different (Table 3). This may be due to a rapid transfer of lead from blood to bone and other tissues in infants. These data emphasize that blood lead values can be misleading in predicting levels of lead exposure by the young, as will be discussed in the next section.

Brain lead to blood lead ratios were significantly greater in 10-day-old infants than in adults after a single exposure to $^{210}Pb(NO_3)_2$ (Table 3), suggesting a more rapid transfer of lead from blood to brain in infants than in adult monkeys. Such data are important in understanding the toxicologic effects of lead on the central nervous system of young nonhuman primates and children.

## IV. BLOOD LEAD LEVELS DURING LONG TERM LEAD EXPOSURE IN NONHUMAN PRIMATES

Lead concentrations in blood are often used as an indicator of body burden of lead and lead exposure, primarily because blood can be conveniently sampled and readily analyzed for lead. It is important to note that blood lead levels may not reflect actual past exposure, particularly if the exposure is intermittent (Vitale et al., 1975). Since lead is more rapidly cleared from blood than other tissues (Momcilovic and Kostial, 1974), blood lead levels may be disproportionately lower than the level of other tissues. In fact, Shapiro et al. (1973) have shown that blood levels may be normal in children whose teeth have high lead levels, indicating a past history of high exposure to lead. As discussed in the previous section of this chapter, the relationship between bone and blood lead levels is substantially different in young animals than adults; consequently, a given blood lead level in an

**Table 3.** Tissue lead concentrations* (mean ± SEM) and tissue lead: whole blood lead ratios 96 hr after oral dosing of monkeys with $^{210}Pb(NO_3)_2$†

| | Age Groups | | | | | |
|---|---|---|---|---|---|---|
| | 10 Days | | 150 Days | | Adults | |
| Tissue | ng Pb/g | Ratio | ng Pb/g | Ratio | ng Pb/g | Ratio |
| Skull | 17.1 ± 0.4 | 15.1 | 8.3 ± 1.4 | 3.7 | 1.7‡ ± 0.8 | 0.6 ¶ |
| Humerus shaft | 13.0 ± 2.6 | 11.7 | 10.9 ± 1.7 | 4.9 | 0.9‡ ± 0.4 | 0.4 ¶ |
| Rib | 11.4 ± 5.8 | 13.4 | 10.6 ± 2.2 | 4.6 | 1.4‡ ± 0.6 | 0.6 ¶ |
| Teeth | 12.9 ± 2.4 | 12.7 | 9.2 ± 0.9 | 4.4 | 0.3‡ ± 0.2 | 0.2 ¶ |
| Occipital cortex | 0.2 ± 0.03 | 0.2 | 0.3 ± 0.06 | 0.1 | 0.1 ± 0.08 | 0.06¶ |
| Whole blood | 1.2 ± 0.03 | — | 2.4 ± 0.6 | — | 2.3 ± 0.9 | — |

* All lead concentrations were calculated on the basis of the specific activity of the lead dose to the monkey and do not include background lead in the various tissues.

† From Willes et al., 1977a.

‡ Significantly different ($p < 0.05$) from ng Pb/g values for all other age groups.

¶ Significantly different ($p < 0.05$) from tissue lead: blood lead ratios for all other age groups.

infant may indicate a higher body burden of lead than in an adult with a comparable blood lead level.

Intermittent pulmonary exposure of adult baboons to lead carbonate in doses ranging from 50 to 135 mg Pb/kg body wt for 39 to 362 days resulted in blood lead levels between 100 and 1000 $\mu$g Pb/dl blood (Hopkins, 1970). Since lead exposure in these studies was not constant throughout the study, it is difficult to establish any relationship between blood lead and lead dose. Interestingly, infant baboons exposed to lead carbonate via the lung had considerably lower blood lead levels than adults, even though the doses were in some cases considerably higher than the adults (Hopkins, 1970; Hopkins and Dayan, 1974). These data suggest that there are considerable differences between infant and adult baboons in either the pulmonary uptake or the metabolism and pharmacokinetics of lead.

Cohen et al. (1974) reported preliminary data on experiments to assess lead toxicity in infant baboons. After 50 days of dosing at 100 $\mu$g Pb/kg body wt/day (as lead acetate), blood lead values were 20 to 30 $\mu$g Pb/dl blood for two infant baboons.

Daily oral exposure of infant monkeys (*Macaca mulatta*) to 1018, 326 or 0 $\mu$g lead/kg body wt (as lead acetate) for their first year of life resulted in mean blood lead levels of 85, 55 and 14 $\mu$g/dl blood, respectively (Bushnell et al., 1977). The blood lead levels in infant monkeys studied by the authors (HPB study) increased from values near 4 $\mu$g Pb/dl blood at birth to mean values of 56, 23, 16, 10 and 4 $\mu$g Pb/dl blood for daily dose rates of 500, 100, 50, 25 and 0 $\mu$g Pb/kg body wt respectively, prior to weaning at 200 days (Fig. 2). The blood lead levels of the highest dose group (2000 $\mu$g PB/kg body wt/day) increased to a mean value of 77 $\mu$g Pb/dl blood prior to weaning at 360 days of treatment (about 470 days of age) (Fig. 2). Thirty days after weaning, the blood lead levels declined to mean values of 50, 40, 15, 10 and 8 $\mu$g Pb/dl blood for doses of 2000, 500, 100, 50 and 25 $\mu$g Pb/kg body wt/day, respectively (Fig. 2). The blood lead levels remained relatively constant for the remainder of the study. Blood lead levels did not change after weaning in the control group.

There is a reasonably consistent relationship between lead dose and blood lead levels for several species of nonhuman primates exposed orally to different doses of lead. The data on blood lead versus lead dose for various nonhuman primate studies are presented together with the authors' data in Fig. 2. Infant Rhesus monkeys (Bushnell et al., 1977) on milk substitute diets developed blood lead levels slightly higher than, but still consistent with, infant Cynomologus monkeys (*Macaca fascicularis*) on similar diets (Fig. 2). When consuming adult primate diets, the relationship between lead dose and blood lead is comparable for baboons (Cohen et al., 1974), young Rhesus adults (Goode et al., 1973, abstract only), adult Cynomologus monkeys (Deknudt et al., 1977) and Cynomologus juveniles (Fig. 2).

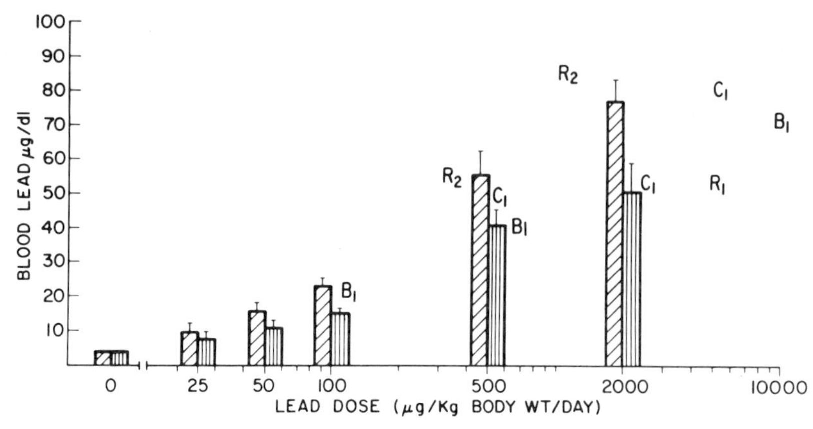


**Figure 2.** Blood lead concentrations of nonhuman primates at various oral doses of lead (log scale) given as lead acetate. Authors' data (HPB study). Crosshatched bars, blood lead levels on milk (S.M.A., Wyeth Ltd., Toronto, Canada) plus primate diet (Teklab Primate Diet, Winfield, Ind.); vertical hatched bars, blood lead levels on primate diet only. $B_1$, baboons on adult diet (Cohen et al., 1974); $C_1$, adult monkeys (*Macaca fascicularis*) (Denudt et al., 1977); $R_1$, adult monkeys (*Macaca mulatta*) (Houser et al., 1973); $R_2$, infant monkeys (*Macaca mulatta*) on milk substitute diet (Bushnell et al., 1977).

Data from young baboons relating blood lead levels to the dose of lead administered were not consistent with data from the five other studies and are not included in Fig. 2; furthermore, the experimental conditions were not comparable. Young baboons consuming adult diets were dosed once weekly with 50 mg Pb/kg body wt as lead acetate (Zook et al., 1976). When this weekly dose is converted to an equivalent daily dose (7142 μg Pb/kg body wt/day), the blood lead levels reported (100 μg Pb/dl blood) were considerably higher than would be predicted from the data collected on other nonhuman primates (Fig. 2). These higher than predicted blood lead levels may have been related to the weekly dosing interval, which could alter the kinetics of lead distribution and excretion and profoundly influence the blood lead values, depending on when the blood samples were collected relative to dosing the baboons with lead.

The consistency of the relationship between data on blood lead levels and lead dose from five completely independent laboratories using three different species of nonhuman primates increases confidence in the relationship between blood lead and lead dose. These data can be utilized to estimate the lead intake required to produce a given blood lead level. Since data on lead intake in human studies are either completely lacking or, at best, crude estimates, blood lead levels from controlled animal studies must be used to delineate a critical lead exposure above which the risk of hazard to health becomes unacceptably high. As mentioned at the beginning of this section, however, the extrapolation of a blood lead level to a lead intake, particularly from an uncontrolled environment (*i.e.*, the population at large), must be done with care.

It has been suggested that blood lead levels greater than 40 μg/dl blood are detrimental to health in children (Chisolm et al., 1974), although there is still much debate over what is an acceptable or safe level of blood lead, especially for children. For example, the US Environmental Protection Agency has proposed that blood lead levels in excess of 30 μg/dl warrant concern (Federal Register, 1978). Extrapolation of data from the log dose-blood lead relationship (Fig. 2) between dose groups suggests that a blood lead level of 40 μg/dl would be attained by the daily ingestion of about 200 μg Pb/kg body wt when consuming milk diets, or about 500 μg Pb/kg body wt/day when not consuming milk diets. A blood lead level of 30 μg Pb/dl blood would correspond to the daily ingestion of about 150 and 400 μg Pb/kg body wt/day on milk and nonmilk diets, respectively (Fig. 2).

## V. BIOCHEMICAL AND HEMATOLOGIC EFFECTS OF LEAD IN NONHUMAN PRIMATES

The effects of lead on heme metabolism are well documented (Gibson et al., 1955; Granick and Mauzerall, 1958) and have been discussed elsewhere in this monograph. This section will describe the effects of lead on various enzymes and products of heme metabolism, on liver function and on hematologic parameters in nonhuman primates.

### A. Free Erythrocyte Protoporphyrins

Inhibition of the enzyme heme synthetase by lead results in an accumulation of the porphyrin substrates of this enzyme and, consequently, an elevation of porphyrins in blood and urine (discussed elsewhere in this monograph). Analytic methodology for porphyrins has progressed to the point that blood porphyrin levels are now used extensively in community screening programs for the diagnosis of excessive lead exposure (Piomelli et al., 1973). In order to utilize blood porphyrin levels as an indicator of lead exposure, the relationship between blood lead, lead dose and blood porphyrins must be understood. It should be emphasized that blood porphyrin levels can be elevated in several other diseases, as well as in lead toxicity. Therefore, in cases where blood porphyrin levels are elevated, blood lead analysis is required to confirm the diagnosis of excessive lead exposure (Piomelli et al., 1973).

Prior to weaning, infant monkeys receiving lead acetate from birth (HPB study) showed elevated free erythrocyte protoporphyrins (FEP) levels at doses as low as 500 μg Pb/kg body wt/day (Fig. 3). Blood lead levels in these infants averaged 40 μg Pb/dl whole blood when consuming adult primate diets (Fig. 2). The elevation in FEP levels in red blood cells was not further elevated at oral doses up to 2000 μg Pb/kg body wt/day (Fig. 3). In addition, FEP levels decreased as the blood lead levels decreased when the infant monkeys were weaned (Fig. 3). The decrease in FEP level lagged a few weeks behind the postweaning fall in blood lead levels.

### B. Delta-Aminolevulenic Acid Dehydrase (ALAD)

ALAD catalyzes the condensation of 2 molecules of ALA to form 1 molecule of porphobilinogen during heme biosynthesis (Gibson et al., 1955; Granick and Mauzerall, 1958). ALAD is inhibited by lead and has been proposed as an indicator of lead exposure in screening for lead-exposed subjects (Goldstein et al., 1975), in spite of the fact that the clinical consequences of depressed ALAD activity in blood are not understood. Analytic procedures for measuring ALAD activity have met with difficulty; consequently, substantial variability has been reported in the correlation between ALAD activity and blood lead level (Cohen et al., 1974; Goldstein et al., 1975). Tomokuni (1974) demonstrated that the pH optimum for ALAD activity decreased following lead exposure. This factor must be considered when attempting to correlate ALAD activity with blood lead values and lead exposure.

Experiments conducted in the authors' laboratory (HPB study) have demonstrated that the effect of lead on ALAD activity in infant monkeys (*Macaca fascicularis*) is pH dependent (Fig. 4). As the pH of the analysis reaction medium is increased from 6.65 to 7.2, the inhibitory effect of lead

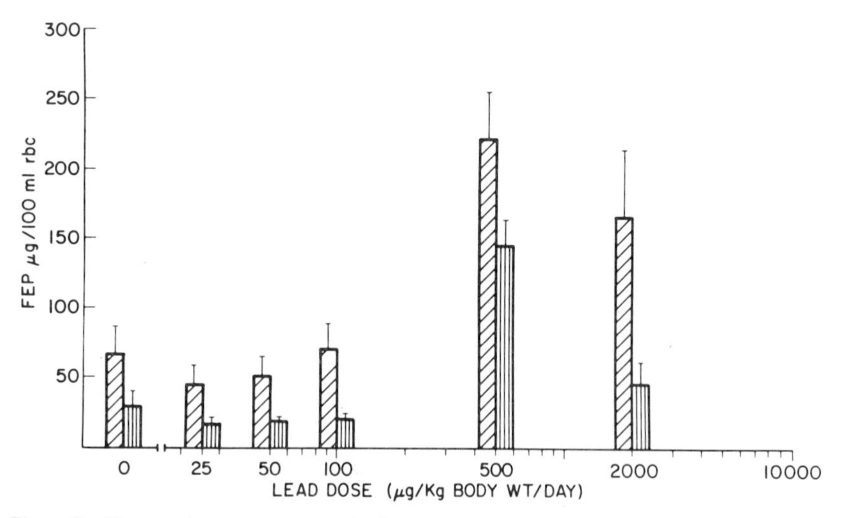


**Figure 3.** Free erythrocyte protoporphyrin concentrations ± SEM (FEP, μg/100 ml blood) of monkeys (*Macaca fascicularis*) at various oral doses of lead (log scale) on milk (S.M.A., Wyeth Ltd., Toronto, Canada) plus primate diet (Teklab Primate Diet, Winfield, Ind.) (crosshatched bars) and on primate diet only (vertical hatched bars).

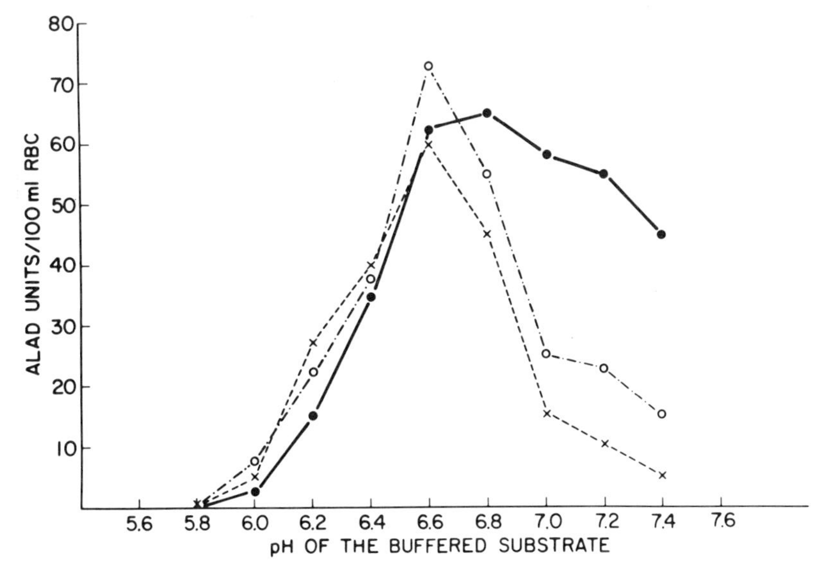


**Figure 4.** Effect of reaction medium pH on ALAD activity of control (filled circle) and lead-treated (x, 500 μg Pb/kg body wt/day; unfilled circle, 100 μg Pb/kg body wt/day) infant monkeys.

on the enzyme increases. By expressing activity as a ratio of the activity at pH 7.2 to the activity at pH 6.65, the effect of pH and lead on the enzyme can be assessed. Also, for reasons not clear at present, this ratio appears to reduce some of the daily variability inherent in measuring ALAD activity in red blood cells.

A significant depression in the ALAD pH 7.2/pH 6.65 ratio was observed at lead doses of 2000, 500, 100, 50 and 25 μg/kg body wt/day (Fig. 5). Monkeys on the lowest daily lead dose studied (25 μg/kg) had a mean blood lead value of 10 and 8 μg/dl blood prior to and after weaning, respectively (Fig. 2). These infant monkeys showed a significant depression in the ALAD ratio compared to those not dosed with lead (Fig. 5).

The depression in ALAD activity occurs rapidly, especially at higher doses of lead. The ALAD ratio was depressed from 0.8 in untreated infant monkeys to 0.25 after 2 weeks of exposure to 500 μg Pb/kg body wt/day. At lower doses of lead, the maximum depression in ALAD activity is reached more slowly over several weeks.

There appears to be some recovery of ALAD activity following the reduction in blood lead levels which occurred following weaning (Fig. 5). These results suggest that the depression of ALAD activity by lead is a reversible effect and that ALAD activity may return to normal after cessation of lead intake and depletion of the body burden of lead.

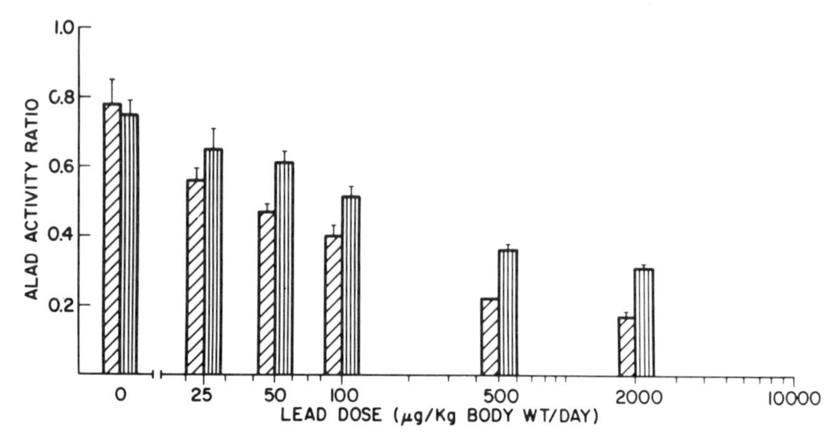


**Figure 5.** Activity ratio of ALAD at pH 7.2 to ALAD activity at pH 6.65 (ALAD activity ratio) ± SEM of infant monkeys on various daily oral doses of lead (log scale) on milk (S.M.A., Wyeth Ltd., Toronto, Canada) plus primate (Teklab Primate Diet, Winfield, Ind.) diet (crosshatched bars) and primate diet only (vertical hatched bars).

## C. Delta-Aminolevulenic Acid (ALA) in Urine

Intratracheal injections of 5 mg of lead carbonate/kg body wt to juvenile baboons produced an immediate elevation in the output of ALA in urine (Cohen et al., 1972). The ALA excretion in urine was generally related to blood lead levels; however, there was considerable variability in the ALA levels observed. Studies conducted by the authors on infant monkeys (*Macaca fascicularis*) exposed to lead acetate from birth (HPB study) also showed that urinary ALA excretion was extremely variable. There did not appear to be a significant elevation in urinary ALA at doses up to 2000 μg Pb/kg body wt/day.

## D. Effect of Lead on Liver Function

Lead has been shown to decrease cytochrome P-450 and inhibit associated drug-metabolizing enzymes in rats (Alvares et al., 1972). Presumably, this effect is due to the inhibition of the synthesis of porphyrins, which form an important structural part of the cytochromes. Mixed function oxidase activity in liver parenchyma can also be assessed by measuring the plasma clearance rate of chemicals such as antipyrine metabolized by these enzymes. The plasma elimination half-time of antipyrine is significantly increased in children exhibiting signs of lead poisoning (Alvares et al., 1976). Some investigators (Meredith et al., 1977) have shown similar effects of lead in adult humans exposed to lead, while others have obtained a less clear association between lead and antipyrine clearance (Fischbein et al., 1977).

The effects of long term lead exposure on antipyrine clearance from plasma have been examined in the HPB study. Monkeys dosed with 2000 μg Pb/kg body wt/day showed a marked inhibition in the elimination of orally administered antipyrine when tested after exposure to lead (Table 4). Blood lead levels averaged 55 μg Pb/dl blood when the tests were conducted. Antipyrine clearance from plasma was not significantly affected at a dose of 500 μg Pb/kg body wt/day (blood lead = 40 μg/dl) (Table 4).

Bromsulphalein (BSP) clearance from blood is another method utilized clinically to assess liver function. This is presumably a different type of liver function than antipyrine, although the metabolism of BSP has not been completely elucidated. BSP is thought to be conjugated with glutathion and excreted by the liver in bile, although other liver enzyme systems may also be involved in BSP clearance.

BSP clearance has also been studied in infant monkeys (HPB study) following long term lead exposure. Compared with control animals, BSP clearance from plasma following iv injection of BSP was significantly reduced in monkeys dosed with 2000 μg Pb/kg body wt/day (Table 5). At lead doses of 500 μg/kg body wt/day, the data suggest a reduced rate of BSP clearance, although the difference in clearance between lead-treated and control monkeys was not statistically significant.

Interestingly, there appears to be a marked difference in the ability of monkeys to clear BSP from plasma as compared to humans. BSP clearance tests in humans usually involve the iv administration of 5 mg BSP/kg body wt. This dose of BSP was cleared from plasma in monkeys within 15 min and the BSP dose in monkeys had to be increased to 20 mg BSP/kg body wt in order to adequately assess the BSP plasma clearance rate in untreated monkeys. Even at a dose of 20 mg BSP/kg body wt, clearance was nearly complete in 35 min, compared to 90 min in humans with one-quarter the

**Table 4.** Plasma antipyrine concentrations (mean ± SEM) after oral administration of antipyrine to control and lead-treated* monkeys (*Macaca fascicularis*)

| Oral lead dose (μg/kg body wt/day) | Time after antipyrine dose† (hr) 2 | 6 | 10 | 14 | 24 |
|---|---|---|---|---|---|
| 0 (3)‡ | 13.6 ± 0.5¶ | 11.7 ± 0.2 | 9.8 ± 0.5 | 7.9 ± 0.4 | 6.4 ± 0.5 |
| 500 (4) | 17.2 ± 0.7 | 11.9 ± 0.9 | 9.1 ± 0.5 | 8.1 ± 0.5 | 7.2 ± 0.5 |
| 2000 (3) | 10.7 ± 0.8 | 10.3 ± 0.4 | 13.7 ± 1.1** | 11.6 ± 0.5** | 10.3 ± 0.3** |

* Lead treatment for 2.5 to 3 years.

† Antipyrine dose, 18 mg/kg body wt given orally.

‡ Number of monkeys tested.

¶ Antipyrine concentration, μg/ml plasma ± SEM.

** Statistically different from control group ($p < 0.05$)

**Table 5.** Plasma bromsulphalein (BSP) concentrations (mean ± SEM) after intravenous administration of BSP to control and lead-treated* monkeys (*Macaca fascicularis*)

| Oral lead dose (μg/kg body wt/day) | Time after BSP dose† (min) | | | | | |
|---|---|---|---|---|---|---|
| | 7 | 15 | 20 | 25 | 30 | 35 |
| 0 (3)‡ | 4.8 ± 0.2¶ | 3.9 ± 0.2 | 2.3 ± 0.2 | 1.2 ± 0.2 | 0.6 ± 0.1 | 0.5 ± 0.1 |
| 500 (4) | 5.2 ± 0.3 | 4.3 ± 0.1 | 3.4 ± 0.2 | 2.1 ± 0.2 | 1.4 ± 0.1 | 1.0 ± 0.1 |
| 2000 (3) | 5.2 ± 0.1 | 5.0 ± 0.1 | 4.5 ± 0.3 | 4.1 ± 0.3** | 3.4 ± 0.1** | 2.8 ± 0.1** |

* Lead treatment for 2.5 to 3 years.

† BSP dose, 20 mg/kg body wt given intravenously.

‡ Number of monkeys tested.

¶ BSP concentration in plasma, μg/ml ± SEM.

** Statistically different from control group ($p < 0.05$)

BSP dose. This demonstrates that *Macaca fascicularis* monkeys clear BSP much more quickly than humans and suggests a difference in certain metabolic functions of the liver of monkeys compared to humans. Antipyrine, on the other hand, was cleared from plasma in untreated monkeys at a rate similar to that of humans.

These data on liver function as assessed by antipyrine and BSP clearance from plasma show that lead acetate, at doses of 2000 μg Pb/kg body wt/day, inhibits the elimination of these compounds by the liver. This would indicate that, at this level of lead exposure, the activity of mixed function oxidases and conjugation enzyme systems in liver is inhibited. The implications of these changes in the assessment of health effects of lead are unclear.

## E. Hematologic Effects of Lead

Lead has been shown to produce changes in a variety of hematologic parameters such as hematocrit, blood hemoglobin concentration, packed cell volume and blood cell morphology. These effects are thought to occur by interference with porphyrin and heme synthesis, interference with red blood cell maturation and shortening of red blood cell life (Albahary, 1972). Abnormal morphology of red cells, known as red blood cell stippling, is also associated with lead exposure (Albahary, 1972). Zielhuis (1971) summarized data on lead exposure in adult humans and correlated blood lead levels with hemoglobin concentration in blood. In adults, blood hemoglobin concentrations were not significantly affected at blood lead levels less than 80 μg Pb/dl blood; only slight decreases in hemoglobin could be expected with blood lead levels up to 120 μg Pb/dl blood. Based on extrapolation from data on nonhuman primates relating blood lead levels to lead intake (Fig.

2), a lead intake of approximately 20 mg Pb/kg body wt/day would be required by adults to produce blood lead levels of 80 μg Pb/dl blood. This figure would be reduced to about 2 mg Pb/kg body wt/day for individuals on milk diets (Fig. 2).

Consequently, hematologic changes in nonhuman primates would only be expected with relatively high blood lead levels. Indeed, anemia has been observed in cases of acute lead poisoning (Hopkins, 1970; Hopkins and Dayan, 1974) or after iv lead dosing (Vermande-Van Eck et al., 1960) in nonhuman primates. Also, exposure of juvenile monkeys to the equivalent of 143 mg Pb/kg body wt/day resulted in a decrease in packed cell volume (Zook et al., 1976). Blood lead levels in these juveniles ranged from 700 to 1200 μg Pb/dl blood, or 9 to 15 times the lowest blood lead levels associated with anemia in human adults.

Exposure of infant baboons to lead hydroxycarbonate (4.9 mg Pb/kg body wt/day; 138 μg Pb/dl blood) or powdered metallic lead (up to 45 mg Pb/kg body wt/day; 400 μg Pb/dl blood) for 4 to 16 months was not associated with any decrease in hemoglobin concentration or hematocrit (Kneip et al., 1976). Similarly, infant monkeys (HPB study and Bushnell et al., 1977) exposed to lower daily doses of lead acetate did not show any significant changes in hemoglobin concentrations in blood or hematocrit.

Lead exposure has also been associated with the appearance of basophilic stippled red blood cells (Zook et al., 1971; Albahary, 1972). Lead is thought to damage the mitochondria and ribosomes of erythrocytes and reticulocytes; this results in greater uptake of dyes by basophilic components of affected red blood cells. Repeated morphologic assessment of red blood cells from infant monkeys (HPB study) demonstrated that basophilic stippled red blood cells are observed more frequently in monkeys exposed to lead acetate at dose levels as low as 100 μg Pb/kg body wt/day. However, basophilic stippled red blood cells are not consistently observed, even at lead doses as high as 2000 μg/kg body wt/day. A lead-treated monkey may show basophilic stippled red blood cells in one blood sample, while a sample collected a few weeks later can be normal in spite of continued lead exposure. Therefore, the presence or absence of basophilic stippled red blood cells is not a reliable indicator of lead exposure in nonhuman primates; this is similar to the situation in humans (Albahary, 1972) and dogs (Zook et al., 1971).

Although it appears that relatively high blood lead levels must be attained before changes in hematocrit or blood hemoglobin concentrations occur, anemia can produce pronounced changes in blood lead levels, FEP concentrations and ALAD activity. Kneip et al. (1976), working with infant baboons, demonstrated that anemia produced by repeated blood withdrawal resulted in higher blood lead levels and greater changes in FEP and ALAD than that observed in lead-treated, but otherwise normal, animals. The authors caution that normal hematopoiesis must be maintained when studying lead toxicity.

Lead has been associated with the development of chromosomal abnormalities in lymphocytes from monkeys treated with lead. Although such effects are not usually considered under the title of hematology, they will be discussed here as a matter of convenience. Monkeys (*Macaca fascicularis*) exposed to 6 mg Pb/kg body wt/day on low calcium diets showed a significant increase in the frequency of severe chromosomal abnormalities (Deknudt et al., 1977). The severity of the abnormalities and their frequency were lower following lead treatment on calcium sufficient diets, but were still increased over control monkeys. Similar effects of lead on chromosomal abberations have been observed in humans (Deknudt et al., 1973; Bauchinger et al., 1976). These data emphasize again the importance of nutritional considerations when assessing lead toxicity. Also, the long term consequences of chromosomal abnormalities could be extremely serious; further investigation of such effects appears warranted.

## VI. CLINICAL AND NEUROLOGIC EFFECTS OF LEAD ON NONHUMAN PRIMATES

### A. Clinical Effects

Both the overt physical and neurologic effects of lead in nonhuman primates are rather nonspecific; they are only observed with high level acute exposure. Monkeys exposed to 1000 mg Pb/week developed Burtonian lines in the gingiva, showed weight loss or decreased weight gain and pallor and developed extreme weakness and anorexia prior to death (Zook et al., 1976). Similar signs of overt lead toxicity have been reported in baboons (Hopkins, 1970). Lower levels of lead exposure (up to 10 mg/kg body wt/day) do not produce any such signs of lead intoxication (Goode et al., 1973; Allen et al., 1974; Cohen et al., 1974; Bushnell et al., 1977; HPB study).

Since lead exposure in young children has been associated with signs of toxicity in the central nervous system (Lin-Fu, 1972), the occurrence of effects on the central nervous system in infant nonhuman primates is of particular interest. Both baboons and monkeys develop seizures and convulsions in the terminal phases of acute lead.toxicity (Hopkins, 1970; Houser and Frank, 1970; Cohen et al., 1972). Blindness has also been observed in monkeys following high level lead exposure (10 to 50 mg Pb/kg body wt/day) (Cohen et al., 1972).

At lower levels of lead exposure, which do not result in a moribund state, changes in neurologic signs are less definitive. An increased incidence of neurologic deficits was observed in monkeys (HPB study) dosed at 500 μg Pb/kg body wt/day prior to 1 year of exposure. Depressed hopping and placing reactions, as well as depressed superficial pain responses, were observed with increased frequency in treated infants compared to controls; however, these deficits were not consistently present in any one monkey and disappeared with time. The qualitative and subjective nature of such neurologic tests makes a definitive assessment of their significance difficult unless the neurologic signs are clear cut and consistently observed.

### B. Electrophysiologic Effects

As another indicator of central nervous system function, the effects of lead on electroencephalograms (EEG) were assessed in monkeys (HPB study) exposed to 500 μg Pb/kg body wt/day. Electroencephalograms were recorded using surface-type EEG electrodes; the data were acquired and stored with a computerized data acquisition system. The data were subsequently analyzed using power spectral analysis techniques. The results showed an 8 to 15% reduction in the proportion of the total power of the EEG power spectra below 7 Hz in the frontal regions of the cerebral cortex of lead-treated monkeys, relative to controls. Further analysis of these data are required and the functional consequences of such changes in electrical activity from the brain following lead treatment are unknown at present.

The effects of lead on peripheral nerve conduction velocities is equivocal at present. Some studies demonstrate a significant correlation between a reduction in peripheral nerve conduction velocity and blood lead values in occupationally exposed humans (Seppalainen and Hernberg, 1972), while others do not show a correlation (Verberk, 1976). Baboons (*Papio anubis*) exposed to lead carbonate intratracheally at doses from 50 to 135 mg Pb/kg body for 39 to 362 days did not show any abnormalities in the peripheral nervous system, as assessed by nerve conduction velocity, electromyography and histopathologic examination of peripheral neurones (Hopkins, 1970). Ulnar nerve conduction velocity measurements in monkeys (HPB study) exposed to 2000 μg Pb/kg body wt/day did not reveal any treatment-related differences compared to control monkeys (Table 6).

### C. Histopathologic Effects

The histopathologic effects of lead on the central nervous system of nonhuman primates has been studied following acute exposure to high levels of inorganic lead. Ferraro and Hernandez (1932b) reported diffuse involvement of nerve cells in various layers of the cerebral cortex, as well as other brain regions. The lesions were described as a "severe type of nerve cell disease" characterized by granular disintegration of the Nissl substance and diffuse pallor of the cytoplasm (Ferraro and Hernandez, 1932b). In baboons (*Papio anubis*) acutely exposed to lead carbonate intratracheally, severe

**Table 6.** Ulnar nerve conduction velocity in lead-treated and control monkeys (*Macaca fascicularis*)

| Lead dose (μg/kg/day) | Number of animals | Age (days ± SE) | Conduction velocity average fibers* (m/sec ± SEM) |
|---|---|---|---|
| 2000† | 3 | 866 ± 18.3 | 82.9 ± 5.2 |
| 0 | 2 | 804 ± 15.5 | 84.4 ± 0.2 |
| 500‡ | 4 | 1225 ± 45.5 | 67.3 ± 5.3 |
| 0 | 4 | 1204 ± 51.2 | 79.5 ± 5.3 |

* Corrected to 37°C.

† Dosed from 100 days of age.

‡ Dosed from birth.

edema was observed in the white matter with severe swelling of the oligodendroglia (Hopkins and Dayan, 1974). Local proliferation of astrocytes was also observed, together with a loss of neurones and coarse sponginess of the neuropil.

There are no reports on the histopathologic effects observed in the brain following long term low level lead exposure where overt signs of toxicity are not observed.

In summary, overt physical and neurologic signs of lead intoxication in nonhuman primates are only evident at higher levels of lead exposure. More subtle and less well established neurologic and electrophysiologic effects are observed at moderate exposure levels, although peripheral nerve conduction velocity in nonhuman primates does not appear to be significantly affected by lead.

## VII. BEHAVIORAL EFFECTS OF LEAD IN NONHUMAN PRIMATES

It is well established that acute or subacute ingestion of lead by children results in encephalopathy, convulsions, and mental retardation (Thurston et al., 1955; Perlstein and Attala, 1966). Damage to the developing brain by chronic low level intake of lead is more difficult to determine in the human population, as it must be inferred from behavioral tests in children, the interpretation of which is often subjective. There is also a problem in matching between control and treated subjects (Damstra, 1977), since results may be subject to influence by nutritional and socioeconomic factors, age, sex, education and previous history of lead exposure. Effects of general intelligence, perceptual, visual perceptual and fine motor skills have been found by some investigators (Beattie et al., 1975; de la Burdé and Choate, 1975; Landrigan et al., 1975), but not by others (Baloh et al., 1975; Lansdown et al., 1974). Conflicting results have even been reported from the same population of children (McNeil et al., 1975; Whitworth, 1974). A controlled laboratory environment is of great advantage, therefore, in assessing the subtle behavioral effects of lead.

### A. Learning

Behavioral assessment of four monkeys exposed to 500 μg/kg/day of lead and their controls (HPB study) was begun when the infants were approximately 1½ to 2 years of age. Lead-treated monkeys were impaired on a series of reversals of a two-choice nonspatial form discrimination (Rice and Willes, 1979). The stimuli were two three-dimensional red forms, a cube and a triangle. After the monkeys learned to push the cube off one of two small food wells to obtain a raisin, two sessions (50 trials/session) were run with only the cube present. Placement over the right and left food wells was alternated randomly. The triangle was then introduced as the negative stimulus over the previously uncovered food well. Sessions consisted of five blocks of ten trials each; subjects were considered to have learned the form discrimination when they made no more than one incorrect response in a ten-trial block. The correct stimulus was then changed from the cube to the triangle, and the monkey tested until it again performed to criterion. The correct stimulus was reversed again for a total of 20 reversals. There were 150 extra or "overtraining" trials on the previous positive stimulus between reversals 5 and 6, and 500 extra trials between reversals 12 and 13.

The lead-treated monkeys learned the original acquisition (reversal 0) in a fewer number of trials and with fewer errors than the controls (Fig. 6). This appeared to be due to the relative reluctance of the lead-treated monkeys to explore the novel negative stimulus. After reversal 1, the control monkeys performed significantly better on the reversal problem than the treated monkeys, up to the second series of overtraining trials (reversal 13). Both treated and control monkeys were disrupted by the same magnitude by the first series of overtraining trials. The second series of overtraining trials influenced performance on successive reversals for both control and treated monkeys. Performance was better on reversals in which the positive stimulus (cube) was the same as that used on the overtraining trials.

A similar set of experiments has been done in Rhesus monkeys exposed to lead from birth to 1 year of age (Bushnell, 1978). Lead doses were adjusted so that blood lead levels were maintained at 80 μg/dl (high lead) or 50 μg/dl (low lead). Monkeys in the high lead group were retarded in learning the first of a series of reversals in a two-choice reversal learning-set paradigm. Both groups of monkeys took more total trials to complete the series of seven reversals, as well as more total postperservative errors (errors made between the first correct response and acquisition of criterion on any reversal). Similar results were obtained in two subsequent series of experiments in the same laboratory.

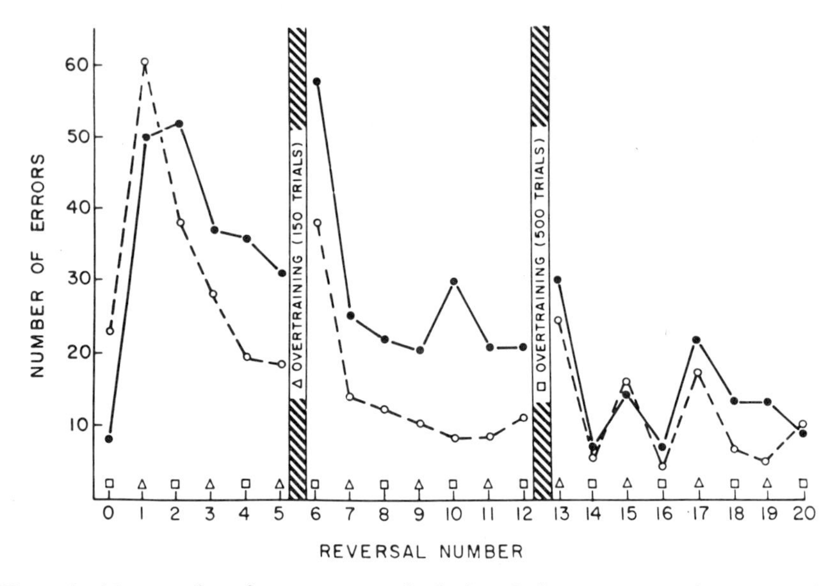


**Figure 6.** Mean number of errors to reversal criterion. Ordinate represents the mean number of total errors before reaching criterion on a reversal for control (unfilled circle) and treated (filled circle) monkeys. Striped bars represent the points in the experiment where "overtraining" occurred.

This same group of monkeys performed on a two-choice discrimination in which they had to choose a planar closed ring rather than one of several rings with gaps of various sizes (Bushnell et al., 1977). Monkeys were tested at four luminance levels. The performance of the high-lead group was severely impaired at the two lowest luminance values. Although it cannot be concluded from this experiment that this represents a visual deficit, there was clearly a differential effect of luminance level on the behavior of the high lead group which deserves further exploration.

## B. Locomotion and Social Behavior

An elevated lead body burden has been implicated in hyperactivity in children, based on retrospective studies (David et al., 1976), although this is not a consistent finding (Whitworth et al., 1974). Hyperactivity, irritability and distractability have also been observed as late sequelae in children with overt lead poisoning (Jenkins and Mellins, 1957; Thurston et al., 1955). There is much literature on the effects of perinatal lead exposure on activity in rodents, but uncontrolled factors such as food intake and nutritional status of the mother or offspring often confound the results. It is known that the nutritional state can affect activity in rodents (Slob et al., 1973). In rodent studies where weight and food intake of treated and control animals are controlled, negative results have been obtained on the effect of lead treatment on activity.

Using nursery-reared primates, Bushnell (1978) found lead-treated monkeys to be more active when tested in a photometer at 18 months of age; however, these results were not reproducable in two subsequent studies. In addition, monkeys in all three experiments were normal or hypoactive in their home cage and in social groups. They also exhibited impaired social behavior, as evidenced by suppression of play and assertive behaviors and increased social clinging.

We have also found no evidence of increased locomotor activity in monkeys treated with 500 $\mu$g Pb/kg body wt/day. The monkeys described in the previous experiment (HPB study) were placed in pairs (two treated and two control pairs) in their exercise cages and monitored for 23 hr/day for 3 weeks. Number of photo cell crossings from treated pairs was not different from control pairs, nor did the treated monkeys show a differential response to d-amphetamine. These results suggest that while lead does not reliably increase locomotor activity in the primate, it may have important effects on social behavior.

### C. Schedule-Controlled Behavior

"Hyperactivity" in children is not defined as merely an increase in locomotion, but involves more subtle changes in behavior such as distractability, fidgeting, repetitive engagement in nonproductive activities, etc. In order to monitor "activity" other than locomotion, the authors studied the pattern of lever pressing on an operant schedule of reinforcement, the fixed interval (FI), on monkeys dosed with 500 $\mu$g Pb/kg body wt/day (HPB study). Monkeys were required to make one response after 8 min had elapsed in order to receive a fruit juice reinforcement. Responses before the 8-min interval had elapsed were recorded but had no scheduled consequences. Thus the FI schedule neither positively nor negatively reinforces any particular rate of responding, since the reinforcement density is not appreciably affected by different rates or patterns of responding, within reasonable limits. In this study, each FI was followed by a 90-second time-out (TO) period, during which responses were never reinforced. A red light behind the response button signaled that the FI was in effect. The button was unlit during the TO.

When a subject is first exposed to the FI schedule, the response pattern tends to be linear or skewed toward more responding at the beginning of the interval, often with relatively long pauses. Over the course of several sessions, FI performance becomes characterized by an initial pause, followed by a gradually accelerating rate of response, terminating in reinforcement.

There were obvious differences between lead-treated and control monkeys in the acquisition of the FI performance, as well as a subtle but lasting differences in the stable performance. During the first 20 to 30 1-hr sessions, lead-treated monkeys had a higher rate of responding and a concomitant decrease in the median time between responses (inter-response time or IRT). TO responding was increased over that of controls during approximately the first 10 sessions. During the early sessions, three of the four lead-treated monkeys responded in "bursts," as indicated by the step-like pattern in the cumulative response records (Fig. 7). This "bursting" response pattern is reflected in the IRT distributions of the treated monkeys compared to the controls (Fig. 8). Three of the treated monkeys had a very high absolute frequency of short IRT's (below 1 second), while the fourth had a high frequency of responses between 3 and 7 seconds. The distribution of the control monkeys were relatively flat between 0 and 7 seconds, with absolute frequencies below those of any of the treated monkeys. Thus both

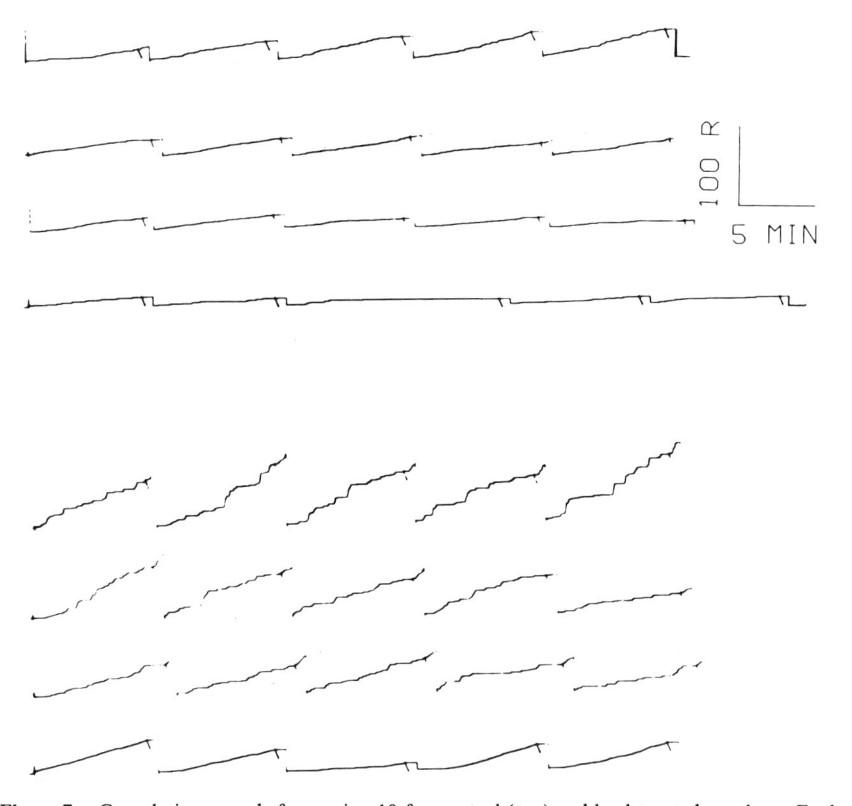


**Figure 7.** Cumulative records for session 10 for control (top) and lead-treated monkeys. Each lever press stepped the pen vertically, while time is represented horizontally. The reinforced response in each fixed interval (FI) was signaled by downward deflection of the pen. The pen reset to baseline at the end of each time-out (TO) period. The response rate in both the FI and TO periods is greater than controls for three of the four lead-treated monkeys. The step-like bursts of response are evident. Note the long pauses in the records of two of the control monkeys. Calibration scale shows pen deflection for 100 responses (R) and 5 minutes elapsed time.

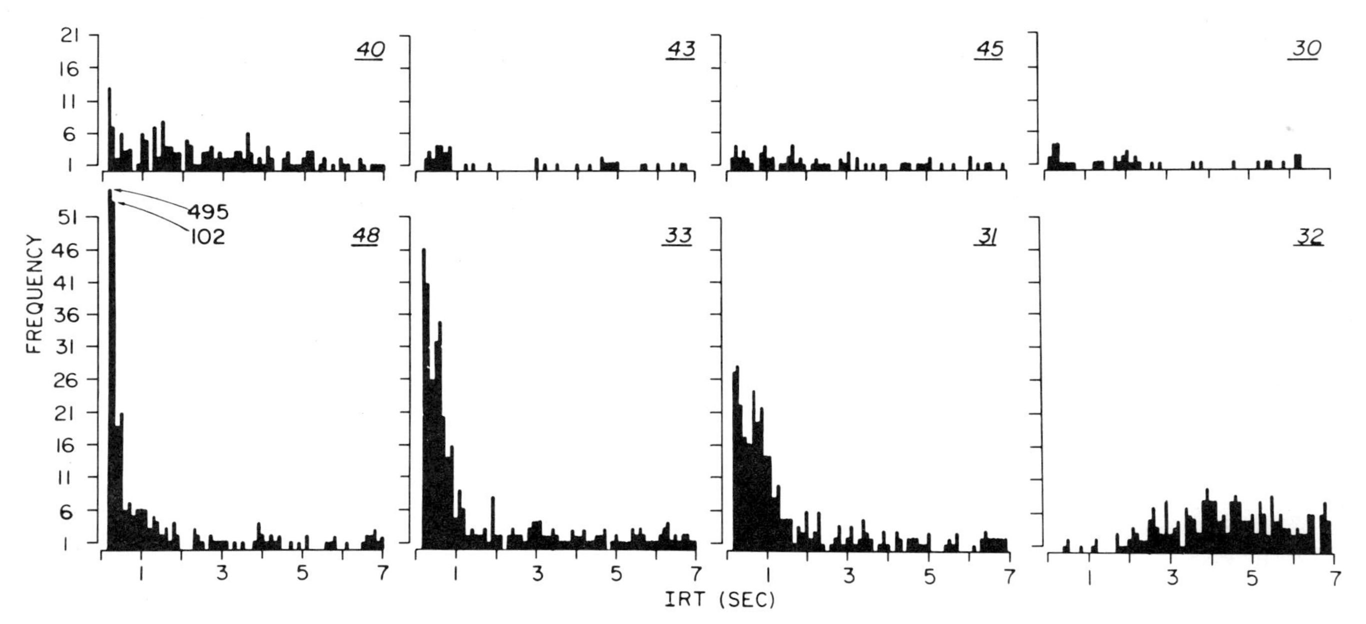


**Figure 8.** Interresponse time (IRT) absolute frequency distribution histograms for control (top) and lead-treated monkeys (500 $\mu$g Pb/kg body wt/day) for session 7. Bin width is 100 msec, with all IRT over 7 seconds eliminated from the graph. The lead-treated monkeys in general had a much higher frequency of short IRT, and a distribution skewed toward very short IRT.

the pattern of responding and absolute frequency of responding differed between control and treated groups. After the FI performance had stabilized, there were no differences between treated and control monkeys in most of the characteristics of FI performance (pause, run rate, quarter-life). Yet the IRT distributions of the two groups remained distinguishable. Whether this difference in "activity" is a generalized phenomenon in this group of monkeys awaits further investigation, but it suggests that the lead-treated monkeys respond differently to their external environment in ways that could influence their performance on many types of tasks.

### D. Comments on Behavioral Effects of Lead in Nonhuman Primates

As a general concluding comment on the behavioral effects of lead on nonhuman primates, it is evident from these studies that blood lead levels of 60 to 80 μg/dl early in life result in clear cut behavioral impairment. The uniformity of results between two independent laboratories using different species of Macaques is gratifying and increases confidence in the validity and generality of the data. Behavioral changes were observed when the monkey's blood lead levels were either normal (Bushnell et al., 1977; Bushnell, 1978) or in the range of 20 to 40 μg/dl (HPB study). These data demonstrate that exposure to relatively low blood lead levels early in life has lasting consequences for the organism.

## VIII. SUMMARY AND CONCLUSIONS

The available data on effects of lead on nonhuman primates can be summarized as follows.

1. Infant nonhuman primates absorb appreciably more lead from the gastrointestinal tract (60 to 70%) than adults (3 to 7%). The majority of the extra lead absorbed by infants is distributed primarily to bone and cartilaginous tissue. The brain Pb to blood Pb ratio is higher in infants, indicating an increased transfer of lead into the infant brain.
2. The difference in lead absorption between infants and adults appears to be associated with a change from milk to solid diets.
3. There is good agreement between the results of five independent laboratories in the relationship between blood lead and lead dose. These data indicate that a blood lead level of 40 μg Pb/dl blood would result from an oral intake of 200 μg Pb/kg body wt/day prior to weaning or from an oral intake of about 500 μg Pb/kg body wt/day postweaning.
4. A decrease in blood ALAD activity is associated with lead exposure in nonhuman primates. A dose of 25 μg Pb/kg body wt/day (12 μg Pb/dl blood preweaning; 8 μg Pb/dl blood postweaning) was associated with a significant depression in the ALAD 6.65/ALAD 7.2 activity ratio.

The ALAD activity was not further depressed at doses above 500 μg Pb/kg body wt/day (60 μg Pb/dl blood preweaning; 40 μg Pb/dl blood postweaning).

5. An increase in blood FEP level is associated with lead exposure in nonhuman primates. This increase in FEP was not statistically significant at doses below 500 μg Pb/kg body wt/day (60 μg Pb/dl blood preweaning; 40 μg Pb/dl blood postweaning).
6. Urinary ALA levels are increased following lead exposure in nonhuman primates, but only at doses above 2000 μg Pb/kg body wt/day.
7. Lead exposure is associated with changes in liver function in nonhuman primates, but only at doses of 2000 μg Pb/kg body wt/day or greater. Both antipyrine and bromsulphalein clearance rates were significantly reduced in monkeys exposed to this dose of lead. These data suggest that lead alters the activity of various drug-metabolizing enzymes in nonhuman primate liver.
8. Changes in hemoglobin concentrations, hematocrit and packed cell volume are only observed at very high levels of lead exposure. Basophilic stippled red blood cells are observed following lead exposure, but their occurrence is intermittent. There is some indication that lead exposure may result in damage to DNA in lymphocytes at doses of 6 mg Pb/kg body wt/day.
9. Unequivocal overt physical and neurologic signs of lead intoxication are only observed after high levels of lead exposure (10 to 50 mg Pb/kg body wt/day), which usually result in death. At lower levels of lead exposure (500 to 2000 μg Pb/kg body wt/day), the neurologic signs of lead intoxication are equivocal and not reproducible.
10. Relatively low levels of lead exposure are associated with changes in learning ability and operant behavior in monkeys exposed as infants. Behavioral effects have been observed at the lowest lead dose tested (500 μg Pb/kg body wt/day; 60 μg Pb/dl blood preweaning; 40 μg Pb/dl blood postweaning).

In conclusion, significant effects on learning ability and operant behavior are evident in nonhuman primates exposed to 500 μg Pb/kg body wt/day from birth (blood Pb, 60 μg/dl preweaning; 40 μg/dl postweaning). Biochemical effects of lead (depressed ALAD activity) are evident at doses as low as 25 μg Pb/kg body wt/day (blood Pb, 12 μg/dl preweaning; 8 μg/dl postweaning). The significance of depressed ALAD activity in blood as a health risk is unclear at present, since the resultant functional consequences are unknown. However, the fact that significant behavioral and biochemical effects and the simultaneous occurrence of other less definitive signs of toxicity (intermittent neurologic deficits, possible changes in EEG patterns) are evident at doses of 500 μg Pb/kg body wt/day indicates that this level of lead intake results in a significant health risk. Blood lead levels at this intake are about 55 μg Pb/dl, preweaning, and 40 μg Pb/dl, postweaning. In light of these observations, the acceptable level of lead exposure in the human population requires re-evaluation.

## ACKNOWLEDGMENTS

The authors wish to thank Mrs. P. Kressler-Willes, Mrs. L. Chenier and the staff of the Animal Resources Division, HPB, for their expertise and diligence in the operation of the infant primate nursery and data collection during the HPB study. The expert technical input of Mr. E. Lok and Mr. R. Tanner is also acknowledged. We are grateful to Dr. S. M. Charbonneau and Dr. C. Miller for the time and effort expended in the critical review of this chapter.

## REFERENCES

Albahary, C. (1972). *Am. J. Med. 52:*367.

Alexander, F.W. (1974). *Environ. Health Perspect. 7:*155.

Allen, J.R., McWey, P.J., and Suomi, S. (1974). *Environ. Health Perspect. 7:*239.

Alvares, A.P., Fischbein, A., Sassa, S., Anderson, K.E., and Kappas, A. (1976). *Clin. Pharmacol. Ther. 19:*183.

Alvares, A.P., Leigh, S., Cohn, J., and Kappas, A. (1972). *J. Exp. Medicine 135:*1406.

Baloh, R., Sturon, R., Green, B., and Glesi, G. (1975). *Arch. Neurol. 32:*326.

Barltrop, D., and Khoo, H.E. (1975). *Postgrad. Med. J. 51:*795.

Bauchinger, M., Schmid, E., Ein-brodt, H.J., and Dresp, J. (1976). *Mutation Res. 40:*57.

Beattie, A., Moore, M., Goldberg, O., Findlayson, M., Graham, J., Mackie, E., Maui, J., McLauren, D., Murdock, R., and Stewart, G. (1975). *Lancet 3:*589.

Bushnell, P.J. (1978). Ph.D. Thesis, Dept. of Psychology, University of Wisconsin, Madison, Wisc.

Bushnell, P.J., Bowman, R.E., Allen, J.R., and Marlar, R.J. (1977). *Science 196:*333.

Chisolm, J.J., Jr., Mellits, D.E., Keel, J.G., and Barrett, M.B. (1974). *Environ. Health Perspect. 7:*7.

Clasen, R.A., Hartmann, J.F., Coogan, P.S., Pandolfi, S., Laing, I., and Becker, R.A. (1974). *Environ. Health Perspect. 7:*175.

Cohen, N., Kneip, T.J., Goldstein, D.H., and Muchmore, E.A.S. (1972). *J. Med. Prim. 1:*142.

Cohen, N., Kneip, T.J., Rulon, V., and Goldstein, D.H. (1974). *Environ. Health Perspect. 7:*161.

Damstra, T. (1977). *Environ. Health Perspect.* 19:297.

David, O., Hoffman, S., and Sverd, J. (1976). *Psychopharmac. Bull. 12:*11.

de la Burdé, B., and Choate, M. (1975). *J. Pediat.* 87:638.

Deknudt, G., Colle, A., and Gerber, G.B. (1977). *Mutation Res. 45:*77.

Deknudt, G., Leonard, A., and Guanou, B. (1973). *Environ. Physics. Biochem. 3:*132.

Federal Register. (1978). *43:*46246.

Ferraro, A., and Hernandez, R. (1932a). *Psychiat. Quarterly 6:*121.

Ferraro, A., and Hernandez, R. (1932b). *Psychiat. Quarterly 6:*319.
Fischbein, A., Alvares, A.P., Anderson, K.E., Sassa, S., and Kappas, A. (1977). *J. Toxicol. Environ. Health 3:*431.
Fisher, L.E. (1954). *J. Am. Vet. Med. Assoc. 125:*478.
Forbes, G.B., and Reina, J.C. (1972). *J. Nutr. 102:*647.
Gibson, K.D., Neuberger, A., and Scottil, J.J. (1955). *J. Biochem 61:*618.
Goldstein, D.H., Kneip, T.J., Rulons, V., and Cohen, N. (1975). *J. Occup. Med. 17:*157.
Goode, J.W., Johnson, S., and Calandra, J.C. (1973). *Toxicol. Appl. Pharmacol. 25:*465.
Goyer, R.A., and Mahaffery, R.R. (1972). *Environ. Health Perspect. 2:*73.
Granick, S., and Mauzerall, D. (1958). *J. Biol. Chem. 232:*1119.
Hausman, R., Sturtevant, R.A., and Wilson, W.J. (1961). *J. Forensic Sci. 6:*180.
Heywood, R., James, R.W., Sortwell, R.J., Prentice, D.E., and Barry, P.S.I. (1978). *Toxicol. Letters 2:*187.
Hopkins, A. (1970). *Brit. J. Industr. Med. 27:*130.
Hopkins, A.P., and Dayan, A.D. (1974). *Brit. J. Indust. Med. 31:*128.
Houser, W.D., and Frank, N. (1970). *J. Am. Vet. Med. Assoc. 157:*1919.
Jenkins, C., and Mellins, R. (1957). *AMA Arch. Neurol. Psychiat. 77:*70.
Kello, D., and Kostial, K. (1973). *Environ. Res. 6:*355.
Kneip, T.J., Mallon, R.P., and Rulon, V.P. (1976). *Am. Ind. Hyg. Assoc. J. 37:*578.
Kostial, K., Kello, D., Jugo, S., Rabar, I., and Maljkovic, T. (1978). *Envir. Health Perspect. 25:*81.
Landrigan, P., Whitworth, R., Baloh, R., Stachling, N., Barthel, W., and Roseublum, B. (1975). *Lancet 1:*708.
Lansdown, R., Shephard, J., Clayton, B., Delves, H., Graham, P., and Turner, W. (1974). *Lancet 3:*538.
Lin-Fu, J. (1972). *New England J. Med. 286:*702.
McNeil, J., Ptasnik, J., and Croft, B. (1975). *Arch. za Higijenu Rada in Toksikologijie 26* (suppl):97.
Meredith, P.A., Campbell, B.C., Moore, M.R., and Goldberg, A. (1977). *Europ. J. Clin. Pharmacol. 12:*235.
Momcilovic, B., and Kostial, K. (1974). *Environ. Res. 8:*214.
Perlstein, M., and Attala, R. (1966). *Clin. Pediat. 5:*292.
Piomelli, S., Davidow, B., Guinee, V.F., Young, P., and Gay, G. (1973). *Pediatrics 51:*254.
Rice, D., and Willes, R. (1979). *J. Environ. Pathol. Toxicol.* In press.
Seppalainen, A.M., and Hernberg, S. (1972). *Brit. J. Industr. Med. 29:*443.
Shapiro, I.M., Dobkin, B., Tuncay, O.C., and Neddlesman, H.D. (1973). *Clin. Chim. Acta 46:*119.
Slob, A., Snow, C., and de Natris-Mathot, E. (1973). *Devel. Psychobiol. 6:*177.
Stephens, R., and Waldron, H.A. (1975). *Food Cosmet. Toxicol. 13:*555.
Thurston, D., Middlekamp, J., and Mason, E. (1955). *J. Pediat. 47:*413.
Tomokuni, K. (1974). *Arch. Environ. Health 29:*274.
Verberk, M.M. (1976). *Int. Arch. Occup. Environ. Hlth. 38:*141.
Vermande-Van Eck, G.J., and Meigs, J.W. (1960). *Fertil. Steril. 11:*223.
Vitale, L.F., Joselow, M.M., Wedeen, R.P., and Pawlow, M. (1975). *J. Occup. Med. 17:*155.
Waldron, H.A. (1975). *Prev. Med. 4:*135.
Whitworth, R., Rosenblum, B., Dickerson, M., and Baloh, R. (1974). *Center for Disease Control Morbidity and Mortality Weekly Report 23:*157.

Willes, R.F., Kressler, P.L., and Truelove, J.F. (1977a). *Lab. Animal Sci. 27:*90.
Willes, R.F., Lok, E., Truelove, J.F., and Sundaram, A. (1977b). *J. Toxicol. Environ. Health 3:*95.
Zielhuis, R.L. (1971). *Arch. Environ. Health 23:*299.
Zook, B.C., London, W.T., Silver, S.L., and Sauer, R.M. (1976). *J. Med. Primatol. 5:*23.
Zook, B.C., McConnell, G., and Gilmore, C.E. (1971). *Am. J. Vet. Med. Assoc. 157:*2092.

# Behavioral Neurotoxicity of Lead

*Kathryn M. Jason and Carol K. Kellogg*

TABLE OF CONTENTS



## I. INTRODUCTION

Of the hazards associated with exposure to lead, behavioral consequences are often the most insidious and the most difficult to diagnose. The relationship between chronic, low level exposure to lead and such nonspecific symptoms as depression, insomnia, irritability, memory impairment or clumsiness is not always recognized. Blood lead (PbB) level, the most common diagnostic tool for evaluating exposure, does not necessarily correlate well with the behavioral symptoms of lead toxicity, since lead in blood only reflects current exposure. This problem is especially evident in cases of childhood lead poisoning, when delayed behavioral and/or intellectual effects of early lead exposure may not be recognized until the child enters school.

With the advent of increasingly reliable techniques in behavioral toxicology, subtle neurobehavioral effects of lead can now be quantified in individuals in whom PbB levels indicate no excessive exposure. Several recent animal studies have revealed such less obvious toxicologic effects of chronic exposure to low levels of lead. As a result of this new evidence, a re-evaluation of the definition of lead toxicity is taking place. This chapter will examine many of the reported neurobehavioral effects of lead poisoning in humans and will present and evaluate animal models currently used to study the behavioral neurotoxicity of lead.

## II. BEHAVIORAL CONSEQUENCES OF HUMAN EXPOSURE TO LEAD

### A. Encephalopathy and Neuropathy Attributed to Lead

Until relatively recently, a diagnosis of lead poisoning was reserved for persons exhibiting severe, acute encephalopathy, the symptoms of which have long been recognized. The syndrome may begin with the onset of intractable seizures and lead to coma and death within a short period of time. In other cases, frequent episodes of vomiting, drowsiness, altered consciousness (stupor), and profound ataxia may be observed, after which the patient becomes comatose. Nonspecific symptoms such as muscle weakness or tremor, numbness, paralysis, or persistent headache, or depression may precede the more severe symptoms. The clinical findings in acute encephalopathy are not unique to lead poisoning, but resemble symptoms of cerebral edema and increased intracranial pressure of unknown origin. In the absence of a known history of exposure to lead, the diagnosis of lead poisoning may be overlooked. This problem is especially evident in the case of people who drink moonshine whiskey and persons exposed to lead through beverages that have come into contact with improperly lead-glazed earthenware.

Chronic encephalopathy resulting in progressive mental deterioration is seen primarily in children. Perlstein and Attala (1966) reported that recurrent seizures and mental retardation occurred in 82% of children with a history of severe lead encephalopathy. The symptoms may resemble a variety of degenerative diseases of the central nervous system (CNS). Children with pica, who eat nonfood lead-containing items such as paint chips from deteriorating houses, may exhibit a steady loss of motor control or speech followed by convulsive disorders. Since the diagnosis of acute lead encephalopathy is often missed, the chronic syndrome may also be the result of several instances of acute encephalopathy which were not attributed to lead poisoning at the time of their occurrence.

While encephalopathy occurs primarily in children, peripheral neuropathy due to lead exposure is more common in adults (Waldron and Stofen, 1974). A predominance of motor disorders, usually with little or no sensory involvement, is the clinical finding in these cases. Three forms of lead-related neuropathy have been identified, all with a common symptom of weakness. In one form, pain and tenderness in the trunk muscles accompanies the acute abdominal colic of lead poisoning. Weakness of the muscles of the extremities may also occur.

The most commonly reported type of peripheral neuropathy involves weakness, primarily of the extensor muscles of the dominant hand and foot. This syndrome is not usually reported to be painful. Extensor muscle weakness is accompanied by reduced muscle stretch reflexes and decreased motor nerve conduction velocity in peroneal and/or radial nerve (Feldman et al., 1977). Such measures may serve as diagnostic tools. More severe extensor weakness can be recognized as the "wrist drop" or "foot drop" of lead palsy. The peripheral neuropathy described here may occur after acute exposure to excessively high levels of lead or, more commonly, after prolonged exposure to moderately increased lead levels. In severe cases, spasticity, rigidity, or tremor may occur, along with clumsy gait or movement disorders resembling those of the extrapyramidal syndrome.

The third (and most severe) form of lead neuropathy resembles other degenerative diseases of the nervous system (Campbell et al., 1970). Muscle fasciculations and atrophy may be present. Lead has long been suspected of having an etiologic role in multiple sclerosis and amyotrophic lateral sclerosis, but the nature of lead's involvement in those diseases is controversial (Damstra, 1977).

### B. Chronic Occupational Exposure to Lead

The manifestations of lead poisoning can be severe, as the foregoing summary has demonstrated. However, the scope of this chapter is not to elaborate on the severe consequences of excessive lead exposure, but rather to discuss less severe (but functionally important) effects of lead on which environmental scientists have only recently begun to focus. Such subtle effects fall into two major categories; these are those principally occurring in adults and those that are most common in children. The exposure of adults occurs primarily in the workplace. Lead smelters (especially secondary smelters) and storage battery manufacturing plants where lead casings are produced are among the most hazardous of working environments in terms of lead exposure.

As recently as 1972, a comprehensive report on lead stated that no studies had been reported concerning behavioral changes in workers exposed to lead concentrations below those which produced overt symptoms of toxicity (National Research Council, 1972). Since then, investigators have started to evaluate systematically workers known to be exposed to lead, with the goal of determining whether present standards for acceptable lead exposure levels are still reasonable in light of the sophisticated measurements for behavioral changes that are now available. Sensory and motor deficits, neuromuscular effects and, occasionally, deficiencies in intellectual functioning can now be evaluated along with the subjective reports of workers. It may be possible, then, to correlate behavioral findings with biologic measurements such as PbB level, delta-aminolevulinic acid (ALA) or zinc protoporphyrin (ZPP) levels used to assess lead exposure.

The studies of Seppäläinen and her coworkers in Finland (Seppäläinen and Hernberg, 1972; Seppäläinen et al., 1975) indicate that definite changes occur in lead-exposed workers who report no physical complaints. These investigators compared workers at a storage battery plant with age- and sex-matched blue collar workers not occupationally exposed to lead. The battery plant workers had spent 1 to 17 years in departments where lead exposure was considered slight or moderate. None had a PbB level greater than 70 μg/100 ml. This level is commonly accepted as safe, because clinical symptoms rarely occur at PbB less than 80 μg/100 ml. In this population, they found that maximal conduction velocities of the median and ulnar nerves of the arm were slowed in the exposed workers. A marked decrease was also found in the conduction velocity of slower fibers (CVSF) of the ulnar nerve. The data on CVSF from these two studies demonstrated a dose-response effect, in that a lead-poisoned group had lowest CVSF values, controls had the highest values and the exposed but not poisoned group had intermediate CVSF values. These findings of damage in the nerves of the upper arm in persons chronically exposed to low levels of lead correlate well with clinical observations that lead palsy is primarily a disease of the arm.

Several behavioral and sensory measures of lead toxicity were reported by Repko and associates (Repko et al., 1975) in a large scale behavioral evaluation of lead-exposed workers. These investigators studied 428 workers, 316 of whom were engaged in various aspects of the manufacture of storage (lead-acid) batteries, while the remainder (controls) had no history of industrial exposure to inorganic lead. A battery of 12 tests to assess intellectual functions (learning and memory), sensory-perceptual functions (visual and auditory acuity, watch-keeping, vigilance, attention and pattern perception), neuromuscular functions (psychophysiologic and psychomotor), and psychologic functions (affect) were administered. The results of those tests were correlated with biomedical indices of body burden of lead such as PbB, urine lead (PbU), blood delta-aminolevulinic acid dehydrase activity (ALAD), urine delta-aminolevulinic acid (ALA) and urine coproporphyrin (CPU), all of which were altered in the lead-exposed group.

The largest difference between lead-exposed and control subjects occurred in measures of neuromuscular functions. Decreased function was demonstrated by increased occurrence of tremors in the lead group and by increases in response variability and latency in eye-hand coordination tests, particularly with the preferred hand. These changes were found with increasing body burdens of lead and were significantly correlated with decreased ALAD. In addition, the changes began to occur at PbB levels between 70 and 79 μg/100 ml. Muscular endurance data also suggested decreased function in the lead group, but data on muscular strength showed decreased strength at the lower PbB levels (39 to 69 μg/100 ml) and increased strength above that level. The investigators suggest that the inverse relationship between muscular strength and endurance may be a result of overcompensation in the lead-exposed group; they may have over-exerted themselves in the measure of strength (possibly due to previous experience with weakness), but were then unable to maintain endurance.

Marked differences between lead-exposed and control workers on measures of psychologic functions were also reported. The lead-exposed group showed significantly greater hostility, depression, and general dysphoria. Again, the largest increases occurred in the range of 70 to 79 μg/100 ml of PbB. However, the authors caution that the causative effect of lead cannot be proved, since the findings could also be the result of other conditions, such as stress factors in the work environment, which were not taken into consideration.

Of the sensory functions evaluated, the effects of lead were most evident in auditory function tests. Hearing losses were reported in all frequency ranges, were correlated with decreased ALAD and were most marked at high body burdens of lead. Initial changes in auditory functions occurred at 70 to 79 μg/100 ml of PbB. Hearing losses at high and low frequencies correlated with PbB and PbU levels, respectively, and auditory pathology was suggested by findings of tone decay correlated with increased PbB and CPU. However, visual dysfunctions did not occur and there were no adverse effects on watch keeping, vigilance or attentive functions at the body burdens of lead obtained in this study.

There was no quantitative relationship between body burden of lead and the type of intellectual functioning measured by Repko et al. (1975). The data did suggest, however, that either a decrease in learning capacity or an increase in intellectual fatigue interacted with arithmetic computation performance in lead-exposed, but not control, workers.

In another recent set of studies, Valcuikas et al. (1978a,b) employed five behavioral tests to determine CNS dysfunction in a group of 90 secondary lead smelter workers compared to 25 steel workers living in the same community. Zinc protoporphyrin levels (ZPP) and PbB levels were biologic indices used to assess lead exposure. Lead workers were impaired on performance of three of the behavioral measures of intellectual function (Block Design, Embedded Figures, Digit Symbol), but the Santa Ana Dexterity Test for both hands or preferred hand did not differentiate the groups. In addition, ZPP levels were significantly correlated with scores on those tests where deficits were found and PbB level was correlated with scores on the Block Design and Embedded Figures tests. These investigators have also reported that there were no differences in scores on performance tests between exposed workers who did or did not demonstrate symptoms such as fatigue, nervousness, and sleep disturbance (Lilis et al., 1977a,b; Valcuikas et al., 1978a).

Although the number of studies of occupational lead exposure is still relatively small, the available data demonstrate that several neurobehavioral deficits occur in workers exposed to lead at levels which do not produce clinical symptoms. The studies of Repko et al. (1975) indicate some intellectual deficits in storage battery plant employees. Quantitative changes were not found in that study, however, and it was suggested that, to detect differences in intellectual functioning, more sensitive tests should be employed. The tests administered by Valcuikas et al. (1978a and 1978b) and Lilis et al. (1977a and 1978b), which demonstrated significant differences between lead smelter workers and controls in tests of CNS functioning, might be more appropriate. Of the sensory-perceptual functions assessed by Repko et al. (1975), the only significant finding was that of auditory dysfunction in the lead-exposed workers. The data obtained from other measurements in that category led the authors to suggest improved methods such as testing for the detection threshold for light and dark adaptation to evaluate changes in visual acuity, and the use of latency measures in tests of perceptual deficits. The most marked neurobehavioral effects of lead reported in these studies are the deficits in neuromuscular functions. The lead-related increases in latency measures of eye-hand coordination found by Repko et al. are consistent with the decreased nerve conduction velocities reported by Seppäläinen and Hernberg (1972) and Seppäläinen et al. (1975).

All of the psychologic, neuromuscular, sensory-perceptual and intellectual changes that these studies ascribe to lead occurred at what have been considered relatively low PbB levels. Most of the changes reported by Repko et al. (1975) occurred initially at 70 to 79 μg/100 ml PbB; the decreased nerve conduction velocities reported by Seppäläinen et al. (1975) occurred at PbB levels below 70 μg/100 ml. Of the lead smelter workers tested by Valcuikas et al. (1978a), 79% had PbB levels of less than 60 μg/100 ml and 99% had levels less than 80 μg/100 ml; 94% of control PbB levels were under 40 μg/100 ml and all were less than 60 μg/100 ml. Repko et al. (1975) measured several biologic indices of lead body burden and proposed that ALAD was a better indicator of neurobehavioral lead toxicity than PbB level. Valcuikas et al. (1978a,b) suggest that ZPP level can also be an important indicator of CNS dysfunction resulting from lead toxicity.

### C. Childhood Lead Exposure

In adults the body burden of lead in workers can be related to lead level in the air, but the most commonly recognized mode of childhood lead exposure is by ingestion. Many young children ingest nonfood items, a habit known as pica. Historically, peeling and chipping lead-based paint in old buildings has been an important contributory source of lead for ingestion, especially in poor neighborhoods of inner cities. However, awareness of the problem has cut the incidence of childhood lead encephalopathy through the enactment of new laws limiting the amount of permissible lead in indoor paint and through routine blood lead screening programs. More recently, it has been determined that children also ingest lead that is present environmentally. For example, children's PbB levels correlated positively with proximity to secondary smelters in El Paso (Landrigan et al., 1975a) and Toronto (Roberts et al., 1974); both studies suggested ingestion of lead-contaminated dirt and dust as the primary route of exposure. Beattie et al. (1975) reported that children living in homes with high levels of lead in tap water had PbB levels significantly higher than control children living where water lead levels were not excessive.

Since lead encephalopathy was a well documented phenomenon, the earliest studies of behavioral and intellectual effects of lead focused on the question of whether there were long lasting sequelae in children hospitalized for lead poisoning. Later investigators examined behavioral changes in asymptomatic children with elevated blood lead or a history of lead exposure.

An early longitudinal study of lead-poisoned children was carried out by Byers and Lord (1943). Twenty children hospitalized with a diagnosis of lead poisoning were followed for several years. Intelligence, perceptual-motor, and memory testing revealed that the children had lasting difficulties indicative of minimal brain damage; 19 of the 20 children were reported to be doing poorly in school. In a large study by Perlstein and Attala (1966), 425 children with diagnosed lead poisoning were followed over a 10-year period. The route of exposure for over 95% of the subjects was ingestion of loose plaster with lead-containing paint from dilapidated housing. Neurologic sequelae were reported in 39% of the children, with mental retardation and recurrent seizures as the most common findings. Sequelae occurred in 80% of children originally presenting with severe encephalopathy and in 67% of patients originally presenting with seizures, but without increased intracranial pressure. More striking than these findings, however, is the remark that, in 61% of the patients in this study, there was "complete recovery from lead intoxication" (Perlstein and Attala, 1966). Such a statement is suspect, since it imples that frank neurologic sequelae (such as retardation and/or seizures) alone are reasonable criteria for establishing the existence of lasting effects of childhood lead poisoning. In this study, lesser neurologic and behavioral consequences which could be functionally significant were not defined as sequelae.

Since these studies were carried out, the definition for excessive lead exposure has changed and screening programs have become more common. Pueschal et al. (1972) used elevated hair lead levels (over 100 $\mu g/g$), marked differences in lead content of distal and proximal hair segments or a positive history of ingestion of lead-containing materials, in the absence of overt symptoms, as criteria for examining lead in venous blood in preschool children. Blood lead levels greater than 50 μg/100 ml or a chelation urinary output of over 500 μg lead/24 hr were considered indicative of increased lead burden. Symptoms of CNS involvement, such as irritability and clumsiness, and the incidence of fine motor dysfunction were significantly more prevalent in the children with increased lead burden than in a matched control group of healthy children from the same environment who demonstrated normal PbB levels. Neurologic and/or motor impairment was still evident in 25% of the patients available for follow-up one and one-half years later.

Another study compared children with a history of asymptomatic lead exposure to controls in performance on several psychologic tests (de la Burde and Choate, 1975). The 67 exposed children had no clinical symptoms, but did have a history of eating plaster and paint between 1 and 3 years of age, with positive CPU and PbB levels greater than 40 μg/100 ml and/or radiologic indications of lead in bone at that age. In the group of 70 control children drawn from the same population, CPU was normal. The latter group was selected to exclude children with a history of lead intake, since radiographic studies and PbB levels could not be obtained. When tested at 7 years of age, neurologic exams revealed twice as many deficits in exposed children as compared to controls. At 8 years of age, the number of abnormalities in exposed children increased to four times those found in control children. In addition, the children exposed to lead demonstrated significant deficits in global IQ, associative abilities, and visual and fine motor coordination at 7 years. Abnormal behavior during testing was eight times as frequent in exposed children as in controls. More children in the lead-exposed group were reported as having severe behavior problems at home (lying, stealing, running away, fire setting); learning and behavior problems leading to school failure were also more frequent in the exposed than the control group.

In an attempt to clarify the allowable limits of lead in children's environments, Albert et al. (1974) compared children with a history of clinical lead poisoning to children with no history of overt poisoning who had evidence of increased cumulative exposure to lead and to children with no indication of increased lead exposure and no incidence of lead poisoning. In addition to PbB levels, lead in deciduous teeth was used as a measure of cumulative exposure to lead. The 371 children in this study were tested for PbB between 1.5 and 2.5 years of age; the interval from blood test to follow-up ranged from 3 to 11 years (averaging 6.7 years). The subjects were divided into categories of those hospitalized for lead poisoning and therefore receiving chelation therapy and those who were not diagnosed as lead-poisoned and did not receive chelation therapy. The subjects in the treated (poisoned) group were subdivided into those cases with or without encephalopathy. The untreated (nonpoisoned) group was divided into those with PbB levels greater than 60 $\mu g/100$ ml, those with less than 60 $\mu g/100$ ml lead in blood with high tooth lead and controls, who had less than 60 $\mu g/100$ ml lead in blood and low tooth lead. The relative difference in tooth lead between the last two groups represented a factor of five in cumulative lead intake.

Both of the treated groups and the high PbB group had significantly higher incidences of diagnosed mental disorders, including retardation, organic brain syndrome, seizures and behavior disorders, than were found in the control group. Only the subjects with encephalopathy were significantly different from controls on full scale IQ and perceptual-motor tests, although the performance of the high PbB group tended to be consistently lower than that of controls on those measures.

Data on school performance reflected significantly greater proportions of children in both of the treated groups and the high PbB group in special classes, due to mental retardation or psychological disorders, than children in the control group. In addition, those groups had higher proportions of children with psychological referrals and with behavioral, attention and concentration problems than the control group. These findings are in agreement with those of other investigators, in that neurologic symptoms ranging from mental retardation to behavioral problems in school may be apparent several years after overt lead poisoning and may also be related to a finding of excessive lead in blood in early life.

Evidence of long-term effects of lead encephalopathy and increased lead burden reported in retrospective studies has led to increased behavioral and psychologic testing of children with documented exposure to excessive lead in their environments. Correlation of PbB levels with the incidence of behavioral problems in children have also been carried out and recent studies have focused on the effects of increased lead burden alone in asymptomatic children.

A relationship between increased lead exposure and behavioral problems in children was first suggested by David et al. (1972). These investigators measured blood and post-chelation urine lead in hyperactive children and controls. The hyperactive children were divided into groups where a possible or probable cause (other than lead) was presumed, a group with a history of lead poisoning, and a group with no known probable or possible etiology. Hyperactivity was determined on the basis of a questionnaire filled out by the child's teacher, parent, and doctor. The most striking finding was that PbB and PbU levels obtained for children with hyperactivity of unknown origin and also for those with a history of lead poisoning were significantly greater than controls. Children with hyperactivity of probable known origin did not have increased lead burdens relative to controls. The PbB levels reported by David et al. (1972) were all low compared to those in most other studies (averaging less than 30 $\mu g/100$ ml in all groups), yet the results suggest that even low level chronic exposure to lead during early life may be responsible for some cases of hyperactivity.

Based on this evidence, David et al. (1976) used chelation therapy in the treatment of hyperkinetic children demonstrating PbB and PbU levels in an elevated but "nontoxic" range (25 to 40 $\mu g/100$ ml in blood). Six of the children had known probable causes for hyperactive behavior; these six showed little profit from chelation treatment. However, in the seven children with no known cause for hyperkinesis, chelation therapy resulted in considerable behavioral improvement, according to ratings of parents, teachers, and physicians. Although these studies do not elaborate on the role of lead in producing childhood hyperkinesis, they indicate that lead burden should be considered as a factor in these children, particularly in the absence of other known causes for the disorder. They also argue for a significant modification in the lower limits considered to be indicative of lead poisoning in children, since the responders to chelation averaged less than 30 $\mu g/100$ ml lead in blood at the beginning of the study.

Hyperactivity was also significantly more prevalent in the histories of lead-exposed children examined by Baloh et al. (1975). In this case, asymptomatic exposed children with PbB levels greater than 40 $\mu g/100$ ml were matched with respect to age, sex, race and socioeconomic status to controls with PbB levels less than 30 $\mu g/100$ ml. There were no significant differences between the groups on the quantitative neurologic or psychologic tests administered, but there was a trend toward poorer scores in the elevated lead group on tests of fine motor ability.

Another area of investigation into the effects of childhood lead exposure is the relationship between PbB levels and cognitive and intellectual functioning in asymptomatic children. Perino and Ernhart (1974) examined black preschoolers with low (less than 30 $\mu g/100$ ml) or moderate (40 to 70 $\mu g/100$ ml) PbB levels on tests of general cognitive ability, and verbal, perceptual-performance, quantitative, memory, and motor abilities. They reported significantly lower cognitive ability and verbal and perceptual-performance scores in the group with the higher PbB level, but no differences in scores on the motor ability and memory scales. When the influences of age, parental intelligence and birth weight were taken into account, PbB level still contributed significantly to the children's performance on the general cognitive ability, verbal and perceptual-performance scales.

The studies of behavioral and intellectual effects of lead discussed here have generally presumed that the method of exposure to lead was ingestion of peeling paint and/or plaster in substandard housing. With the enforcement of laws which regulate the amount of lead allowable in house paint and legislation encouraging improvements in substandard housing, the prevalence of lead intoxication from this route may be decreasing. However, there is increasing recognition that children living in high lead environments may be ingesting excess lead from sources other than paint and plaster.

In a study of children living near an El Paso lead smelter, Landrigan et al. (1975a) found that children's PbB levels were correlated with the proximity of their homes to the smelter. They reported that chronic ingestion and inhalation of particulate lead were the primary routes of exposure. Children were evaluated on a variety of neurologic measures and intelligence tests and medical histories were compared (Landrigan et al., 1975b). Those children with PbB levels greater than 40 μg/100 ml had significantly lower performance IQ scores and significant slowing in a finger-wrist tapping test. The elevated lead group included greater proportions of children with histories of pica, abdominal colic, clumsiness, irritability and convulsions than the control group, but these differences did not reach significance. The positive findings in this study are in contrast to negative results reported in an earlier study of children living near a smelter (Lansdown et al., 1974). However, Landrigan et al. (1975b) point out that the earlier study used distance from the smelter rather than PbB level for assignment of groups.

A longitudinal design was employed to evaluate development and behavior in children living near a lead battery manufacturing plant (Ratcliffe, 1977). Psychologic and fine-motor tests and a behavioral questionnaire were administered to children 4 to 5½ years old, approximately 3 years after an initial blood lead level had been obtained. None of the children exhibited symptoms of lead poisoning. The 47 children participating in this study were divided into two groups according to the PbB levels obtained when they were 2 years old. Scores of children with levels up to and including 35 μg/100 ml were compared to scores of children who had greater than 35 μg/100 ml lead in blood. In this study, no significant differences between groups were reported for developmental or behavioral scores. Ratcliffe (1977) points out the difficulty of testing preschool children and the possibility that the tests were not sufficiently sensitive to detect differences attributable to lead. Also, interpretation of this study is made difficult, since total exposure of the children to lead is unknown. There is no evidence that the children's environmental exposure to lead changed between the time of lead testing and behavioral and psychological evaluation and no analyses of PbB levels were available at the time of follow-up.

In addition to the reports of sequelae of childhood lead poisoning and the attempts to evaluate behavior and intellectual ability in asymptomatic children, correlative studies have linked lead exposure to a variety of neurologic deficits in children. Beattie et al. (1975) reported that PbB levels were significantly higher in a group of mentally retarded children than in controls, and found that the retarded children lived in homes where the lead content of tap water was significantly elevated. A group of learning disabled children studied by Pihl and Parkes (1977) had significantly increased amounts of lead in hair, and Cohen et al. (1976) reported that autistic children had higher PbB levels than nonautistic psychotic children and controls. This last study noted that pica is often associated with severely atypical development such as that found in autistic children.

Interpretation of studies that correlate lead burden with mental retardation must consider variables such as pica. This abnormal behavior itself may account for excessive lead intake and therefore may be indirectly responsible for the neurologic symptoms reported. Kotok et al. (1977) studied preschool children, all of whom had histories of ingestion of nonfood items, and reported no significant differences in cognitive function between children with less than 40 $\mu g/100$ ml and a group demonstrating over 60 $\mu g/100$ ml lead in blood. The authors point out that these findings do not prove the lack of neurologic damage due to lead. They stress that early detection and prompt abatement of the environmental sources may be effective in preventing neurologic damage in children exposed to lead.

The literature findings vary widely regarding the intellectual and behavioral deficits associated with lead exposure; it is likely that most of the differences in results can be attributed to differences in study design and the selection of experimental and control subjects. Many problems beset these studies, since it is particularly difficult to control variables such as type and length of exposure. Even so, the overwhelming conclusion of the investigators is that adverse behavioral and intellectual effects occur with increased lead burdens in so-called asymptomatic children and that the acceptable limits for lead exposure need to be lowered.

As recently as 1966, it was suggested that a diagnosis of childhood lead poisoning be considered and treatment initiated when PbB level exceeded 60 $\mu g/100$ ml and at least two other symptoms were present. Additionally, it was proposed that a child with 60 $\mu g/100$ ml lead in blood and a history of pica without additional signs or symptoms be classified as "possibly" lead poisoned (Jacobziner, 1966). Many of the more recent studies consider PbB levels over 40 $\mu g/100$ ml to be indicative of excessive exposure and some authors suggest that allowable levels should be set even lower (David et al., 1976). One of the problems with setting standards is that, in many cases, the techniques for assessing deficits are not sensitive enough to detect behavioral differences that exist or it may not be possible to sample representative behaviors. Lin-Fu (1972) has pointed out that many children reported to be asymptomatic may fall into that category simply because people don't know what to look for; even today, many mild cases of lead intoxication probably go undiagnosed.

## III. ANIMAL MODELS FOR THE STUDY OF BEHAVIORAL EFFECTS OF LEAD

### A. Effects of Organic and Inorganic Lead in Adults

Just as more studies have been directed toward defining the consequences of lead exposure during childhood than in adults, most experimental work in animals has focused on developing a model for childhood lead poisoning. The number of studies investigating behavioral effects of lead in adult animals is still rather small. Making comparisons among the studies using animals is quite difficult. The dosage of lead administered is given as ppm lead, percentage of lead in water or food or as mg/kg/day administered by injection. We have given the dosage of lead used in the various studies as stated by the authors. For purposes of comparison, however, the following guidelines are suggested. A 100 ppm solution of lead is equivalent to 0.01% lead or to 0.1 mg/ml. Assuming a 300-g male rat drinks 35 ml of water per day which contains 100 ppm lead, he would be getting approximately 11.67 mg/kg/day of lead. If he consumed 25 g of food containing 100 ppm lead, he would get around 7.5 mg/kg/day of lead. Whereas human exposure to lead is described in terms of PbB lead, few animal studies report any biologic index of lead exposure. Absorption of lead is a complex process and may differ according to species and route of exposure. The animal studies must be considered in light of these limitations.

Very few studies have dealt with the effects of organic lead on the nervous system, even though organic lead compounds are highly toxic and are known to produce many specifically neurologic effects, such as hallucinations and convulsions (Weiss, 1978). Those studies which have examined the effects of organic lead point out some of the problems which are related to dosing and to the design of behavioral testing methodologies.

Bullock et al. (1966) reported that the administration of tetraethyllead (TEL) did not affect the performance of rats in a water T-maze. The animals had received a total of 15 mg/kg TEL either before training began or during training and all of the lead-treated animals exhibited neurologic impairment such as tremors, ataxia and convulsions. Such impairment may well obscure any effects of lead on the animals' ability to learn, and these results suggest that the T-maze is not a sensitive indicator of the behavioral neurotoxicity of lead.

In another study, Avery et al. (1974) investigated the effect of TEL on several complex behavioral tasks using intragastric administration. They found markedly decreased bar-press response rates after a single dose of 10 mg/kg of TEL. In an operant discrimination task, administration of single doses of 8 or 10 mg/kg TEL had significantly negative effects on performance; the treated animals were also deficient on reversal testing and retention of reversal. When the same doses were administered over 10 days, these effects were not seen. The authors interpret their results as demonstrating that TEL impairs learning and memory in the rat. Interestingly, they state that the deficits did not appear to be related to an effect of TEL on emotionality, since neither avoidance nor open field behavior was significantly affected by the treatment. They do not report any findings of ataxia or tremors which could occur at the dose levels employed in this study. The conclusions are open to question, since many performance deficits could be accounted for by the effects of TEL on motor ability alone. Also, the fact that the cumulative dosing did not affect performance as drastically as acute doses suggests that the findings may be related more to poisoning than to effects on the central nervous functions involved in learning and memory.

To date, studies of organic lead have not addressed the less obvious effects of chronic exposure. This is a subject which should be investigated in light of the fact that we do not have good data on the long term effects of solvent abuse, such as inhalation of leaded gasoline (Seshia et al., 1978).

Information on the effects of chronic exposure to inorganic lead is more readily available, as well as more germane to the understanding of clinical findings in workers occupationally exposed to lead. Atmospheric exposure to lead has been investigated by Russian workers using classical conditioning paradigms. In one study of conditioned motor responses in rats, 1.5 to 2 months of daily exposure to a high atmospheric concentration of lead oxide (11 $\mu g/m^3$) produced disturbed reflexes; severity of impairment increased over the rest of the 6-month exposure period (Gusev, 1961). The reflexes returned to baseline after 10 to 23 days off the lead exposure regimen. Six months of exposure to a lower dose of lead (1.13 $\mu g/m^3$) did not impair the animals on the force and latency of response measures employed. In a similar study, Shalamberidze (1962) reported disturbances in conditioned reflexes in rats receiving 6 months of daily exposure to 48.3 $\mu g/m^3$ of lead in the form of lead sulfide ore dust.

In one of the few studies conducted in the US dealing with environmental lead and conditioning (Weir and Hine, 1970), goldfish exposed to a low level of lead nitrate in the water (0.07 ppm) for 24 or 48 hr were impaired on performance of a shock avoidance task. This concentration of lead is minute in comparison to the lethal dose for the animals and is close to that found in drinkable water.

Information on the effects of lead on natural behaviors is also sparse. Xintaras et al. (1967) reported that in rats chronic lead exposure can alter the rapid eye movement (REM) phase of sleep, but more information regarding lead's effect on these types of behavior is not currently available.

The findings from these few studies indicate the need for a greater commitment to research into the areas of reflexive and natural behaviors. The results of classical conditioning studies and investigations of atmospheric lead will be especially important since most workers are exposed to airborne lead and the early clinical manifestations of lead poisoning include decreased motor nerve conduction velocity (Seppäläinen and Hernberg, 1972; Seppäläinen et al., 1975) and altered neuromuscular and psychologic function (Repko et al., 1975).

Most of the behavioral investigations of lead toxicity in adult animals have employed acute or chronic injection of some form of inorganic lead; some have exposed animals to lead via food or drinking water. The utility of these studies is primarily related to the fact that they can direct the course of future research in this area.

Several investigators have examined maze learning in animals exposed to lead, some with negative results. Brown et al. (1971) reported that a single dose of 100 mg/kg of lead acetate did not affect learning and memory of rats in a water escape maze, even though the dose used was lethal to some animals. Another study found that injection of rats with 12 mg/kg of lead daily for 37 days had no effect on learning in a Hebb-Williams maze (Snowden, 1973). In contrast, exposure for 2 weeks to 10 mg/ml lead acetate in drinking water decreased weight gain in mice and also decreased speed in a swimming maze (Ogilvie, 1977). Fewer lead-treated than control mice were able to perform the task; the treated animals exhibited significantly more errors and needed more trials to reach criterion. Despite the weight differences, there were no overt signs of toxicity and a learning deficit along with a motor disability was proposed to account for the differences. The effects of the motor deficit alone should not be discounted in view of the findings by Spyker et al. (1972) that swimming behavior can be very sensitive to disruption by toxic agents. Ogilvie (1978) also employed a maze task for mice given a single injection of 100 mg/kg lead acetate after acquisition of the task. In this case, lead-treated and control mice were not different when tested for retention of the task 2 weeks later.

In agreement with the findings of Bullock et al. (1966) using organic lead, these studies demonstrate that maze learning is a relatively insensitive indicator of the CNS effects of lead. The deficits seen in some reports are more likely due to neuromuscular impairment than to effects on CNS mechanisms involved in the learning process. There is a need to design studies of behavioral tasks which adequately separate toxic effects related to information processing from those due to motor deficits. Studies employing mazes fail to meet this criterion.

Some of the recent studies of lead exposure have attempted to address this problem by using operant behavioral training techniques. These methods have been suggested as rather sensitive indicators of behavioral toxicity, even to lower levels of lead exposure than have typically been used (Evans and Weiss, 1978). Shapiro et al. (1973) trained rats to bar press on a multiple variable interval 30" extinction schedule and reported that acute injections of 16 mg/kg of lead acetate resulted in increased variability between subjects and shorter intervals between responses. In a study using pigeons, Barthalmus et al. (1977) administered 6.25 to 25 mg/kg of lead acetate daily by gastric intubation for up to 70 days. The finding of decreased responding on a multiple fixed ratio-fixed interval schedule was related to dose level. In some cases, however, the changes were associated with toxicity, particularly at the highest dose level. When lead treatment was discontinued, dramatic increases in responding occurred, particularly under the fixed interval component of the schedule.

Few other studies have evaluated operant behavior in adult animals, although some have examined the effect of chronic lead exposure initiated at weaning. Padich and Zenick (1977) were unable to detect postweaning effects of lead acetate ingestion (750 mg/kg daily via restricted watering) on a fixed ratio schedule of reinforcement. Cory-Slechta and Thompson (1979) also examined the effects of postweaning lead exposure, but used a fixed interval schedule. They found greater intersubject variability, increases in rate and shorter latencies to first response in rats exposed to 50 or 300 ppm lead in drinking water. In the low dose group, responding gradually returned to control levels after cessation of exposure. Decreased rates and longer latencies were observed with exposure to 1000 ppm lead, but some of these animals exhibited signs of toxicosis. According to these investigators, the increased variability itself may be an indicator of behavioral toxicity. This type of finding was also reported by Van Gelder et al. (1973) in a study in which ewes exposed daily to 100 mg/kg of lead for 9 weeks showed poorer performance and significantly increased between-session variability on an auditory signal detection test.

The number of useful behavioral paradigms for the study of chronic low level lead exposure is still small. The work already completed indicates that doses in the toxic range are still being employed in studies which attempt to evaluate subtle (but functionally relevant) changes. The literature suggests that of the behavioral paradigms available, interval schedules are more sensitive than ratio schedules to evaluate lead-related behavioral toxicity. Future research in this area should continue to explore these operant techniques and attempt to use lower lead levels for chronic exposure regimens. The development of models for atmospheric exposure to lead, such as those used in some of the classical conditioning experiments, should also be encouraged to approximate more closely the conditions of occupational lead exposure.

## B. Developmental Exposure to Lead

Since the most common modes of lead exposure and symptoms of lead poisoning in children differ from those of adults, the clinical data were a challenge to research methodology to devise an animal model for investigating the developmental effects of low-level lead exposure on behavior and CNS functioning. Preferably, lead should be introduced early in life and should be ingested. Animals could then be exposed to lead during the critical period of brain growth and maturation in a nonstressful manner.

Pentschew and Garro (1966) were the first to demonstrate that lead introduced into a lactating rat's diet can produce neurologic sequelae in the offspring with no obvious effects on the mother. On a 4% lead carbonate maternal diet, the offspring were retarded in growth and developed paraplegia during the fourth week of life. Exposure of this nature produces encephalopathy and gross histologic changes in brain, including cerebellar hemorrhages (Pentschew and Garro, 1966; Rosenblum and Johnson, 1968; Michaelson, 1973). Krigman et al. (1974) reported anatomical changes upon light and electron microscopic examination of brains from 30-day-old rats exposed to 4% lead carbonate, first through mother's milk and then weaned to the mother's diet. Although these studies provide a model of childhood lead poisoning, the behavioral findings of encephalopathy and other gross changes do not address the problem of subtle behavioral manifestations of early low-level lead exposure.

Since the original studies of Pentschew and Garro (1966), models for childhood lead exposure have utilized several dosing procedures. Some investigators have studied offspring when lead treatment of the dams was initiated as early as 10 weeks prior to mating and carried through gestation and postnatal life to weaning (Padich and Zenick, 1977; Zenick et al., 1978). Other studies initiated treatment 40 days before mating and continued it through adulthood of the offspring (Reiter et al., 1975). In some cases, the premating lead exposure was of both male and female parents (Brady et al., 1975; Reiter et al., 1975; Driscoll and Stenger, 1976). Most studies, however, have focused on lead treatment of the offspring at some time during the postnatal period.

Direct exposure of pups to lead was examined by Sobotka and Cook (1974), Overman (1977) and Jason and Kellogg (1977), who performed daily oral intubations of rat pups, and by Brown (1975), who provided lead to pups by daily injections. The advantage of these procedures is elimination of problems of altered quality or quantity of maternal milk, or abnormal maternal behavior due to lead exposure of the mother, but these modes of exposure are the most stressful for the neonates.

Indirect methods for treatment of offspring via the mother's milk are more common. Brown (1975) and Krehbiel et al. (1976) treated nursing rats by gavage; others have used indirect exposure through the dam's water (Silbergeld and Goldberg, 1973; 1974; 1975; Hastings et al., 1977) or food (Sauerhoff and Michaelson, 1973; Michaelson and Sauerhoff, 1974; Kostas et al., 1976 and 1978). These methods do not involve stress for the neonate and are more practical to carry out than direct dosing of the pups. However, the effects of lead on the dam (as discussed above) cannot be controlled. Most studies begin treatment of the dam at parturition and continue exposing the offspring beyond weaning. Although these procedures may maximize exposure, they do not parallel the exposure of children to lead during a limited segment of life.

There have also been problems associated with growth and development of neonates treated with lead via maternal milk. Michaelson (1973) found that animals raised on a diet where the mothers's food consisted of 5% lead carbonate exhibited retarded growth and developed ataxia and paraplegia during the latter part of the third week of life. By adjusting the mother's diet to one containing 25 ppm lead at 16 days after parturition, the ataxia and paraplegia could be avoided, but the growth lag was not eliminated (Michaelson and Sauerhoff, 1974). The degree of growth retardation in lead-treated animals correlates with lead exposure level (Silbergeld and Goldberg, 1973; Kostas et al., 1976). These problems of undernutrition and the resultant growth retardation in young rats exposed to lead were considered by Loch et al. (1978), who reported that some of the behavioral changes attributed to lead could be accounted for by undernutrition alone. A few studies have taken this variable into account by pair-feeding control animals (Sauerhoff and Michaelson, 1973; Reiter, 1977; Zenick et al., 1978).

Finally, an additional problem of indirect lead exposure studies is that the actual dosage reaching the pups is seldom known. Bornschein et al. (1977) have reported a method for estimating the lead intake of pups from mother's milk. Their theoretical estimates of lead body burden fell within 5% of the actual lead body burden of neonatal rats.

Behavioral analyses of the effects of early lead exposure in animals have tended to examine levels of locomotor activity, in response to the observations of David et al. (1972) indicating a relationship between hyperkinesis and lead exposure in children. The results of activity studies have not been entirely consistent, with reports of increases, decreases, or no change in locomotor activity occurring after neonatal lead exposure. The differences can be attributed to differences in test environment, as well as age at exposure and dosage of lead.

In several studies in which early lead exposure reportedly increased activity, growth and development were also retarded. Silbergeld and Goldberg (1973) observed significantly higher activity in lead-treated mice than in controls when the animals were tested individually in an activity meter at 40 to 60 days of age. The mice received lead via dams drinking 2,

5, or 10 mg/ml lead in water and were weaned to the lead concentrations of the dam's water. Michaelson and colleagues also noted elevated motor activity following neonatal lead exposure in rats tested for one 24-hr light-dark cycle at 23 to 25 days of age (Sauerhoff and Michaelson, 1973; Michaelson and Sauerhoff, 1974; Golter and Michaelson, 1975). The lactating dams were fed a chow consisting of 4% lead carbonate. At 17 days, the diet was changed to one containing 25 to 40 ppm lead carbonate. Reportedly, direct dosing of pups with 1.09 mg lead/day from birth to 16 days, followed by a diet of 40 ppm in food, eliminated effects on growth rate but did not elicit significant increases in motor activity in all instances. Kostas et al. (1976) also reported increased activity of adult rats exposed to lead from birth to 35 days of age when tested in a Y-maze or a tilt box, but not in a running wheel. The lactating rats were placed on diets of 5%, 0.5%, or 0.05% lead acetate in food. When the pups were 21 days of age, the diet was reduced to one containing 25, 2.5, or 0.25 ppm lead. At 35 days, the pups were placed on normal chow.

Increased activity has been reported in rats given 10, 30, or 90 mg/kg lead acetate directly (via intubation) on postnatal days 3 through 21 (Overman, 1977). No growth retardation was observed, but the animals exhibited increased motor activity when tested in jiggle cages postweaning. Deficits were also reported in motor coordination revealed by testing on a rotarod.

However, in the majority of studies in which the animals were directly exposed to lead for a limited period following parturition and in which no growth retardation was observed, no differences have been noted in locomotor activity between lead-exposed and control animals when tested postweaning. Sobotka and Cook (1974) found no differences in animals individually tested at 24 to 28 days in a photoactometer. Exposed animals had been given doses of lead acetate up to 81 mg/kg daily on postnatal days 3 through 21. Similarly, oral administration of up to 75 mg/kg lead acetate on days 2 through 14 failed to produce activity differences from controls at 35 days of age (Jason and Kellogg, in preparation). We have also analyzed blood and brain levels of lead (Jason and Kellogg, 1977) and found that, whereas PbB levels measured at 35 days had decreased by 90% following cessation of exposure, brain lead had decreased only 50%. Substantial amounts of lead were still present in the brain.

Brown (1975) also reported unchanged spontaneous motor activity and exploratory behavior in 7-week-old rats nursed by dams treated with 35 mg/kg lead acetate by gavage on postnatal days 1 to 10, 11 to 21, or 1 to 21. Intubation of dams twice daily to administer up to 1.25 g/kg lead acetate per day during lactation, followed by placing the animals on diets up to 37.5 ppm lead, did not alter the 23-hr activity levels of animals tested between 24 and 27 days of age, as compared to pair-fed controls (Krehbiel et al., 1976). In an additional study, running wheel activity, measured at 30 days, was not altered in animals nursed by dams receiving up to 0.1% lead acetate in drinking water through day 21 (Hastings et al., 1977). However, in the latter study, the control animals demonstrated more habituation than lead-treated animals.

With chronic lead treatment beginning at or before conception and continuing into adulthood of the offspring, reduction in activity has been observed. Reiter et al. (1975) exposed male and female rats to low doses of lead (5 or 50 ppm lead in drinking water) beginning 40 days prior to mating. Pregnant females were continued on these regimens throughout gestation and lactation and pups were weaned to the same lead solutions. No weight differences occurred in the offspring, but both doses of lead produced significant reductions in locomotor activity of 120-day-old males tested in a residential maze. Exposure to $10^{-2}$ M lead acetate in drinking water (approximately 4 mg/ml) beginning at the time of mating, with pregnant and nursing mothers and their offspring continued on that dose, resulted in 20% weight decreases and decreased locomotor activity in an open field in 31-day-old offspring (Driscoll and Stenger, 1976). However, behavior of rats in an open field is often considered to be more a measure of emotionality than of general motor activity. This finding may not be directly related to the results of other locomotor activity studies.

In addition to studies which measure activity on one or several occasions postweaning, some investigators have examined the developmental pattern of locomotor activity taking place preweaning. Others have attempted to manipulate locomotor activity using pharmacologic methods.

Reiter (1977) exposed neonatal rats to lead by providing a 5% lead carbonate diet to nursing mothers from parturition to postnatal day 16 and exposing offspring to 50 ppm lead in water thereafter. Jiggle cage activity was measured in animals at various ages. Acute exploratory activity of individual lead-treated subjects was significantly higher than activity of pair-fed controls at 13, 16, and 29 days, but not at 44 days. When chronic lead treatment began 40 days prior to mating, offspring exhibited delays in eye opening and development of the righting reflex, but development of the startle reflex was unaffected (Reiter et al., 1975).

In our study (Jason and Kellogg, 1977) in which rat pups were given lead acetate via intubation on days 2 to 14, we also observed only a transient difference in activity between lead-exposed and control groups. The behavior of the control animals was consistent with literature reports (Campbell et al., 1969; Melberg et al., 1976) with peak activity levels occurring at 15 days, followed by a significant decrease in activity between 15 and 18 days. Activity levels in the lead-treated animals were maximal also at 15 days, but this peak was sustained and the decrease did not occur until after

18 days. This difference in development of the locomotor pattern resulted in a significant difference between control and lead-exposed animals only at 18 days. Whether the prolonged peak in the treated animals is due to high levels of lead present in the brain following the cessation of daily exposure or whether it represents a true difference in the development of this behavior is yet to be determined. However, in the study previously discussed (Reiter, 1977), activity levels were higher in exposed animals during a comparable time period but then decreased to control values even though the animals were maintained on lead diets. The presence of lead may not be the determining factor in the lead-related changes in development of activity.

Early investigations of drug responses in an animal model of childhood lead exposure showed that, in lead-treated mice, amphetamine and methylphenidate suppressed hyperactivity and phenobarbital increased motor activity (Silbergeld and Goldberg, 1974). These drugs are reported to have the same effects on activity in some children with minimal brain dysfunction (Wender, 1971). An additional study of lead-treated mice reported that the animals responded to several cholinergic and catecholaminergic drugs differently than controls (Silbergeld and Goldberg, 1975). These pharmacologic findings, and the epidemiologic evidence of David et al. (1972), led to speculation that neonatal lead exposure in animals could be a model for the study of hyperkinesis and minimal brain dysfunction (MBD) in children. Although involvement of lead in MBD cannot be discounted, these studies must be carefully scrutinized since growth and development of treated animals was retarded. Loch et al. (1978) reported that developmental growth retardation subsequent to undernutrition alters general motor activity and responses to amphetamine in mice in a manner similar to that reported by Silbergeld and Goldberg (1974) for lead-treated animals.

However, altered locomotor responses to amphetamine have been noted in animals where lead treatment did not alter weight gain. Neonatal exposure to lead eliminates or attenuates amphetamine-induced increases in activity at doses which effectively increase activity in control animals (Sobotka and Cook, 1974; Kostas et al., 1978). We have verified these findings in our own laboratory (Jason and Kellogg, in preparation). Lead treatment also diminished amphetamine-induced activity when chronic low level lead treatment was initiated prior to mating (Reiter et al., 1975).

Considering the amount of literature related to lead-induced changes in locomotor activity, very few conclusions can be drawn from the results. Generally, neonatal lead exposure with accompanying undernutrition and growth retardation results in increased spontaneous activity, while low level lead treatment limited to the lactation period shows no effect and chronic lead treatment from preconception through adulthood seems to produce decreased motor activity in offspring. In addition, delays in the development of some reflexes and of adult-like locomotor behavior have been found and responsiveness to amphetamine is altered in lead-exposed animals, even in cases where growth retardation does not occur.

Problems related to measuring motor activity are numerous; they include lack of specificity of the behavior observed and factors related to the effects of the test environment (Evans and Weiss, 1978). In motor activity studies of lead-exposed animals, it is clear that differing test environments were employed and that the aspects of motor behavior measured in the various studies differed. Certainly, the results of these studies in developing animals demonstrate that gross measurement of activity alone is not a reliable indicator of behavioral changes that can be attributed to lead.

Several investigators have examined more complex tasks in neonatally exposed subjects. A number of behavioral paradigms, including mazes, discrimination tasks and shock-avoidance situations, have been used to evaluate the effects of lead on learning and emotional behavior.

Young adult animals (8 to 10 weeks old) exhibited learning deficits in a T-maze after either direct or indirect neonatal lead exposure (Brown, 1975). Exposure of the nursing dam to either 1.0 mg/ml lead acetate in drinking water or 25 mg/kg lead acetate per day by gavage during the suckling period resulted in deficits in T-maze performance of the offspring. To determine the influence of age on sensitivity of the neonates to lead, nursing rats were exposed to 35 mg/kg lead acetate by gavage on days 1 to 10, or 11 to 21, postpartum. The earlier treatment produced deficits in T-maze performance, but no changes were found in the performance of animals exposed on days 11 to 21. Treatment on days 11 to 21 produced deficits only with a maternal dose that induced decreases in body weight of the offspring (140 mg/kd/day). Direct treatment of pups with 5 mg/kg lead acetate by ip injection on days 1 through 10 also produced the learning deficit (Brown, 1975).

Sobotka and Cook (1974) reported that postnatal lead exposure caused a performance deficit in a two-way shuttle avoidance situation. Pretreatment with 3.0 mg/kg d,1-amphetamine improved the performance of lead-treated subjects on the task and produced a slight depression in the performance of controls. Lead-treated rats also demonstrated impairment on an operant habit reversal task (Sobotka et al., 1975). Stress responses, as measured by step-out latency in a passive avoidance situation, and learning tasks involving one-way shuttle avoidance or single bar press avoidance-escape responding, were not affected by lead (Sobotka et al., 1975).

In contrast to these behavioral findings, Driscoll and Stenger (1976) found that continuous postnatal exposure to lead, leading to weight reductions, produced a significantly greater number of errors during acquisition of a visual discrimination problem, no differences in reversal performance and superior, rather than inferior, performance by lead-treated animals in two-way avoidance responding. These results, together with the finding of decreased exploratory behavior in an open field, were interpreted as indicating that relatively high level neonatal lead exposure may increase responsiveness to aversive stimuli.

Overmann (1977), using low level direct exposure to lead during the suckling period, found that acquisition and extinction of passive avoidance were not altered by lead treatment. However, both acquisition and extinction of active avoidance were affected; this is consistent with the findings of Sobotka and Cook (1974) and Driscoll and Stenger (1976). Again, a higher level of fear and emotionality among lead-treated subjects was postulated (Overmann, 1977). The lead treatment in this study had no effect on simple learning of an E-maze, visual discrimination in the maze, or visual acuity, but impairment was noted in the exposed animals on reversal learning and they were unable to inhibit responding in an operant situation.

Hastings et al. (1977) also suggested that neonatal exposure to a low level of lead may affect emotional behavior. However, less emotionality in lead-treated subjects was hypothesized, since pairs of treated animals in this study displayed less aggressive behavior in shock-elicited aggression tests than controls. In this case, neither acquisition nor reversal of a brightness discrimination task was altered by lead treatment.

The influence of pre- and neonatal exposure on discrimination learning and operant response patterns has been investigated by Zenick and colleagues (Padich and Zenick, 1977; Zenich et al., 1978 and 1979). Lead exposure of dams to 750 mg/kg lead acetate per day via restricted watering began at 21 days of age and continued throughout breeding, gestation and the suckling period until the pups were weaned. These investigators reported that the offspring exhibited significantly more errors on both brightness and shape discrimination tasks in a water escape T-maze (Zenick et al., 1978). Swimming times were shorter in lead-exposed animals and the authors suggest that the deficits observed could be attributed to the failure of the exposed animals to attend to relevant discriminative cues.

Padich and Zenick (1977) examined fixed ratio responding; Zenick et al. (1979) considered fixed interval responding in rats exposed to lead through weaning as described above. At weaning, half of the treated and control pups were placed on the lead concentrations in water used for dams and the other half were provided with distilled water. On both tasks, effects of lead were demonstrated only in the group receiving lead throughout the study. That group received fewer reinforcements across sessions and took longer to emit 20 responses on the fixed ratio schedule (Padich and Zenick, 1977). They also received fewer reinforcements across sessions on the fixed interval schedule (Zenick et al., 1979).

In addition to changes in performance and emotional behavior and some aspects of learning, there is evidence that neonatal lead exposure can affect nonlearned behaviors such as spontaneous alternation in rats (Kostas et al., 1976; 1978). Neonatal lead treatment until 35 days of age resulted in a significantly lower percentage of alternation in 70-day-old rats (Kostas et al., 1976). Pretreatment with 1.5 or 3.0 mg/kg of d-amphetamine reduced the percentage of alternation in control animals, but in lead-treated subjects both doses increased percentage alternation to the level of control animals receiving saline (Kostas et al., 1978).

As discussed previously, severe behavioral toxicity usually occurs when neonatal lead exposure is sufficient to cause weight reduction. Many studies, however, have reported lead-related behavioral changes when no overt toxicity was produced. Some indications of behavioral neurotoxicity were found when exposure occurred during a restricted period (Brown, 1975), whereas in other instances changes occurred only when the lead exposure extended over two generations (Padich and Zenick, 1977; Zenick et al., 1978 and 1979). Although none of the behavioral paradigms used to examine neonatal lead toxicity have resulted in consistent findings, a few generalizations can be made.

Maze learning in lead-treated offspring is impaired in some instances, yet Overmann (1977) was unable to find any effects of lead on learning when a small amount of lead was administered directly to neonates. The effect of lead on discrimination is also unclear. In rats, short term low dose levels of lead produced no effect on visual discrimination and visual acuity (Overmann, 1977) or brightness discrimination (Hastings et al., 1977), whereas higher level exposure (Driscoll and Stenger, 1976) or long term exposure (Zenick et al., 1978) resulted in deficits on similar tasks. A study using lambs (Carson et al., 1973) demonstrated that prenatal exposure resulting in a maternal PbB level of 34 $\mu$g/100 ml produced impairment on a visual discrimination task. Deficits in reversal of discrimination have been found using tactile discrimination in a maze (Overmann, 1977) and in an operant situation (Sobotka et al., 1975) when similar dosing schedules were used. However, reversal of visual discrimination in a Y-maze was unaffected by lead (Driscoll and Stenger, 1976). Again, the results of the latter study may have been obscured by the effects of undernutrition. The available data indicate that neither maze learning nor most simple discrimination tasks can be considered reliable measures of the behavioral effects of neonatal lead exposure and that reversal tasks are more sensitive indicators of behavioral changes resulting from lead treatment.

The evidence obtained using shock avoidance paradigms suggests that neonatal lead treatment has some effect on emotionality which can be measured in active avoidance situations. Sobotka and Cook (1974) and Overmann (1977) have found that lead treatment impairs two-way shuttle avoidance responding but has no effect on passive avoidance. Animals exposed to higher doses (Driscoll and Stenger, 1976) performed better on that task than did controls. Together with the observation of less shock-induced aggression in lead-exposed subjects (Hastings et al., 1977), these results suggest that more specific effects of lead may be found using aversive stimuli in active tasks than with the use of positive rewards.

Operant testing situations have revealed that lead-exposed rats are deficient on tasks requiring response inhibition (Overmann, 1977) and on both fixed ratio and fixed interval schedules of reinforcement (Padich and Zenick, 1977; Zenick et al., 1979). These findings suggest that studies of operant response patterns, particularly with schedules involving waiting periods, may be useful in defining effects of neonatal lead treatment on behavior.

The altered responsiveness to amphetamine after neonatal lead exposure that was observed in activity studies is also evident in studies of more complex behaviors (Sobotka and Cook, 1974; Kostas et al., 1978). Long term behavioral effects of early exposure to drugs or toxins may be manifest most readily when the animal is subjected to environmental challenges.

### C. Evaluation of the Animal Models

In terms of behavioral neurotoxicity, an appropriate animal model for occupational exposure to lead has not yet been defined. Animal studies have generally exposed subjects by injection or by adding lead to the diet, but chronic environmental exposure of animals to lead, the most common mode of occupational exposure, is rare. Only a few classical conditioning experiments have dealt with this problem, providing a start in the investigation of altered reflexes and of motor deficits related to chronic lead exposure (Gusev, 1961; Shalamberidze, 1962). Reports of increased variability of response patterns in lead-exposed animals (Shapiro et al., 1973; Cory-Slechta and Thompson, 1979) correlate with the findings in humans of increased variability in coordination tests (Repko et al., 1975). The observations of Xintaras et al. (1967) that chronic lead exposure in rats alters REM-phase patterning may be related to the fact that insomnia is one sign of lead poisoning in humans. However, no good correlative work in the area of sensory deficits, fine motor impairment, or altered emotional (psychological) responsiveness have been performed using adult animals, even though occupational lead exposure produces behavioral changes in those categories (Repko et al., 1975; Valcuikas et al., 1978a and 1978b).

An animal model for childhood lead exposure has been more fully explored. In most cases, those studies employing dosing procedures that produce no overt toxicity and do not interfere with growth of the animal are the best available models of "subclinical" childhood lead poisoning. Such studies have established that low level lead exposure can produce behavioral effects in the young even when PbB levels are low or have returned to normal following cessation of exposure (Carson et al., 1973; Brown, 1975; Sobotka and Cook, 1974; Jason and Kellogg, in preparation). Few animal studies have limited exposure of animals to lead to a period equivalent to that associated with the greatest period of lead ingestion in children. The neural maturation occurring during the second 2 postnatal weeks in the rat correlates best with human CNS development during the first 3 years of life. Behavioral studies should consider species differences in neural development.

There has been little behavioral investigation of attentional deficits, but impaired performance of neonatally exposed animals on habit reversal (Sobotka et al., 1975), reversal in discrimination tasks (Sobotka et al., 1975; Zenick et al., 1978) and deficits in response inhibition (Overmann, 1977) indicates that developmental exposure to lead may alter attentional behavior. Problems of attention and the inability to inhibit responses are some of the behavioral deficits attributed to children with MBD (Wender, 1971). The animal models also report a number of instances of altered emotional responsiveness which may be related to psychological problems in children with MBD or to lead-intoxicated children such as those observed by de la Burde and Choate (1975). Although animal studies of neonatal lead exposure do not totally support or refute the hypothesis of lead's involvement in MBD or hyperkinesis (David et al., 1972, 1976; Silbergeld and Goldberg, 1973 and 1974), evidence that low level lead exposure during development can affect behavior on a variety of measures suggests that the problem of childhood lead poisoning requires more study.

## IV. SUMMARY AND CONCLUSIONS

Since human studies of occupational lead exposure are still in the early stages, there is an obvious need for more widespread behavioral analysis of complex tasks, such as those used by Repko et al. (1975) and Valcuikas et al. (1978b). The studies so far completed reveal that biologic indicators other than blood levels are more reliable predictors of the behavioral manifestations of lead toxicity. Attention should be directed to testing for changes in reflexive behavior using classical conditioning paradigms; procedures for the evaluation of sensory deficits must be designed to detect subtle effects such as changes in auditory or visual thresholds or acuity. Further investigation of psychological effects in lead-exposed workers is important. Long term behavioral evaluations must be carried out, since toxic effects may only be measurable several years after exposure ceases. It is likely that the toxic effects of lead are potentiated in the presence of other agents (Novakova, 1969); therefore, behavioral evaluation of workers simultaneously exposed to lead and other toxic substances is needed to determine standards for occupational exposure limits.

The most important direction for animal models of occupational lead exposure in the near future will be the development of a model of chronic environmental exposure to lead which can approximate the human situation. Once this is accomplished, many of the behavioral paradigms already available can be used to examine behavioral toxicity. Since mazes and simple learning tasks generally do not detect lead-related deficits even when overt poisoning occurs, future studies should include more complex learning problems, such as reversal tasks, and should evaluate patterns of responding on operant schedules. Classical and operant conditioning studies for examining sensory functioning, especially of the auditory system, should be stressed in light of the findings by Repko et al. (1975). Evaluation of performance, endurance, and muscular strength should also be encouraged in an attempt to correlate the animal model with human studies that report muscular weakness as an early sign of lead toxicity related to occupational exposure (Repko et al., 1975; Seppäläinen et al., 1975; Valcuikas et al., 1978a and 1978b). Increases in response variability have been reported in lead-exposed workers (Repko et al., 1975) and animals (Shapiro et al., 1973; Cory-Slechta and Thompson, 1979), and further examination of this phenomenon may aid in the detection of subtle effects of lead.

Epidemiologic studies in children, along with the animal studies to date, indicate that PbB levels are not a good index of early lead exposure. Lead in hair or deciduous teeth or postchelation urinary lead may be better indicators of a child's body burden of lead. Children with suspected excessive lead exposure should also be evaluated for urine ALA and ZPP levels.

Childhood hyperactivity, which has been correlated with lead exposure (David et al., 1972; 1976) is a very nonspecific term. Increased understanding of the behavioral effects of early lead exposure in children can be obtained only when more precise assessments of behavior are done. Barocas and Weiss (1974) suggest that techniques derived from applied behavioral analysis such as quantification of specifically defined behaviors may prove to be reliable indicators of the behavioral manifestations of childhood lead exposure. These techniques may also be applicable to the measurement of psychological or emotional disturbances in children.

The animal data indicate that early evaluation of lead-related changes in sensory systems in children should be carried out. In addition, reports of maturational delays in the appearance of some behaviors in animals suggest that future epidemiologic studies in children should employ longitudinal designs to detect any changes in the development of behavior patterns associated with childhood lead exposure.

Animal studies of neonatal exposure to lead should focus on refinement of the devices used to detect subtle effects. The available data indicate that evaluation of complex learning and discrimination, attention to relevant environmental cues and the ability to inhibit responding will add to our understanding of developmental neurotoxicity of lead. The examination of response rates and latencies and of patterns of operant schedules is an area still virtually unexplored with this model. Quantification of responses in aversive or stressful situations and behavioral responses to pharmacologic manipulations, such as those employed by Sobotka and Cook (1974) or Kostas et al. (1978), should be carried out; these may be the types of challenges that produce measurable behavioral effects in neonatally lead-exposed animals.

To date, few studies of neonatal animals have concentrated on the effects of lead on sensory functioning outside of discrimination paradigms; no information is available on the effects of lead on the auditory system. Data are needed on the incidence of altered visual and auditory acuity and threshold detection levels in neonatally lead-exposed animals. The studies demonstrating altered maturation of behavior patterns after developmental exposure to lead should also be expanded; it may be necessary to use higher order mammals than rodents to investigate these behaviors.

The relative youth of behavioral toxicity as a discipline is evidenced by the fact that nearly all of the studies of behavioral neurotoxicity of lead have been performed during the last 10 years. The goal of behavioral research in terms of human health is to aid in the establishment of safe exposure limits for toxic substances. Behavioral changes can be used as early sensitive indicators of altered CNS functioning. Behavioral investigations of lead exposure have established that, with increasingly sensitive test devices, changes occur at exposure levels well below those that have been associated with overt toxicity in adults and in the young. Although more research is needed to fully describe the effects of lead on behavior, all of the results to date emphasize the need for downward revision of safe exposure levels and vigorous behavioral evaluation of children and adults suspected of excessive exposure to lead.

## REFERENCES


Albert, R.E., Shore, R.E., Sayers, A.J., Strehlow, C., Kneip, T.J., Pasternak, B.S., Friedhoff, A.J., Coran, F., and Cimino, J.A. (1974). *Environ. Hlth. Persp. 7:*33.

Avery, D.D., Cross, H.A., and Schroeder, T. (1974). *Pharmac. Bioch. Behav. 2:*473.

Baloh, R., Sturm, R., Green, B., and Gleser, G. (1975). *Arch. Neurol. 32:*326.

Barocas, R., and Weiss, B. (1974). *Environ. Hlth. Persp. 7:*47.

Barthalmus, G.T., Leander, J.D., McMillan, D.E., Mushak, P., and Krigman, M.R. (1977). *Toxicol. Appl. Pharmacol. 42:*271.

Beattie, A.D., Moore, M.R., Goldberg, A., Finlayson, M.J.W., Graham, J.F., Mackie, E.M., Main, J.C., McLaren, D.A., Murdoch, R.M., and Stewart, G.T. (1975). *Lancet 1:*589.

Bornshein, R.L., Fox, D.A., and Michaelson, I.A. (1977). *Toxicol. Appl. Pharmacol. 40:*577.

Brady, K., Herrera, Y., and Zenick, H. (1975). *Pharmac. Bioch. Behav. 3:*561.
Brown, D.R. (1975). *Toxicol. Appl. Pharmacol. 32:*628.
Brown, S., Dragann, N., and Vogel, W.H. (1971). *Arch. Environ. Hlth. 22:*370.
Bullock, J.D., Wey, R.J., Zaia, J.A., Zarembok, I., and Schroeder, H.A. (1966). *Arch. Environ. Hlth. 13:*21.
Byers, R.K., and Lord, E.E. (1943). *Am. J. Dis. Child. 66:*471.
Campbell, A.M.G., Williams, E.R., and Barltrop, D. (1970). *J. Neurol. Neurosurg. Psychiat. 33:*877.
Campbell, B.A., Lytle, L.D., and Fibiger, H.C. (1969). *Science 166:*637.
Carson, T.L., Van Gelder, G.A., Karas, G.C., and Buck, W.B. (1974). *Arch. Environ. Hlth. 29:*154.
Cohen, D.J., Johnson, W.T., and Caparulo, B.K. (1976). *Am. J. Dis. Child. 130:*47.
Cory-Slechta, D.A. and Thompson, T. (1979). *Toxicol. Appl. Pharmacol. 47:*151.
Damstra, T. (1977). *Environ. Hlth. Persp. 19:*297.
David, O.J., Clark, J., and Voeller, K. (1972). *Lancet 2:*900.
David, O.J., Hoffman, S.P., Sverd, J., Clark, J., and Voeller, K. (1976). *Am. J. Psychiat. 133:*1155.
de la Burde, B., and Choate, M.S. (1975). *J. Peds. 87:*638.
Driscoll, J.W., and Stenger, S.E. (1976). *Pharmac. Bioch. Behav. 4:*411.
Evans, H.L., and Weiss, B. (1978). In "Contemporary Research in Behavioral Pharmacology," D.E. Blackman and D.J. Singer (Eds.), p. 499. Plenum Press, New York.
Feldman, R.G., Hayes, M.K., Younes, R., and Aldrich, F.D. (1977). *Arch. Neurol. 34:*481.
Golter, M., and Michaelson, I.A. (1975). *Science 187:*359.
Gusev, M.I. (1961). V. A. Ryazanov (Ed.), In "Limits of Allowable Concentration of Atmospheric Pollutants," Book 4, p. 5. Translated from the Russian, U.S. Dept. of Commerce, Off. Tech. Serv., Washington, D.C.
Hastings, L., Cooper, G.P., Bornschein, R.L., and Michaelson, I.A. (1977). *Pharmac. Bioch. Behav. 7:*37.
Jacobziner, H. (1966). *Clin. Peds. 5:*277.
Jason, K.M., and Kellogg, C.K. (1977). *Fed. Proc. 36:*1008.
Kostas, J., McFarland, D.J., and Drew, W.G. (1976). *Pharmacol. 14:*435.
Kostas, J., McFarland, D.J., and Drew, W.G. (1978). *Pharmacol. 16:*226.
Kotok, D., Kotok, R., and Heriot, J.T. (1977). *Am. J. Dis. Child. 131:*791.
Krehbiel, D., Davis, G.A., LeRoy, L.M., and Bowman, R.E. (1976). *Environ. Hlth. Persp. 18:*147.
Krigman, M., Kruse, M.J., Trayton, T.D., Wilson, M.H., Newell, H.R., and Hogan, E.L. (1974). *J. Neuropath. Exp. Neurol. 33:*671.
Landrigan, P.J., Gehlbach, S.H., Rosenblum, B.F., Shoults, J.M., Candeleria, R.M., Barthel, W.F., Liddle, J.A., Smrek, A.L., Staehling, N.W., and Sanders, J.F. (1975a). *N. Eng. J. Med. 292:*123.
Landrigan, P.J., Whitworth, R.H., Baloh, R.W., Staehling, N.W., Barthel, W.F., and Rosenblum, B.F. (1975b). *Lancet 2:*708.
Landsdown, R.G., Clayton, B.E., Graham, P.J., Shepherd, J., Delves, H.T., and Turner, W.C. (1974). *Lancet 1:*538.
Lilis, R., Blumberg, W.E., Eisinger, J., Fishbein, A., Diamond, S., Anderson, H.A., and Selikoff, I.J. (1977a). *Arch. Environ. Hlth. 32:*256.
Lilis, R., Fishbein, A., Eisinger, J., Blumberg, W.E., Diamond, S., Anderson, H.A., Rom, W., Rice, C., Sarkosi, L., Kon, S., and Selikoff, I.J. (1977b). *Environ. Res. 14:*285.

Lin-Fu, J.S. (1972). *N. Engl. J. Med. 286:*702.
Loch, R.K., Rafales, L.S., Michaelson, I.A., and Bornschein, R.L. (1978). *Life Sci. 22:*1963.
Melberg, P.E., Alhenius, S., Engel, J., and Lundborg, P. (1976). *Psychopharmacol. 49:*119.
Michaelson, I.A. (1973). *Toxicol. Appl. Pharmacol. 26:*539.
National Research Council. Comm. on Biologic Effects of Atmospheric Pollutants. (1972). In "Lead: Airborne Lead in Perspective," 330 pp, p. 158. National Academy of Sciences, Washington, D.C.
Novakova, S. (1969). *Hyg. Sanit. 34:*96.
Ogilvie, D.M. (1977). *Can. J. Zool. 55:*771.
Ogilvie, D.M. (1978). *Bull. Environ. Contam. Toxicol. 19:*143.
Overmann, S.R. (1977). *Toxicol. Appl. Pharmacol. 41:*459.
Padich, R., and Zenick, H. (1977). *Pharmac. Bioch. Behav. 6:*371.
Pentschew, A., and Garro, F. (1966). *Arch. Neuropathol. 6:*266.
Perino, J., and Ernhart, C.F. (1974). *J. Learn. Disab. 7:*616.
Perlstein, M.A., and Attala, R. (1966). *Clin. Peds. 5:*292.
Pihl, R.O., and Parkes, M. (1977). *Science 198:*204.
Pueschel, S.M., Kopito, L., and Schwachman, H. (1972). *J. Am. Med. Assoc. 222:*462.
Ratcliffe, J.M. (1977). *Br. J. Prev. Soc. Med. 31:*258.
Reiter, L.W. (1977). *J. Occup. Med. 19:*201.
Reiter, L.W., Anderson, G.E., Laskey, J.W., and Cahill, D.F. (1975). *Environ. Hlth. Persp. 12:*119.
Repko, J.D., Morgan, B.B., and Nicholson, J. (1975). "Behavioral Effects of Occupational Exposure to Lead," 239 pp., p. 1. HEW Publication (NIOSH) 75-164. Washington, D.C.
Roberts, T.M., Hutchinson, T.C., Paciga, J., Chattopadhyay, A., Jervis, R.E., Van Loon, J., and Parkinson, D.K. (1974). *Science 186:*1120.
Rosenblum, W.I., and Johnson, M.G. (1968). *Arch. Pathol. 85:*640.
Sauerhoff, M.W., and Michaelson, I.A. (1973). *Science 182:*1022.
Seppäläinen, A.M., and Hernberg, S. (1972). *Br. J. Ind. Med. 29:*443.
Seppäläinen, A.M., Tola, S., Hernberg, S., and Kock, B. (1975). *Arch. Environ. Hlth. 30:*180.
Seshia, S.S., Rajani, K.R., Boeckx, R.L., and Chow, P.N. (1978). *Devel. Med. Child. Neurol. 20:*323.
Shalamberidze, O.P. (1962). V.A. Ryazanov (Ed.), In "Limits of Allowable Concentration of Atmospheric Pollutants," Book 5, p. 29. Translated from Russian, U.S. Dept. of Commerce, Off. Tech. Serv., Washington, D.C.
Shapiro, M.M., Tritschler, J.M., and Ulm, R.A. (1973). *Bull. Psychon. Soc. 2:*94.
Silbergeld, E.K., and Goldberg, A.M. (1973). *Life Sci. 13:*1275.
Silbergeld, E.K., and Goldberg, A.M. (1974). *Exp. Neurol. 42:*146.
Silbergeld, E.K., and Goldberg, A.M. (1975). *Neuropharmacol. 14:*431.
Snowden, C.T. (1973). *Pharmac. Bioch. Behav. 1:*599.
Sobotka, T.J., Brodie, R.E., and Cook, M.P. (1975). *Toxicol. 5:*175.
Sobotka, T.J., and Cook, M.P. (1974). *Am. J. Ment. Def. 79:*5.
Spyker, J.M., Sparber, S.B., and Goldberg, A.M. (1972). *Science 177:*621.
Valcuikas, J.A., Lilis, R., Eisinger, J., Blumberg, W.E., Fishbein, A., and Selikoff, I.J. (1978a). *Int. Arch. Occup. Environ. Hlth. 41:*217.
Valcuikas, J.A., Lilis, R., Fishbein, A., and Selikoff, I.J. (1978b). *Science 201:*465.

Van Gelder, G.A., Carson, T., Smith, R.M., and Buck, W.B. (1973). *Clin. Toxicol.* *6:*405.
Waldron, H.A., and Stofen, D. (1974). "Subclinical Lead Poisoning," 224 pp., p. 96. Academic Press, London.
Weir, P.A., and Hine, C.H. (1970). *Arch. Environ. Hlth.* *20:*45.
Weiss, B. (1978). *Fed. Proc.* *37:*22.
Wender, P. (1971). "Minimal Brain Dysfunction in Children," 242 pp, p. 87. Wiley Interscience, New York.
Xintaras, C., Sobecki, M.F., and Ulrich, C.E. (1967). *Toxicol. Appl. Pharmacol.* *10:*384.
Zenick, H., Padich, R., Tokarek, T., and Aragon, P. (1978). *Pharmac. Bioch. Behav.* *8:*347.
Zenick, H., Rodriguez, W., Ward, J., and Elkington, B. (1979). *Devel. Psychobiol.* *12:*509.

# Neurochemical Correlates of Lead Toxicity

*Pavel D. Hrdina, Israel Hanin and Thea C. Dubas*

**TABLE OF CONTENTS**



## I. INTRODUCTION

The deleterious effect of lead toxicity on the physical condition of exposed individuals has been well documented. Apart from this evidence, there is also a potential for significant neurologic and behavioral sequalae of subclinical exposure to lead in animals, including man, through its action on central neurotransmitter systems. As will be documented in this chapter, lead exposure has a profound effect on certain neurotransmitter mechanisms in mammals. When one combines this information with the increasing evidence supporting a definitive role for central neurotransmitter function in the etiology of a variety of neurologic and psychiatric disease states (see review by Hanin, 1978), it is not difficult to speculate that lead exposure may, in some heretofore unexplored manner, also be responsible for certain movement and behavior-related disorders.

Although clinical data are relatively sparse, a significant amount of information has been obtained in experimental animals with respect to the effect of lead exposure on central as well as peripheral neurotransmitter function.

In this chapter we have attempted to provide an overview of the effects that lead could induce on a variety of neurochemical correlates. Whenever possible, we have made reference to comprehensive reviews which will provide more extensive documentation and discussion of the particular subject to the interested reader.

## II. EFFECTS OF LEAD ON BRAIN DEVELOPMENT AND CHEMICAL COMPONENTS OF NERVOUS TISSUE

### A. Neuronal Growth, Protein and DNA and RNA Content

Lead exerts a detrimental effect upon the morphologic and biochemical features of cerebral ontogenesis. Cerebral growth is impaired in both rats and mice exposed to lead postnatally (Maker et al., 1973; Michaelson, 1973; Krigman and Hogan, 1974). The mass of both cerebral gray and white matter is reduced due to the retardation in neuronal growth and maturation, leading to hypomyelination and a reduction in the number of synapses per neuron. Synaptogenesis, however, is neither delayed nor perturbed, but is reduced by the limited development of neuronal dendritic fields; similarly, the hypomyelination is primarily related to retarded neuronal growth and maturation and not to any defect in the myelinating glia or delay in the initiation of myelination (Krigman and Hogan, 1974). Whole brain deoxyribonucleic acid (DNA) content (used as a rough indicator of the number of brain cells) does not appear to be altered (Maker et al., 1973; Krigman and Hogan, 1974), although brain weights were reduced in lead-treated animals. However, postnatal lead exposure in rats has been found to produce a significant reduction (10 to 20%) in the DNA content of the cerebellum; on the other hand, no change was noted in the concentration (mg/gm) of DNA, ribonucleic acid (RNA) or protein in the cerebellum and cerebral cortex. There was also no change in the content (mg/tissue) of DNA, RNA and protein in the cerebral cortex, as well as in the cerebellar content of RNA and protein (Michaelson, 1973). These observations suggest that lead may specifically inhibit new cell formation in this particular region of the brain.

### B. Myelin Formation

As a consequence of lead-induced hypomyelination secondary to retarded neuronal growth and maturation, the brain content of neuronal- and myelin-associated lipids and fatty acids (phospholipids, galactolipids, plasmalogens, cholesterol, gangliosides, cerebrosides, sulfatides and ceramides) is significantly reduced (Maker et al., 1973; Krigman and Hogan, 1974). Although reduced in content, ultrastructural and detailed molecular subspecies studies indicate that myelin formed in the face of lead intoxication is indistinguishable from that of the normal rat (Maker et al., 1973; Krigman and Hogan, 1974). Composition of the phospholipids, galactolipids and gangliosides showed no appreciable difference between lead-treated and control groups, although the monosialo-ganglioside $Gm_1$ appeared to be more prominent in the experimental animal (Krigman and Hogan, 1974). It is the neuron and perhaps the myelinating glial cells which appear to be most compromised by exposure to this heavy metal.

### C. Chemical Components of Neuronal Metabolism

Retardation in the biochemical maturation of the brain in terms of development of glucose metabolism (reduction in the conversion of glucose carbon into amino acids associated with the tricarboxylic acid cycle, *i.e.*, decreased tissue glucose utilization) and metabolic compartmentation (different labeling pattern of glutamine versus glutamate when glucose and acetate are the labeled precursors) have also been reported to occur in various brain areas (cortex, cerebellum, whole brain) of rats exposed to lead from birth (Patel et al., 1974a and 1974b). Histochemically, reversible changes appear in the brain phosphatases of lead-treated rats; these consist of a slight reduction in the activity of alkaline phosphatase in the vessels and an increase in the activity and changed distribution of acid phosphatase in the neurons (Brun and Brunk, 1967). The increased activity of the latter enzyme may be compatible with reversible degenerative processes in the neuron; inhibition of alkaline phosphatase, which is assumed to partake in glucose resorption in the brain, may be an important link in the chain of events leading to interference with nerve function, since the brain is almost entirely dependent upon glucose metabolism as its prime source of energy.

Lead also interferes with the $K^+$-stimulated respiration of rat cerebral cortex slices in the presence of glucose or lactate, but not when pyruvate serves as substrate (Bull et al., 1975). Inhibition of cytoplasmic NAD(P)H oxidation by brain mitochondria provides additional evidence that the neurotoxic effects of lead may be mediated directly at the neuronal level. This is further substantiated by reports indicating the lead is sequestered in synaptosomal mitochondria where it may influence energy and $Ca^{++}$ metabolism and neurotransmitter release (Silbergeld et al., 1977). The interference of lead with essential trace metals (Cu, Fe, Zn) which are part of the prosthetic group of enzymes in mitochondria can affect vital metabolic pathways (respiratory chain, oxidative phosphorylation, ATP synthetase complex), with resultant loss of structural integrity of nerve tissue and functional disorders, as recognized after exposure of animals and humans to lead (Niklowitz, 1974). Lead-induced Cu deficiency, for instance, inhibits the synthesis of phospholipids by affecting mitochondrial cytochrome oxidase. It is clear, therefore, that lead produces primary abnormalities in oxidative energy metabolism in the immature brain. In fact, the initial stimulation and then progressive inhibition of NAD-linked respiration and cytochrome oxidase activity appears earlier than any other reported biochemical or morphologic effect of lead (Holtzman and Shen Hsu, 1976). This effect may be part of or, more likely, results in the more general effect on glucose metabolism in the developing brain described above. Some believe that the lead-induced delay in neurochemical development of the nervous system may be the basis for observed neurophysiologic (Fox et al., 1977) and behavioral (Maker et al., 1973; Reiter et al., 1975) alterations.

## D. Enzyme Systems

Lead is toxic to many enzyme systems (Vallee and Ulmer, 1972), but only a few of these are sensitive to the relatively low concentrations of lead normally encountered during environmental exposure. Such enzyme systems include $\delta$-aminolevulinic acid dehydrase, $Na^{+}$-$K^{+}$-$Mg^{++}$-ATPase and lipoamine dehydrogenase. Inhibition of the first enzyme leads to the accumulation of $\delta$-aminolevulinic acid, which may play a role in the pathogenesis of neurologic complications. $Na^{+}$-$K^{+}$-$Mg^{++}$-activated ATPase is required for cation transport; lipoamine dehydrogenase is part of the pyruvate-decarboxylase system, defects in which may lead to profound central nervous system (CNS) dysfunction (*e.g.*, in Wernicke's syndrome). To date, however, no lead-induced abnormalities in the functioning of these enzymes have yet been reported (Goldstein and Diamond, 1974). There is one enzyme system which may prove to be a crucial factor in understanding the biochemical mechanism underlying lead neurotoxicity. The metal has been reported to be a potent inhibitor of brain adenylate cyclase and it has been suggested that interference with cAMP metabolism and its associated synaptic mechanisms may play an important role in some of the neurochemical (specifically catecholaminergic) and behavioral manifestations of lead toxicity (Nathanson and Bloom, 1975, 1976). The effects of lead on the levels and turnover rates of the various putative neurotransmitters and on their synthesizing and catabolising enzymes will be discussed in the following sections.

## III. EFFECTS OF LEAD ON MONOAMINERGIC SYSTEMS

### A. Norepinephrine, Dopamine and Serotonin

Since the first report (Sauerhoff and Michaelson, 1973) of possible involvement of brain catecholamines in lead-induced hyperactivity, several researchers have investigated the neurochemical changes in brains of animals treated chronically with this toxic metal. The results of these reports are summarized in Tables 1 and 2. In all but one (Hrdina et al., 1976) of these studies, the animals were exposed to lead during early stages of their postnatal development. No changes in brain norepinephrine (NE) levels were reported in four of these studies; increases in whole brain or discrete brain regions ranging from 13 to 60% were reported in five of these studies. In three of these reports, increases in whole brain (Golter and Michaelson, 1975), forebrain (Silbergeld and Goldberg, 1975) or midbrain (Dubas and Hrdina, 1978) NE levels were found to be associated with enhanced spontaneous locomotor activity of lead-treated animals. In contrast, Jason and Kellogg (1977) noted a decline in spontaneous activity of 15-days-old rats which had increased levels of brainstem NE.

Data on depamine (DA) levels in brain tissue of lead-exposed rodents are somewhat less controversial, as shown in Table 2. Most workers were unable to show any alterations of DA levels in whole brain or discrete regions of animals exposed to lead. However, three laboratories reported lowered DA concentration in brains of lead-treated rats which also showed altered spontaneous locomotor activity (Sauerhoff and Michaelson, 1973; Jason and Kellogg, 1977; Dubas and Hrdina, 1978).

Steady-state levels of 5-hydroxytryptamine (5-HT) were found to be either unaltered (Sobotka et al., 1975; Hrdina et al., 1976) or decreased (Dubas and Hrdina, 1978) by lead treatment in rat whole brain, as well as in rat brain regions.

As is apparent from Tables 1 and 2, the findings on the effect of inorganic lead on the levels of various brain biogenic amines in experimental animals are by no means concordant. Moreover, attempts to correlate the observed lead-induced hyperactivity in small rodents with changes in the content of some putative central neurotransmitters in the brain have failed to yield consistent results. The difficulties in comparing or interpreting the results of studies summarized in Tables 1 and 2 are due to several factors which may be, at least in part, responsible for apparent discrepancies. These

**Table 1.** Effect of chronic lead exposure on brain norepinephrine levels

| Species and age (reference) | Dose (level), route and time of exposure | Brain region | Result | Locomotor activity |
|---|---|---|---|---|
| Rats, 29 days old (Sauerhoff & Michaelson, (1973) | 40 ppm in milk and/or water on days 1–28 | Whole brain | →* | ↑ |
| Rats, 33 days old (Golter & Michaelson, 1975) | 40 ppm in diet | Whole brain | 13% ↑ | ↑ |
| Rats, 30–32 days old (Grant et al., 1976) | 25, 100 or 200 μg Pb/g/day on days 3–25 | Cortex, striatum, hypothalamus, brainstem | → | → |
| Rats, 21–22 days old (Sobotka et al., 1975) | 9, 27 or 81 mg Pb acetate/kg/day orally on days 1–21 | Cortex, brainstem cerebellum | → | → |
| Rats, 15 days old (Jason & Kellogg, 1977) | 25 or 75 mg Pb acetate/kg/day orally on days 2–14 | Brainstem | ↑ | ↓ |
| Rats, 56 days old (Dubas et al., 1978) | 2% Pb acetate in drinking water until weaning, then 20, 40 or 80 ppm | Midbrain,<br>striatum,<br>hypothalamus | 60% ↑<br>20–30% ↓<br>24% ↓ or → | — |
| Rats, 56 days old (Dubas & Hrdina, 1978) | 50 μg Pb/pup/day orally on days 1–21, then 80 ppm in drinking water | Cortex,<br>midbrain<br>striatum,<br>hypothalamus | 25% ↓<br>16% ↑<br>27% ↓<br>→ | ↑ |
| Rats, adult (Hrdina et al., 1976) | 1 mg/kg/day for 45 days | Brainstem | 27% ↓ | — |
| Mice, 44 days old (Schumann et al., 1977) | 2 or 10 mg Pb/ml of drinking water to mother or offspring since birth | Whole brain | → | — |
| Mice, 40–90 days old (Silbergeld & Goldberg, 1975) | 5 mg Pb acetate/ml of drinking water to mother and offspring | Forebrain | 32% ↑ | ↑ |

* Symbols: ↑, increase; ↓, decrease; →, no change; —, not given

**Table 2.** Effect of chronic lead exposure on brain dopamine levels

| Species and age (reference) | Dose (level), Route and time of exposure | Brain region | Result | Locomotor activity |
|---|---|---|---|---|
| Rats, 33 days old (Golter & Michaelson, 1975) | 40 ppm in diet | Whole brain | →* | ↑ |
| Rats, 29 days old (Sauerhoff & Michaelson, 1973) | 40 ppm in milk and/or water on days 1–28 | Whole brain | 20% ↓ | ↑ |
| Rats, 30–32 days old (Grant et al., 1976) | 25, 100 or 200 μg/Pb/g/day on days 3–25 | Cortex, striatum, hypothalamus, brainstem | → | → |
| Rats, 21–22 days old (Sobotka et al., 1975) | 9, 27 or 81 mg Pb acetate/kg/day orally on days 1–21 | Cortex, brainstem, cerebellum | → | → |
| Rats, 15 days old (Jason & Kellogg, 1977) | 75 mg/kg/day orally on days 2–14 | Striatum | 20% ↓ | ↓ |
| Rats, 56 days old (Dubas & Hrdina, 1978) | 50 μg Pb/pup/day orally on days 1–21, then 80 ppm in drinking water | Cortex,<br>striatum,<br>midbrain<br>hypothalamus | 17% ↓<br>→<br>17% ↓<br>20% ↓ | ↑ |
| Mice, 44 days old (Schumann et al., 1977) | 2 or 10 mg Pb/ml of drinking water to mothers and offspring since birth | Whole brain | → | — |
| Mice, 40–90 days old (Silbergeld & Goldberg, 1975) | 5 mg Pb acetate/ml of drinking water to mother or offspring | Forebrain | → | ↑ |

* Symbols: ↑, increased; ↓, decreased; →, no change; —, not given

include 1) different doses (exposure levels) of lead used in various studies, 2) differences in duration of exposure to lead, 3) time of performing biochemical or behavioral tests, 4) lack of information on actual brain content of lead achieved by various routes and levels of exposure, 5) lack of time course studies of the neurochemical parameters in question, and 6) the fact that the experimental animals in some of the earlier studies (Silbergeld and Goldberg, 1975; Golter and Michaelson, 1975) did show a significant retardation in growth. The possibility that malnutrition may have influenced the neurochemical changes (Loch et al., 1976) cannot therefore be excluded.

### B. Correlation of Neurochemical and Behavioral Changes

Recently, Dubas and Hrdina (1978) have examined the regional levels and turnover of brain biogenic amines in developing rats at various times of exposure to lead and simultaneously determined the time course of changes in spontaneous locomotor activity. In this study, the pups were given 50 μg of lead as Pb acetate daily from day 1 until day 21. A fixed dose per pup rather than per body weight was chosen because in certain aspects it resembles more closely the conditions of environmental exposure where the "external dose" of the metal is relatively constant. After weaning, the pups were switched to drinking water containing 80 ppm lead. Considering the difference in absorption of lead between suckling pups and after weaning and the amount of water consumed, one can estimate that the average daily amount of lead absorbed (the "internal dose") during the first 4 weeks of exposure was 4.2, 2.2, 1.4 and 1.6 mg/kg. Under these conditions, there was virtually no difference in weight gain and appearance of key developmental landmarks between the lead-treated rats and their coetaneous controls.

Analysis of the time course of spontaneous locomotor activity in developing rats (Fig. 1) shows age-related increases until 10 weeks of age in both control and experimental animals. The lead-treated rats exhibited higher locomotor activity than age-matched controls at 6, 8 and 10 weeks of age, but not at 12 weeks. These results clearly show that lead-induced hypermotility was of a transient nature, since at 12 weeks of age the locomotor activity of both control and lead-treated animals was in the same range. In animals which were withdrawn from lead treatment at the age of 8 weeks, the enhanced motor activity returned to normal range even sooner (at 10 weeks). These results agree with some earlier reports (Golter and Michaelson, 1975; Overman, 1977) on hyperactivity produced in rats by similar exposure to lead. However, they also emphasize the crucial importance of measuring the temporal course of changes in locomotor activity in behavioral toxicology studies.

Data on regional content of brain NE, DA, 5-HT and 5-hydroxyindoleacetic acid (5-HIAA) in 8-week-old hyperactive rats (Fig. 2) demonstrate that significant changes do occur in the levels of these biogenic

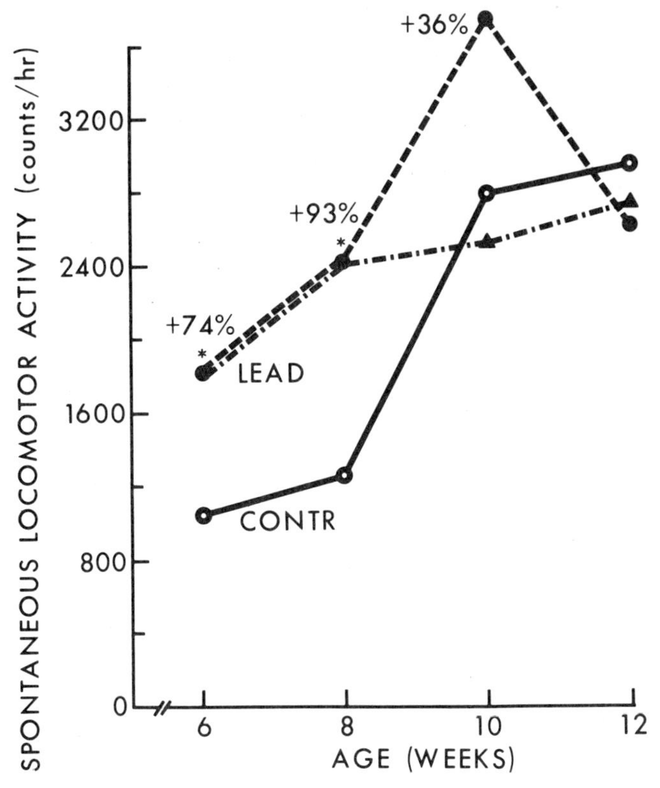


**Figure 1.** Effect of lead on the time course of spontaneous locomotor activity in developing rats. Pups were given 50 $\mu$g of lead as Pb acetate daily, from day 1 to day 21, and were switched after weaning to drinking water containing 80 ppm lead. Solid bottom line, control rats given equimolar solution of sodium acetate; middle broken line, rats treated with lead until 8 weeks of age and then withdrawn from the metal; top broken line, lead-treated rats. The numbers represent percent increase in spontaneous locomotor activity in comparison with age-matched controls and belong to the uppermost curve. Points represent mean values of four groups each containing six rats. Asterisk indicates a value significantly different ($p < 0.05$) from values obtained in controls.

amines during the phase of lead-induced hyperactivity. NE concentrations were found to be decreased in the cortex and striatum, enhanced in midbrain and unchanged in the hypothalamic region. In contrast, levels of DA were not altered in the striatum, but were significantly decreased in cortex, midbrain and hypothalamic region. Finally, the 5-HT levels in cortex and hypothalamus were also lower than in controls; the decreases in 5-HT were accompanied by a corresponding decline in the concentration of 5-HIAA.

In contrast to the alterations in regional content of NE and DA in brains of 8-week-old hyperactive rats, almost no significant changes were found in regional levels of these catecholamines in rats whose locomotor

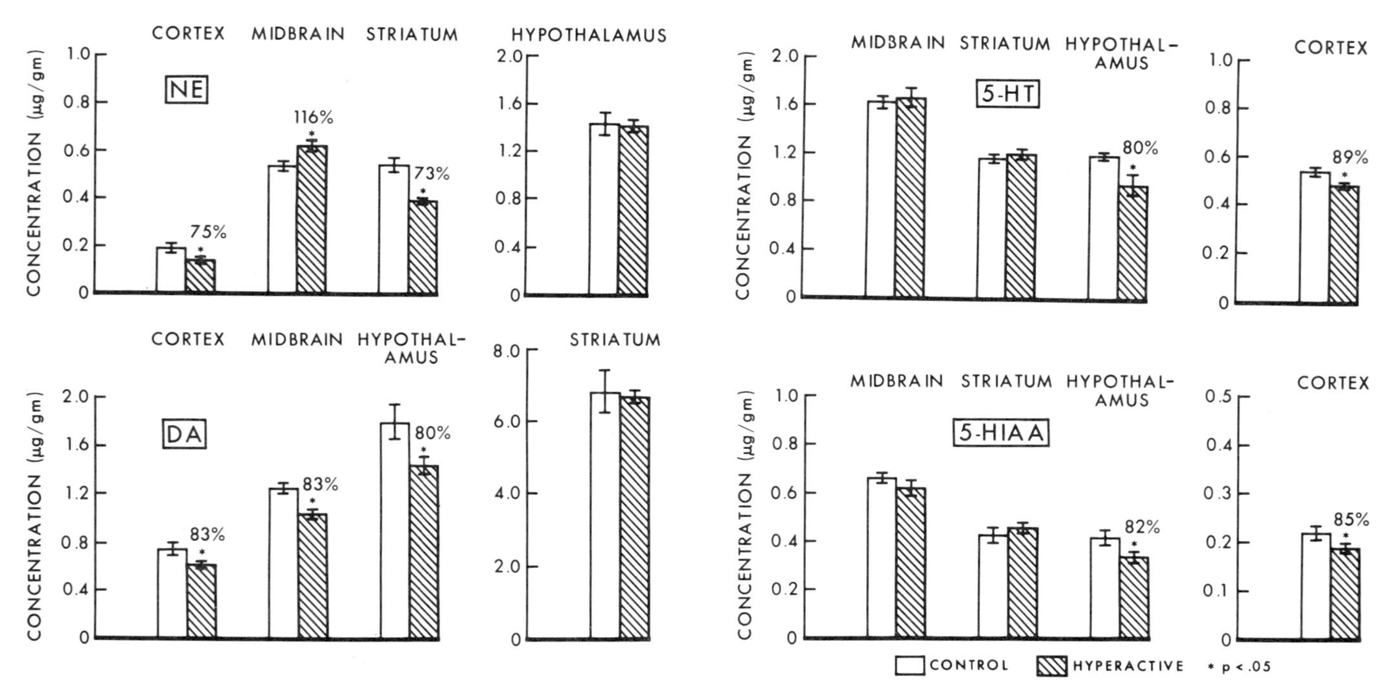


**Figure 2.** Concentration of NE, 5-HT and 5-HIAA in cortex, midbrain, striatum and hypothalamus of 8-week-old rats exposed to lead from birth and showing signs of hypermotility (hatched columns) and in age-matched controls (open columns). Columns represent mean values ± SEM of eight animals in each group. Experimental rats received 50 $\mu$g of lead/pup from day 1 to day 21 and were switched after weaning to drinking water containing 80 ppm of lead. Asterisk indicates a value significantly different ($p < 0.05$) from the corresponding control value.

activity had returned to near normal range regardless of whether they continued receiving lead until 12 weeks or were withdrawn from the metal at 8 weeks of age (Table 3). The changes in the regional brain content of indole compounds (5-HT and 5-HIAA), however, persisted in rats whose locomotor activity had subsided to within normal range at 10 or 12 weeks of age.

Both NE and DA have been implicated in regulation of motor activity in rodents (Geyer et al., 1972; Thornburg and Moore, 1973; Gordon and Schellenberger, 1974; Herman, 1975). Although the neurochemical data on the effects of chronic lead exposure on central monoaminergic systems are far from being conclusive, there is a growing body of evidence suggesting that alterations in functioning of brain NE and DA may play a role in neurotoxic effects of lead, including transient hyperactivity (Sauerhoff and Michaelson, 1973; Golter and Michaelson, 1975; Jason and Kellogg, 1977; Dubas et al., 1978; Silbergeld and Goldberg, 1975; Dubas and Hrdina, 1978). The finding that alterations in regional levels of NE and DA are associated with hyperactivity in lead-exposed rats, whereas decreases in 5-

**Table 3.** Summary of regional changes in brain NE, DA, 5-HT and 5-HIAA levels in rats chronically exposed to lead from birth

| | | | Levels (% control) and direction of changes | | | | | | | |
|---|---|---|---|---|---|---|---|---|---|---|
| Brain region | Experimental group* | | NE | | DA | | 5-HT | | 5-HIAA | |
| Cortex | Hyperactive | (8 wk) | 75† | ↓ | 83† | ↓ | 89† | ↓ | 85† | ↓ |
| | Normal | (12 wk) | 91 | 0 | 110 | 0 | 87† | ↓ | 120† | ↓ |
| | Withdrawn | (10 wk) | 95 | 0 | 107 | 0 | 87† | ↓ | 110 | 0 |
| Midbrain | Hyperactive | (8 wk) | 116† | ↑ | 83† | ↓ | 102 | 0 | 94 | 0 |
| | Normal | (12 wk) | 97 | 0 | 95 | 0 | 101 | 0 | 95 | 0 |
| | Withdrawn | (10 wk) | 92 | 0 | 88† | ↓ | 100 | 0 | 91 | 0 |
| Striatum | Hyperactive | (8 wk) | 73† | ↓ | 99 | 0 | 104 | 0 | 107 | 0 |
| | Normal | (12 wk) | 100 | 0 | 103 | 0 | 67† | ↓ | 84† | ↓ |
| | Withdrawn | (10 wk) | 92 | 0 | 90 | 0 | 77† | ↓ | 84† | ↓ |
| Hypothalamus | Hyperactive | (8 wk) | 99 | 0 | 80† | ↓ | 86† | ↓ | 82† | ↓ |
| | Normal | (12 wk) | 104 | 0 | 97 | 0 | 97 | 0 | 77† | ↓ |
| | Withdrawn | (10 wk) | 107 | 0 | 97 | 0 | 86† | ↓ | 86† | ↓ |

* All experimental rats were treated from Day 1 until Day 21 with 50 μg of Pb/pup and were switched after weaning to drinking water containing 80 ppm Pb. Hyperactive, group of animals showing increased locomotor activity and killed at 8 weeks of age; normal, group of animals showing hyperactivity at 8 weeks, but whose locomotor activity at 12 weeks was within control range despite continued lead treatment; withdrawn, animals withdrawn from lead treatment at 8 weeks and showing no hypermotility at 10 weeks, when they were sacrificed. Levels given in percent of age-matched controls.

† Significant difference versus controls ($p < 0.05$)

HT metabolism were virtually the same in hyperactive rats and animals whose locomotor activity had returned to normal level (Dubas and Hrdina, 1978), would suggest that changes in 5-HT metabolism are not likely to be directly related to changes in locomotor activity induced by lead treatment.

One has to consider, of course, the possibility that the reported lead-induced changes in levels of brain biogenic amines may be due to a non-specific toxic effect of the metal on nervous tissue. However, if it were so, one would expect to find a more or less uniform pattern in these neurochemical changes; this certainly is not the case with lead-induced alterations. In addition, differences also seem to exist between the neurochemical effects of lead in developing and adult brain, as well as between the effect of various toxic metals upon the levels of NE and 5-HT in the brain of rats (Hrdina et al., 1976).

Alterations in the functional activity of a putative neurotransmitter may not be adequately reflected in its steady-state levels. Lead may influence neurotransmission by affecting one or more of the several processes involved in the synthesis, storage, release, action or disposition of putative neurotransmitters. In contrast to earlier reports on increased NE turnover in the whole brain of rats exposed chronically to lead (Michaelson et al., 1974), we were unable to find any changes in this functional parameter for NE or DA in selected brain regions of rats treated with lead (Table 4). Other authors also failed to observe any changes in synthesis rate (Schuman et al., 1977) or synaptosomal uptake (Silbergeld and Goldberg, 1975) of NE in brain of lead-exposed animals. On the other hand, active synaptosomal transport of DA was found to be decreased (Silbergeld and Goldberg, 1975), whereas levels of its metabolite, homovanillic acid (HVA), increased in brain tissue by lead treatment (Silbergeld and Chisolm, 1976). This would seem to agree with some reports of decreased DA levels resulting from lead treatment (Table 2). Again, however, some workers were unable to detect any changes in rate of brain DA synthesis (Schumann et al., 1977) or turnover (Michaelson et al., 1974) in lead-exposed animals.

It is thus fair to conclude that the findings on the effects of lead exposure on brain monoamine levels and metabolism are equivocal and difficult to interpret. Some of the reasons for apparent discrepancies have already been discussed. Results of some of the early studies could have been confounded by growth retardation, probably resulting from malnourishment due to large doses of the metal employed. Their significance thus remains unclear. More recent studies in which precaution was taken not to impair the postnatal growth and development of experimental animals have demonstrated that low doses of lead do produce transient hypermotility associated with concurrent, but also transient, changes in regional levels of norepinephrine and dopamine, but not of serotonin. From these experiments, however, no conclusion can be made as to whether the observed

**Table 4.** Turnover rate of norepinephrine (NE) and dopamine (DA) in some brain regions of control and lead-treated rats*

| | Norepinephrine | | | |
|---|---|---|---|---|
| | Midbrain | | Cortex | |
| | k(hr$^{-1}$) | TR($\mu$g/g/hr) | k(hr$^{-1}$) | TR($\mu$g/g/hr) |
| Control | 0.37† | 0.29‡ | 0.41 | 0.12 |
| Lead | 0.35 | 0.27 | 0.37 | 0.11 |
| | Dopamine | | | |
| | Midbrain | | Striatum | |
| | k(hr$^{-1}$) | TR($\mu$g/g/hr) | k(hr$^{-1}$) | TR($\mu$g/g/hr) |
| Control | 0.30 | 0.37 | 0.23 | 1.65 |
| Lead | 0.37 | 0.38 | 0.21 | 1.62 |

* Rats were exposed orally to 50 $\mu$g/pup of lead (as Pb acetate) from birth until day 21 and then to 80 ppm in their drinking water. The animals were sacrificed at 6 weeks of age.

† Fractional rate constant (k) was calculated from the mean slope of amine decline after inhibition of its synthesis by $\alpha$-methyl-p-tyrosine ($\alpha$MPT). Seven animals from each group were used to determine mean tissue concentration of NE and DA at times 0, 0.5, 1 and 2 hr after administration of $\alpha$MPT.

‡ Turnover rate (TR) of NE and DA was calculated by multiplying *k* by the initial (at 0 time) tissue concentration of the amine.

changes in catecholamine levels were due to a direct effect of lead or were secondary to lead-induced activity changes brought about by distinctly different mechanism. Clearly, more work is needed to elucidate the mechanism of lead-induced alterations in monoaminergic neurotransmission and their relationship to any of neurological or behavioral symptoms resulting from exposure to this heavy metal.

## IV. EFFECTS OF LEAD ON CHOLINERGIC SYSTEMS

### A. Acetylcholine and Choline Levels

Available reports on the effect of lead on levels of acetylcholine (Ach) and choline (Ch) have been conducted primarily in brain samples. Results have not been consistent. As is shown in Table 5, Ach levels following lead administration were either found to remain constant or to be increased in specific brain areas. Ch levels, on the other hand, either were shown to be unaffected by lead administration or, as demonstrated in one case, were diminished in rat midbrain following chronic lead administration.

Such variability is probably due to differences in the approaches used in the various laboratories to expose the experimental animals to lead, as well as in the methodologic techniques used to kill the experimental animals and to analyze the tissue extracts for Ach and Ch content. In particular, the

**Table 5.** Effect of chronic lead exposure on brain acethycholine (ACh) and choline (Ch) levels and cholinergic metabolic enzyme activity*

| Compound measured | Species | Brain Region | Results | References |
|---|---|---|---|---|
| ACh | Mouse | Forebrain | No change | Silbergeld & Goldberg, 1975<br>Carroll et al., 1977 |
| | Rat | Cerebellum, cortex hippocampus, medulla-pons, midbrain, striatum | No change | Modak et al., 1975a, 1975b |
| | Rat (neonate) | Cortex, hippocampus, midbrain, striatum | No change | Shih & Hanin, 1977, 1978a |
| | Rat | Diencephalon | 20% increase | Modak et al., 1975a, 1975b<br>Hrdina et al., 1976 |
| | Rat | Cortex | 32–48% increase | |
| Ch | Mouse | Forebrain | No change | Carroll et al., 1977 |
| | Rat (neonate) | Cortex hippocampus, striatum | No change | Shih & Hanin, 1977, 1978a |
| | Rat (neonate) | Midbrain | 30% decrease | Shih & Hanin, 1977, 1978a |

| | | | | |
|---|---|---|---|---|
| ChAT | Mouse | Forebrain | No change | Carroll et al., 1977 |
| | Rat | Cerebellum diencephalon, striatum | No change | Modak et al., 1975a, 1975b |
| | Rat | Cortex, hippocampus, medulla-pons | 14–18% increase | Modak et al., 1975a, 1975b |
| AChE | Mouse | Forebrain | No change | Carroll et al., 1977 |
| | Rat | Brainstem | No change | Sobotka et al., 1975 |
| | Rat | Cortex | No change | Hrdina et al., 1976 |
| | Rat | Cerebellum, cortex hippocampus, pineal, striatum | No change | Modak et al., 1975a, 1976b |
| | Rat | Cerebellum, cortex telencephalon | Decrease by up to 30% | Sokotka et al., 1975 |
| | Rat | Diencephalon, midbrain, medulla-pons | 9–12% decrease | Modak et al., 1975a, 1976b |
| | Man | Serum | Decrease | DeBruin, 1971 |
| BuChE | Rat | Cerebellum | No change or slight increase | Sobotka et al., 1975 |
| | Rat | Brainstem, telencephalon | Decrease | Sobotka et al., 1975 |
| ChPK | Rat | Forebrain | No change or slight increase | Carroll et al., 1977 |

* Modified from Shih and Hanin, 1978a.

mechanism used for killing the experimental animals prior to analysis of Ach and Ch content in their brains is extremely significant, because there are rapid postmortem alterations in levels of both these substances (Dross and Kewitz, 1972). Ideally, one should be able to inactivate instantaneously *in situ* all postmortem degradative changes, in order to obtain a reflection of levels of ACh and Ch in the animal's brain at the time of its death. Several approaches have been utilized over the past few years; better techniques are even now in the process of further development (Takahashi and Aprison, 1964; Stavinoha et al., 1973; Veech et al., 1973; Giudotti et al., 1974).

## B. Metabolic Enzymes

The effect of lead on four endogenous enzymes which affect levels of ACh and Ch *in vivo* has been studied by a number of investigators. These enzymes include 1) choline acetyltransferase (ChAT), which is responsible for the synthesis *in vivo* of ACh from Ch; 2) acetylcholinesterase (AChE), which hydrolyzes ACh *in vivo* and converts it back to Ch; 3) butyrylcholinesterase (BuChE), otherwise known as "pseudocholinesterase," an enzyme found endogenously, but the functional role of which is presently not clear; and 4) choline phosphokinase (ChPK), which is responsible for the conversion of Ch to phosphorylcholine *in vivo*.

A summary of reported results to date is included in Table 5. Note that here, as in the case of the levels of Ch and ACh, not all investigators obtained similar effects of lead exposure on brain enzyme activity. This again may be attributed to differences in treatment and exposure of the animals to lead in the various laboratories. At the same time, it is interesting to note that, where changes in enzyme activity were found, there appeared to be differences in such activity when measured in various brain areas. These data point to a selectivity of action of lead in discrete brain regions of rats and mice, rather than to an overall uniform effect of lead exposure on the entire brain.

## C. Metabolism and Turnover of Acetylcholine

Measured levels of Ch and ACh in a tissue provide only one index of the metabolic state of the particular cholinergic system. Other indices of metabolism at the cholinergic nerve terminal include 1) turnover studies, 2) analysis of the dynamics of availability of Ch for ACh synthesis as measured by high and low affinity transport of Ch and 3) spontaneous, as well as externally induced, release of ACh from the nerve terminal and extent of a subsequent modification of ACh release-mediated physiologic response.

Ideally, a combination of all these parameters, *including* an analysis of Ch and ACh levels, should provide a reliable picture of the metabolic status of the system in question. Table 6 summarizes available information regarding the effect of lead exposure on the above mentioned parameters.

This phase of research to date has yielded the most promising and least contradictory data regarding the effect of lead exposure on the cholinergic system. It would appear to be quite conclusive, from the results shown in Table 6, that lead administration has a consistent inhibitory influence on neuronal cholinergic metabolism. This effect is evident *in vitro*, as well as *in vivo*, and in the central, as well as in the peripheral, nervous system.

### D. Possible Role of Cholinergic Mechanisms in Neurotoxic Effects of Lead

It is evident from the information presented in the three previous sections that lead exposure has some definitive consequence on cholinergic nerve activity in mammalian as well a nonmammalian systems. The reported electrophysiologic as well as neuropharmacologic studies on neuromuscular systems (Table 6) have shown conclusively that lead exposure will induce a functional deficit in these systems. When one couples this information with evidence for segmental demyelination and axonal degeneration of peripheral motor nerves (Fullerton, 1966; Lampert and Schochet, 1968), it would not be surprising if lead toxicity were to be associated with an induced pathology in neuromuscular function. In fact, it has been documented clinically that children who have been exposed to high doses of lead exhibit impairment in fine motor coordination (De la Burde and Choate, 1972) and a significant degree of reduction in motor nerve conduction velocities (Feldman et al., 1973).

The possible role of cholinergic mechanisms in the central nervous system as a result of lead neurotoxicity is less understood. As we have noted in the previous sections, this is partially due to the difficulty in interpretation of reported data. Nevertheless, strong evidence has been presented in the literature which has implied the involvement of lead-induced cholinergic alterations in brain in the induction of minimal brain dysfunction (juvenile hyperactivity) in man (Silbergeld and Goldberg, this volume). Moreover, it has been suggested by Shih and coinvestigators (1977) that chronic lead exposure will eventually result in the development of supersensitivity of central cholinergic receptor sites. This effect may have no physiological consequences on its own. However, because of the intricate interaction and balance of several neurotransmitter systems in the brain *in vivo* (Hanin, 1978), development of such supersensitivity of one neurotransmitter system could induce an imbalance in the others, with consequent deleterious physiologic effects. This picture is complicated even further when one considers that lead exposure probably affects systems other than those that involve ACh, some of which may be part of the above network of interacting neurotransmitter systems.

**Table 6.** Effects of chronic lead exposure on various parameters of acetylcholine metabolism*

| Parameter studied | Species | Tissue | Results | References |
|---|---|---|---|---|
| Acetylcholine | | | | |
| Turnover rate *in vivo* | Rat | Cortex, hippocampus, midbrain, striatum | 33–54% decrease | Shih & Hanin, 1977; 1978b |
| Spontaneous release | Mouse | Cortical minces | 40% increase | Carroll et al., 1977 |
| Potassium-induced release | Mouse | Cortical minces | 16–30% decrease | Carroll et al., 1977 |
| Preganglionic stimulation-induced release | Cat | Perfused superior cervical ganglion | Decrease | Kostial & Vouk, 1957 |
| Size of endplate potential (epp) and frequency of miniature epps | Frog | Isolated sciatic nerve-sartorius muscle preparation | Reduction in size of epp; increased frequency of mepps | Manalis & Cooper, 1973 |
| Force and latency of muscle contraction following nerve stimulation | Rat; mouse | Phrenic nerve-hemidiaphragm preparation | Reduction in force; increase in latency | Silbergeld et al., 1974a 1974b |
| Choline | | | | |
| High affinity transport | Mouse | Forebrain synaptosomes | 50% decrease | Silbergeld & Goldberg, 1975 |
| | Mouse | Cortical minces | No change | Carroll et al., 1977 |
| Low affinity transport | Mouse | Forebrain synaptosomes | No change | Silbergeld & Goldberg, 1975 |
| | Mouse | Cortical minces | No change | Carroll et al., 1977 |
| Spontaneous release | Mouse | Cortical minces | No change | Carroll et al., 1977 |
| Potassium-induced release | Mouse | Cortical minces | 34–57% decrease | Carroll et al., 1977 |

* Modified from Shih and Hanin, 1978a

## V. EFFECTS OF LEAD ON OTHER NEUROTRANSMITTER SYSTEMS

A number of studies have also been conducted with the goal of attempting to correlate the effect of chronic lead exposure on the putative neurotransmitter systems, in addition to the monoaminergic and the cholinergic systems. The following is a summary of the results obtained pertaining to the cyclic AMP (cAMP) and $\gamma$-aminobutyric acid (GABA) systems.

### A. cAMP

Data on the effect of lead on this substance are very sparse. Nevertheless, it would appear from available information that chronic lead exposure will not affect the rate or extent of norepinephrine-stimulated cAMP production in hearts of neonatal rats (Williams et al., 1977). Lead exposure will, on the other hand, inhibit the activity of brain adenylate cyclase, the enzyme responsible for cAMP synthesis *in vivo* (Nathanson and Bloom, 1976).

### B. GABA

Studies on the GABAergic systems have indicated that lead induces no change in levels of rat whole brain GABA (Michaelson and Sauerhoff, 1974) or in levels of GABA in individual rat brain regions, including brainstem and cortex (Sobotka and Cook, 1974; Sobotka et al., 1975; Piepho et al., 1976). There would appear, however, to be a discrepancy in the literature regarding the effect of lead exposure on GABA levels in rat cerebellum. While one group of investigators observed no change in GABA levels (Sobotka and Cook, 1974; Sobotka et al., 1975), another group reported a decrease in rat cerebellar GABA levels following chronic lead exposure (Piepho et al., 1976). This discrepancy has yet to be resolved. Studies on high affinity transport of GABA in forebrain synaptosomes indicated no change following chronic lead exposure in mice (Silbergeld and Goldberg, 1975).

## VI. IONIC MECHANISMS IN NEUROTOXIC EFFECTS OF LEAD

Lead has been shown to decrease radiolabeled calcium (Ca) uptake by bullfrog sympathetic ganglia *in vitro* (Kober and Cooper, 1976), but to enhance *in vitro* radiolabeled Ca entry into rat caudate synaptosomes (Silbergeld, 1977). Thus, lead interacts with and interferes with the normal process of Ca transport across various tissue membranes.

Since the Ca ion is intimately involved in the process of ACh and dopamine synthesis and/or release, it should not be surprising if the reported effects of lead on neurotransmitter systems were found to be mediated, at least in part, by a competition with ionic mechanisms which are responsible for the release of these neurotransmitters *in vivo*. An analysis of such an interaction would explain the discrepant effect of lead on ACh versus dopamine (which we have already documented in this chapter), in which the release of the former neurotransmitter substance is reduced, while that of the latter is increased.

Silbergeld and Adler (1978) studied this phenomenon in detail and appear to have provided an explanation for this discrepancy. Their data indicate that the inhibitory effect of inorganic lead on ACh release may be due to a blockade of Ca binding to the membrane, resulting in an inhibition of Ca-dependent Ch uptake into and subsequent release of ACh from the nerve terminal. On the other hand, the observed potentiation of dopamine release by lead may occur as a result of a different mechanism involving, not only lead and Ca, but intracellular sodium as well. The interaction of these three ions in the proposed scheme is such that sodium-induced release of Ca from intrasynaptosomal mitochondria is inhibited by lead. This is believed to result in a compensatory increase of transmembrane flux of exogeneous Ca leading to increased exocytotic release of dopamine into the synaptic cleft (Silbergeld and Adler, 1978).

Obviously more work is required in this area; nevertheless, the above data are intriguing in that they begin to explain some of the earlier reported findings on the effect of lead toxicity on several neurotransmitter mechanisms.

## VII. LEVELS AND LOCALIZATION OF LEAD IN THE CENTRAL NERVOUS SYSTEM

There is some indication that in young organisms exposed to lead the brain may be the "critical" organ (Norberg, 1976). The "critical" organ is defined as that particular organ which first attains the "critical" concentration of the metal under specified circumstances of exposure and for a given population. But the organ which accumulates the metal most is not necessarily the "critical" organ. For example, after exposure to lead the highest concentration of the metal may be attained in the bone without any apparent effect. "Critical" concentration in the "critical" organ results in "critical" effect, which is generally recognized as an adverse effect with some consequences for the health of the whole organism. This is particularly important in the case of the central nervous system.

### A. Whole Brain and Regional Levels After Different Ways of Exposure

Table 7 summarizes the data on brain lead levels included in reports published during the recent decade. It shows that, depending on the dose and length of exposure, brain levels of lead in experimental animals can vary from as low as 0.35 ppm to as high as 5.75 ppm.

Small amount of lead can be detected even in normal brain tissue of man or experimental animals (Table 7). It is of interest that in normal laboratory rats (Wistar strain) the concentration of this heavy metal in hippocampus was found to be about seven times higher than that seen in the whole brain (Fjerdingstad et al., 1974). It is not clear why this preferential accumulation occurs and whether it is of any pathophysiologic consequence.

It has been reported that, in lead-treated rats showing hyperactivity, increased aggressiveness and stereotype behavior, the levels of the heavy metal in brain were several times higher than those seen in age-matched controls (Sauerhoff and Michaelson, 1973; Golter and Michaelson, 1975). However, relatively high levels in brain already had been attained by the 5th day of onset of exposure; the levels increased by only 33% after an additional 3 weeks of lead treatment (Sauerhoff and Michaelson, 1973). Grant et al. (1976) observed dose-dependent increases in the concentration of lead in brains of rats exposed to varying levels of the metal (25 to 200 $\mu$g/g body wt/day) without noting any apparent signs of behavioral toxicity. Similarly, Krehbiel et al. (1976) reported a lack of neurotoxic effects in rats which had a brain concentration of lead as high as 1.77 ppm. This is not surprising in view of the fact that brain lead levels of 0.44 ppm were found in normal laboratory rats without any experimental exposure to the metal (Fjerdingstad et al., 1974). On the other hand, guinea pigs whose brains contained only 0.427 ppm lead have shown frequent seizures after a 5-day exposure to toxic doses of the metal (O'Tuama et al., 1976).

Regional distribution of lead in brains of rats exposed to the heavy metal right from birth is shown in Table 8. At 3 weeks, the highest concentration of lead was found in the cerebellum and pons-medulla. This is in agreement with the observation of Goldstein et al. (1974) that in rat pups the brain region showing most damage after exposure to lead is the cerebellum. There is also some evidence that cerebellar capillaries may be more sensitive to toxic effects of lead than the capillaries in other parts of the brain (Goldstein and Diamond, 1974). This could be, at least in part, a cause of increased penetration of the metal into cerebellar tissue. There was some redistribution of lead in various brain regions at 8 weeks of age, when the animals also showed a significant increase in spontaneous locomotor activity. It is of interest that in addition to the pons-medulla, the highest content of the heavy metal was found in hippocampus, hypothalamus and striatum, regions of the brain believed to play a fundamental role in the elaboration of behavior and regulation of motor activity.

At present, however, not even a tentative conclusion can be made regarding the relationship of neurotoxic effects seen in experimental animals and the "critical" concentration in brain or brain regions needed to produce such effects. It would seem that the appearance of the neurotoxic symptoms depends on at least three factors. These are 1) age at which the organism is

**Table 7.** Concentration of lead in whole brain or discrete brain regions found in humans or experimental animals after various routes and levels of exposure

| Species (reference) | Brain region | Dose (level), route and time of exposure | Pb concentration, ppm, fresh tissue | | Symptoms |
|---|---|---|---|---|---|
| | | | Controls | Treated (exposed) | |
| Man (Zaworski & Oyasu, 1973) | Right frontal lobe | Material from autopsy (N = 191) | 0.5 ± 0.68 (0–7.9) | —* | — |
| Man (Barry & Mossman, 1970) | Cortex | Heavy lead exposure<br>Occupational exposure<br>Nonoccupational exposure | — | 4.17 (N = 1)<br>0.17 (0.06–0.26) (N = 3)<br>0.14 (0.02–0.73) (N = 14) | - |
| Rabbits, 30 days old (Lorenzo & Gewirtz, 1977) | Whole brain | 1 mg Pb nitrate/day on days 1–30 | 0.10 ± 0.05 (N = 7) | 0.35 ± 0.06 (N = 6) | |
| Guinea-pig, adult (Bouldin & Krigman, 1975) | Whole brain | 155 mg Pb carbonate/day for 5 days | 0.117 ± 0.08 (N = 4) | 3.32 ± 1.6 (N = 4) | Seizures |
| Guinea-pig, adult (O'Tuama et al., 1976) | Whole brain | 155 mg Pb carbonate/day for 5 days | 0.095 ± 0.016 (N = 3) | 0.427 ± 0.067 (N = 3) | Frequent seizures |
| Rats, 29 days old (Sauerhoff & Michaelson, 1973) | Whole brain | 40 ppm in milk and/or water on days 1–28 | 0.1 ± 0.01 (N = 6) | 0.88 ± 0.11 (N = 6) | Increased activity |

| | | | | | |
|---|---|---|---|---|---|
| Rats, 25 days old Michaelson & Sauerhoff, 1974) | Whole brain | 25 ppm in milk and/or food on days 1–24 | 0.1 | 0.5 | Increased activity |
| Rats, 27 days old (Goldstein et al., 1974) | Whole brain | 4% Pb carbonate in mother's diet on days 12–26 | — | 5.75 ± 1.25 (N = 3) | — |
| Rats, 29 days old (Krehbiel et al., 1976) | Whole brain | 0.83 g Pb acetate/kg/day orally to mother from day 2 through weaning | 1.05 ± 0.09 (N = 11) | 1.77 ± 0.19 (N = 6) | → |
| Rats, 45 days old (Fjerdingstad et al., 1974) | Whole brain | Normal food | 0.443 ± 0.052 (N = 6) | — | — |
| | hippocampus | | 1.65 ± 0.20 (N = 10) | — | |
| Rats, 30 days old (Grant et al., 1976) | Whole brain | 25 μg Pb/g/day<br>100 μg Pb/g/day<br>200 μg Pb/g/day on days 3–25 | 0.15 ± 0.07 | 0.38 ± 0.06<br>0.51 ± 0.08<br>0.68 ± 0.11 | → |
| Rats, 180 days old (Krigman et al., 1974) | Whole brain | 4% Pb carbonate in diet since birth | 0.28 | 5.2 | — |
| Mice, adult (Seth et al., 1976) | Whole brain | 8 mg/kg on days 1, 3 and 5; killed at day 11 | 1.2 ± 0.05 (N = 6) | 2.9 ± 0.10 (N = 6) | — |

* Symbols: →, no change in locomotor activity; —, not given

**Table 8.** Regional distribution of lead in brain of lead-exposed rats

| Brain region | Lead content (ppm wet wt) | |
|---|---|---|
| | 3 weeks | 8 weeks |
| Cortex | 1.28* | 1.77 |
| Striatum | 2.11 | 2.65 |
| Hypothalamus | 3.76 | 2.88 |
| Midbrain | n.d. | 1.27 |
| Hippocampus | 2.62 | 3.48 |
| Cerebellum | 4.59 | 2.12 |
| Pons-medulla | 4.21 | 3.51 |

* Mean value of duplicate sample pooled from four brains. Rats were exposed orally to 50 μg/pup of lead as Pb acetate from birth until day 21 and then to 80 ppm lead in their drinking water. Lead content was measured by flame atomic absorption spectroscopy. Lead levels in respective regions from control animals were not detectable by the method used.

exposed to the metal; it appears that in young organism the absorption from the gastrointestinal tract is more complete and the blood-brain barrier is more vulnerable, thus allowing for faster build-up of lead levels in brain tissue; 2) levels and duration of exposure; and 3) species differences.

Accurate measurement of lead in brain tissue, particularly in small samples from various brain regions, depends to a great extent on the availability of a sensitive and reproducible method of detection. The wide range of values reported for brain lead in control animals (see Table 7) would indicate that more work is required to establish a comparable basis for metal analysis in brain tissue.

### B. Penetration of Lead Into Brain and Relationship of Blood and Brain Lead Levels

Little information is available on the relationship between lead levels in blood and those in the brain. The penetration of this heavy metal into brain tissue may depend on the function of the blood-brain barrier, which is believed to play an important role in the development of lead encephalopathy (Goldstein et al., 1974). Several investigators have attempted to answer the question whether there is a threshold for the entry of lead into the brain. It was found that, after acute administration of lead to adult experimental animals (guinea pigs), the deposition of lead in the brain is proportional to its concentration in the blood (Goldstein and Diamond, 1974; Savolainen and Kilpio, 1977). However, the interdependency of blood and brain lead contents changes with the time of exposure and the dose accumulated; after a certain period of time, the blood lead appears to enter the brain more freely, which is reflected in higher brain-blood ratios. Savolainen and Kilpio (1977) have explained this by "saturation" of the permeability barrier. It thus appears that instead of a simple threshold for lead entry into the brain, there may be a dose- and equilibrium-dependent saturation point of the barrier, beyond which the metal penetrates into brain tissue more rapidly, thus resulting in a high brain-blood ratio.

Willes et al. (1977) have shown that an important factor in penetration of lead into the brain is the age at which the metal is administered. They found that the brain-blood ratio of lead 96 hr after oral dosing with $^{210}Pb(NO_3)_2$ was more than three times higher in 10-day-old monkeys than in adult ones. Studies in young rats (Goldstein and Diamond, 1974) have shown that after a single dose the brain concentration of lead remains virtually unchanged for several days, even though blood levels fall to one seventh the initial concentration. It thus seems likely that progressive accumulation of lead in brain can occur without marked changes in blood concentration when subjects are exposed to lead sporadically. Consequently, blood lead levels are not likely to be a reliable index of the concentration of lead in the brain.

In order to understand the mode of neurotoxic action of lead, it is important to know precisely the cellular site of its action. Despite much research, it is not known whether lead acts primarily on neurons, glial tissue or brain capillaries. Information on subcellular localization of lead in the brain is rather limited. Krigman et al.(1974) reported the following mean relative distribution of lead (percent of total lead) in subcellular brain fractions of 6-month-old rats exposed to a diet of 4% Pb carbonate since birth: mitochondrial supernatant, 16.5%; mitochondria, 34.4%; nuclear fraction, 30.8%; synaptosomes, 3.4%; and myelin, 2.4%. By using X-ray elemental analysis, Silbergeld et al. (1977) were able to demonstrate that Pb rapidly penetrates into isolated brain synaptosomes and is selectively accumulated by mitochondria, where it may influence energy metabolism. Indeed, it has been shown by Bull et al. (1975) that this heavy metal inhibits cytoplasmic NAD(P)H oxidation by brain mitochondria. Further research is obviously needed to elucidate the nature of lead action on subcellular particles of nervous tissue.

## VIII. SUMMARY AND CONCLUSIONS

Inorganic lead has varied but profound effects on neurochemical components of the living system. It is safe to conclude from the data presented in this chapter that the observed toxic effects of lead appear to be specific for particular structures and/or physiologic components of mammalian as well as nonmammalian nervous systems. Because of a lack of uniformity in the approaches which have been used by investigators to expose their experimental animals to lead, neurochemical data obtained to date are not concordant from laboratory to laboratory. Among factors which could be responsible for the apparent discrepancies are 1) differences in dose (exposure) levels and in duration of exposure, 2) differences in age or time at which various neurochemical or behavioral parameters are measured, or 3) species differences. Nevertheless, certain definitive trends are gradually emerging which are providing us with a continually sharper focus on the mechanisms by which lead intoxication exerts its influence on a wide range of neurochemical correlates *in vivo*.

The important observation that undernutrition can be responsible for alterations in central neurotransmitter function which parallel those observed in animals exposed to lead (Michaelson, this volume) make it essential that one should consider the effect of undernutrition on the reported results whenever lead-exposure data are reported. Nevertheless, as documented in this chapter, carefully controlled experiments have been conducted with this consideration in mind which have verified the fact that lead exposure *per se* will also induce perfound effects on neurotransmitter function.

Neurotransmitter mechanisms play an essential role in a variety of neurological as well as psychiatric disease states. A perturbation of these mechanisms by a toxic agent such as lead could therefore result in significant deleterious neurologic and behavioural sequelae in an individual. Consequently, it is essential that more research be conducted with the goal of further elucidating neurochemical correlates of lead toxicity. In view of the prevalence of lead in the atmosphere which surrounds us, this urgency would particularly pertain to the study of the effects of subclinical low level lead exposure in biological systems, including man.

## REFERENCES

Barry, P.S.I., and Mossman, D.B. (1970). *Brit. J. Industr. Med. 27:*339.

Brun, A., and Brunk, U. (1967). *Acta Path. Microbiol. Scand. 70:*531.

Bull, R.J., Stanaszek, P.M., O'Neill, T.T., and Lutkenhoff, S.P. (1975). *Environ. Hlth. Perspect. 12:*89.

Carroll, P.T., Silbergeld, E.K., and Goldberg, A.M. (1977). *Biochem. Pharmacol. 26:*397.

De La Burde, B., and Choate, M.S., Jr. (1972). *J. Pediatr. 81:*1088.

DeBruin, A. (1971). *Arch. Environ. Health 23:*249.

Dross, K., and Kewitz, H. (1972). *Naunyn-Schmied. Arch. Pharmacol. 274:*91.

Dubas, T.C., and Hrdina, P.D. (1978). *J. Env. Path. Toxicol. 2:*473.

Dubas, T.C., Stevenson, A., Singhal, R.L., and Hrdina, P.D. (1978). *Toxicol. 9:*185.

Feldman, R.G., Haddow, J., Kopito, L., and Schwachman, H. (1973). *Am. J. Dis. Child. 125:*39.

Fjerdingstad, E.J., Danscher, G., and Fjerdingstad, E. (1974). *Brain Res. 80:*350.

Fox, D.A., Lewdowski, J.P., and Cooper, G.P. (1977). *Toxicol. Appl. Pharmacol. 40:*449.

Fullerton, P.M. (1966). *J. Neuropathol. Exp. Neurol. 25:*214.

Geyer, M.A., Segal, D.S., and Mandell, A.J. (1972). *Physiol. Behav. 8:*653.

Goldstein, G.W. (1977). *Brain Res. 136:*185.

Goldstein, G.W., and Diamond, T. (1974). E. Plum (Ed.), In "Brain Dysfunction and Metabolic Disorders", p. 293. Raven Press, New York.
Goldstein, G.W., Asbury, A.K., and Diamond, I. (1974). *Arch. Neurol. 31:*382.
Golter, M., and Michaelson, I.A. (1975). *Science 187:*359.
Gordon, J.H., and Shellenberger, M.K. (1974). *Neuropharmacol. 13:*129.
Grant, L.D., Breese, G., Howard, J.L., Krigman, M.R., and Mushak, P. (1976). *Fed. Proc. 35:*503.
Guidotti, A., Cheney, D.L., Trabucchi, M., Doteuchi, M., Wang, C., and Hawkins, R.A. (1974). *Neuropharmacol. 13:*1115.
Hanin, I. (1978). *Environ. Hlth. Perspect. 26:*135.
Herman, Z.S. (1975). *Brit. J. Pharmacol. 55:*351.
Holtzman, D., and Shen Hsu, J. (1976). *Pediat. Res. 10:*70.
Hrdina, P.D., Peters, D.A.V., and Singhal, R.L. (1976). *Res. Comm. Chem. Path. Pharmacol. 15:*488.
Jason, K., and Kellogg, C. (1977). *Fed. Proc. 36:*1008.
Kober, T.E., and Cooper, G.P. (1976). *Nature 262:*704.
Kostial, K., and Vouk, V.B. (1957). *Brit. J. Pharmacol. 12:*219.
Krehbiel, D., Davis, G.A., Leroy, L.M., and Bowman, R.E. (1976). *Environ. Hlth. Perspect. 18:*147.
Krigman, M.R., and Hogan, E.L. (1974). *Environ. Hlth. Perspect. 7:*187.
Krigman, M.R., Traylor, D.T., Hogan, E.L., and Mushak, P. (1974). *J. Neuropathol. Exp. Neurol. 33:*562.
Lampert, P.W., and Schochet, S.S. (1968). *J. Neuropathol. Exp. Neurol. 27:*527.
Loch, R., Bornschein, R.L., and Michaelson, I.A. (1976). *Pharmacologist 18:*124.
Maker, H.S., Lehrer, G.M., Silides, D.J., Weissbarth, S., and Weiss, C. (1973). *Trans. Am. Neurol. Assoc. 98:*281.
Manalis, R.S., and Cooper, G.P. (1973). *Nature 243:*354.
Michaelson, I.A. (1973). *Toxicol. Appl. Pharmacol. 26:*539.
Michaelson, I.A., (1979). R.L. Singhal and J.A. Thomas (Eds.), In "Lead Toxicity," Urban and Schwarzenberg, Baltimore, Md.
Michaelson, I.A., and Sauerhoff, M.W. (1974). *Environ. Hlth. Perspect. 7:*201.
Michaelson, I.A., Greenland, R.D., and Roth, W. (1974). *Pharmacologist 16:*250.
Modak, A.T., Weintraub, S.T., and Stavinoha, W.B. (1975a). *Pharmacologist 17:*212.
Modak, A.T., Weintraub, S.T., and Stavinoha, W.B. (1957b). *Toxicol. Appl. Pharmacol. 34:*340.
Momcilovic, B. (1974). *Arch. Hig. Rada. 35:*359.
Nathanson, J.A., and Bloom, F.E. (1975). *Nature 255:*419.
Nathanson, J.A., and Bloom, F.E. (1976). *Mol. Pharmacol. 12:*390.
Niklowitz, W.J. (1974). *Environ. Res. 8:*17.
O'Tuama, L.A., Kim, C.S., Gatzy, J.T., Krigman, M.R., and Mushak, P. (1976). *Toxicol. Pharmacol. 36:*1.
Overmann, S.R. (1977). *Toxicol. Appl. Pharmacol. 41:*459.
Piepho, R.W., Ryan, C.F., and Lacz, J.P. (1976). *Pharmacologist 18:*125.
Reiter, L.W., Anderson, G.E., Laskey, J.W., and Cahill, D.F. (1975). *Environ. Hlth, Perspect. 12:*119.
Sauerhoff, M.W., and Michaelson, I.A. (1973). *Science 182:*1022.
Savolainen, H., and Kilpiö, J. (1977). *Scand. J. Work. Environ. Health 3:*104.
Schumann, A.M., Dewey, W.L., Borzelleca, J.F., and Alphin, R.S. (1977). *Fed. Proc. 36:*405.
Seth, T.D., Agarwal, L.N., Satija, N.K., and Hasan, M.Z. (1976). *Bull. Environ. Contamination Toxicol. 16*(2):190.

Shih, T.M., and Hanin, I. (1977). *Fed. Proc. 36:*977.
Shih, T.M., and Hanin, I. (1978a). *Life Sci. 23:*877.
Shih, T.M., and Hanin, I. (1978b). *Psychopharmacology 58:*263.
Shih, T.M., Khachaturian, Z.S., and Hanin, I. (1977). *Psychopharmacology 55:*187.
Silbergeld, E.K. (1977). *Life Sci. 20:*309.
Silbergeld, E.K., and Adler, H.S. (1978). *Brain Res. 148:*451.
Silbergeld, E.K., and Chisolm, J.J., Jr. (1976). *Science 192:*153.
Silbergeld, E.K., and Goldberg, A.M. (1973). *Life Sci. 13:*1275.
Silbergeld, E.K., and Goldberg, A.M. (1974). *Exp. Neurol. 42:*146.
Silbergeld, E.K., and Goldberg, A.M. (1975). *Neuropharmacology 14:*431.
Silbergeld, E.K., and Goldberg, A.M. (1979). R.L. Singhal and J.A. Thomas (Eds.), In "Lead Toxicity," Urban and Schwarzenberg, Baltimore, Md.
Silbergeld, E.K., Adler, H.S., and Costa, J.L. (1977). *Res. Comm. Chem. Pathol. Pharmacol. 17:*715.
Silbergeld, E.K., Fales, J.T., and Goldberg, A.M. (1974a). *Nature 247:*49.
Silbergeld, E.K., Fales, J.T., and Goldberg, A.M. (1974b). *Neuropharmacology 13:*795.
Sobotka, T.J., and Cook, M.P. (1974). *Am. J. Ment. Defic. 79:*5.
Sobotka, T.J., Brodie, R.E., and Cook, M.P. (1975). *Toxicology 5:*175.
Stavinoha, W.B., Weintraub, S.T., and Modak, A.T. (1973). *J. Neurochem. 20:*361.
Takahashi, R., and Aprison, M.H. (1964). *J. Neurochem. 11:*887.
Thomas, J.A., Dallenbach, F.D., and Thomas, M. (1971). *Virchows Arch. Path. Anat. 352:*61.
Thornburg, J.E., and Moore, K.E. (1973). *Neuropharmacology 12:*853.
Vallee, B.L., and Ulmer, D.D. (1972). *Ann. Rev. Med. 23:*91.
Veech, R.L., Harris, R.L., Veloso, D., and Veech, E.H. (1973). *J. Neurochem. 20:*183.
Williams, B.J., Griffith, V.H., III, Albrecht, C.M., Pirch, J.H., and Hejtmancik, M.R., Jr. (1977). *Toxicol. Appl. Pharmacol. 40:*407.
Willes, R.F., Lok, E., Truelove, J.F., and Sunderam, A. (1977). *J. Toxicol. Environ. Health 3:*395.
Zaworski, R.E., and Ryoichi, O. (1973). *Arch. Environ. Health 27:*383.

# An Appraisal of Rodent Studies on the Behavioral Toxicity of Lead

## The Role of Nutritional Status

*I. Arthur Michaelson*

TABLE OF CONTENTS

## I. INTRODUCTION

The toxicity of lead as a clinical entity in man and animals has been known for centuries; more recently the relationship between lead poisoning in children and their socioeconomic status has been recognized. Goyer and Mahaffey (1972) have reviewed both constitutional and environmental factors which either enhance or reduce susceptibility to the toxic effects of lead. The intimate relationship of nutritional factors and predisposition to raised body burden of lead is now recognized along with age, season of the year, etc. In many young children with prolonged exposure to lead, the low dietary intake of iron and/or calcium is not rare. It is an unfortunate circumstance that the human diet frequently contains lower than recommended amounts of calcium and iron. At the experimental level, Barltrop and Khoo (1975a,b) used adult male rats to identify calcium and phosphate as the two minerals responsible for major effects on lead uptake. When calcium and phosphate were reduced or omitted from the diet, the lead content of all the organs increased.

From many reports it can be inferred that the composition of the diet of an animal or man is intimately related to the intensity of the toxic effects which may ensue after exposure to lead. The characteristics of absorption and retention (toxicokinetics) of lead are influenced by a number of dietary constituents. Furthermore, the toxicologic effects of dietary lead depend greatly on the amount and variety of substances other than lead in the diet. Nutritional factors in relation to lead toxicity are described elsewhere and will not be discussed here, other than to emphasize that the biologic effects of experimental lead toxicity can be decreased by availability of excess nutrient. It is the intent to deal with an appraisal of behavioral toxicity of lead and the role of nutritional status, and to show that available information has not been taken into consideration in evaluating the results from experiments on anatomical, neurochemical, pharmacologic and behavioral effects following lead exposure in rodents. Effects of excess nutrients and the relevance of dietary oversupplementation when using commercial rodent diets in relationship to low level lead toxicity will be discussed. Diet as a source of variability and undernutrition as a frequent covariable suggests a need for experimental reappraisal where diet as well as growth and development are considered. Behavioral measures following undernutrition alone are described. In addition, the effect of early undernutrition on neurochemistry, locomotor activity and learning is reviewed; the conclusion is reached that many of the reported findings on effects of lead on neurochemistry and behavior should be viewed with healthy skepticism until such time when data from adequately controlled experimentation are provided. This does not negate the established fact that lead is a neurotoxin. The intent is to demonstrate that much of the published work lacks adequate control for experimental covariables which are, in themselves, of questionable relevance to the human problem, *i.e.*, dietary oversupplementation or protein-caloric undernutrition.

## II. EFFECT OF EXCESS NUTRIENT ON LEAD TOXICITY

A number of experiments have shown that essential minerals such as calcium can protect against lead poisoning when fed at higher levels than those recommended as daily requirements for that species. The same is true for iron. Recently, Conrad and Barton (1978) described their method of evaluating factors affecting the absorption and excretion of lead in the rat. It was found that animals fed an iron-deficient diet absorbed significantly more lead (21.3 ± 2.3%) than did the control rats (11.2 ± .4%). Iron-loaded animals absorbed slightly less lead (7.71 ± 1.2 versus 11.2 ± 1.4%, $p < 0.05$). In another series of experiments, various amounts of iron as $FeCl_2$ were added to a test dose of radioactive lead and absorption was quantified from isolated duodenal loops *in situ*. Diminished absorption of lead was observed in animals which received the test dose containing increased amounts of iron. In testing various dietary constituents, it was found that other elements such as zinc and calcium at equimolar concentrations likewise decreased the absorption of radioactive lead. It was concluded that iron probably competes with lead for absorptive sites in the intestinal mucosa. Conrad and Barton (1978) believe that lead and iron share a common absorptive receptor. Kochen and Greener (1975) demonstrated a decreased intestinal absorption of lead in iron-supplemented rats. They point out that ferritin limits the absorption and facilitates the storage of unneeded iron. Their study indicated that ferritin may fulfill similar functions with lead by binding the metal. Interaction of lead with ferritin results in a displacement of hydrogen ion and precipitation of a complex with a maximum binding capacity of 400 atoms of lead per molecule of ferritin. At reaction concentrations below those required to produce an insoluble complex, binding of $^{210}Pb$ to purified ferritin was shown by the crystalization and/or immunoprecipitation of isotope containing ferritin. Similar binding to ferritin *in vivo* was demonstrated by the isolation of isotope containing ferritin from rat liver following oral administration of $^{210}Pb$; 50% of liver lead was found to be ferritin bound. Intestinal absorption of lead was diminished in iron-treated rats, as compared to iron-deficient rats. The decrease in lead absorption after iron supplementation was associated with decreased toxicity, indicated by improvement in weight gain and survival. The binding of lead by ferritin in intestingal mucosa cells and their exfoliation may explain the decreased absorption of lead following iron supplementation. Oral administration of ferritin similarly diminished lead absorption. Halliday et al. (1976) have shown that there are at least two biologically important iron-containing fractions in the rat intestinal mucosal cell during iron absorption in addition to ferritin II. These fractions appear to be involved in the transfer of iron from the lumen to the serosal surface, both in the normal and in the iron-deficient state. Analogous studies with lead and iron-binding protein (FeBP) to that of lead and calcium-binding protein (CaBP) would be informative.

Klauder et al. (1972) found that a low copper diet (0.5 ppm) fed to rats for 8 weeks exaggerated the toxic effects of dietary lead (5000 ppm) as compared to a diet with higher copper content (2.5 ppm) and 5000 ppm lead. Evidence included elevated erythrocyte lead concentrations, depressed hemoglobin and hematocrit and impaired animal growth. Although experimentally the mechanism of lead-copper interaction was unclear, it was suggested that the two metals compete for absorption in the gastrointestinal tract. Petering (1974) further reported that dietary deficiency of copper not only increased the toxic manifestations of lead, but also resulted in an increased retention of lead in kidney and liver. The addition of small increments of copper to the semipurified, copper-deficient rat diet significantly reduced the adverse effects of lead on growth and hematopoietic parameters. An inverse relationship between erythrocyte lead and plasma ceruloplasmic levels was also found. When rats were given 500 ppm lead in copper-deficient diet, they developed internal exposures which were three times that of rats ingesting the same dose of lead given in a copper and iron-fortified diet (Klauder and Petering, 1975).

Some essential minerals can protect against lead poisoning when fed in excess of requirements by interaction at absorptive at well as the macromolecular active sites. A zinc-lead interaction has been evidenced at the absorptive as well as the enzymatic site. Earlier studies describe a protective effect by relatively large doses of zinc on lead toxicity in the young horse (Willoughby et al., 1972). But this interaction of lead and zinc was not studied extensively. Subsequently, others have demonstrated the interaction of zinc and lead at the enzymatic level. $\delta$-Aminolevulinic acid dehydratase (ALAD; E.C.4.2.1.24), an enzyme of the heme synthetic pathway which is exceptionally sensitive to inorganic lead, has been reported to be activated *in vitro* by zinc (Abdulla and Haeger-Aronsen, 1971). These same authors demonstrated the antagonistic effect of zinc on the inhibition of ALAD by lead (Abdulla and Aronsen, 1973). Finelli et al. (1974) found that both synthesis and activity of blood and liver ALAD are dependent on the dietary levels of zinc. Moreover, the inhibition of ALAD by lead could be reversed *in vitro* (Abdulla and Haeger-Aronsen, 1973; Finelli et al., 1975) and *in vivo* (Finelli et al., 1975; Haeger-Aronsen et al., 1976) by zinc. The effects of lead on blood clotting factors in rats were also shown to be dependent on the dietary levels of zinc (Finelli and El-Gazzar, 1977).

Using the rat, Cerklewski and Forbes (1976) examined the influence of dietary zinc (8, 20 and 35 ppm) on dietary lead. Their study indicated that, as dietary zinc increased, the severity of lead toxicity decreased, as evidenced by decreased lead concentrations in blood, liver, kidney and tibia, by decreased excretion of urinary ALA, by decreased accumulation of free erythrocyte prophyrins, by decreased inhibition of kidney ALAD activity and by a decrease in apparent lead absorption. Injected zinc did not afford protection against lead toxicity. It was suggested that the protective effect of excess dietary zinc on lead toxicity is largely mediated by an inhibition of lead absorption at the level of the gastrointestinal tract. The overall effect of excessive dietary zinc in rats fed 200 ppm of lead was to reduce the toxic effects to those typically observed at the level of 50 ppm of lead. That antagonism between zinc and lead may exist at the intestinal absorption sites has also been suggested by El-Gazzar et al. (1978). However, one cannot exclude the interaction of these metals in tissues at the enzyme level. Thawley et al. (1978) found that zinc administration resulted in the reduction of urinary ALA (U-ALA) excretion to near normal levels in lead-intoxicated rats. The effect of zinc in reducing U-ALA levels was observed when zinc was administered both orally and parenterally. Since this was observed in the absence of any significant reduction in blood lead levels, it was suggested that the effect of zinc occurs at a physiologic level beyond gastrointestinal absorption. In the light of past research experience with metal-metal interaction for interacting sites, it is not unreasonable to suggest that zinc and lead both have affinities for similar sites and/or mechanisms at both the level of gastrointestinal absorption and intracellular macromolecules such as enzymes. The mechanism of zinc absorption and its control has recently been illuminated by the work of Evans and associates. They have described a low molecular weight zinc-binding protein factor in the intestinal lumen (Hahn and Evans, 1973), intestinal mucosa (Evans and Hahn, 1974) and pancreas of rats and pancreatic secretions from dog (Evans et al., 1975).

Metabolic interactions between zinc and cadmium, zinc and iron, zinc and chromium and zinc and copper have been discussed by Underwood (1977). Such interactions are to be expected among elements that share common chemical parameters and compete for common metabolic sites. Competitions by lead for zinc binding protein comparable to the studies reported by Barton et al. (1978) on calcium binding protein have not been reported.

One of the more exhaustive studies on the influence of nutritional factors on the absorption of lead has been conducted by Barltrop and his associates (Barltrop and Khoo, 1975a and 1975b) using both intact animals and ligated gut loop preparation *in situ*. This was a systematic study of the effect of individual dietary factors on lead absorption from the gastrointestinal tract, using 30 to 32 day old rats fed a diet containing 750 ppm Pb, tagged with radioactively labeled Pb. In essence, low protein, high fat and low mineral diets increased, whereas high mineral diets decreased, blood lead concentrations. Low fat, low fiber, high fiber, low vitamin and high vitamin diets had no effect on lead absorption. Low mineral diets have been shown to increase lead uptake in all organs studied. The greatest increase was observed in the blood, which had a 17.7-fold enhancement of lead uptake. In their studies to identify the individual minerals responsible for increasing lead absorption from the gut, each mineral was omitted sequentially from the diet. Diets deficient in major components, *i.e.*, calcium, phosphate, magnesium, sodium, potassium and chloride, resulted in a marked increase in lead absorption. There was an 8- to 18-fold increase in the lead content of kidney, femur, liver, blood and carcass. However, exclusion of minor components, *i.e.*, iron, maganese, copper, zinc, iodine and molybdenum from the diet had no effect of lead absorption from the gut. In the study by Barltrop and Khoo (1975a and 1975b) on the influence of nutritional factors on lead absorption, synthetic diets of known composition were compounded to contain 0.07% lead as lead chloride labeled with $^{203}Pb$. Rats were exposed to this radiolabeled lead-containing diet for periods of 48 hr. The dietary intake was then measured and the capacity for the absorption of lead was measured. Although a 48-hr exposure via the gastrointestinal tract may not duplicate the real-life situation, the question being posed is not compromised by such experimental design. The turnover time of the various metal binding proteins is relatively short and intestinal epithelial cells are replaced every 72 to 96 hr. Conrad and Barton (1978) have shown that absorption of lead occurs almost solely from the small intestine, maximally from the duodenum; maximal mucosal uptake occurs in the rat within 30 min of the administration of a test dose. On the other hand, absorption into the carcass increases slowly over a 2-hr period before it reaches a plateau.

Systematic exclusion of minor component(s) from the mineral mix, in order to identify the factor(s) responsible for enhanced uptake of lead, showed that sodium- potassium- and chloride-deficient diets did not affect absorption. Diets without calcium, but complete in other factors, resulted in a 3.5- to 7.4-fold increase in retention. Furthermore, diets without phosphate also enhanced lead absorption, but to a lesser degree than in calcium-deficient diets. However, omission of both calcium and phosphate resulted in an additive effect relative to increased lead retention by all tissues. The exclusion of magnesium from the diet resulted in a significant increase in lead uptake by kidney, liver and blood. This complements the findings by Fine et al. (1976) in their study of the effect of magnesium on the intestinal absorption of lead in dogs, which showed that there was more evidence of lead toxicity in the group not ingesting magnesium.

Although dietary deficiencies have clinical correlates, oversupplementation has relevance to animal experimental conditions. In Barltrop's 1975a,b study, animals fed extra minerals had decreased uptake of lead in all their tissues. Kidney and blood lead uptake was decreased by 70% while the values for femur and liver were decreased by 90 and 80%, respectively. Diets with four times the normal concentration of calcium, phosphate, magnesium, sodium, potassium and chloride resulted in a 60 to 80% decrease in lead absorption in all tissues examined. In light of earlier reports, it is surprising that collectively increasing iron, manganese, copper, zinc, iodine and molybdenum to four times the normal concentration had no effect on kidney, femur, blood and carcass, while liver lead retention was significantly enhanced.

Diets containing twice the recommended concentration of calcium or phosphate had no effect on lead retention. However, when doses of both elements were doubled simultaneously, lead retention was halved in all organs studied except in blood. Increasing calcium and phosphate content in the diet, individually or simultaneously, to four times the normal value reduced lead uptake in all the organs. A fourfold dietary calcium increase decreased lead uptake in the liver, blood and whole body by 50%; there were decreases in kidney by 30% and femur by 60%. Diet containing four times the normal concentration of phosphate resulted in a 60% decrease in femur, liver and blood, a 70% decrease in kidneys and a 50% decrease in the whole body. Increasing calcium and phosphate content four times simultaneously decreased lead retention by 50 to 80% in all organs except blood. The results from a number of investigators show that nutritional factors have a marked effect on the acute absorption of lead from the gastrointestinal tract and retention by various organs of the body. Whereas Barltrop and associates have identified calcium and phosphate as the two principal minerals modifying lead absorption, the prevailing view is that deficiencies and excesses of these (as well as iron, copper and zinc) likewise influence absorption, distribution and retention. It is therefore essential to consider nutritional status and the quality and quantity of diet being consumed in studies seeking answers to the biological impact of ingested lead.

## III. ANIMAL MODELS FOR THE EFFECTS OF LEAD ON CENTRAL NERVOUS SYSTEM FUNCTION

Although there has been long standing recognition of lead as a clinically important poison, concerns about the biologic impact of low level environmental lead is of more recent origin. Furthermore, despite the recognition of potential interactions of nutritional status with lead uptake and deposition, researchers have not given this entity adequate consideration. The use of man as an experimental animal in lead poisoning is a hazardous occupation, especially when it concerns dose-response relationships of the acute or chronic form of toxicity. There are ethical considerations, especially in children, since the exposure to low levels in early life is already suspect in leading to brain damage. Accordingly, most human studies have been metabolic balance studies or retrospective studies of an epidemiologic nature.

In recent years, there has been a growing awareness of the important contributions which animal models can make to the advancement of knowledge in biomedical science. Animal models have become increasingly important as a means of investigating processes occurring in humans. The use of animals is, of course, much more practical and realistic in testing for adverse toxicologic reaction. Animals permit the use of experimental procedures and intervention studies not possible in humans. Furthermore, because of their short life span, meaningful measures may be made in a few months in animals which might spread over many years in the lives of humans. This is especially pertinent to the question being posed as to whether early low level lead exposure in infancy leads to subsequent deficits in brain function. It would be ideal if the animal model were analogous to the disease state under study, such as plumbism. Unfortunately, the mode of central nervous system dysfunction following either high or low levels of lead exposure is unknown. Furthermore, there is uncertainty as to whether low level lead exposure produces cerebral dysfunction as manifest in cognitive and behavioral disorders. The prevailing hypothesis is that low level lead is detrimental to normal brain function. Thus, an animal model which provides symptomatology similar to that suspected of occurring in humans may offer valuable insight into the etiology of the disorder. Considering that segment of the human population we are concerned about, what is an appropriate model for lead-mediated cerebral dysfunction? It has been suggested that the appropriate model should satisfy the following two criteria. 1) Specific symptoms and cardinal features of the disorder under question should be replicated in the animal model. Such features may include those suspect of being caused by lead, including hyperactivity as well as cognitive dysfunction. 2) The pathogenesis of the disorder should bear some relationship to the pathogenesis of the suspected lead-related disorder in children, with insult taking place during early development and deficits being evident shortly thereafter at ages comparable to those of children. One would expect that experimental studies carried out in the laboratory would provide opportunity for defining exposure levels in relation to concomitant nutritional components with some precision. Unfortunately, appropriate experimental designs and consideration of these interactions are frequently lacking. A recent search of the literature for those animal experimental studies which have examined some aspect of the interaction of lead and the central nervous system revealed some disturbing omissions. Eighty-eight reports dealing with rats and 22 on mice were selected because interpretation of data arising from them have stimulated the oft-quoted view, fitting the hypothesis, that low level lead results in cerebral dysfunction. In reviewing the reports dealing with rodents, certain items which might conceivably contribute to the experimental results were identified; these are sex, strain, animal supplier, diet and blood- and brain-lead levels. These are tabulated in Table 1.

**Table 1.** Protocols for studying the effect of lead on rodent CNS function

| | Rat | | Mouse | |
|---|---|---|---|---|
| | (N) | (%) | (N) | (%) |
| Number of reports | 80 | 100 | 22 | 100 |
| Species named | 80 | 100 | 22 | 100 |
| Sex identified | 29 | 36 | 10 | 45 |
| Strain identified | 56 | 70 | 16 | 73 |
| Animal supplier | 23 | 29 | 11 | 50 |
| Diet mentioned | 32 | 40 | 5 | 23 |
| Diet identified | 20 | 25 | 2 | 9 |
| Blood-lead levels | 28 | 35 | 3 | 14 |
| Brain-lead levels | 23 | 29 | 2 | 9 |

Nutritional status, consumptive activity and daily and cumulative exposure are considered equally important; however, they were not tabulated, because information in the original article frequently is lacking. Less than 25% of the articles mentioned the type of food given and/or frequency of feeding. There should be no need to emphasize the importance of the dietary status of animals as a major variable in experimental design. Dietary factors can be regarded as both 1) a previously uncontrolled variable which can confound the interpretation of experimental findings, and 2) a potential independent variable which adds a critically important dimension to the development of an animal model of experimental lead exposure. It has already been pointed out that numerous dietary constituents have been shown to influence lead metabolism. The list is extensive and includes protein and fat content, trace metals (*e.g.*, iron, copper, zinc, magnesium, etc.), Vitamin D and essential minerals (*e.g.*, calcium and phosphate). Absolute concentrations as well as concentrations relative to other dietary components can alter the toxic effects which may ensue following exposure to lead. It therefore serves no useful purpose to publish nonspecific descriptors such as stock diet, commercial diet, commercial pellets, pelleted diet, laboratory chow, powdered chow, etc., as has frequently been the case. Even identification of the source presents problems considering the váriability in content from any one supplier (Table 3).

## IV. DIET AS A SOURCE OF VARIABILITY IN ANIMAL MODELS OF LEAD EXPOSURE

The American Institute of Nutrition (Bieri et al., 1977) has recommended that rat and mouse diets should not contain quantities of vitamins and minerals highly in excess of the requirements for rats and mice set forth by the Committee of Animal Nutrition, National Research Council (NRC) (1972). Commercially available animal chows are characteristically oversupplemented with essential nutrients. Often they exceed by five- to tenfold the levels recommended by the NRC (1972) for growth and development. The concentration of any particular dietary constituent may vary considerably between different manufacturers, as well as between different lots of chow obtained at different points in time from the same manufacturer. However, the concentrations seldom fall below the NRC (1972) recommendations. The extent of variability between different suppliers and the extent of oversupplementation is demonstrated in Table 2. In this table are listed several dietary constituents known to influence lead metabolism.

The magnitude of the difference in metal content of commercial rodent diet and that recommended by the NRC (1972) and more recently by the American Institute of Nutrition (AIN) (Bieri et al., 1977) is striking.

It is evident from the above considerations that rodents consuming commerical chows are recipients of a diet heavily fortified with essential nutrients. Such fortified diets are seldom encountered in the human population. In fact, there is good documentation that a substantial portion of our population is deficient in one or more of those nutrients considered essential for normal health. The parameters of greatest concern with respect to the

**Table 2.** Comparison of trace metal content of diets recommended by American Institute of Nutrition (AIN) and National Research Council (NRC) with three commercial preparations

| Metal | AIN 76™ | NAS-NRC | Charles River 4RF | | Purina 5001 | | Teklad | |
|---|---|---|---|---|---|---|---|---|
| Protein (%) | | 20 | 26.3 | (1.3)* | 23.4 | (1.2) | 26.2 | (1.3) |
| Fat (%) | | 5.0 | 7.1 | (1.4) | 4.5 | (0.9) | 6.0 | (1.2) |
| Fiber (%) | 5.0 | nr† | 3.4 | (0.7) | 5.2 | (1.0) | 4.2 | (0.9) |
| Calcium (%) | 0.52 | 0.5 | 1.1 | (2.2) | 1.2 | (2.6) | 2.0 | (3.9) |
| Phosphorus (%) | 0.40 | 0.4 | 0.7 | (1.8) | 0.9 | (2.4) | 1.3 | (3.3) |
| Sodium (%) | 0.10 | 0.05 | 0.4 | (8.2) | 0.4 | (9.8) | 0.7 | (14.2) |
| Chlorine (%) | 0.16 | 0.05 | nr | | 0.5 | (10) | 0.8 | (15.8) |
| Potassium (%) | 0.36 | 0.18 | 1.1 | (6.0) | 1.0 | (4.6) | 0.8 | (4.6) |
| Magnesium (%) | 0.05 | 0.04 | 0.2 | (4.5) | 0.2 | (6.5) | 0.2 | (5.5) |
| Sulfate (%) | 0.1 | nr | nr | | nr | | nr | |
| Cobalt, ppm | nr | nr | nr | | 0.4 | | 1.3 | |
| Copper, ppm | 6 | 5 | 15.5 | (3.1) | 18.0 | (3.6) | 13.1 | (2.6) |
| Iodine, ppm | 0.2 | 0.15 | 1.4 | (9.3) | 1.7 | (11.3) | 3.0 | (19.7) |
| Iron, ppm | 35 | 35 | 370.0 | (10.5) | 198.0 | (5.7) | 369.8 | (10.6) |
| Manganese, ppm | 54 | 50 | 46.7 | (0.9) | 51.0 | (1.0) | 149.3 | (3.0) |
| Zinc, ppm | 30 | 12 | nr | | 58 | (4.8) | 37.8 | (3.2) |
| Chromium, ppm | 2.0 | nr | nr | | nr | | nr | |
| Fluorine, ppm | nr | nr | nr | | 35 | | nr | |
| Selenium, ppm | 0.1 | 0.04 | nr | | nr | | nr | |
| Vitamin D, IU | 1000 | 1000 | 1090 | (1.1) | 5300 | (5) | 5142 | (5.1) |

* Figures in parentheses represent ratio relative to NAS-NRC recommendations.

† nr, not reported.

lead problem are most likely iron and calcium. The recent Ten State Nutritional Survey (1972) indicated that, in children under 3 years of age, 50% were deficient in iron intake irrespective of ethnic background or socioeconomic status. Approximately 40% of this population had an iron intake of less than one-half the Recommended Dietary Allowance. In this context, one questions the experimental rationale of employing oversupplemented nutritional regimens in research on biologic effects of lead when modeling for relevant human concerns in a population which is nutritionally deficient in specific interacting components. This would be most appropriate if the experimental intent were to determine protective or remedial effects of supplementation. Unfortunately, this has not been the purpose behind most of the studies on neurologic consequence of early lead exposure.

There are two additional criticisms about the use of commercial chows in studies of low level lead toxicity. These are 1) the variability between advertised content of nutrient and that actually found in different lots obtained over the course of several years and 2) contamination with heavy metals (Table 3).

Both Schroeder et al. (1965) and, more recently, Fox et al. (1976) examined the heavy metal contaminants of standard commercial animal chows. The latter report indicated that rat chows contained from 0.1 to 1.5 ppm of lead. Table 3 indicates that as much as 1.7 ppm of Pb has been found in rat rations. An adult rat consuming 20 g of diet per day would therefore ingest as much as 34 $\mu$g Pb. Assuming the rat weighs 200 g, its ingested dose could be as high as 170 $\mu$g Pb/kg/day. A 70 kg man ingests approximately 350 $\mu$g Pb/day or 5 $\mu$g Pb/kg/day. Whereas humans absorb approximately 10% of an ingested dose, rodents have been reported to absorb less than 1%. On a weight basis, the rodent will unintentionally be exposed to six times as much lead as man. The fact that ordinary laboratory chows do contain lead as well as cadmium may be extremely important for those results from long term laboratory animal experiments designed to measure subtle biologic effects, especially where the frame of reference is control animals which have been unwittingly exposed to intoxicant.

## V. NUTRITIONAL CONSEQUENCE OF LEAD EXPOSURE REGIMENS

Prior to 1970, the literature on cerebral dysfunction in animals resulting from lead was limited, largely due to inability to produce unequivocal neurologic changes in laboratory animals comparable to human clinical cases of lead encephalopathy. Adult animals show resistance to the central nervous system effects of lead poisoning. At the histopathologic level, the study of lead encephalopathy had been hampered by the repeated failure to produce the histopathologic changes characteristic of the human disease (as described by Blackman (1937) and others) by administration of lead to

**Table 3.** Summary of food analysis done on Ralston Purina's rat/mouse Chow 5001 from 8/76 to 4/78; includes 22 different batches over a 22-month period*

| Metal | Measure | AIN Standard | Low | Average | High |
|---|---|---|---|---|---|
| Phosphorus | % | 0.4 | 0.677 | 0.71 | 0.78 |
| Potassium | % | 0.36 | 1.20 | 1.29 | 1.36 |
| Magnesium | % | 0.05 | 0.168 | 0.34 | 1.74 |
| Sodium | % | 0.10 | 0.29 | 0.41 | 0.59 |
| Arsenic | ppm | — | 0.29 | 0.381 | 0.51 |
| Cadmium | ppm | — | 0.09 | 0.114 | 0.15 |
| Lead | ppm | — | 0.26 | 0.507 | 1.68 |
| Manganese | ppm | 54 | 45.2 | 58.9 | 112.6 |
| Mercury | ppm | — | <0.05 | <0.05 | <0.05 |
| Selenium | ppm | 0.10 | 0.21 | 0.33 | 0.51 |
| Aluminum | ppm | — | 118 | 162 | 192 |
| Barium | ppm | — | 7.93 | 10.96 | 16.2 |
| Iron | ppm | 35 | 253 | 305.6 | 387 |
| Strontium | ppm | — | 18.2 | 25.2 | 31.7 |
| Boron | ppm | — | 13.4 | 14.6 | 15.6 |
| Copper | ppm | 6 | 8.2 | 10.8 | 14.4 |
| Zinc | ppm | 30 | 54.5 | 57.4 | 65.2 |
| Chromium | ppm | 2-0 | 2.11 | 2.86 | 3.81 |
| Malathion | ppm | — | 0 | 0.07 | 0.13 |
| Estrogen | ppb | — | 0 | 2.3 | <4 |
| PCB's ppb | ppb | — | 10 | 36 | 185 |
| DDT | ppb | — | <5 | 9.4 | 24 |
| DDD | ppb | — | <5 | 6.3 | 13 |
| DDE | ppb | — | <5 | 7.2 | 22 |
| TDE | ppb | — | <5 | 5.2 | 7 |
| Dieldrin | ppb | — | <5 | <5 | <5 |
| Endrin | ppb | — | <5 | <5 | <5 |
| Heptachlor epoxide | ppb | — | <5 | 5.1 | 8 |

* From W.E. Coleman and R.G. Tardiff, in press

adult animals. This situation has dramatically changed in recent years following the report by Pentschew and Garro (1966) who described a technique for producing lead encephalopathy in neonatal suckling rodents by exposing the lactating dam to a diet containing as much as 38,000 ppm lead. It was found that the experimental sucklings exhibited morphologic alterations similar to those occurring in children with lead encephalopathy. It was further suggested that lesions of the central nervous system of neonatal rats were produced by transmission of the lead fed to the lactating dam to the suckling young via maternal milk. The milk contained 40 ppm lead. Pentschew and Garro (1966) reported that the developing neonate showed no early adverse effects of being nourished by lead-containing milk other than a profound retardation in body growth. The model developed by Pentschew and Garro (1966) has come to serve as a prototype for studies of the CNS effects of lead in the developing young rat. A similar approach has been employed in the neonatal mouse by Rosenblum and Johnson (1968), who examined neuropathologic changes in suckling mice by adding lead to the maternal diet in an attempt to reproduce the finding of Pentschew and Garro (1966) in rat. Silbergeld and Goldberg (1973) reported lead-induced hyperactivity in mice. In this study, lead acetate (2730 ppm lead) was presented to lactating dams. In following the protocols described by Pentschew and Garro (1966) for rats and that of Silbergeld and Goldberg (1973) for mice, the lactating dam consumes relatively large amounts of lead and the neonate is weaned and maintained on comparable amounts of lead until completion of the study.

Replication of the Pentschew and Garro (1966) study on neonatal rodent induced encephalopathy has been reported by a large number of investigators. Replication of the behavioral and neurochemical measures reported in rat and mouse have been inconsistent. In all instances, there is an attendant retardation in growth. In spite of this consistent finding, the availability of a lead-related cerebral dysfunction in an animal model has encouraged its employment in behavioral and neurochemical studies.

### A. Rat

Michaelson and Sauerhoff (1974) were the first to quantitate the food consumption by the lactating dam when presented with 27,300 ppm of lead in diet relative to those on control diet. Figure 1 illustrates that the experimental lactating dam presented with this amount of lead in her diet eats 30% less food and experiences a precipitous fall in body weight, which remains below 90% of that on day zero.

The rate of body growth is significantly retarded in the lead-exposed sucklings relative to controls. Whereas the normal animals gain approximately 3 g a day, the experimental animals gain an average of 1.2 a day. Figure 2 illustrates the body weight changes with age in both experimental and control neonatal rats. The average body weight of the lead-exposed group at 20 days of age was comparable to that of a 9-day-old control suckling from a normal control lactating dam. The diminished food consumption (possibly due to taste aversion) by the lactating dam ingesting 38,000 ppm of lead and the slowed growth of offspring relative to control groups implicated nutritional influences among the test group. Although experimental proof was lacking, growth retardation in the suckling was related to diminished milk production by undernourished dams and not necessarily related to lead in the diet of the neonates. Paired feeding studies demonstrated the ability to match the growth rate of the lead experimental

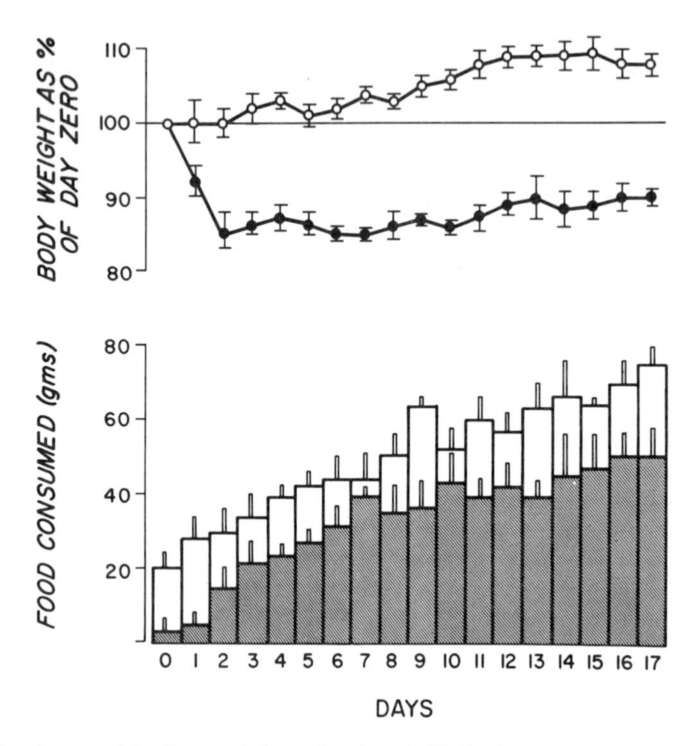


**Figure 1.** Amount (g) of normal (open bars) and 5% lead acetate (shaded bars) food consumed per day. Relative percentage body weight of control (open circles) and lead-exposed (filled circles) lactating rats. Day zero defined as 100%. Each value is the mean ± SD of five animals.

groups by matched food restriction of controls (Fig. 3). Golter and Michaelson (1975) also showed that daily intragastric administration of 2.5 to 25 times the amount lead found in milk (40 ppm) had no effect on growth (Fig. 4). It is evident that lead *per se* does not result in retarded growth in the young. The only plausible explanation supported by experimental evidence is the diminished food intake by the lactating dam. The effect of lead on the nutritional quality of milk produced by lead-exposed lactating dams has not been adequately researched.

With regard to the lactating dam, it was found that the average daily food consumption when presented with 37,300 ppm lead was 35 g for the first 17 days following parturition. Thereafter, the daily and cumulative consumption of lead *per se* was 955.5 mg and 16.24 g, respectively. If the nursing dam weighed 300 g, then exposure would be approximatly 3.15 g Pb/kg/day. This attests to the remarkable resistance of the *adult* rodent to lead poisoning. If absorption in the adult rodent is less than 1% (Kostial et al., 1971), 31.5 mg/kg/day continues to represent an extremely high lead exposure. The exposure procedure described by Pentschew and Garro (1966) and followed by a number of investigators permits direct access to the 27,300 to 38,800 ppm lead-containing diet at approximately 18 days of age when rat pups normally gain access to solid food. The 618 to 1100-fold increase in lead exposure when moving from leaded milk to maternal diet is the primary contributing factor in the precipitous toxic signs of encephalomyelopathy. Exposure to such high levels of lead during a critical time of brain development leads to clinical signs of poisoning (*i.e.*, cerebral hemorrhage, paralysis) at 20 to 30 days of age. Clasen et al. (1974) emphasized that the production of experimental disease in the infant rat

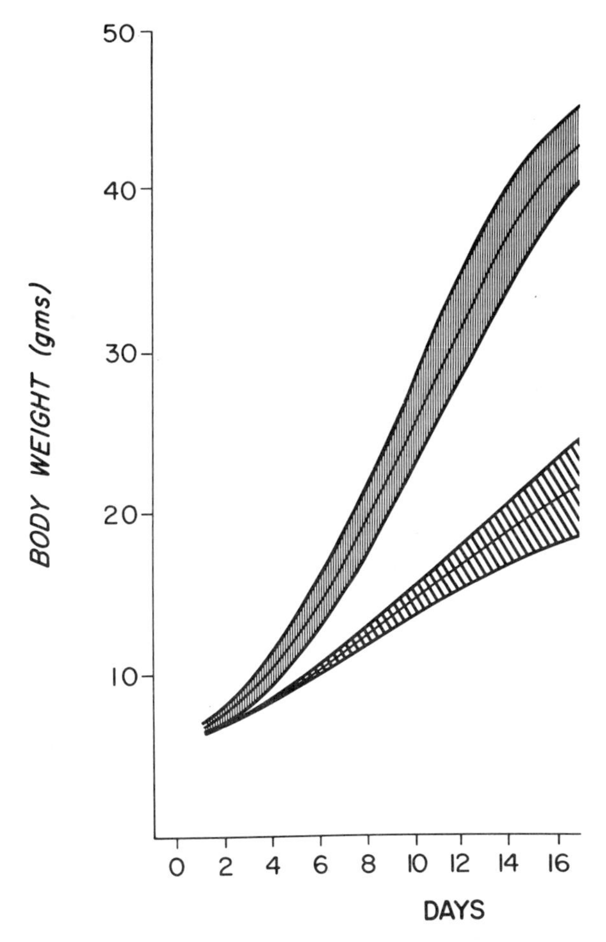


**Figure 2.** Growth (g) of neonatal rats suckling dams on normal (upper curve) and 5% lead acetate (lower curve) diets.

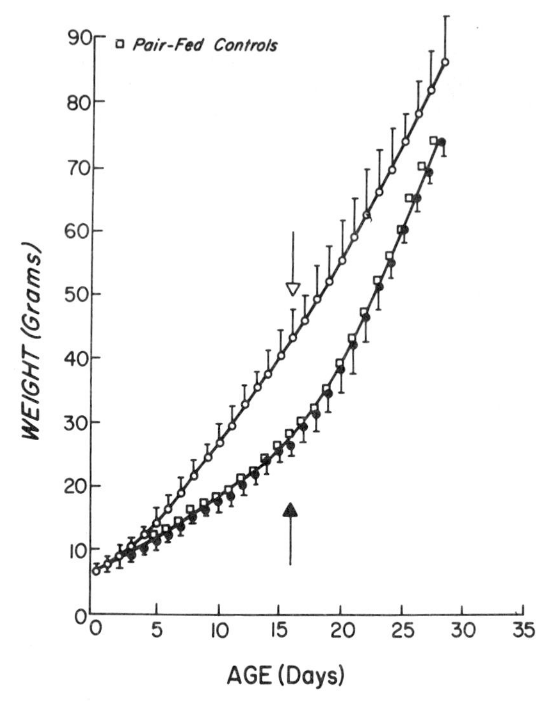


**Figure 3.** Growth (g) of young rats suckling dams on normal (open circles) and 27,000 ppm lead (filled circles) diets, followed by weaning to normal (↓) or 25 ppm lead (↑) diets. Control young rats (open squares) and suckling dams pair fed the same amount of food consumed by experimental dams. The vertical bars indicate ± SD of five animals.

requires that exposure to significant amounts of lead be instituted at a critical time of development. One is unable to produce the disease by giving lead after the age of 20 days. What has not been fully elucidated is the contribution of the severe undernutrition to the pathogenesis of the lead-induced encephalopathy. Krigman et al. (1977) have been able to achieve encephalopathies without serious weight loss by intragastric feeding of lead, but the concentrations have been quite high. In order to circumvent confounding due to two exposure regimens (40 ppm Pb in milk followed by 27,000 ppm in diet), Michaelson and Sauerhoff (1974) offered 31,000 ppm lead to lactating dams and allowed sucklings consuming 25 to 40 ppm lead-containing milk to be weaned to comparable lead containing diets. These animals did not develop the usual encephalopathy, but were reported to exhibit abnormal behavior. It must be emphasized that these animals were still growth retarded. A similar approach to indirect exposure via the lactating dam and weaning to a relatively lower concentration of lead has been used by a number of investigators in behavioral and neurochemical studies of early lead exposure, but unfortunately no attention has been paid to the fact that undernutrition is a complicating accompaniment.

## B. Mice

The impetus for much of the experimental work on lead-induced behavioral disorders in mice stems from reports originating from Silbergeld and Goldberg (1973). The animal model of chronic lead poisoning developed for their studies followed a design similar to that of Pentschew and Garro (1966) in the rat and that of Rosenblum and Johnson (1968) in the mouse. Upon parturition, dams were given lead in drinking water. In most of the reports from this group of researchers, 2,731 ppm lead was employed and

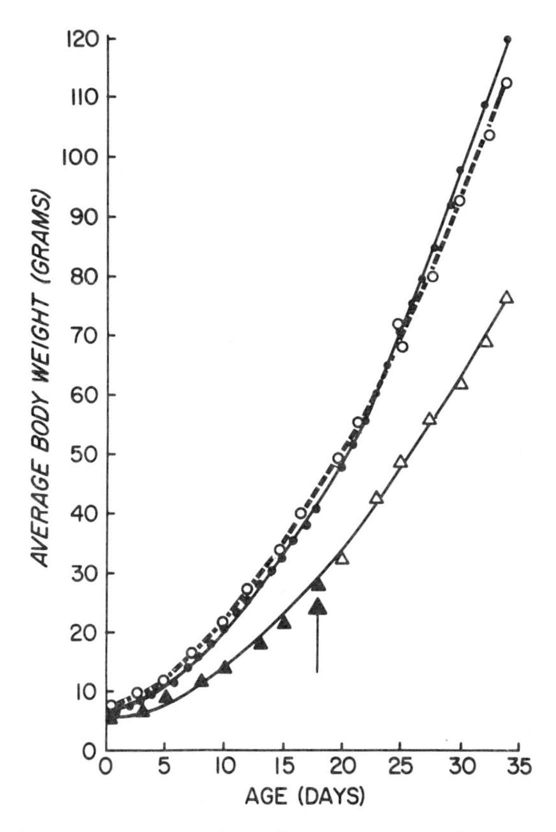


**Figure 4.** Growth of young rats receiving daily oral doses of sodium acetate and weaned to normal food (open circles) and daily oral doses of 1.0 mg of lead and weaned to a diet containing 40 ppm lead (filled circles) as compared to neonates suckling lactating dams eating 27,300 ppm lead (filled triangles) and weaned (at arrow) to a diet containing 40 ppm lead (open triangles).

sucklings weaned to and maintained on the same solution until testing. The reported finding of a lead-induced hyperactivity in mice has attracted considerable attention. In none of the reports was the volume of fluid or amount of food ingested by the lactating dam or postweaning animals described. Because of our long term interest in detailing the effects of lead on the central nervous system, we carried out an in-depth study of the experimental conditions. We have monitored the volume of fluid and quantity of food consumed by lactating mice presented with 5 mg/ml lead acetate (2,730 ppm lead) in drinking water (Fig. 5).

Panel A shows the solid food consumption by control dams (hatched bars) and the somewhat smaller intake of food by those drinking 2,730 ppm

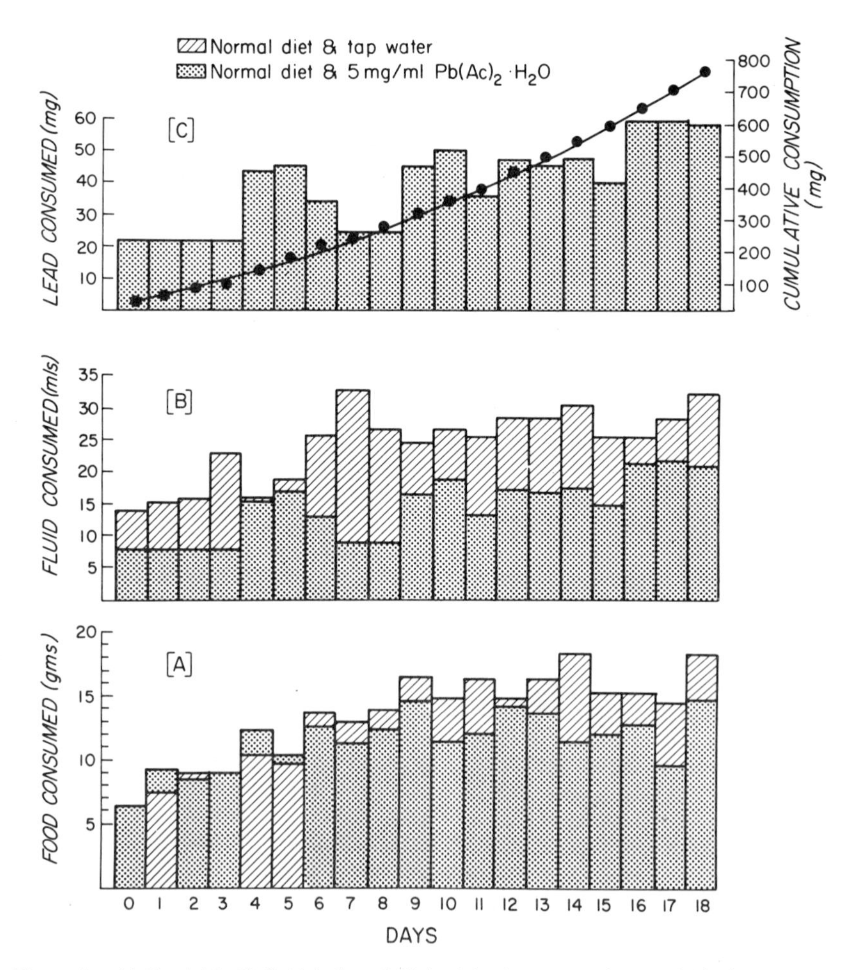


**Figure 5.** A) Food (g), B) fluid (ml) and C) lead (mg) consumption per 24 hr in control and lead-exposed lactating mice (Michaelson, unpublished).

lead-containing water (dotted bars). Panel B shows the fluid consumption. There is a very significant difference in the volume of fluid intake. It is important to point out that the moisture and fat content of diets used by the respective investigators could influence the water needs of the animals; hence, the findings may not be directly transferable. However, the diets were not identified.

Panel C shows the daily lead consumption (mg) as represented by the height of the bars and the cumulative ingestion by the connected solid circles. By the 18th day of lactation, the nursing mouse has taken in 800 mg of lead or about 24 g/kg body weight. As observed in our earlier experience with rats, ingestion of considerable quantities of lead by the dam does not result in overt signs of lead poisoning. The disruption in normal consumptive activity in lactating mice drinking 0.5% $Pb(Ac)_2$ in water is shown in Fig. 6. Under normal dietary conditions, the fluid to food consumption ratio for rodents is approximately 1.8. When presented with drinking water containing 5 mg/ml lead acetate, this ratio decreases as much as 40% to 1.1. On the basis of previous experimental experience with rat, it is reasonable to suggest that it is this aspect of maternal consumption behavior which contributes to the depression in body weight of lead-exposed suckling mice (Fig. 7).

## VI. BEHAVIORAL ALTERATIONS AS A CONSEQUENCE OF UNDERNUTRITION

We have evaluated the effects of various levels of undernutrition during development on later levels of locomotor activity and response to centrally acting drugs (Loch et al., 1978). Mice were raised in either small litters of 8 pups (control animals) or large litters of 16 pups in order to produce growth rates approximating those which occurred in those experiments employing 5 mg/ml lead acetate (2,730 ppm lead) in drinking water (Silbergeld and Goldberg, 1973). Growth rates of fully nourished and undernourished animals approximated those reported in the literature under similar conditions of lead exposure (Fig. 8).

Mice were tested for activity when they were 35 to 36 days of age (Table 4). Baseline activity was recorded for 2 hr, following which each animal received a 10 mg/kg ip injection of d-amphetamine. Activity under the influence of the drug was monitored for an additional 2 hr. The spontaneous activity of all animals decreased throughout the initial 2-hr period of testing. However, the rate of decrease depended upon *nutritional status;* growth-retarded animals displayed a slower rate of decline over the 2-hr period than did normal-sized animals. This decreased rate of habituation occurred primarily in male growth-retarded mice (Table 4). Following the ip injection of d-ampetamine (10 mg/kg), activity in all animals increased to a maximum within 30 min after injection and then declined over the next 90

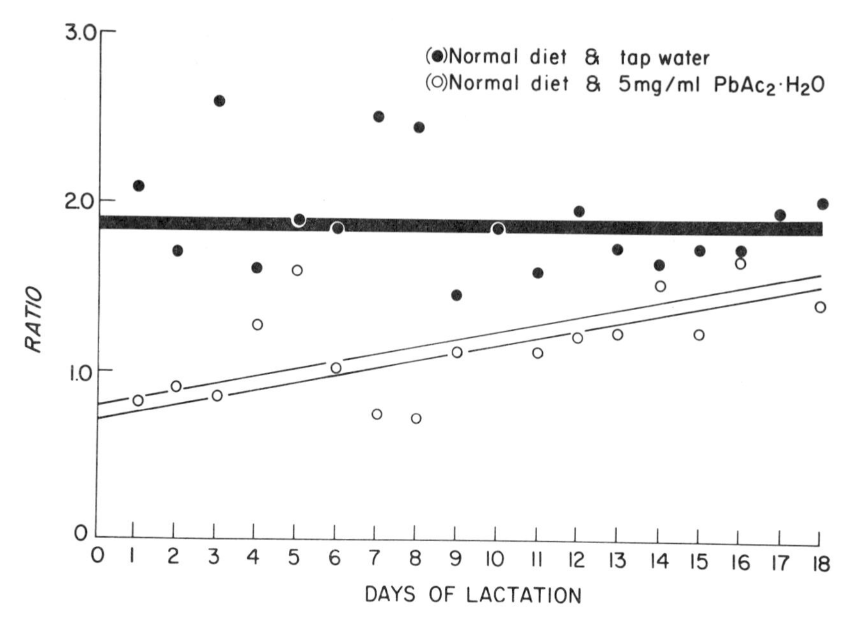


**Figure 6.** Ratio of grams of fluid to grams of food consumed by lactating mice.

min of testing. Growth-retarded mice, as a group, increased their overall activity to a lesser extent than normal-sized mice. Growth-retarded males showed the smallest increases in activity. The *rate* of decline in activity also differed between normal and growth-retarded mice, the latter having a more rapid decline than the former.

Mice from both large (16-pup) and small (8-pup) litters were combined for an analysis of trend. Two analyses were performed. One correlated body weight to the activity which occurred during the first hour after the d-amphetamine injection; another correlated body weight to the activity which occurred during the second hour after drug injection. Scattergrams of body weight and activity are illustrated in Figure 9. A statistically significant relationship was found between body weight and change in activity for both the first hour and for the second hour after injection.

There is sufficient evidence from a number of independent investigators that weight loss (whether it be termed undernutrition, malnutrition, growth retardation or small body size for age) varies systematically with the lead exposure history of an animal. These results demonstrate that growth retardation can be responsible for altering an animal's behavioral response both before and following drug treatment. Consequently, it is not possible to establish with any degree of confidence whether lead exposure or nutritional status (or both) is the causal agent(s) responsible for the behavioral changes which have been reported following lead administration to a lactating dam

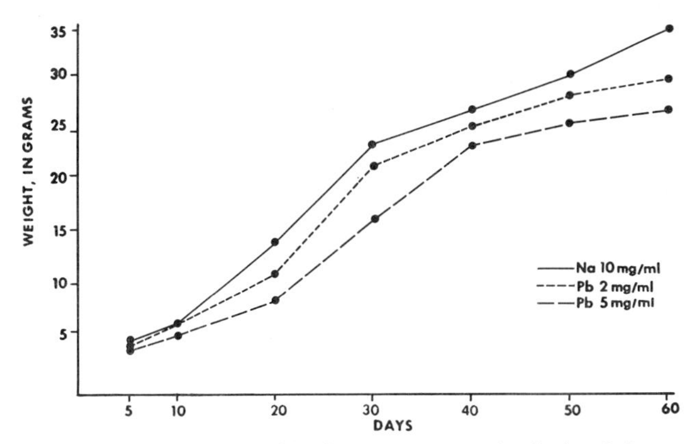


**Figure 7.** Effects of increasing doses of lead acetate on growth of mice. Points are means for 30 to 40 animals. Differences in weight for all doses and controls are significant from day 20 to day 60 ($p < 0.01$), except for controls and 2 mg/ml animals on day 40 (Silbergeld and Goldberg, 1973).

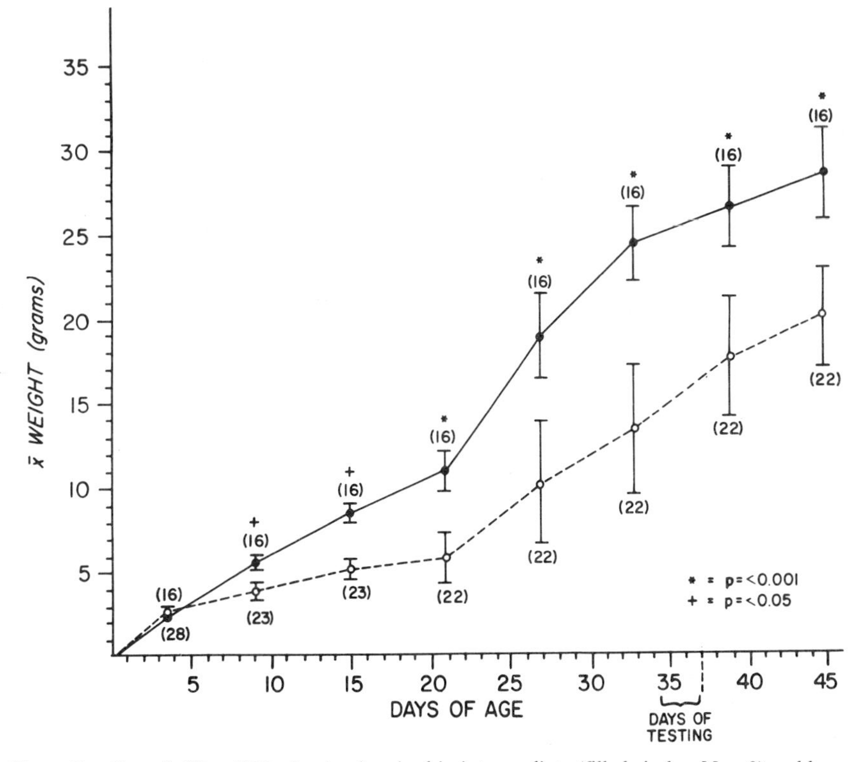


**Figure 8.** Growth (X ± SD) of animals raised in intermediate (filled circles, N = 8) and large (open circles, N = 16) litters. Parentheses indicate total number of animals contributing to each data point.

**Table 4.** Predrug activity in CD-1 mice raised in large (16-pup) and small (8-pup) litters

| Time (minutes) | Litter Size: Large | | Litter Size: Small | | Ratio of large to small |
|---|---|---|---|---|---|
| | $\bar{x}$ | SD | $\bar{x}$ | SD | |
| 15 | 20.09 | 2.68 | 22.26 | 2.74 | 0.90 |
| 30 | 15.82 | 2.59 | 18.62 | 2.47 | 0.85 |
| 45 | 14.08 | 3.62 | 16.24 | 3.40 | 0.87 |
| 60 | 13.46 | 5.20 | 14.11 | 3..46 | 0.95 |
| 75 | 11.11 | 5.75 | 10.83 | 6.51 | 1.03 |
| 90 | 8.87 | 6.82 | 7.04 | 6.31 | 1.26 |
| 105 | 7.83 | 7.93 | 2.74 | 4.41 | 2.86 |
| 120 | 7.14 | 7.04 | 1.38 | 2.76 | 5.17 |

(Rafales et al., 1979). It is evident that weight retardation, and not lead *per se*, may at least partially account for some of the behavioral effects which have been reported in rodents poisoned with lead via suckling contaminated milk. Lead exposure does not have to alter the growth rate of suckling neonates to produce nutritional deficits. At two levels of lead exposure which do not alter the food or water consumption of lactating dams (0.02 and 0.20% lead acetate in drinking water), there is a depression in the zinc and copper concentration of milk to the extent of 33 and 28%, respectively, during the early states of lactation. The critical role of these essential trace elements in brain development and function (Dodge et al., 1975) suggests that such deficits may contribute to behavioral and neurochemical alterations later in life. In addition, there is a considerable collection of respectable scientific literature which demonstrates the long term effect of early undernutrition on brain development and function.

## VII. ROLE OF EARLY UNDERNUTRITION ON NEUROCHEMISTRY, LOCOMOTOR ACTIVITY AND LEARNING

In reviewing the published reports on effects of lead on brain functions, undernutrition is a frequent associative component of the experimental results. This factor has not been taken into account in evaluating the results from such experiments. It therefore seems appropriate to examine the effects of undernutrition alone on these same parameters. Admittedly, there is a high degree of selection in order to emphasize the objectives of this review.

### A. Experimental Methods for Producing Early Undernutrition

The most common method used in altering nutritional status of neonatal rodents is to vary the number of suckling pups receiving milk from a single lactating dam. The normal rat litter consists of 8 to 12 pups. Arbitrarily, a nursing group of 10 pups per dam has come to be considered a normal size litter. Nutritional deprivation can be imposed by increasing the size of the nursing group to 16 or 18 animals. Conversely, overnutrition can be achieved by decreasing the size to 3 animals per lactating dam. A complicating factor in this design is that, in addition to altering the nutritional status of the developing pups, the quality of maternal stimulation is also affected.

Another approach has been protein and/or caloric restriction in the lactating dam, reducing the quantity of milk produced without altering its composition. Others have employed restricted nursing time by allowing the animals to suckle for limited periods of time during the day, thereby reducing the quantity of milk consumed.

All of these methods produce a total caloric restriction as well as a restriction in individual nutrients, the most important of which is probably

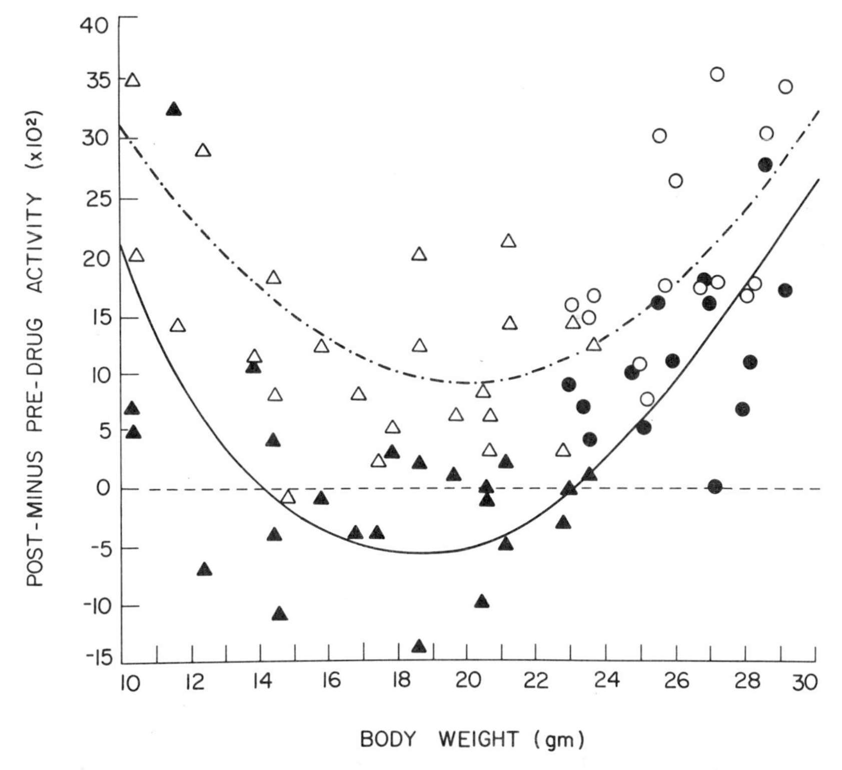


**Figure 9.** Relative locomotor activity in mice 1 hr (open symbols) and 2 hr (filled symbols) after d-amphetamine (10 mg/kg), as a function of body weight (triangles represent mice from growth-retarded litters; circles represent mice from normal size litters).

protein. All the methods have a comparable effect on brain growth; hence no distinction will be made between various methods employed.

### B. Undernutrition and Brain Size

Employing the large and small litter technique, Widdowson and McCance (1960) demonstrated that the growth rate of the nursing pups was inversely proportional to the number of animals in the group. Moreover, it was demonstrated that the weight of the brain was reduced in undernourished animals and, conversely, increased in overnourished animals. It was further shown that, no matter what the state of nutrition after weaning, the undernourished animals never attained normal size and their brains never recovered normal weight. This has also been found to be true in other species. However, this may or may not be true in some of the lead studies which involved early growth retardation.

Dickerson et al. (1967) found that complete recovery in either body size or brain size could not be obtained even with maximal nutritional rehabilitation following growth retardation produced in the neonatal period in pigs.

Many years earlier, Jackson and Steward (1920) had shown that, if undernutrition was imposed later during the growing period of the rat, nutritional rehabilitation could restore normal body and brain weights, although initial growth retardation resulted. The factor that determines whether an animal recovers appears to be the age at which nutritional restriction occurs. The earlier the undernutrition, the less likely will recovery occur after the stress is discontinued. Generally, fundamental characteristics of early growth differ from those in later growth to an extent which allows the relatively older animal to recover from undernutrition. A possible explanation is that early organ growth is mainly due to cell division and a real increase in cell numbers. On the other hand, later organ growth is mainly due to cellular hypertrophy in which existing individual cells become larger.

Winick and Noble (1966) repeated the original experiments reported by Widdowson and McCance (1960) and compared these data with those from animals undernourished at two later times during the growth period. They concluded that if nutritional deprivation is imposed during the proliferative phase of growth, the *rate* of cell division is slowed and the final number of cells is reduced. Most importantly, this change is permanent and cannot be reversed once the normal time for cell division has passed. In contrast to this early phase, if undernutrition is imposed during the period when cells are normally enlarging, the stress will curtail the enlargement but subsequent rehabilitation will enable cells to resume their normal size.

These studies by Winick and Noble (1966) demonstrate unequivocably that total numbers of brain cells are permanently reduced by undernourishing the rat during the first 21 days of life (the proliferative stage); even vigorous rehabilitation thereafter fails to alleviate this reduction in cell number. More precisely, overnourished neonatal rats in litters of three have increased numbers of brain cells compared with animals raised in litters of ten. Cell division can be manipulated in either direction by changing the state of nutrtion during specific stages of the proliferative stage. Winick et al. (1968) found that undernutrition for the first 9 days of life produced a deficit in brain cell number which could be overcome entirely by nutritional augmentation for the remaining 12 days before weaning. This undoubtedly relates to the critical period of spurts in growth rate during early periods of development (Dobbing, 1968). Most importantly, undernutrition slows the rate of cell division, but cells do continue to divide for the same period of time as in the normal animal. Further division in undernourished and well nourished rodents ceases at 21 days of age.

### C. Cellularity, Cortical Thickness and Cellular Composition

The cerebral cortex has a laminar structure which is essentially similar in all mammals. A number of independent studies have shown that pre- and postnatal nutritional deprivation in the rat produces a decrease in the thickness of the cerebral cortex (Bass et al., 1970; Cragg, 1972; Clark et al., 1973).

In undernourished rats at 10 days of age, the laminar layering is poorly defined relative to normal animals (Bass et al., 1970). Neuronal density counts at 24 and 50 days of age reveal normal but more compressed patterning (Cragg, 1972). Clark et al. (1973) found that in undernourished rats at 10 days, the deficit in cortical thickness rostrally is twice that to be expected from the total cerebral weight deficit.

Neurons and glia of the CNS are both derived from the germinal ependymal zone. The sequence of proliferation, migration and differentiation has been described by Bass (1971). The acquisition of neurons occurs before the glial complement is attained. Altman and Das (1966), employing autoradiography, showed that the majority of cerebral neurons are present in the rat before birth. By contrast, in the rat cerebellum, the majority of neurons (excluding Purkinje cells) are formed postnatally. In view of this anatomical development, it is not surprising that the type of cell affected under the influence of inadequate nutrition is intimately related to the timing during development when the stress of undernutrition is applied.

Since neuronal proliferation in the rat cerebrum is prenatal, decreased brain weight resulting from prenatal undernutrition would be expected to be associated with neuronal deficits (Zamenhof et al., 1968). Accordingly, a postnatal undernutrition would not be expected to interfere significantly with cerebral neuronal number. Dobbing et al. (1971) produced postnatal growth retardation by raising rats in large litters. A deficit in neuronal density was found only at those levels corresponding to the internal and external granular layers, which are areas affected by postnatal neurogenesis. Complementary observations were made by Cragg (1972), who imposed postnatal nutritional deprivation in rats. It was found that, although there was a 15% deficiency in cerebral cortical thickness, there was a concomitant 33% increase in neuronal density.

The findings of Dobbing et al. (1971) and those of Cragg (1972) support the view that postnatal undernutrition does not affect those nerve *cells* produced prior to birth. In contrast to the cerebrum, neurogenesis occurs postnatally in the cerebellum. In this region, Dobbing et al. (1971) showed that postnatal nutritional deprivation produces a profound deficit in cerebellar neuronal numbers. This observation was made when the rodents were adult and the deficit was not alleviated by subsequent nutritional rehabilitation. Neville and Chase (1971) found that it was the cells of the internal granular and molecular layers and not the Purkinje cells which were involved in the persistent cerebellar neuronal deficit following rehabilitation.

With regard to glia, the following opinion prevails. Since cerebral neuronal proliferation is primarily prenatal, while glial proliferation is postnatal, a postnatal nutritional deprivation should principally affect glial cell numbers. Indeed, this appears to be the case. Decreased glial cell numbers following postnatal undernutrition have been observed in both the cerebral cortex (Bass et al., 1970; Dobbing et al., 1971; Cragg, 1972; Siassi and Siassi, 1973) and in the cerebellum (Dogging et al., 1971; Clos et al., 1973 and 1977).

It has been demonstrated by Winick (1969) and by Zamenhof (1971a, 1971b and 1971c) that pre- and postnatal caloric and/or protein deprivation result in a decreased total deoxyribonucleic acid (DNA) content of rat brain. More specifically, Winick and Noble (1966) found that caloric restriction imposed postweaning, after the time of maximum brain growth, does not result in deficits of brain DNA content. If, however, caloric restriction is introduced to the lactating dam during neonate suckling, there is an ultimate deficit in brain DNA of the young.

## D. Regional Patterns of Cellular Growth with Undernutrition

Regional patterns of cellular growth are also modified by nutritional deficits during the suckling period. The cerebellum, where the rate of cell division is most rapid, is affected earliest (by 8 days of life) and most severely. The cerebrum, where cell division occurs at a slower rate, is affected later (at 14 days of life) and less severely. The effects produced include a reduced rate of cell division in both cerebellum and cerebrum, as well as a reduction in overall pattern of protein synthesis of various lipids. In addition to these effects on areas of rapid cell division, the increase in DNA content which normally appears in the hippocampus between 14 and 17 days of life in rat is delayed and perhaps even partially prevented. Those regions in which the rate of cell division is highest are affected earliest and most strenuously; cell migration is also curtailed. Regional patterns of lipid synthesis and the effects of undernutrition on these patterns have not been thoroughly studied, as in the case of cell formation. However, there is a consensus of opinion which states that those areas where myelination is most rapid are most susceptible to the adverse effects of early malnutrition.

### E. Undernutrition and Myelin Formation

When a preweanling animal is restricted in its access to the maternal milk supply, a general decrease in brain lipid ensues. Deficits in the constituent cerebroside have been found by Culley and Mertz (1965) to be most marked, suggesting a retardation of myelination. In addition, Dobbing and Widdowson (1965) have reported a deficit in cholesterol. Bass et al. (1970) hold the view that the myelin deficiency may be due to a reduced number of oligodendrocytes; Winick (1970) suggests that lipid synthesis within each cell may also be lowered.

Not only are there changes in myelin lipid with early undernutrition, but protein components of myelin change as well. Wiggins et al. (1974) have shown that, by the end of the third week of life, neonatal nutritional deprivation of rats produces a 30% reduction in the rate of substrate incorporation into myelin protein. The greater percentage of high molecular weight protein found in these animals suggests that myelin at day 20 is of a more immature composition than that of age-matched nutritionally adequate controls.

The slowing of axonal proliferation in undernourished animals decreases the axonic surface area available for myelination; hence the myelin lipid deficiency. In addition, it has been shown by Sima (1974) that restriction of the maternal dietary intake results in a decreased number of myelin lamellae in the offspring, particularly around axons in some lumbar, dorsal and ventral roots, as well as in the optic nerve. With more defined restriction of undernutrition to gestation (*in utero* effects), permanent impairment of myelin deposition in sensory dorsal root fibers ensues.

One would expect that impaired myelin formation would lead to decrements in efficiency of nerve conduction and transmission due to impaired laminar insulation of nerve fibers. It has been established that the frequency distribution of myelin sheath thickness normally exhibits maximal peaks at 41 to 50 lamellae and at 101 to 110 lamellae per axon. Sima (1974) has demonstrated that undernourished rats display peaks at 31 to 40 and 91 to 100 lamellae in dorsal roots. Significantly, the decreased myelin sheath thickness is not directly proportional to the decrease in axonal circumference. Myelin deficiency could account, in part, for some of the neurophysiologic symptoms (*i.e.*, delayed transmission times) observed in undernourished animals and humans.

The effects of nutritional deprivation on individual cell types have not been vigorously researched, but three types of studies have been conducted on brains of animals undernourished during periods of rapid growth. These are 1) histologic evaluations employing specific staining techniques, 2) histochemical techniques attempting to differentiate effects on patterns of ontogeny of specific enzymes and 3) radioautographic procedures to determine the effect of undernutrition on division of particular cell types.

Histologic changes have been observed in the central nervous system of a variety of species (rats, pigs and dogs) raised after weaning on protein-deficient diets (Platt, 1962; Platt et al., 1964). Both neurons and glia in spinal cord and medulla degenerate. These degenerative changes persist even after intensive rehabilitation and are even more severe when nutritional deprivation is imposed at an earlier age.

Histochemical changes in brains of rats subjected to early undernutrition have been described by Zeman and Stanbrough (1969). The appearance of a variety of enzymes is delayed and the quality is reduced.

Radioautographic studies show that, in neonatal rats undernourished during the first 10 days of life, only glial cell division is inhibited in cerebrum, since neuronal cell division ceases prior to birth. On the other hand, in cerebellum, the rate of cell division of external granular cells, internal granular cells and molecular cells is reduced. In addition, the rate of cell division in neurons under both the third and lateral ventricles is decreased. This reduction in neurons under the lateral ventricle explains, in part, the reduced DNA content in hippocampus 5 days later, since these are the cells that are destined to migrate into the hippocampus.

A considerable degree of predictability exists which states that 1) postnatal undernutrition will affect both cell division and myelination in the developing rat; 2) the vulnerable periods coincide with the maximum rate of DNA and myelin synthesis; 3) all regions are vulnerable, but the timing of their vulnerability varies, depending on the maximum rate of synthesis in that particular region; and 4) all cell types are affected if they are dividing at the time the undernutrition occurs.

## F. Undernutrition and Synaptic Organization

Maximal neuronal proliferation in rat cerebral cortex occurs between birth and 25 days of age (Bass, 1971). The total sequence involves axonal and dendritic growth as well as the formation of synaptic connections. The increased neuronal packing observed in undernourished animals suggests that this outgrowth has not taken place. Such an alteration of neuronal processes itself suggests an altered pattern of interconnections between nerve cells. Using electron microscopic techniques, Cragg (1972) observed that the number of synaptic terminals associated with one neuron in layer IV of the undernourished rat was decreased by 38 to 41% relative to rats of adequate nutritional status. Others have used biochemical markers such as choline acetyltransferase (CAT), a constituent of the synapse, as an index of synaptic population. Gaetani et al. (1975) assayed CAT in the synaptic ganglia of mice raised in large litters and found lesser amounts than normal. Using light microscopy, Salas et al. (1974) found a reduced number of dendritic spines and a reduction in the diameter of dendritic processes near their point of origin from the soma in postnatal nutritional deprivation. The electron microscopic study by Hogan et al. (1973) reported findings similar to those of Cragg (1972), with both decreased synaptic numbers and fewer synapses per neuron in response to a postnatal nutritional insult. Many of these observations have been confirmed in the subsequent studies of Gambetti et al. (1974), who found a decreased density of synapses in the undernourished cerebral cortex.

Synapses can be assigned to two general categories with the opposed functions of inhibition and excitation. The assignment can be based on accepted morphologic characteristics and ultimate physiological functionality on ratios of different synaptic types within a particular layer. It has come to be accepted that axodendritic synapses are formed before axosomatic synapses (Voeller et al., 1963; Bunge et al., 1967; Schwartz et al., 1968; Molliver et al., 1973). It becomes axiomatic that, if disruptions of the normal pattern of synaptic formation occur, then the development of normal functional patterns of inhibition and excitement may be impaired. This in turn may create a modification of the normal interneuronal processing of signals and subsequent modification of the behavioral processes mediated by these connections. Dyson and Jones (1976) have investigated changes in synaptic development which may result from developmental undernutrition. They have categorized the synaptic population of the rat cerebral cortical molecular layer into five successive divisions, "A" through "E," where "A" represents a typical adult type junction and "E" an immature stage of synaptic formation. This scheme provides for a continuum of developmental transition from "E" to "A" with the stages "B" through "D" representing intermediate stages. According to their classification, it was found that at 7 days of age "A" (adult) constituted only 1% of the synaptic population, whereas by 16 weeks this proportion increased to 26%. Concomitantly, the percentage of "E" (immature) type junctions decreases from 42% at day 7 to 5% in the 16 week old adult.

The major finding has been that this synaptic distribution pattern is significantly different in rats which have been undernourished during gestation and lactation. For instance, at 20 days of age, less than 2% of the junctions in undernourished rats have the mature "A"-type characteristics, in contrast to the 10% complement of "A"-type junctions in normally fed rats by this stage of development. The only conclusion which can be drawn is that there is a delay in synaptic development in undernourished growth-retarded rodents. Adult animals rehabilitated from day 35 postnatal to adulthood reveal that the synaptic distribution of the previously nutritionally inadequate animals is of a much more immature nature than that displayed by nutritionally adequate control animals.

## G. Undernutrition and Peripheral Nerves

The effect of undernutrition on development and function of peripheral nerves has not been researched as extensively as that of the CNS. It has been reported by Sima (1974) and Sima and Sourander (1974) that undernutrition initiated early in life of rats reduces the caliber of nerve fibers in the sciatic nerve and roots. This reduction is most marked in the thicker fibers. Myelin deposition is also impaired in the ventral root fibers and within fibers of other peripheral nerves such as the optic nerve. The quantitative effect of early nutritional deprivation varies with the particular nerve examined. In the ventral root fibers, myelin deposition is curtailed more than axonal expansion. In the dorsal root fibers, the reduction in myelination and radial axonal growth is proportional. In the optic nerve, the effect of axonal growth is greater than on myelination. Reduction of the caliber and reduced myelin content in the dorsal root fibers persists even after prolonged nutritional rehabilitation. Partial recovery was observed in both sciatic and optic nerves. Since these studies demonstrate that undernutrition early in life can induce changes in peripheral nerves similar to those described in the CNS, the report by Silbergeld et al. (1974) on the effect of inorganic lead on the phrenic nerve-hemidiaphram preparations from obviously growth-retarded mice requires careful interpretation. On the other hand, experimental investigations on the phrenic nerve in undernourished rats (Lewis, 1971) have not revealed any differences in nerve conduction velocity between control and undernourished rats. In the lead study cited, the frame of reference was normal mice; this is not an adequate control.

## H. Undernutrition and Neurochemistry

The observation that severe undernutrition early in life is associated with structural changes in brain composition, *i.e.*, decreased size, number of cells and DNA, protein and lipid content, has already been discussed. In general, these effects are more pronounced, longer lasting and less amenable to rehabilitation the earlier and more prolonged the nutritional inadequacy during development. These well established observations affirm that the cellular and structural integrity of the brain is critical to proper brain function. They further support the view that brain cellularity and connectivities and hence, functionality, are most vulnerable to stresses of dietary restrictions during the critical period of development when cellular proliferation and brain growth is most rapid. However, the appropriate number of brain cells (*i.e.*, neurons and glia) and the correct connectivities among them are not necessarily the only determinants in establishing normal functionality. Brain neurons must be capable of acquiring and/or synthesizing and releasing neurotransmitter molecules. Chemical function in neuronal activity is a thoroughly accepted phenomenon. The effect of diet and undernutrition on the capacity of neurochemical functionality has been the subject of much research, especially inasmuch as all the known or putative neurotransmitters are themselves constituents of the diet (amino acids such as glycine, aspartate or glutamate) or chemically related metabolites of precursors in the diet (serotonin from tryptophan, noradrenaline and dopamine from phenylalanine or tyrosine, $\gamma$-amino butyric acid from glutamine, histamine from histidine and acetylcholine from choline). The working hypothesis is that if nutritional inadequacies limit the input to the body of nutrients essential to the formation of neurotransmitters, then the availability of these precursors to the brain may become rate limiting; thus, the synthesis and utilization of neurotransmitter may be reduced. This is then translated to chemical functionality as it relates to maturation of performance in intelligence, sensory and motor function. The cholinergic system has been demonstrated to be involved in the control of behavior (Russel, 1969). The catecholamine-containing cells in the brain have also been implicated in a variety of visceral and neuroendocrine functions (Wurtman, 1970), some aspects of learning (Ketty, 1971), control of mood (Shildkraut, 1967) and motor function (Sourkes et al., 1969). In the same vein, some specific functions which have become associated with serotonin-containing neurons include control of sleep (Hery et al., 1970), affect and mood (Bowers, 1970), control of body temperature (Bechman and Eiseman, 1970), secretions of the anterior pituitary gland (Wurtman, 1971) and sensitivity to painful stimuli (Harvey and Junger, 1973).

Table 5 lists those reports which have examined the effects of lead on various aspects of neurotransmitter metabolism in animals experiencing undernutrition as reflected in growth retardation. This should be contrasted with Table 6, which tabulates those studies on the influence of early undernutrition on brain monoamines, their metabolites and precursor amino acids.

**1. Brain cholinergic system** The steady state concentration of acetylcholine (ACh) in cholinergic neurons is relatively constant under physiologic conditions. This is due to conditions in which the rate of synthesis of ACh *de novo* adjusts to changes in the rate of release. The rate of synthesis is regulated by mechanisms that depend upon the concentration of ACh in the neuron. The dynamic balance of synthesis and release, under normal conditions, maintains brain concentrations of ACh within a relatively narrow range. The concentration of ACh is more carefully regulated as compared to the catecholaminergic or indolergic neuronal systems. The rate of synthesis of ACh in brain is at least 40 and as much as 600 times

**Table 5.** Reports on lead effects on neurochemistry in presence of undernutrition

| Reference | Species | Period of exposure PM* | G | L | PW | Exposure protocols | Cholinergic | Catechol-aminergic | Indol-ergic | GABA-nergic | Mitochondrial respiration | Histaminergic |
|---|---|---|---|---|---|---|---|---|---|---|---|---|
| Krall et al. (1972) | Rat | — | — | X | X | 4% $PbCO_3$ dam's diet, pups weaned to same | | X | | | | |
| Sauerhoff & Michaelson (1973) | Rat | — | — | X | X | 5% $Pb(Ac)_2$ dam's diet, pups weaned to same | | X | X | X | | |
| Sauerhoff & Michaelson (1973) | Rat | — | — | X | X | 4% $PbCO_3$ dam's diet, weaned to 40 ppm day 15 | | X | | | | |
| Silbergeld & Goldberg (1974) | Mice | — | — | X | X | 5 mg $Pb(Ac)_2$/ml in dam's drinking water, pups weaned to same | X | X | | | | |
| Michaelson et al. (1974) | Rats | — | — | X | X | 4% $PbCO_3$ dam's diet, pups weaned to 400 ppm | | X | | | | |
| Michaelson & Sauerhoff (1974) | Rat | — | — | X | X | 5% $Pb(Ac)_2$ dam's diet, pups weaned to 40 ppm day 16 | | X | X | X | | |
| Silbergeld & Goldberg (1975) | Mouse | — | — | X | X | 5 mg $Pb(Ac)_2$/ml in dam's drinking water, pups weaned to same | X | X | X | X | | |
| Golter & Michaelson (1975) | Rat | — | — | X | X | 5% $Pb(Ac)_2$ dam's diet, weaned to 40 ppm Pb | | X | | | | |
| Modak et al. (1975) | Rats | — | — | X | X | 10 mg $Pb(Ac)_2$/ml in dam's drinking water, pups weaned to same, 60 days | X | | | | | |
| Modak et al. (1975) | Rats | — | — | X | X | 10 mg $Pb(Ac)_2$/ml in dam's drinking water, pups weaned to same 60 days | X | | | | | |

| | | | | | | | | | | | |
|---|---|---|---|---|---|---|---|---|---|---|---|
| Silbergeld et al. (1975) | Mouse | — | — | X | X | 5 mg $Pb(Ac)_2$/ml in dam's drinking water, pups weaned to same | X | X | | | |
| Carroll et al. (1975) | Mouse | — | — | X | X | 5 mg ($Pb(Ac)_2$/ ml dam's drinking water, pups weaned to same | X | | | | |
| Modak et al. (1976) | Mouse | — | — | X | X | 2.5, 5, 10 mg $Pb(Ac)_2$/ ml dam's drinking water, pups weaned to same 30 days | X | | | | |
| Bull et al. (1975) | Rats | — | — | — | X | 3, 12, 60 mg Pb/Kg total dose/2 wks ip | | | | X | |
| Holtzman and Hsu (1976) | Rats | — | — | X | X | 4% $PbCO_3$ at 2 wk postpartum | | | | X | |
| Silbergeld & Chisolm (1976) | Mice | — | — | X | X | 5 mg $Pb(Ac)_2$/ml dam's drinking water, weaned to same | | X | | | |
| Shih & Hanin (1977) | Rat | — | — | X | X | 4% $Pb(Ac)_2$ dam's diet, pups weaned to 40 ppm | X | | | | |
| Modak et al. (1977) | Mouse | — | — | X | X | 2.5, 5, 10 mg/$Pb(Ac)_2$/ ml in dam's drinking water, pups weaned to same 30 days | X | | | | |
| Carroll et al. (1977) | Mice | — | — | X | X | 2, 5, 10 mg/$Pb(Ac)_2$/ ml in dam's drinking water, pups weaned to same | X | | | | |
| Gerber et al. (1978) | Rats | — | — | X | X | 10 mg $Pb(Ac)_2$ g dam's diet, pups weaned to same 6–12 mos. | | X | X | | X |
| Shih and Hanin (1978) | Rats | — | — | X | X | 4% $pbCO_3$, dam's diet, pups weaned to 40 ppm Pd, days 17–51 | X | | | | |

* PM, premating; G, gestation; L, lactation or neonatal; PW, postweaning

**Table 6.** Influence of early undernutrition on brain monoamines and their

| Authors | Species | Dietary conditions | Period of undernutrition PM* | G | L | PM |
|---|---|---|---|---|---|---|
| Miller et al. (1977) | Rat | PR† | X | X | X | X |
| Stern el al. (1975) | Rat | PR | X | X | X | X |
| Stern et al. (1975) | Rats | PR | X | X | X | X |
| Ahmad and Rahman (1975) | Rats | PR+UL | X | X | X | X |
| Stern et al. (1974) | Rats | PR | X | X | X | X |
| Miller et al. (1977) | Rat | PR | X | X | X | X |
| Miller et al. (1977) | Rat | PR | X | X | X | X |
| Ramanamurthy (1977) | Rat | PR | — | X | X | X |
| Kalyanasundarm (1976) | Rats | PR | — | X | X | X |
| Shoemaker & Wurtman (1973) | Rats | PR | — | X | X | — |
| Lee and Dubos (1972) | Mice | PR | — | X | X | — |
| Shoemaker and Wurtman (1971) | Rats | PR | — | X | X | — |
| Dickerson & Pao (1975) | Rats | PR | — | X | X | — |
| Hernandez (1976) | Rat | UL | — | — | X | X |
| Sereni et al. (1966) | Rats | UL | — | — | X | X |
| Hernandez (1973) | Rats | UL | — | — | X | X |
| Sobotka et al. (1974) | Rats | PR | — | — | X | X |
| Enwonwu & Worthington (1974) | Rats | PR | — | — | — | X |
| Dickerson and Pao (1975) | Rats | PR | — | — | — | X |

* PM, premating; G, gestational; L, lactation or neonatal; PM, postweaning

† PR, protein restriction; UL, unequal litter size

greater (Haubrich and Chippendale, 1977) than the rate of turnover of either noradrenaline (0.6 to 0.8 nmol/g/gr), dopamine (1.4 to 2.8 nmol/g/hr) or serotonin (1.1 to 2.2 nmol/g/hr) (Neff et al., 1971).

At the biochemical and cellular level, ACh is synthesized from the condensation of acetylcoenzyme-A and choline (Ch) via the catalytic reaction involving the enzyme choline acetyltransferase (CAT). Choline acetyltransferase is synthesized within the cell bodies and localized within the nerve terminals of the cholinergic neurons. It is the only known enzyme capable of catalyzing this reaction. The CAT reaction is readily reversible and therefore suggests that the synthesis of ACh in the cholinergic neurons could be influenced by mass action. Choline acetyltransferase does not appear to be the rate-limiting step in the synthesis of acetylcholine. It is generally believed that CAT is present in excess and that the bioavailability of substrate, choline, is rate limiting in the synthesis of ACh *in vivo*. It is axiomatic that for the concentration of a substrate to be an important factor in regulating the rate of synthesis of a product, its concentration in the vicinity of the synthetic enzyme must be below its $K_m$. Choline, as a precursor to acetylcholine, is not highly concentrated within cholinergic neurons. The concentration of free choline in brain of rodents killed by microwave irradiation has been reported to be 33 $\mu$M, corrected for 80% tissue water (Stavinova and Weintraub, 1974). On the other hand, the $K_m$ of choline for soluble CAT from brain tissue is greater than 400 $\mu$M.

metabolites and precursor amino acids

| 5 HT | 5HIAA | Tryptophan | NA | DA | Tyrosine | Phenylalanine | Histidine | Histamine |
|---|---|---|---|---|---|---|---|---|
| X↑ | X↑ | X↑ | — | | | — | — | — |
| X↑ | X↑ | — | X↑→ | — | | — | — | — |
| X↑ | X↑ | — | X↑→ | — | | — | — | — |
| X→ | — | — | X→ | X→ | | — | — | — |
| X↑ | X↑ | — | X↑ | — | | — | — | — |
| — | — | — | — | — | | Utilization ↑ | — | — |
| — | — | Utilization ↓ | — | — | | — | — | — |
| X↓ | — | — | X↓ | X↓ | | — | — | X↑ |
| — | — | X↓ | — | — | — | — | — | — |
| — | — | — | X↓ | X↓ | X→ | — | — | — |
| — | — | — | X↓ | X↓ | — | — | — | — |
| — | — | — | X↓ | X↓ | X→ | — | — | — |
| X→ | — | — | X→ | — | X↑ | — | — | — |
| X↑ | — | — | X→ | — | | — | — | — |
| X↓ | — | — | X↓ | — | — | — | — | — |
| X→ | — | — | — | — | — | — | — | — |
| X↑ | X↑ | — | X↑→ | X↑→ | | — | — | — |
| — | — | — | — | — | | — | ↑ | ↑ |
| X↓ | — | X↓ | — | — | — | — | — | — |

Thereafter, the intracellular level of free choline is less than one-tenth its $K_m$ for the enzyme. Hence, relatively small changes in the intracellular concentration of free choline can be expected to alter the rate of synthesis of ACh. Consumption of diet supplemented with choline chloride (Cohen and Wurtman, 1976) or a single meal containing lecithin (Hirsch and Wurtman, 1978) increases the concentration of serum and brain Ch and brain ACh in rats. Increases in brain ACh can be partially blocked by pretreatment with hemicholinium-3, which inhibits the synthesis of ACh by competitive blockade of choline transport into neurons. These studies indicate that choline is the rate-limiting substrate in the *in vivo* synthesis of ACh and that the carrier responsible for transport of choline into cholinergic neurons under physiologic conditions is not saturated with choline *in vivo*. Because choline is a rate-limiting substrate in the synthesis of ACh, it is important to understand the mechanisms that regulate the amount of choline available for acetylation within the neuron. This is especially true in those instances where the effect of lead on cholinergic function has been studied in obviously undernourished animals. Free choline used in the synthesis of ACh is primarily derived from dietary choline or choline-containing dietary constituents.

Choline is transported into cholinergic neurons by a carrier that has a high affinity for choline. Kinetic studies have revealed two distinct transport systems. The affinity of choline for the high affinity transport system of brain ($K_m$ = 1 to 6 $\mu$M) is at least six times greater than it is for the low affinity system, with the former being associated exclusively with cholinergic neurons. It is now believed that the rate of transport of choline by the high affinity carrier is the rate-limiting step in the synthesis of ACh.

**2. Brain serotonin synthesis** Serotonin (5-hydroxytryptamine, 5-HT), a putative transmitter, is synthetized from the essential amino acid L-tryptophan by a two-step reaction involving the hydroxylation of tryptophan to 5-hydroxytryptophan (5-HTP) catalyzed by the enzyme tryptophan hydroxylase. This is then followed by the decarboxylation of 5-HTP to the putative neurotransmitter 5-HT by the enzyme aromatic L-amino acid decarboxylase. The principal deactivated metabolite of serotonin, 5-hydroxyindoleacetic acid (5HIAA), is formed by the enzymatic oxidative deamination of 5-HT to an aldehyde intermediate (catalyzed by monoamine oxidase, MAO) with subsequent oxidation by aldehyde dehydrogenase to 5-HIAA. The rate-limiting step in serotonin formation is tryptophan hydroxylase in the hydroxylation of substrate tryptophan, which to a great extent depends upon brain tryptophan bioavailability. The dependence of 5-HT biosynthesis on tryptophan availability probably arises from the unusually high substrate $K_m$ that characterizes the enzyme tryptophan hydroxylase (Lovenberg, et al., 1968; Moir and Eccleston, 1968). It seems likely that this enzyme normally functions in an unsaturated state. Hence, physiologic increases in intraneuronal tryptophan could drive the hydroxylation of the amino acid and, ultimately, its conversion to 5-HT. There are at least two observations which support the view that the rate of tryptophan hydroxylation *in vivo* depends upon precursor availability (Fernstrom, 1976). 1) The $K_M$ (for tryptophan) for tryptophan hydroxylase ($3 \times 10^{-4}$ M) using the $DMPH_4$ cofactor (Jequier et al., 1969) was found to be considerably greater than the concentration of tryptophan in whole brain ($2 \times 10^{-5}$ to $5 \times 10^{-5}$M, $5 \times 10^{-5}$; Fernstrom and Wurtman, 1971). 2) A single injection of a large dose of L-tryptophan was shown to elevate brain 5-HT and 5-HIAA concentrations within 30 to 60 min (Ashcroft et al., 1965; Moir and Eccleston, 1968). According to Kaufman (1974), the $K_M$ of the tryptophan hydroxylase, using the natural cofactor biopterin ($BH_4$) is closer to $5 \times 10^{-5}$ M). This is still consistent with the hypothesis that tryptophan hydroxylase is not normally saturated with its substrate *in vivo*; *i.e.*, an increase or decrease in brain tryptophan levels can alter the degree of enzyme saturation and thus the rate of tryptophan hydroxylation.

**3. Brain catecholamine synthesis** The catecholamines noradrenaline (NA) and dopamine (DA) are located in brain, exclusively within neurons (Dahlstrom and Fuxe, 1964; Fuxe, 1965), where they probably function as neurotransmitters. The catecholamine (CA) neurotransmitters DA and NA are synthesized in the brain from L-tyrosine and/or phenylalanine (Udenfriend and Zaltzman-Nirenberg, 1963; Shiman et al., 1971). The first, and rate-limiting, step in their formation involves the hydroxylation of the aromatic ring, which is catalyzed by the enzyme tyrosine hydroxylase. The product of this reaction is dihydroxyphenylalanine (DOPA), which is decarboxylated to DA in a reaction mediated by aromatic L-amino acid decarboxylase. Noradrenaline (NA) is formed directly from DA via $\beta$-hydroxylation at the C-7 position.

The rate of CA synthesis is controlled via modulation of the rate-limiting enzyme tyrosine hydroxylase. End product inhibition has been demonstrated, as has long term control of synthesis rate via alterations in the amount of hydroxylase protein. In addition, the rate of catechol formation also depends upon the bioavailability of substrate tyrosine.

## I. Undernutrition and Neurotransmitters

Inasmuch as neurotransmitters, or those bioactive substances suspected of having a role in neurotransmission, are essential for normal central nervous system function, the effect of nutritional stress on those and other chemicals uniquely important to brain function have been studied.

### 1. Biogenic amines

One of the earliest reports on the effect of neonatal undernutrition on catecholamines and indolamines in rat brain was by Sereni et al. (1966). Employing the technique of unequal litter size (4 versus 16) as originally described by Widdowson and McCance (1960) they examined the concentrations of NA and 5-HT in young rats undernourished from birth to 21 days of life, the usual weaning age. The concentration of NA and 5-HT (micrograms fresh brain tissue) was determined at 6, 8, 14, 35 and 45 days of age. Underfeeding slowed the accumulation of NA and 5-HT so that statistically significant differences were evident at 6 and 8 days of age. Subsequently, no significant differences in concentration of NA and 5-HT among the group of animals were detectable. When measuring NA in the brains of undernourished and control animals at 6, 8 and 35 days of age, it was reported that NA concentrations were significantly lower at 6 and 8 days of age in undernourished animals. When nutritional deprivation was extended to 35 days, no significant difference in NA concentration was found between the two groups. If the data of Sereni et al. (1966) are recalculated to express NA content per brain at 35 days, then the undernourished animals contain 17.5% less NA (0.33 $\mu$g versus 0.40 $\mu$g) relative to controls. When the total amount of NA per brain is calculated, the ratio of NA in control versus undernourished animals is approximately 1.60 at 6 days of age and 1.49 at 35 days of age. The effect of undernutrition is therefore retained. This is because the rate of brain growth in the undernourished animals relative to control animals decreases with age, with a correspondingly larger difference in brain weights between the two groups.

Shoemaker and Wurtman (1971 and 1973) nutritionally deprived rats during gestation and through the postnatal period in their study of the effects of undernutrition on a specific property of brain neurons, catecholamine metabolism. Fourteen-day-old timed pregnant rats were fed an inadequate 8% or adequate 24% protein diet which was continued during lactation and suckling. The technique used to impair the nutrition of neonatal rats is based on an early observation by Macomber (1933); it is possibly due to the reduced cystine intake (Nelson and Evans, 1958). Furthermore, it has been demonstrated that the low protein diet consumed by the dam decreases the quantity of milk produced, but does not alter its composition (cf. Shoemaker and Wurtman, 1971). Milk produced by protein-deprived dams contains the same percentage of protein as that of adequately fed dams, but is reduced in quantity. The content of brain NA was lower at 12 days of age, but did not reach statistical significance. On the other hand, both NA and DA were significantly depressed in rats suckling protein-restricted (PR) dams at 24 days of age. To determine whether dopamine was lost only from the neurons that store it or whether it was lost from noradrenergic neurons as well, DA levels were measured in the corpus striatum (Shoemaker and Wurtman, 1973). The DA deficit in this small brain region (24%) accounted for about 60% of the deficit in the whole brain. Concentrations of whole brain tyrosine, a precursor for NA, were similar in both protein-deprived and normal animals.

When normal neonates are crossfostered to PR lactating dams, no change in NA occurs, but a significant depression in DA results at 24 days after birth. When calculated on the basis of brain weight, the concentrations *per se* were not altered, but total amounts were low because of the low brain weights.

As noted above, one effect of early undernutrition on the developing brain is retardation of the accumulation of myelin (Dobbing and Widdowson, 1965), with this deficiency contributing to low brain weight in undernourished rodents. Therefore, expressing the *concentration* of a substance such as a neurotransmitter (*i.e.*, NA which is found only in a fraction of brain neurons) as per unit weight of tissue might obscure functionally significant changes caused by undernutrition in noradrenergic neurons of the CNS. Similarly, changes in the number of cells caused by perinatal undernutrition (Winick and Noble, 1965 and 1966) might occur among cell types which do not contain NA, such as oligodendrocytes or neurons with cell bodies originating in the cerebral cortex. Therefore, expressing the NA content in brains of undernourished animals as the amount per cell might obscure significant changes among noradrenergic neurons. For these reasons, Shoemaker and Wurtman (1971 and 1973) and others have chosen to express data as per whole organ rather than per gram of brain or per cell.

There is some debate as to whether the proper method of expressing transmitter levels in brain of growth retarded rodents is $\mu g/g$ or $\mu g$/total brain. Stern et al. (1975b) agree that this is an important question since, beginning at about day 11, the brains from undernourished rats weigh less than normal. Shoemaker and Wurtman (1971 and 1973) have already pointed out that, because the brain weight changes, it is possible for "increases" in transmitter content to be observed in undernourished rats simply as a result of decrease in the weight of non-neuronal components of the brain. Shoemaker and Wurtman (1971) argued that a substantial portion of the decrease in brain weight in undernourished rats resulted from decreased myelination. However, while many studies report significant impairment of myelination in undernourished rats (Benton et al., 1966; Dobbing, 1966) it is the view of Stern et al. (1975b) that it seems unlikely that this factor would produce an appreciable change in brain weight. They cite Benton et al. (1966) as showing that the brain lipid content (of which myelin is but one) accounts for only 3 to 6% of the brain weight of rats; total brain lipid is decreased in malnourished rats by only 10 to 15%. Thus, at best, a decrease in myelin in the brains of malnourished rats would produce a 0.45 to 0.90% decrease in brain weight and could not, therefore, appreciably contribute to the 100 to 200% changes in monoamine concentrations seen by Stern et al. (1975b) in undernourished rats when expressing their data as $\mu g/g$. It is Stern's contention that if the NA data from growth-retarded rat brains of Shoemaker and Wurtman (1971 and 1973) are calculated on a $\mu g/g$ basis, rather than $\mu g$/brain, the whole brain NA levels are not lowered, but are increased by 20% in undernourished rats. Also bearing on the question of whether the increases in monoamines as seen by Stern et al. (1974, 1975a and 1975b) are due to the use of $\mu g/g$ values is the observation of increased brain concentrations of 5-HT, 5HIAA and NA at birth, an age when brain weights are reported to be almost identical in pups born from nutritionally deprived or adequate dams. Stern et al. (1975b) further commented that this logic clearly demonstrates that feeding of a low protein diet increases the amine concentration in the developing brain, irrespective of whether data are expressed as $\mu g/g$ or $\mu g$/brain.

Lee and Dubos (1972) studied brain catecholamine metabolism in undernourished mice. These were progeny of protein-restricted mice of dams fed a diet containing 20% wheat gluten as the sole source of protein. Dams were placed on the gluten diet from day 14 of pregnancy and through lactation to day 21 after delivery of young. After weaning all animals were transferred to nutritionally adequate diets. At 3 months of age, the NA and DA content of brain were significantly depressed in mice raised by undernourished dams. The results were similar when expressed as either total content or concentration per unit weight basis.

Hernandez (1973) studied serotonin metabolism in young rats undernourished from birth using unequal litter size (6 versus 16 as described by Widdowson and Kennedy, 1962). At weaning, the nutritionally stressed animals were continued at 50% of the normal caloric requirements. The concentration of brain 5-HT was not significantly different from controls at 5, 12 and 21 days of age. A tendency towards increases at 12 days and decreases at 21 days in undernourished animals was noticed but failed to achieve statistical significance. Interpretation is clouded by variation of as much as 21.6% around the standard error of the mean values.

Sobotka et al. (1974) also examined the effect of early undernutrition on NA, DA, 5-HT, and 5-HIAA. An effective state of undernutrition was imposed on newborn rat pups throughout their postnatal period of development by feeding lactating dams a 12% versus 24% casein diet. Only one time interval was examined; this was 22 days of age. The primary neurochemical effect of neonatal undernutrition appeared to involve the serotonergic system. Brain stem concentration of 5-HT (+40%) and its primary metabolite, 5-HIAA (+102%), were both significantly increased above control values in the 22-day-old weanling. No changes were observed in either NA or DA concentrations. Since there was a statistically significant depression in tissue weights of telencephalon and brain stem, it is not entirely clear whether stronger trends would be evident if calculated in amounts per organ rather than concentration. Restricting the time of undernutrition to the postnatal period is similar to the conditions of studies by Shoemaker and Wurtman (1971 and 1973) and Sereni et al. (1966). However, whereas depression in catecholamines was noted in these latter studies, investigation by Sobotka et al. (1974) could not confirm this. On the other hand, Stern et al. (1974) suggest a reduced rate of NE and 5-HT synthesis in the brains of protein-deprived rats (8% versus 25%). throughout development. This view was based upon the response of brain biogenic amines to a monoamine inhibition tissue depleter and physiologic stress.

Stern et al. (1974) used a more protracted period of nutritional restriction than previously attempted by feeding protein deficient diets (8% versus 25% casein) to female rats for 4 to 5 weeks prior to mating with normal males. The diets were maintained through gestation and lactation; animals were weaned to the same diet. Brain 5-HT, 5-HIAA and NA levels in five brain regions were measured at 140 to 150 days of age. Brain levels of 5-HT, 5-HIAA and NA were significantly increased in the protein-restricted animals irrespective of whether the data are expressed as concentration (μg/g) or content (μg/brain). At the time of testing, the brain weights of PR rats were 90% of that of normal rats. With regard to NA findings, this study by Stern et al. (1974) is in stark contrast with the earlier reports by

Shoemaker and Wurtman (1971 and 1973) and by Sobotka et al. (1974), in which NA levels in 2- to 24-day-old protein-deprived rats were reduced by 20% or remained unchanged. The differences may be due to the extreme age differences in the test animals. With respect to the alterations in brain indoleamines, there is close agreement with the findings of Sobotka et al. (1974). Subsequent reports by Stern et al. (1975a and 1975b) employing the same experimental design yielded the same findings. The ontogenetic development of 5-HT, 5-HIAA and NA in protein-deprived rats from birth to 300 days of age was examined (Stern et al., 1975b). In undernourished rats, brain concentrations of 5-HT and 5-HIAA remained elevated up to 300 days of age, with the largest effect (200% increase) occurring in the subtelencephalic brain regions. At most ages, the increase in brain NA levels in malnourished rats was less pronounced than for indoles.

To produce protein undernutrition *in utero*, Ahmad and Rahman (1975) fed low protein (8%) diets to female rats for 1 week before breeding. The offspring were exposed to different degrees of "energy" undernutrition during suckling by varying the number of pups per lactating dam (unequal litter size), as well as by feeding the lactating dam a low protein diet. The weanlings were fed diets at different levels of protein restriction to impose different degrees of postweaning protein inadequacies up to 42 days of age. The concentrations of DA, NA and 5-HT, when expressed as micrograms per gram wet weight of the brain of the rats exposed to different degrees of nutritional restrictions during gestation and suckling and after weaning, remained unchanged as compared with control animals. The contents of monoamines per whole brain varied proportionately with the weight of the brain.

Dickerson and Pao (1975) fed protein-deficient diet (7% versus 21%) to pregnant rats, starting with the fifth day of gestation through 21 days of lactation. Animals were examined for certain brain free amino acids, NA, 5-HT and a number of enzymes within 10 hr of birth and again at 21 days of age. In spite of the fact that the degree of growth retardation became more pronounced as the nutritional restriction was continued through the suckling period, the *concentration* of NE and 5-HT remained unchanged. On the other hand, the concentration of tyrosine was dramatically higher in each part of the brain of the undernourished animals. If these data are recalculated on the bases of total content relative to weights of brain parts reported, one finds a considerable depression in catecholamine content in brain of undernourished animals. The same holds true for 5-HT and supports the earlier view of Shoemaker and Wurtman (1971, 1973). The report by Kalyanasundarum (1976) is also of interest in this context. Starting on the fourth day of pregnancy, rats were given a protein-deficient diet (7.5% versus 20%) which was continued throughout gestation and lactation. At 21 days of age, the young were weaned onto the respective diets of their dams.

Brain tryptophan levels were measured at 7, 14, 21, 28 and 35 days after birth. During the first week of life, brain tryptophan levels in undernourished rats were significantly higher than controls; thereafter (days 14, 21, 28, 35), they were significantly lower. Although 5-HT and 5-HIAA were not assayed, the finding that tryptophan levels are lowered by dietary and caloric deficiency suggests that brain 5-HT levels and/or turnover may also be affected. This is consistent with the earlier reports of lowered brain content of 5-HT in protein-deprived neonatal rats due to lack of substrate availability.

Since the main development of brain monoaminergic processes and connections occurs postnatally (Aghajanian and Bloom, 1967; Lorizon, 1969), Hernandez (1976) tested this period of fast outgrowth for its relative vulnerability to undernutrition. Nutritional restriction was developed using unequal litter size. Within 24 hr after birth, the newborn rats were mixed and redistributed into litters of different sizes. For the undernourished group, litters of 16 pups per dam were formed. The numbers in normal size litter were not indicated. After weaning, the animals from the undernourished group were divided into two subgroups; one was nutritionally rehabilitated and the other was kept nutritionally restricted at 50% of the normal caloric requirements. At the ages of 10, 21, 40 and 105 days, the animals were killed for analysis. Brains were divided into two parts, the posterior brain (pons plus medulla mesencephalon and the thalamohypothalamus complex) and the anterior brain (cortex, caudate hippocampus and remaining structures). Serotonin (5-HT) and noradrenaline (NA) analyses were only determined at 10 and 21 days. In malnourished brains, a significant increase of 5-HT levels was observed in both the posterior and anterior brain sections only at 21 days of age. No changes in NA levels were found. When serotonin content is expressed, not as concentration per gram of wet brain tissue, but as total amount, there is a tendency toward higher levels in malnourished brains; however, the difference fails to achieve statistical significance.

Ramanamurthy (1977) examined the levels of NA, DA, 5-HT and histamine in brain at different ages of development in the rat and the effect of protein caloric restriction on these levels. Protein-deficient diets (7.5% versus 20%) were given to rats through gestation and lactation; pups were weaned to the same diet. Serotonin, dopamine and noradrenaline were measured in brains of fetuses at 10, 15, 17 and 19 days of gestation and at birth and 7, 14, 21, 28 and 35 days of life. Serotonin, noradrenaline and DA could not be detected in fetal brains before the gestational age of 15 days. From then on there was a gradual and steady increase in the levels of all three amines until the 28th day postnatal. Maternal nutritional deprivation did not lead to significant alterations in these amine levels in the brain of fetuses *in utero* or up to the end of 7 days after birth. However, 14 days after birth they were found to be significantly *lower* and the difference continued up to 35 days of age. Histamine was measured from birth onward at 7-day intervals until 28 days of age. Levels of histamine were significantly higher in the brains of those pups born to and reared by dams fed a low protein diet, as compared to those in control groups at all stages of brain development. Similar findings of raised brain histamine have been reported in more mature undernourished animals (Enwonu and Worthington, 1973).

Miller et al. (1977a) have studied the question of tryptophan availability in relation to elevated brain tryptophan, 5-HT and 5-HIAA, as well as phenylalanine utilization (Miller et al., 1977) and tryptophan utilization (Miller et al., 1977b) in protein-deprived rats. Developmental changes in tryptophan, serotonin (5-HT) and 5-hydroxyindoleacetic acid (5-HIAA) in a number of brain regions were examined in normal and deprived rats from birth to 30 days of age (Miller et al., 1977a). Female rats were fed isocaloric (4.3 Kcal/g) diets containing low protein (8% casein) and normal amounts (25% casein) for 5 weeks prior to mating. At birth, each litter was culled to 8 pups and randomized with other litters fed the same diet and born on the same day. After weaning at 21 days, the animals were continued on their respective nutritionally inadequate diets until killed for chemical analysis of indole metabolism. Significantly elevated brain tryptophan, 5-HT and 5-HIAA were found at most ages examined. The regional anatomical analysis of 5-HT, 5-HIAA and tryptophan showed that they were more highly concentrated in the diencephalon, midbrain and pons-medulla regions in the undernourished rats than in controls. The study demonstrated that rearing rats on a diet low in protein but adequate in all other respects significantly elevates brain tryptophan and its related amine concentrations, probably as a consequence of developmental alterations in plasma tryptophan availability.

Miller et al. (1977a) suggest that the increased amount of brain tryptophan, 5-HT and 5-HIAA seen in the undernourished animals can be directly related to the increased amounts of free plasma tryptophan available for their brain metabolism. Undernourished rats may be shunting more tryptophan to the brain at the expense of the periphery. A similar suggestion has been made by Shoemaker and Wurtman (1971 and 1973) for tyrosine which was concentrated in the brains of undernourished rats at the expense of the rest of the body. Miller et al. (1977 and 1977b) have also shown a significant increase in $^{14}$C-phenylalanine, but a decrease in $^{14}$C-tryptophan uptake and incorporation into brain protein of undernourished rats.

## 2. Noradrenaline turnover

In their study of the effect of undernutrition on the metabolism of catecholamines, Shoemaker and Wurtman (1973) examined NA turnover. The turnover of brain NA was estimated by determining the $t_{0.5}$ for ($^3$H)-NA disappearance between 10 and 180 min after its intracisternal injection, calculated a rate constant and multiplying the rate constant by the pool size of endogenous NA. It was found that nutritional deprivation lengthened the $t_{0.5}$ and slowed the turnover in the main storage compartment of brain NA. Lee and Dubos (1972) injected ($^{14}$C)-tyrosine and ($^{3}$H)-NA intracisternally in undernourished mice and followed the incorporation of radioactivity into the brain tissue. Incorporation of ($^{14}$C)-tyrosine into the brain tissue was rapid and reached a maximum 30 min after injection. After 1 hr, there was a slow but progressive fall in the incorporation. The progeny of undernourished dams showed significantly higher incorporation at 1 hr after injection. The authors suggest that their findings revealed that perinatal undernutrition has an effect similar to that of increased activity of sympathetic neurons in that both the biosynthesis of catecholamine and the catabolic activity are changed. During the period of ½ to 6 hr after intraventricular injection of ($^{3}$H)-NA, the labeled catecholamine was metabolized in such a way that 41 to 61% of NA, 11 to 15% of normetanephrine, 2.5 to 4.0% of deaminated metabolites and 23 to 25% of O-methylated deaminated metabolites were found in the brain. However, it is cautioned that their findings do not establish whether the high levels of radioactivity in brain of undernourished animals are due to increased accumulation of metabolites produced from the catabolic activity of ($^{3}$H)-NA or to the slow metabolic turnover or utilization of catecholamine. It should be recalled that, like those of Shoemaker and Wurtman (1971 and 1973), these mice raised from dams fed a low protein diet have low levels of NA and DA; Unlike those of Shoemaker and Wurtman (1971 and 1973), they have depressed enzymatic activities in tyrosine hydroxylase. The only plausible explanation offered is the two different methods of maternal dietary deprivation, namely poor quality protein (wheat glutein) versus a type of protein deficiency which depresses milk production (without affecting quality).

### 3. Cholinergic neurochemistry

Those research reports dealing with effects of early undernutrition or growth retardation on cholinergic neurochemistry of the central nervous system are tabulated in Table 7. Acetylcholine (ACh) plays a key role in the transmission of nerve impulses (Nachmanson, 1961). It has been suggested that the concentration of ACh in the brain varies inversely with the functional activity of the brain. Acetylcholine has been found to increase in anesthesia and sleep, in deep narcosis and after the administration of CNS depressants. On the other hand, the concentration of ACh in the brain has been found to decrease during convulsions, electrical stimulation of the brain and emotional excitement. Russell (1969) has reviewed the relationship of cholinergic transmission to some aspects of behavioral function. Ontogenetically, the concentration of ACh in brains of rodents increases with age up to 100 days or more (Crossland, 1951).

*Acetylcholine (ACh)*

It would appear that the report by Rajalakshmi et al. (1974) is the only study on the effects of preweaning and postweaning undernutrition on acetylcholine levels in rat brain. The bioassay of ACh levels of whole brain of rats raised in unequal litters showed that, while the concentration of ACh ($\mu$g ACh/g) remained unaffected by neonatal undernutrition, protein deprivation following weaning led to decreases in ACh concentrations. The maintenance of cholinergic transmitter levels may be the result of a number of complex interacting factors which are difficult to reconcile with this one report. The early work by Adlard and Dobbing (1970) on neonatal undernutrition showed depressed acetylcholinesterase (AChE) activity during the early weeks of life, but enhanced AChE activity at 16 weeks of life. Furthermore, the use of unequal litter size raises the question of how environmental and/or neonatal stimulation offsets these results. The studies by Eckhert et al. (1975) and others have shown that postnatal stimulation (*e.g.*, handling) has definite effects on cholinergic enzyme activity in undernourished rats.

*Cholineacetyltransferase (CAT) and Acetylcholinesterase*

These enzymes are responsible for the synthesis and inactivation of the neurotransmitter acetylcholine (ACh). The synthetesizing enzyme (CAT) is mainly localized in cholinergic nerve endings, whereas a significant percentage of acetylcholinesterase (AChE) activity is localized in the membranes of nerve terminals (Whittaker, 1965). Therefore, the assay of CAT and/or AChE mainly provides information about those synapses whose transmission involves the cholinergic system.

Sereni et al. (1966) examined the activity levels of AChE in brains of developing rats subjected to undernutrition by the method of unequal litter sizes (4 or more versus 16). Following weaning at 21 days of age, the undernutrition was continued by providing one-half the amount of food eaten per day by the well nourished control animals. At 21 days of age, the experimental animals weighed approximately one-half as much as the control group. The weights of those undernourished until 35 days of age were about one-third those of control animals of the same age, whereas the weights of those animals fed ad libitum from 21 to 35 days of age fell about halfway between those of control animals and those of animals deprived for the entire period. Significant differences in brain growth were observed among the control and experimental groups at every age studied. The activity of AChE was estimated at 6, 8, 14, 21, 35 and 45 days of age in animals undernourished until 21 days of age. Underfeeding slowed the activity of AChE in the brain so that statistically significant differences were evident until 14 days of age for AChE. Subsequently, no significant differences in enzymatic activity among groups of animals were detectable. Upon subsequent nutritional rehabilitation, the enzyme activity per milligram

**Table 7.** Influence of early undernutrition on cholinergic neurochemistry

| Reference | Species | Dietary conditions | Period of undernutrition | | | |
|---|---|---|---|---|---|---|
| | | | PM* | G | L | PW |
| Sereni et al. (1966) | Rat | UL† | — | — | X | X |
| Adlard & Dobbing (1970) | Rat | PCR | — | X | X | — |
| Im et al (1971) | Rat | PCR | — | — | X | X |
| Im et al. (1971) | Rat | PCR | — | — | X | X |
| Adlard & Dobbing (1971) | Rat | PCR | — | X | X | — |
| Adlard & Dobbing (1971) | Rat | PCR | — | X | X | X |
| Adlard & Dobbing (1971) | Rat | PCR | — | X | X | — |
| Adlard & Dobbing (1972) | Rat | PCR | — | X | X | — |
| Adlard & Dobbing (1972) | Rat | PCR | — | X | X | — |
| | | | | X | X | X |
| Gambetti et al. (1972) | Rat | PCR | — | X | X | — |
| Eckert et al. (1973) | Rat | | — | X | — | — |
| | | | — | — | X | — |

| Tissue preparation | ACh | CAT | AChE | ChE | BuChE |
|---|---|---|---|---|---|
| Whole brain minus cerebellum | — | — | d‡6–14↓<br>d21–45→ | — | — |
| Whole brain | — | — | | — | — |
| PCR, 0–7 wk, test at 6–9 mos. whole brain | — | — | — | ↑6–9 mos. | — |
| | | | | specific activity | |
| PCR, 0–7 wk, test 1, 3, 7, 39 | — | — | — | 1W→ | — |
| wk, whole brain | — | — | — | 3W→ | — |
| | — | — | — | 7W↑ | — |
| | — | — | — | 39W↑ | — |
| PR, $G_{d7}$, $L_{d21}$, test 16 wk, whole brain | — | — | W16 ↑14% | — | — |
| PR, $G_{d7}$, $L_{d21}$, $PW_{W21}$ | — | — | d21 ↓16% | — | — |
| (crude mitochondrial) Forebrain | — | — | W12 ↑20% | — | — |
| | — | — | d21 ↓8% | — | — |
| (crude mitochondria) Brainstem | — | — | W12 ↑19% | — | — |
| Olfactory Bulb | — | — | d21 ↓(?%) | — | — |
| Cerebellum | — | — | d21→ | — | — |
| PR, $G_{d7}$, $L_{d21}$, test d21 whole brain | — | — | d21 ↓11% | — | — |
| PR, $G_{d7}$, $L_{d21}$, Forebrain | — | — | d21 ↓16% | — | — |
| (crude mitochondria) Cerebellum | — | — | d21→ | — | — |
| Brainstem | — | — | d21 ↓8% | — | — |
| Olfactory | — | — | d21 ↓22% | — | — |
| whole brain | — | — | — | — | →(?%) |
| PR, $G_{d7}$, $L_{d21}$, Forebrain | — | — | W12 ↑11% | — | — |
| Cerebellum | — | — | W12 ↑11% | — | — |
| Brainstem | — | — | W12 ↑8% | — | — |
| PR, $G_{d7}$, $PW_{12W\ Ks}$ Forebrain | — | — | W12 ↑16% | — | — |
| Cerebellum | — | — | W12 ↑10% | — | — |
| Brainstem | — | — | W12 ↑13% | — | — |
| PR, $G_{d14}$, $L_{d21}$ | — | d12→ | d12 ↓26% | — | — |
| Cortex | — | d24→ | d24→ | — | — |
| Midbrain | — | ↓d49<br>↓d49 | ↑d49<br>↓d49 | — | — |

**Table 7.**—*continued*

| Reference | Species | Dietary conditions | Period of undernutrition | | | |
|---|---|---|---|---|---|---|
| | | | PM* | G | L | PW |
| Eckhert et al. (1974) | Rat | PR | — | X | — | — |
| | | | — | — | X | — |
| Coupain & Tyzbir (1974) | Rat | PR | X | X | X | X |
| Sobotka et al. (1974) | Rat | PR | — | — | X | — |
| Rajalakshmi et al. (1974) | Rat | UL | — | — | X | X |
| | | PR | — | — | — | X |
| Ahmed & Rahman (1975) | Rat | Pb | — | X | X | X |
| | | UL | | | | |
| Eckhert et al. (1975) | Rat | PR | — | — | X | — |
| Im et al. (1976) | Rat | PR | — | — | X | X |
| Eckhert et al. (1976) | Rat | PCR | — | — | X | — |
| Eckhert et al. (1976) | Rat | PCR | — | X | — | — |
| | | | — | — | X | — |
| | | | — | — | — | X |

* PM, premating; G, gestation; L, lactation or neonatal; PW, postweaning

† Dietary restriction: UL, unequal litter size; PCR, protein colonial restriction; PR, protein restriction.

‡ d, day of age; W, week of age, ns, not significant

**Table 7.**—*continued*

| Tissue preparation | ACh | CAT | AChE | ChE | BuChE |
|---|---|---|---|---|---|
| G only, test d49 Cerebrum | | ↓d49 | ↓d49 | — | |
| Brainstem | | ↓ | ↑ | — | |
| Cerebellum | | → | → | — | |
| L only, test d49 Cerebrum | | → | → | — | |
| Brainstem | | ↓ | ↑ | — | |
| Cerebellum | | ↓ | ↑ | — | |
| $PR_{d42}$ whole brain | — | — | ↑ | — | — |
| PR, $L_{d2\text{-}21}$ Telencephalon | — | — | ↓10%(ns) | — | ↓21%(ns) |
| Brainstem | — | — | ↓10%(ns) | — | ↓2%(ns) |
| Cerebellum | — | — | ↓31% | — | ↓7%(ns) |
| Whole brain | d7→<br>d14<br>d21→<br>d28→ | — | — | — | — |
| Whole brain | | — | — | — | — |
| d42 | — | — | d42 | — | — |
| Whole brain | | | → | | |
| d21 whole brain minus cerebellum | — | ↓ | ↑ | — | — |
| Neuron Rich Fraction | — | ↓ | ↑(ns) | — | — |
| Glial Rich Fraction | — | ↓(ns) | ↑(ns) | — | — |
| $PR_{W7}$, test 20 wk, whole brain | — | — | — | ↑ | — |
| $L_{d0\text{-}21}$, test W7 Brainstem | — | ↓ | ↑ | — | → |
| Forebrain | — | — | ↓ | ↓ | — |
| Brainstem | — | — | ↑ | ↑ | — |
| Brainstem | — | ↓ | ↑ | ↑ | — |
| Cerebellum | — | ↓ | — | — | — |
| Forebrain | — | — | ↑ | ↑ | — |
| Brainstem | — | ↓ | ↑ | ↑ | — |
| Cerebellum | — | ↑ | → | → | — |
| Forebrain | — | — | ↑ | ↑ | — |

nitrogen more closely approached the normal range. However, as the *total* brain/nitrogen is reduced, the *total* AChE activity is still deficient.

The difficulty of expressing enzyme activities in a meaningful way is emphasized by the work of Adlard and Dobbing (1971), which showed that the AChE activity per brain is lower in undernourished rats when they are matched to coetaneous controls. By contrast, when the controls are matched for body weight, the undernourished rats exhibit a higher activity level. One of the major contributions to the study of the effects of early undernutrition on cholinergic neurochemistry was made by Adlard et al. (1970) and Adlard and Dobbing (1971a, 1971b, 1971c, 1972a, and 1972b). Their earliest studies showed that the development of whole brain AChE activity in suckling rats was somewhat retarded by underfeeding the dam during pregnancy and lactation. More in keeping with the results of Sereni et al. (1966), the study by Adlard and Dobbing (1971a) demonstrated an 11% reduction in AChE activity (per gram wet weight) of whole brain tissue at 21 days of age in male rats undernourished during fetal life and lactation. Undernutrition was achieved by underfeeding the dam during gestation and lactation such that the young at 21 days of age had a 37% deficit in body weight as compared to control animals. This is comparable to the degree of growth retardation observed in the studies on lead-exposed neonatal rodents when exposure is via lead contaminated maternal milk (Silbergeld and Goldberg, 1973). One of the more interesting observations of Adlard and Dobbing (1971a) was that the activity per brain was less in undernourished animals than in control animals matched for age, but the activity was much higher in growth-retarded animals than controls when matched for body weights.

In order to determine whether deficits in AChE activity persist in adult life, previously undernourished animals were nutritionally rehabilitated after 3 weeks of age (Adlard and Dobbing, 1971). At 16 weeks of age the body weight deficit in rehabilitated animals was 28%, compared with brain weight deficits of 10%. Despite the deficit in brain weight, rehabilitated animals had whole brain AChE activity 14% higher than controls. It was shown that an enzyme whose activity per gram of fresh brain was reduced at the time of weaning by undernutrition during the suckling period exhibits a change in the opposite direction following rehabilitation.

Subcellular fractionation studies by Adlard and Dobbing (1971) using crude mitochondrial fractions (mixtures of myelin, synaptosomes and mitochondria) from rats undernourished during gestation through 12 weeks of age showed highly significant deficits in AChE activity in total forebrain and brain stem at 21 days of age. At 12 weeks of age, deficits persisted in brain stems with complete recovery and enhancement of enzyme activity of forebrain and brain stem. Contemporaneous with this is the study by Gambetti et al. (1972) which employed purified synaptosomal fractions from cerebral cortex of undernourished rats. In this instance there was a significant total deficit of AChE in 12-day-old undernourished rats, although by 24 days this difference was no longer significant.

The contrasting finding by Adlard and Dobbing (1971) and Gambetti et al. (1972) should be viewed in terms of their experimental procedures. The variations between the timing of recovery from the early deficits in AChE activity in these two groups of experiments is probably attributable, in part, to the different brain regions used and, more importantly, to the different purities of the fractions employed. The crude mitochondrial fraction used by Adlard and Dobbing (1971) can be expected to contain mitochondria and myelin in addition to synaptosomes (pinched off nerve endings). By comparison, Gambetti et al. (1972) used a highly purified synaptosomal preparation. Another feature is the method Adlard and Dobbing (1971) used to express their results. Expression of data in terms of fresh weight often yields results that are higher than expected when compared to the content of an enzyme or substrate in a given area or whole brain. This can probably be explained by the relatively pronounced effect of early postnatal undernutrition on glial formation and myelin deposition as compared to other membranes. Therefore, the concentration of a chemical component uniquely localized in relatively unaffected cell (nerve) would appear higher; a more meaningful measure for normalizing data (*i.e.*, mg nitrogen) might be more instructive.

Another active contributor to this area of research is Barnes and his associates (Im et al., 1971a, 1971b, 1972, 1973 and 1976; Eckhert et al., 1973, 1974, 1975, 1976a and 1976b). Initially, they showed that a reduced milk intake prior to weaning, protein restriction for an additional 4 weeks and rehabilitation up to 31 weeks of age led to higher AChE brain activities (Im et al., 1971). These findings are consistent with the regional studies of AChE by Adlard and Dobbing (1972) and comparable to the premature and early weaning study of Himwich et al. (1968). Similar observations have been made in other species. Im et al. (1973) fed low protein diet to piglets for 8 weeks, followed by nutritional rehabilitation. Both underfed and protein-deficient animals demonstrated increased ChE activity in the cerebrum but not in the cerebellum or brain stem. Eckhert et al. (1976) examined regional changes in rat brain CAT and AChE activity resulting from undernutrition imposed during different periods of development, *i.e.*, gestation and the postweaning period as well as during the suckling period alone. The effect on the activity of the individual cholinergic enzymes is complex and not the same in different regions of the brain exposed to undernutrition during different periods of development.

The results of this study, particularly in the differences in AChE activity in the forebrain in prenatally and postnatally undernourished rats, show how pooling different brain regions or combining the effects of undernutrition during the different developmental periods could account for the crossover (Adlard and Dobbing, 1972) of AChE activity between day 21 and 12 weeks of age in the undernourished and control rats. Furthermore, the lower activity of the presynaptic marker CAT in the brain stem and cerebellum of rats undernourished during lactation indicates that cells and cell processes containing CAT are probably less prevalent in the undernourished animals than in the normal control animals. The lower CAT activity in the brain stems of postnatally undernourished rats may reflect an alteration in the biochemical development of the caudate, putamen and cranial motor nuclei where the highest concentrations of the CAT activity are localized in the rat brain stem. Recently Eckhert et al. (1975), using glial-rich and neuronal-rich cell fractions of rat brain obtained from animals undernourished during the suckling period, found decreases in CAT in the neuron-rich fractions and increases in AChE of whole brain. The tendency toward increase of AChE activities in neuron and glial-enriched fractions was not statistically significant. Of considerable interest is the effect of postnatal handling (stimulation) on enzyme activities of undernourished rats during the suckling period. Whereas postnatal undernutrition resulted in a decrease in CAT activity in the neuronal-rich cell fraction of nonstimulated (nonhandled) rats, no change in CAT activity was observed in this cell fraction from handled rats. This bears directly on the interpretation of data derived from those lead exposure studies where newborn sucklings are intubated daily for as many as 21 days. Other relevant observations are those reported by Sobotka et al. (1974), who examined several biochemical parameters in rat brain after a 3-week suckling period in malnourished versus normal litters. In addition to evident alterations in the development of different areas, obviously related to cell number and size and myelination, these workers observed a significant impairement in the maturation of cerebellum as indicated by depressed AChE levels. The same undernourished animals exhibited a marked increase in adrenal weight and corticosterone content, indicating an activation of the pituitary-adrenal system.

It is evident that the experimental conditions in which undernutrition and/or growth retardation are associative factors during early development will have measurable effects on brain growth, cell maturation, synaptogenesis and neurotransmitter metabolism. These dysfunctions, either individually or in combination, inevitably will be translated into physiologic and behavioral performance.

### J. Undernutrition, Spontaneous Activity and Learning

The question which has most concerned investigators is whether or not intellectual capacity is affected by early growth retardation. There is indisputable evidence that early growth restriction results in both deficits and distortions in brain growth and cellular and neurochemical differentiation.

It is therefore not surprising that learning ability, *at the time of the nutritional deprivation*, is deficient (Cowley and Griesel, 1959; Baird et al., 1971). However, the evidence of any impairment outlasting lengthy nutritional rehabilitation is equivocal. The influence of early undernutrition on behavioral development and learning in rodents has recently been reviewed by Leathwood (1978). His conclusion is that, while prenatal and early postnatal undernutrition can lead to permanent deficits and distortions in the physical structure of the brain, behavioral studies using a range of tasks and assessment procedures have produced little or no evidence that perinatal undernutrition *per se* causes any permanent deficits in brain function. Dobbing and Smart (1974) stated that instances of both inferior and superior performance by previously undernourished rodents are to be found in equal abundance. Part of the reason may be that there are differences in other aspects of behavior which affect performance in tests of habit formation, sometimes to the disadvantage of the previously underfed animal, sometimes to its advantage. The methodologic and design problems associated with studies on effects of early developmental undernutrition on postweaned or adult learning in rodents have been addressed by Plaut (1970), Crnic (1976) and Wehmer and Jen (1977).

It has long been the impression that rodents which have experienced early undernutrition are more excitable than normal animals, even after lengthy rehabilitation. They tend to be more "nervous" and more difficult to handle. Recent attempts have been made to quantify these impressions experimentally. The consensus seems to be that in many cases the results are united by a common theme, which states that most of the differences in behavior, in response to food, to unpleasant stimulation and to other animals, are consistent with differences in threshold of arousal. Previously undernourished animals appear to have a lower threshold of arousal. The same amount of stimulation seems to produce a greater response in them than in normal animals. One would predict that they should require less stimulation to elicit specific responses. What has frequently been omitted is the quality of physiologic stress engendered by undernutrition and its influence on hormonal interaction with developing brain and behavior.

Undernutrition is a frequent covariable in studies on lead-related changes in locomotor activity, but it is less often a factor in learning studies. For instance, a substantially lower body weight is almost always noted in mice and rats reported to be hyperactive and display altered response to CNS-active drugs. Invariably, they were compared to age-matched but not weight-matched controls. These animals had been inadvertently subjected to early malnutrition during the course of a lead toxicity study, as well as during a critical period of brain development. It would therefore be appropriate to examine some of the research which has been reported on the effects on spontaneous locomotor activity of early undernutrition alone.

One of the earliest studies on the influence of early developmental undernutrition on behavior associated with and following rehabilitation was by Lát et al. (1961). New born rats were suckled in litters of 3 versus 15 to 20 from the first day of life until weaning at 21 days of age using the technique of uneven litter size as described by Widdowson and McCance (1960). These results showed that accelerating the growth of rats by providing them with an abundance of food in the first week or two of their lives made them more active and inquisitive at each chronological age throughout their growth period. Lát (1956) had reached the same conclusion some years earlier about rats which were growing faster than their litter mates for genetic reasons. However, the study by Lát et al. (1961) indicated that it was possible to alter the psychosomatic constitution of rodents by nutritional means. The authors suggested that the relation between nervous activity and development may really be a matter of physical development rather than age *per se*, for they were able to show that, in those cases where the fast growing and slow growing groups could be compared with regard to size, their activities were approximately the same. This observation appears to have been lost to current researchers in behavioral effects of lead on early development.

Barnes et al. (1968) examined postnatal nutritional deprivation as a determinant of adult behavior toward food consumption and utilization. Dams of newborn rats were given an adequate or protein-deficient diet on the day following parturition. At 3 weeks of age, pups were weaned to either a nutritionally adequate diet *ad libitum* or restricted quantities; a third group of undernourished animals was weaned to a deficient diet up to 7 weeks of age. Under conditions of restricted access of 1 hr a day, it was noted that rats which had been subjected to nutritional deprivation early in life spilled significantly more food than normal animals in the restricted access situation. The authors concluded that the increased food spillage was a manifestation of heightened excitability in stressful circumstances. Evidence from other investigators has accumulated to support this view. Guthrie (1968) assessed the effect of severe undernutrition for varying intervals during the period of rapid development of the central nervous system immediately following birth. Behavior was studied in mature rats which had been rehabilitated with an adequate diet. Undernutrition was achieved by feeding pups in litters of 16 with an undernourished dam which restricted weight gain to less than one-third that of pups nursed in litters of 8. In the postweaning period, malnutrition was sustained by a protein-deficient diet. Running wheel activity was for a 1-hr test period. Those groups of male rats nutritionally deprived throughout the suckling period only or, in addition, for some weeks postweaning ran much more in the wheels than adequately nourished controls. Hence, on the evidence of this study and that of Barnett et al. (1971), the long term activity of rats is significantly increased as a result of early inadequate nutrition. Levitsky and Barnes (1970) showed that growth-retarded rats press a lever to postpone an electric shock at a higher rate than do well fed controls. Rats that had been undernourished made greater responses to noise than controls. The behavioral differences became apparent only when the animals were placed in a moderately stressful situation. This suggested to the authors that one effect of malnutrition during early development was to produce a rather long lasting, if not permanent, lowering of the threshold of the stress-response system. The observation could be loosely described as overreactions in unpleasant situations.

Barnes et al. (1971) recorded the daily activity of three groups of male rats in a residential "plus-maze." Controls were given a complete diet *ad libitum* at all times; undernourished rats received the complete diet in restricted amounts for 8 weeks from the age of 28 days. Low protein rats received a low protein diet for the same period. From 12 weeks all rats were given the complete diet *ad libitum*, until tested at 35 weeks. Each rat was then put alone, for 12 days, in a plus-maze with a central nest box and four passageways radiating from it; food was supplied at the end of one arm and water was supplied in the nest box. On days 4 through 9, each rat had access to the arms for only 1 hr daily. Entries into the passageways and duration of stay were recorded automatically. The low protein group made most visits to the arms, even during the 6 days of restricted access, yet the undernourished rats spent most time in them. Presumably when the undernourished group was out of the nest they were indulging in some kind of "activity," perhaps eating, or gnawing or exploring. Hence, both nutritionally deprived groups can be described as being more active than the controls, although in contrasting ways. The only other observations on the activity of previously underfed rats over a period greater than a few minutes are those of Guthrie (1968) on activity in running wheels, discussed previously.

Pregnant rats fed diets adequate in protein concentration but reduced in total amount produced offspring exhibiting deviant behavior in open-field testing later in life (Simonson et al., 1971). The authors reduced total food intake to 50% during the gestational period only. Male offspring of these dams, as compared to male offspring of dams fed *ad libitum*, displayed increased emotional behavior and decreased exploratory activity. Furthermore, there was a dramatic difference in reaction time. The control animals exited from an enclosure nine times as rapidly as did the experimental animals. Of interest is the fact that the body weights of the two groups were comparable, decreasing the likelihood that these investigators were biased in their observations of the several behaviors studied.

The question of activity in isolation as compared to activity carried out in social interaction was examined by Frankova (1973). Frankova (1973) employed the method of protein deprivation during lactation. Animals of both sexes in the protein-deficient group interacted less with littermates, the dam and the environment during the last half of the suckling period. Males were kept on the same protein-deficient diet until 49 days of age. As a measure of social interaction, the response of the experimental animal to a partner was examined. The nutritionally deprived animals withdrew from the partner rather than approaching, which is the normal behavior; nevertheless, they did exhibit a higher level of exploratory activity when alone, in agreement with earlier observations. Frankova and Barnes (1968a and 1968b) speculated that the differences observed in the normal animal seeking a new social experience and the protein caloric restriction, P-C-R rat avoiding it could mark the onset of pathologic changes in behavior that are manifest later in life in response to different situations and could therefore affect learning performance.

Slob et al. (1973) addressed themselves to the fact that previous studies of the effects of early undernutrition on subsequent behavior in adult rats confounded underfeeding with maternal deprivation or membership in a large litter. They undertook to correct this by arranging for food-deprived rats to receive fulltime maternal care and to live in the same sized litters as well fed controls. This was achieved by placing animals with a nonlactating foster mother for the light period and returned to the lactating dam for the dark period. A second group was daylight fed, reversed relative to the first group. Animals were weaned at 25 days of age and nutritionally rehabilitated for several weeks. Food-deprived animals did not differ from controls in the open-field test of motor coordination and in two learning tasks. However, food-deprived animals were more active than controls in a residential maze.

Randt and Derby (1973) reported behavioral changes resulting from maximal early life undernutrition in mice. Protein-caloric undernutrition throughout gestation and lactation until weaning resulted in a significant reduction in body and brain weights. Mice were weaned to nutritionally adequate diet. Measurement of behavior indicated a higher set-point for arousal in the nutritionally deprived animals, as exemplified by increasing activity when presented with environmental stimulation. Smart (1974) studied the activity and exploratory behavior of adult offspring of undernourished mother rats. The conclusion was that previously undernourished animals behave in a way consistent with a state of heightened excitability, even in situations which are less obviously stressful. There is evidence from several sources that, even months after a period of growth restriction, rats display a higher than normal level of general activity.

Studies on "activities" related to social interaction in rats have been reported by Whatson et al. (1974, 1976). Male rats, growth-retarded until weaning, but well fed afterward, engaged in more social interaction than did control animals. More of it was classified into a category which could be termed "aggressive," Pairs of experimental rats "box" and "fight" more often than do pairs of controls. Castellano and Oliverio (1976) employed unequal litter size to develop a state of undernutrition and growth retardation. Mice were reared in small, intermediate or large litters (4, 8 or 16). All animals were weaned at 21 days to standard laboratory diet. Locomotor activity was measured with a toggle-floor box. The number of crossings from one side to the other of the box was recorded automatically by means of a microswitch connected to the tilting floor of the box. Exploratory activity was measured for 25 min. The exploratory activity of small sized litters was consistently lower than that evident in the growth-retarded large litters. The data showed a clear trend toward an increase of exploratory behavior as the litter size increases. The average activity of the growth-retarded large litters was twice as high as that evident in the small litters. The study on locomotor activity and social behavior of old rats after preweaning undernutrition (Schenck et al., 1978) led the authors to conclude that animals undernourished neonatally stay more active than controls well into old age and that the gender difference in locomotor activity persists as the animals grow older. Similar observations by Loch et al. (1978) on enhanced locomotor activity in mice growth retarded to the same extent as those employed in a number of studies on lead toxicity has been described earlier.

## VIII. SUMMARY AND CONCLUSIONS

The major problem today relative to inorganic lead as an environmental pollutant is the uncertainty concerning possible subtle toxic neurologic effects of lead exposure in children. The presence or absence of such subtle effects is a matter of great uncertainty and controversy.

The overall objectives of present research efforts are to determine 1) what level of lead exposure is minimally toxic to young children who live in environmental milieu suspected of being hazardous, 2) whether there are undesirable effects on the developing nervous system which have not been hitherto detected due to lack of suspicion and lack of cause and effect relationship relative to neurologic aspects of subclinical or asymptomatic lead poisoning, 3) the cellular mechanism by which raised brain lead concentrations produce disruptive effects on neurotransmitter metabolism in function, and 4) the relevance of existing experimental models to human asymptomatic childhood lead poisoning relative to external versus internal exposure, nutritional status and time of onset and duration of exposure.

There is little disagreement that lead has a deleterious effect on brain function. However, the mechanism of molecular toxicity remains obscure. Furthermore, there is little knowledge about the relationship between blood lead concentration and brain lead concentration. Consequently, present attempts to define the critical level for lead in the developing or mature brain are frustrated.

The problems associated with childhood lead exposure have taken on a new focus. This is primarily due to the recent recognition that many otherwise asymptomatic children have had or continue to have elevated blood and/or lead body burdens. The current concern is whether exposure to lead during early life has a detrimental effect on brain development, behavior and intellectual maturation. In the search for answers to this question, many researchers have turned to neurochemical and behavioral measures in animals for assistance. Since it is not possible to manipulate lead exposure in humans in the same way that it can be manipulated in laboratory animals, animal models can be used in protocols designed to answer important questions about neurologic consequences of early low level exposure.

Until quite recently there have been few studies on the cerebral effects of lead in the usually available laboratory animals. The historical circumstance has been that investigators often employed adult members of the various species, the most frequent being the rodent. Adult rodents exhibit extreme resistance to the CNS effects of lead poisoning; they require large doses of lead to produce manifestations of lead poisoning. Pentschew and Garro (1966) demonstrated that lead encephalopathy can be produced in 3- to 4-week-old rats when, beginning at birth, they are suckled by a lactating dam receiving 4 to 5% lead carbonate (21,800 to 27,300 ppm Pb) in her diet and producing milk containing 40 ppm of lead. The model developed by Pentschew and Garro (1966) has come to serve as a prototype for studies of the CNS effects of lead in the developing young. In most instances, lead is transmitted to the neonate via milk from a lactating dam consuming relatively large amounts of lead. The neonate is then allowed to wean spontaneously to, and be maintained on, comparable amounts of lead until completion of the study. The exposure procedure permits direct access to the high lead-containing maternal diet at approximately 18 days of age, when pups gain access to solid food and drinking water. In spite of this, investigators continue to use this experimental design and term it low level lead exposure. Neonatal mice suckling dams consuming 2,731 ppm lead-containing drinking water have been weaned to and maintained on the same level of lead exposure up to the time of testing (Silbergeld and Goldberg, 1973). Unfortunately, the relatively high concentration of lead in the dam's diet (Pentschew and Garro, 1966) and drinking water (Silbergeld and Goldberg, 1973) leads to aversion and disruption in consumptive activity in the rat (Michaelson and Sauerhoff, 1974), as well as the mouse (Morrison et al., 1975). This undoubtedly affects milk production and subsequent growth of progeny. The young experience a 20 to 40% diminution in body weight during the first 3 weeks of life compared to controls of the same age. The relevance of behavioral and neurochemical findings at 40 to 60 days of age in growth-retarded treated animals compared to normal size control animals requires evaluation of the contribution of early malnutrition to the interpretation of experimental findings. Unfortunately, much of the data generated by the studies cited on the interaction of lead, neurochemistry and behavior are of little value due to a lack of clinical perspective both in the definition of the critical issues and in the design of the studies to deal with these issues. With respect to the general experimental design, several deficiencies have been discussed. These are the use of nutritionally oversupplemented commercial diets and their contamination with toxic substances which in themselves have neurologic effects. The most damaging of the confounding variables is maternal undernutrition and/or neonatal growth retardation, which results in an inability to determine specific causes for demonstrated effects.

There should be no doubt that early undernutrition either during fetal or neonatal life or both causes serious deficits in structure, chemistry and some behaviors. A review of recent animal experiments on the effects of lead on brain function shows that this undesirable condition, undernutrition, which is known to give rise to brain defects, has been a covariable with the experimental intent, namely the biological effect of lead. Both experimental maneuvers have adverse effects on brain function. Their combined presence, whether intended or unintended, makes it impossible to ascribe the experimental measure to one condition alone.

## ACKNOWLEDGMENTS

I would like to thank Dr. H. Zenick for his reading of the manuscript and for valuable suggestions. I am also grateful to Evelyn Widner for help with the literature and to Mary Jo Loftus and Gail Gair for patient secretarial assistance. The research work described from the author's laboratory was supported by United States Public Health, National Institutes of Environmental Health Sciences Grants ES-00159 (Center for the Study of the Human Environment); ES-00129 (R. K. Loch), ES-02614 (R. L. Bornschein), ES-05103 (L. S. Rafales) and ES-01077 (I. A. Michaelson).

## REFERENCES

Abdulla, M., and Haeger-Aronsen B. (1971). *Enzymes 12:*708.

Abdulla, M., and Haeger-Aronsen, B. (1973). *Intern. Res. Commun. Systems* (73-8)8-14-1.

Adlard, B.P.F., and Dobbing, J. (1971a). *Brain Res. 28:*97.

Adlard, B.P.F., and Dobbing, J. (1971b). *Proc. Nutr. Soc. 30:*67A.

Adlard, B.P.F., and Dobbing, J. (1971c). *Brain Res. 30:*198.

Adlard, B.P.F., and Dobbing, J. (1972a). *Pediat. Res. 6:*38.

Adlard, B.P.F., and Dobbing, J. (1972b). *Br. J. Nutr. 28:*139.

Adlard, B.P.F., Dobbing, J., and Smart, J.L. (1970). *Biochem. J. 119:*46P.

Aghajanian, G.K., and Bloom, F.E. (1968). *Brain Res. 6:*716.
Ahmad, G., and Rahman, M.A. (1975). *J. Nutr. 105:*1090.
Altman, J., and Das, G.D. (1966). *J. Comp. Neurol. 126:*337.
Ashcroft, G.W., Eccleston, D., and Crawford, T.B. (1965). *J. Neurochem. 12:*483.
Baird, A., Widdowson, E.M., and Cowley, J.J. (1971). *Br. J. Nutr. 25:*391.
Barltrop, D., and Khoo, H.E. (1975a). *Postgrad. Med. J. 51:*795.
Barltrop, D., and Khoo, H.E. (1975b). D.D. Hemphill (Ed.), In "Trace Substances in Environmental Health - IX," P. 369. University of Missouri-Columbia, Columbia, Missouri.
Barnes, R.H., Neely, C.S., Kwong, E., Labadan, B.A., and Frankova, S. (1968). *J. Nurt. 96:*467.
Barnett, S.A., Smart, J.L., and Widdowson, E.M. (1971). *Develop. Psychobiol. 4(1):* 1.
Barton, J.C., Conrad, M.E., Harrison, L., and Nuby, S. (1978). *J. Lab. Clin. Med. 91(3):*366.
Bass, N.H. (1971). R. Paoletti and A.N. Davison (Eds.), In "Chemistry and Brain Development;" p. 413. Plenum Press, New York.
Bass, N.H., Netsky, M.G. and Young, E. (1970). *Arch. Neurol. 23:*289.
Beckman, A.L., and Eisenman, J.S. (1970). *Science 170:*334.
Benton, J.W., Moser, H.W., Dodge, P.R., and Carr, S. (1966). *Pediat. 38(5):*801.
Bieri, J.G., Stoewsand, G.S., Briggs, G.M., Phillips, R.W., Woodward, J.C., and Knapka, J.J. (1977). *J. Nutr. 107(7):*1340.
Blackman, S.S., Jr. (1937). *Bull. Johns Hopkins Hospital. 61:*43.
Bowers, M.B., Jr. (1970). *Neuropharmacol. 9:*599.
Bull, R.J., Stanaszek, P.M., O'Neill, J.J., and Lutkenhoff, S.D. (1975). *Envir. Htlh. Persp. 12:*89.
Bunge, M.B., Bunge, R.P., and Peterson, E.R. (1967). *Brain Res. 6:*728.
Carroll, P.T., Silbergeld, E.K., and Goldberg, A.M. (1975). *Neurosciences Abstracts 1:*250.
Carroll, P.T., Silbergeld, E.K., and Goldberg, A.M. (1977). *Biochem. Pharmacol. 26:*397.
Castellano, C., and Oliverio, A. (1976). *Brain Res. 101:*317.
Cerklewski, F.L., and Forbes, R.M. (1976). *The J. Nutr. 106(5):*689.
Clark, G.M., Zamenhof, S., and Van Marthens, E. (1973). *Brain Res. 54:*397.
Clasen, R.A., Hartmann, J.F., Starr, A.J., Coogan, P.S., Pandolfi, S., Laing, I., Becker, R., and Hass, G.M. (1974). *Am. J. Pathol. 74(2):*215.
Clos, J., Bebiere, A., and Legrand, J. (1973). *Brain Res. 63:*445.
Clos, J., Favre, C., Selme-Matrat, M., and Legrand, J. (1977). *Brain Res. 123:*13.
Cohen, E.L., and Wurtman, R.J. (1976). *Science 191:*561.
Conrad, M.E., and Barton, J.C. (1978). *Gastroenterol. 74(4):*731.
Cowley, J.J., and Criesel, R.D. (1959). *J. Genet. Psychol. 95:*187.
Cragg, B.G. (1972). *Brain 95:*143.
Crnic, L.S. (1976). *Psychol. Bull. 83:*715.
Crossland, J. (1951). *J. Physiol. 114:*318.
Dahlstrom, A., and Fuxe, K. (1964). *Acta Physiol. Scand. 62 (Suppl. 232):*1.
Dickerson, J.W.T., and Pao, S.K. (1975). *Biol. Neonate 25:*114.
Dickerson, J.W.T., Dobbing, J., and McCance, R.A. (1967). *Proc. Roy. Soc. 166:*396.
Dobbing, J. (1966). *Biol. Neonate 9:*132.
Dobbing, J., and Smart, J.L. (1974). *Brit. Med. Bull. 30:*164.
Dobbing, J., and Widdowson, E.M. (1965). Brain 88:357.

Dobbing, J., Hopewell, J.W., and Lynch, A. (1971). *Exper. Neurol. 32:*439.
Dodge, P.R., Prensky, A.L., and Feigin, R.D. (1975). S.J. Holmes (Ed.), In "Nutrition and the Developing Nervous System," p. 243. The C.V. Mosby Co., St. Louis.
Dyson, S.E., and Jones, D.G. (1976). *Progress in Neurobiol. 7:*171.
Eckhert, C., Barnes, R.H., and Levitsky, D.A. (1975). *Am. J. Physiol. 229(6):*1532.
Eckhert, C., Barnes, R.H., and Levitsky, D.A. (1976a). *Brain Res. 101:*372.
Eckhert, C.D., Barnes, R.H., and Levitsky, D.A. (1976b). *J. Neurochem. 27:*277.
Eckhert, C., Levitsky, D., and Barnes, R.H. (1973). *Fed. Proc. 32:*902.
Eckhert, C., Levitsky, D.A., and Barnes, R.H. (1974). *Fed. Proc. 33:*661.
Eckhert, C.D., Levitsky, D.A., and Barnes, R.H. (1975). *Proc. Soc. Exper. Biol. and Med. 149:*860.
El-Gazzar, R.M., Finelli, V.N., Boiano, J., and Petering, H.G. (1978). *Tox. Lett. 1:*227.
Enwonwu, C.O., and Worthington, B.S. (1973). *J. Neurochem. 21:*799.
Evans, G.W., Grace, C.I., and Votava, H.S. (1975). *Am. Physiol. 228:*501.
Evans, G.W., and Hahn, C.J. (1974). M. Friedman (Ed.), In "Protein-Metal Interactions," p. 285. Plenum Press, New York.
Fernstrom, J.D. (1976). *Fed. Proc. 35:*1151.
Fernstrom, J.D., and Wurtman, R.J. (1971). *Science 173:*149.
Fine, B.P., Barth, A., Sheffet, A., and Lavenhar, M.A. (1976). *Environ. Res. 12:*224.
Finelli, V.N., and El-Gazzar, R.M. (1977). *Toxicol. Lett. 1:*33.
Finelli, V.N., Klauder, D.S., Karaffa, M.A., and Petering, H.G. (1975). *Biochem. Biophys. Res. Comm. 65:*303.
Finelli, V.N., Murthy, L., Peikano, W.B., and Petering, H.G. (1974). *Biochem. Biophys. Res. Comm. 60:*1418.
Fox, J.G., Aldrich, F.D., and Boylen, G.W. Jr. (1976). *J. Tox. Environ. Hlth 1:*461.
Frankova, S. (1973). *Develop. Psychobiol. 6(1):*33.
Frankova, S., and Barnes, R.H. (1968a). *J. Nutr. 96:*477.
Frankova, S., and Barnes, R.H. (1968b). *J. Nutr. 96:*485.
Fuxe, K. (1965). *Physiol. Scand. 64 (Suppl. 247):*37.
Gaetani, S., Mengheri, E., Spadoni, M.A., Rossi, A., and Toschi, G., (1975). *Brain Res. 86:*75.
Gembetti, P., Autilio-Gambetti, L., Gonatas, N.K., Shafer, B., and Stieber, A. (1972). *Brain Res. 47:*477.
Gambetti, P., Autilio-Gambetti, L., Rizzuto, N., Shafer, B., and Pfaff, L. (1974). *Exper. Neurol. 43:*464.
Gerber, C.B., Maes, J., Gilliavod, N., and Casale, G. (1978). *Tox. Lett. 2:*51.
Golter, M., and Milchaelson, I.A. (1975). *Science 187:*359.
Goyer, R.A., and Mahaffey, K.R. (1972). *Environ. Hlth Persp. 1:*73.
Guthrie, H.A. (1968). *Physiol Behav. 3:*619.
Haeger-Aronsen, B., Schutz, A., and Abdulla, M. (1976). *Arch. Environ. Hlth 31:*215.
Hahn, C.J., and Evans, G.W. (1973). *Proc. Soc. Exp. Med. 144:*793.
Halliday, J.W., Powell, L.W., and Mack, U. (1976). *Br. J. Haematol. 34:*237.
Harvey, J.A., Schlasberg, A.J., and Yunger, L.M. (1974). *Adv. Biochem. Psychopharmacol. 10:*233.
Haubrich, D.R., and Chippendale, T.J. (1977). *Life Sci. 20:*1465.
Hernandez, R.J. (1973). *Experientia 29:*1487.
Hernandez, R.J. (1976). *Biol. Neonate 30:*181.

Henry, F., Pujol, J.F., Lopez, M., Macon, J., and Glowinski, J. (1970). *Brain Res. 21:*391.

Himwich, W.A., Davis, J.M., and Agrawal, H.C. (1968). J. Wortis (Ed.), In "Recent Advances in Biological Psychiatry," Vol. 10, p. 266. Plenum Press, New York.

Hirsch, M.J., and Wurtman, R.J. (1978). *Science 202:*223.

Hogan, E.L. (1975). *J. Neuropathol. Exp. Neurol. 33(1):*2023.

Holtzman, D., and Shen Hsu, J. (1976). *Pediat. Res. 10:*70.

Im, H.S., Barnes, R.H., and Levitsky, D.A. (1971). *Nature 233:*269.

Im, H.S., Barnes, R.H., and Levitsky, D.A. (1976). *J. Nutr. 106(3):*342.

Im, H.S., Barnes, R.H., Levitsky, D., and Pond, W.G. (1971). *Fed. Proc. 30:*459.

Im, H.S., Barnes, R.H., Levitsky, D.A., and Pond, W.G. (1973). *Brain Res. 63:*461.

Im, H.S., Barnes, R.H., Levitsky, D., Krook, L., and Pond, W.G. (1972). *Fed. Proc. 31:*697.

Jackson, C.M., and Stewart, C.A. (1920). *J. Exp. Zool. 30(1):*97.

Jaquier, E., Robinson, D.S., Lovenberg, W., and Sjoerdsma, A. (1969). *Pharmacol. 18:*1071.

Kalyanasundaram, S. (1976). *J. Neurochem. 27:*1245.

Kaufman, S. (1974). G.E.W. Wolstenholme and D.W. Fitzsimons (Eds.), In "Aromatic Amino Acids in the Brain," p. 86. Associated Scientific Publishers, Amsterdam.

Kety, S.S. (1971). F.O. Schmidt (Ed.), In "The Neurosciences: Second Study Program," p. 324. Rockefeller University Press, New York.

Klauder, D.S., and Petering, H.G. (1975). *Environ. Hlth Persp. 12:*77.

Klauder, D.S., Murthy, L., and Petering, H.G. (1972). D.D. Hemphill (Ed.), In "Trace Substances in Environmental Health - VI," p. 131. University of Missouri-Columbia, Columbia, Missouri.

Kochen, J., and Greener, Y. (1975). *Pediat. Res. 9:*323.

Kostial, K., Simonovic, I., and Pisonic, M. (1971). *Nature 233:*564.

Krall, A.R., Pesavento, C., Harmon, S.J., and Packer, R.M., III. (1972). *Fed. Proc. 31:*655.

Krigman, M.R., Mushak, P., and Bouldin, T.W. (1977). L. Roizin, H. Shiraki and N. Grcevic (Eds.), In "Neurotoxicology," Vol. 1, p. 299. Raven Press, New York.

Lát, J. (1956). *Physiol. Bohemoslov. 5:*38.

Lát, J. Widdowson, E.M., and McCance, R.A. (1961). *Proc. Roy. Soc. B. 153:*347.

Leathwood, P. (1978). G. Gottlieb (Ed.), In "Studies on the Development of Behavior and the Nervous System," Vol. 4, p. 187. Academic Press, New York.

Lee, C.J., and Dubos, R. (1972). *J. Exper. Med. 136:*1031.

Levitsky, D.A., and Barnes, R.H. (1970). *Nature 225:*468.

Loch, R.K., Rafales, L.S., Michaelson, I.A., and Bornschein, R.L. (1978). *Life Sci. 22:*1963.

Loizou, L.A. (1969). *J. Anat. (Lond.) 104:*588.

Lovenberg, W., Jequier, E., and Sjoerdsma, A. (1968). S. Garattini and P.A. Shore (Eds.), In "Advances in Pharmacology," Vol. 6A, p. 21. Academic Press, New York.

Macomber, D. (1933). *N. Engl. J. Med. 209:*1105.

McCance, R.A., and Widdowson, E.M. (1962). *Proc. Roy. Soc. B. 156:*326.

Michaelson, I.A., and Sauerhoff, M.W. (1974a). *Environ. Hlth Persp. 1:*201.

Michaelson, I.A., and Sauerhoff, M.W. (1974b). *Tox. Appl. Pharmacol. 28:*88.

Michaelson, I.A., Greenland, R.D., and Roth, W. (1974). *Fed. Proc. 33:*578.

Miller, M., Leahy, J.P., McConville, F., Morgane, P.J., and Resnick, O. (1977a). *Brain Res. Bull. 2:*347.
Miller, M., Leahy, J.P., McConville, F., Morgane, P.J., and Resnick, O. (1977b). *Brain Res. Bull. 2:*189.
Miller, M., Leahy, J.P., Stern, W.C., Morgane, P.J., and Resnick, O. (1977). *Exper. Neurol. 57:*142.
Modak, A.T., Weintraub, S.T., and Stavinoha, W.B. (1975a). *Tox. Appl. Pharmacol. 34:*340.
Modak, A.T., Weintraub, S.T., and Stavinoha, W.B. (1975b). *Pharmacologist 17(2):*213, Abstract 204.
Modak, A.T., Purdy, R.H., and Stavinoha, W.B. (1977). *Toxicol. Appl. Pharmacol. 41:*151.
Modak, A.T., Weintraub, S.T., Stavinoha, W.B., and Purdy, R.H. (1976). *Tox. Appl. Pharmacol. 37:*133.
Moir, A.T., and Eccleston, D. (1968). *J. Neurochem. 15:*1093.
Molliver, M.E., Kostovic, I., and Van Der Loos, H. (1973). *Brain Res. 50:*403.
Nachmansohn, D. (1961). *Science 134:*1961.
National Academy of Sciences-National Research Council (1972). "Nutrient Requirements of Laboratory Animals," Publ. No. 990. Washington, D.C.
Neff, N.H., Spano, P.F., Gropetti, A., Wang, C.T., and Costa, E. (1971). *J. Pharmacol. Exp. Ther. 176:*701.
Nelson, M., and Evans, H.M. (1958). *Proc. Soc. Exp. Biol. Med. 99:*723.
Neville, H.E., and Chase, H.P. (1971). *Exp. Neurol. 33:*485.
Pentschew, A., and Garro, F. (1966). *Acta Neuropathologica 6:*266.
Petering, H.G. (1974). W.G. Hoekstra, J.W. Suttie, H.E. Ganther and W. Mertz (Eds.), In "Trace Element Metabolism in Animals-2" p. 311. University Park Press, Baltimore, Md.
Platt, B.S. (1962). *Proc. Roy. Soc. B 156:*337.
Platt, B.S., Heard, C.R.C., and Stewart, R.J.C. (1964). H.N. Munro and J.B. Allison (Eds.), In "Mammalian Protein Metabolism," Vol. 2, Chapter 21, p. 445. Academic Press, New York.
Plaut, S.M. (1970). *Develop. Psychobiol. 3(3):*157.
Rafales, L.S., Bornschein, R.L., Michaelson, I.A., Loch, R.K., and Barker, G.F. (1979). *Pharmacol. Biochem. Behav. 10:*95.
Rajalakshmi, R., Kulkarni, A.B., and Ramakrishnan, C.V. (1974). *J. Neurochem. 23:*119.
Ramanamurthy, P.S.V. (1977). *J. Neurochem. 28:*253.
Randt, C.T., and Derby, B.M. (1973). *Arch. Neurol. 28:*167.
Rosenblum, W.I., and Johnson, M.G. (1968). *Arch. Path. 85:*640.
Russell, R.W. (1969). *Fed. Proc. 28(1):*121.
Salas, M., Diaz, S., and Nieto, A. (1974). *Brain Res. 73:*139.
Sauerhoff, M.W., and Michaelson, I.A., (1973a). *Pharmacol. 15:*164.
Sauerhoff, M.W., and Michaelson, I.A. (1974b). *Science 182:*1022.
Schenck, P.E., Koos Slob, A., and Van Der Werff Ten Bosch, J.J. (1978). *Develop. Psychobiol. 11:*205.
Schildkraut, J., and Kety, S. (1967). *Science 156:*21.
Schroeder, H.A., Balassa, J.J., and Vinton, W.H., Jr. (1965). *J. Nutr. 86:*51.
Schwartz, I.R., Pappas, G.D., and Purpura, D.P. (1968). *Expt. Neurol. 22:*394.
Sereni, F., Principi, N., Perletti, L., and Sereni, L.P. (1966). *Biol. Neonate 10:*254.
Shih, T.M., and Hanin, I. (1977). *Fed. Proc. 33(3):*977.

Shih, T.M., and Hanin, I. (1978). *Psychopharmacol. 58:*263.
Shiman, R., Skino, M., and Kaufman, S. (1971). *J. Biol. Chem. 246:*1330.
Shoemaker, W.J., and Wurtman, R.J. (1971). *Science 171:*1017.
Shoemaker, W.J., and Wurtman, R.J. (1973). *J. Nutr. 103:*1537.
Siassi, F., and Siassi, B. (1973). *J. Nutr. 103:*1625.
Silbergeld, E.K. (1973). *Life Sci. 13:*1275.
Silbergeld, E.K., and Chisolm, J.J., Jr. (1976). *Science 192:*153.
Silbergeld, E.K., and Goldberg, A. (1974). *Pharmacologist 16:*249.
Silbergeld, E.K., and Goldberg, A.M. (1975). *Neuropharmacol. 14:*431.
Silbergeld, E.K., Carroll, P.T., and Goldberg, A.M. (1975). *Pharmacologist 17:*212.
Silbergeld, E.K., Fales, J.T., and Goldberg, A.M. (1974). *Neuropharmacol. 13:*795.
Sima, A. (1974). *Acta Physiol. Scand. Suppl. 406:*1.
Sima, A., and Sourander, P. (1974). *Acta Neuropath. 28:*151.
Simonson, M., Stephan, J.K., Hanson, H.M., and Chow, B.F. (1971). *The J. Nutr. 101:*331.
Slob, A.K., Snow, C.E., and Natris-Mathot, E. (1973). *Develop. Psychobiol. 6:*177.
Smart, J.L. (1974). *Develop. Psychobiol. 7:*315.
Sobotka, T.J., Cook, M.P., and Brodie, R.E. (1974). *Brain Res. 65:*443.
Sourkes, T., Potrier, L., and Singh, P. (1969). F.J. Gillingham and I.M.L. Donaldson, (Eds.), In "Third Symposium on Parkinson's Disease," p. 54. Livingstone, Edinburgh.
Stavinoha, W.B., and Weintraub, S.T. (1974). *Science 183:*964.
Stern, W.C., Forbes, W.B., Resnick, O., and Morgane, P.J. (1974). *Brain Res. 79:*375.
Stern, W.C., Miller, M., Forbes, W.B., Morgane, P.J., and Resnick, O. (1975). *Exper. Neurol. 49:*314.
Stern, W.C., Morgane, P.J., Miller, M., and Resnick, O. (1975). *Exper. Neurol. 47:*56.
Stern, W.C., Resnick, O., Miller, M., Forbes, W.B., and Morgane, P.J. (1974). *Fed. Proc. 33:*661.
Ten-State Nutrition Survey (1968–70). DHEW Publication No. (HSM) 72-8133, Atlanta, Ga.
Thawley, D.G., Pratt, S.E., and Selby, L.A. (1978). *Envir. Res. 15:*218.
Udenfriend, S., and Zaltzman-Nirenberg, P. (1963). *Science 142:*394.
Underwood, E.J. (1977). "Trace Elements in Human and Animal Nutrition," 4th Ed., p. 196. Academic Press, New York.
Voeller, K., Pappas, G.D., and Purpura, D.P. (1963). *Expt. Neurol. 7:*107.
Wehmer, F., and Jen, K-L. C. (1977). *Develop. Psychobiol. 11(4):*353.
Whatson, T.S., Smart, J.L., and Dobbing, J. (1974). *Br. J. Nutr. 32:*413.
Whatson, T.S., Smart, J.L., and Dobbing, J. (1976). *Develop. Psychobiol. 9(6):*529.
Whittaker, V.P. (1965). *Prog. Biophys. Mol. Biol. 15:*41.
Widdowson, E.M., and Kennedy, G.C. (1962). *Proc. Roy. Soc. B. 156:*96.
Widdowson, E.M., and McCance, R.A. (1960). *Proc. Roy. Soc. B. 152:*188.
Wiggins, R.C., Benjamins, J.A., Krigman, M.R., and Morell, P. (1974). *Brain Res. 80:*345.
Willoughby, R.A., MacDonald, E., McSherry, B.J., and Brown, G. (1972). *Can. J. Comp. Med. 36:*348.
Winick, M. (1969). *J. Pediat. 74:*667.
Winick, M. (1970). *Fed. Proc. 29:*1510.
Winick, M., and Noble, A. (1965). *Develop. Biol. 12:*451.
Winick, M., and Noble, A. (1966). *J. Nutr. 89:*300.

Wurtman, R.J. (1970). J. Meites (Ed.), In "Hypophysiotrophic Hormones of the Hypothalamus, Assay and Chemistry," p. 184. The Williams & Wilkins Co., Baltimore, Md.

Wurtman, R. (1971). *Neurosciences Res. Program Bull. 9:*172.

Zeman, F.J., and Stanbrough, E.C. (1969). *J. Nutr. 99:*274.

Zamenhof, S., van Marthens, E., and Grauel, L. (1971a). *J. Nutr. 101:*1265.

Zamenhof, S., van Marthens, E., and Grauel, L. (1971b). *Science 172:*850.

Zamenhof, S., van Marthens, E., and Grauel, L. (1971c). *Science 174:*954.

Zamenhof, S., van Marthens, E., and Margolic, F.L. (1968). *Science 160:*322.

# Biochemical and Clinical Effects and Responses as Indicated by Blood Concentration

*Sven Hernberg*

TABLE OF CONTENTS



## I. INTRODUCTION

Thorough knowledge of the relationship between the degree of exposure to a toxic agent, on one hand, and the severity and frequency of the resulting effects, on the other, is a prerequisite for the definition of health-based criteria for safe exposure. This knowledge should be utilized when political decisions are made concerning standards or acceptable levels of tolerance.

The toxicity of lead has been studied intensively for more than a hundred years; one could therefore expect such knowledge to exist. However, although the literature is vast indeed, comparatively few studies have been designed for the evaluation of quantitative relationships between exposure or dose and various toxic effects. In addition, most of these studies have focused on hematologic effects; only during the last few years have dose-related data on neurologic functional impairment been published. This chapter attempts to review the present knowledge of this important field, as well as discuss and clarify some concepts and definitions.

## II. MEASURES OF DOSE AND EXPOSURE

*Dose* is a measure of what has been absorbed in the organism through the lungs and the gastrointestinal tract and sometimes via other routes; it is influenced by individual variations in absorption. A dose can generally be regarded as a pharmacological concept. In the occupational and public health setting, it is not possible to assess dose, because exact estimations would require tedious balance studies that are impracticable outside the test laboratory. What can be measured, although it also requires much effort, is exposure, *i.e.*, the degree of environmental contamination in air, water, food, etc. Exposure however, does not take into consideration individual differences, such as respiratory volume, differences in absorption rate, etc.; hence, the exposure level does not necessarily reflect the individual dose. Consequently, some substitute for dose or individual exposure, or both, is needed for operational purposes.

After absorption, lead is transported to the organs via the bloodstream. It can therefore be assumed that the concentration of lead in circulating blood (PbB) reflects the amount of lead which has entered the organism. However, the situation is confounded by the dynamic equilibrium prevailing between different tissues and blood. Therefore, the PbB level is also a function of the concentration gradients of, and the specific affinities for, lead of the tissues in question.

It has been postulated that there are at least three bodily compartments for lead. One compartment is blood lead, probably having a mean life of 27 days. Still another compartment is soft tissue lead, plus a rapidly exchangeable fraction in bone, with an estimated life of 30 days. A third compartment is composed of a stable bone fraction, whith a mean life of 30 years (Rabinowitz et al., 1973). If the stores (the body burden) are large, their relative impact on the PbB level is great; conversely, if they are small, they influence it to a lesser degree. In other words, the PbB level reflects currently absorbed lead better during the beginning of (occupational) exposure than later in its course. Under the steady state conditions prevailing in the general population, the PbB level is also a useful indicator of recent absorption, especially when an increase in exposure intensity occurs. Later in the course of heavy (meaning occupational or otherwise exceptional) exposure, this level is a poor indicator of current absorption, because the relative impact of lead released from the stores increases. After the termination of such exposure, the PbB level self-evidently reflects only the lead that is being released.

In spite of the above mentioned considerations, which warrant careful interpretation, the PbB level is generally considered to be the best indicator available of so-called recently absorbed lead (Task Group of Metal Toxicity, 1976). Furtermore, it can be assumed that the PbB level is always a good indicator of the biologically active lead which is responsible for the toxic effects at any given period of time. Hence, its use as a dose indicator can be justified, although it undoubtedly creates some confusion for the concept of dose. It should be stressed that the PbB level is not a good indicator of accumulated body burden. However, since a major part of the latter is biologically inactive, this restriction needs not necessarily hamper the use of the PbB level as a reference value for toxic effects.

One important difficulty of a more practical nature arises from the fact that PbB determinations are unfortunately very vulnerable to methodologic errors. In the past, many laboratories have performed them unsatisfactorily. It has been said that even the most experienced laboratories may have had errors of up to 10%; the error rate of less experienced laboratories may have been much higher. For example, in a recent interlaboratory comparison among 66 European laboratories employing varying analytical techniques, differences with a factor of up to ten were noted in the results (Berlin et al., 1974a). Similar examples were reported earlier from the USA (Donovan et al., 1971). These reports have had good effects, since they have prompted intensified interlaboratory method control programs and other measures to increase the precision and accuracy of analytic methods. There are now indications that the analytic level of laboratories taking part in such programs has improved substantially. For example, the results of a Scandinavian intercomparison program initiated in 1971 have shown ever decreasing interlaboratory differences, which must be interpreted to reflect both a primarily good level and improvements intensified by the awareness of continuous external control.

In spite of recent improvements, data from only a minority of even the most recent studies can be fully accepted for relating PbB levels to toxic effects; even then the comparison of results from different laboratories is not always possible. It is obvious that a great number of seemingly conflicting reports have arisen from comparisons between incompatible series. Subsequent sections of this chapter have considered only studies from laboratories with well established analytical accuracy, yet confusion can never be eliminated completely.

## III. EFFECTS AND RESPONSES

The Scientific Committee on the Toxicology of Metal, under the auspices of the Permanent Commission and International Association on Occupational Health, has arranged two task group meetings which have produced consensus reports in which some definitions have been agreed upon (Task Group of Metal Accumulation (TGMA), 1973; Task Group of Metal Toxicity (TGMT), 1976). As has been indicated, the PbB level will be used as a substitute for dose; hence the word "dose" in the definitions given below actually means "blood concentration," rather than dose in its pharmacologic sense.

The term "effect" is used to mean a biologic change caused by exposure or dose. Sometimes this effect can be measured on a graded scale of severity; at other times, it may be possible only to describe a qualitative effect occurring within some range of PbB levels. When data are available for the graded effect, a relationship between PbB levels and the gradiation of the effects in the population can be established. This relation can be defined as the dose-effect relationship. The noneffect level is the level below which current methods fail to detect any changes.

The effects are usually divided into all-or-none and graded effects. The death of a cell is an all-or-none effect, while disturbances of the heme synthesis represent graded effects. However, this is not an absolute distinction. In a physiologic sense, individual fibers or cell components may function only on an all-or-none basis; when a large number of these units react to lead together, however, the summation of their individual actions may yield measurements that one observes as graded effects. Many graded effects can be expressed as percentages of the unimpaired function observed in a general population.

The term "response" is used to express the proportion (prevalence or incidence) of a population that demonstrates a specific effect; its correlation with the PbB level provides the dose-response relationship. The nonresponse level is the level at which the proportion of affected individuals exceeds that of an unexposed reference population. However, one may also use a 5% response level as the reference level (Zielhuis, 1975).

The distinction between effect and response may be illustrated by the following example. While the relationship between the PbB level and the concentration of delta-aminolevulinic acid (ALA) in the urine is a dose-effect relationship, the relationship between different PbB levels (*e.g.*, in categories of 10 $\mu$g/100 ml) and the percentages of individuals with an ALA excretion in excess of certain amount (*e.g.*, 10 mg ALA/1 urine) is a dose-response relationship.

The dose-response relationship is, in its simplest form, usually displayed graphically as an S-shaped curve with the dose on the X axis and the response on the Y axis. If different effects are to be put in the same graphic display, the relationship requires a great number of curves and a multidimensional model. Such a model would be unintelligible; hence the various dose-response relationships will be treated separately. At the end of the chapter a simplified summary will be given.

The term critical effect has been used to denote the earliest adverse or undesirable effect in an organ. The critical effect may be regarded in qualitative (*e.g.*, demyelination of a nerve) or quantitative (*e.g.*, a certain degree of protoporphyrin accumulation in the erythrocytes) terms. This critical effect may or may not be of immediate importance for the health of the whole organism.

At exposure levels that are lower than the one resulting in critical effects, some other effects may occur that do not impair cellular function, but are still evident by means of biochemical or other tests. These effects have been defined as subcritical effects. The biologic meaning of a subcritical effect is sometimes not known; however, in some cases it may be a precursor of a critical effect.

A critical organ can be defined as that particular organ which first shows critical effects under specified circumstances of exposure and for a given population. The organ of greatest accumulation is not necessarily the critical organ. For example, the highest concentrations of lead are reached in bone without any identifiable effect; hence, the bone cannot be the critical organ. The Task Group of Metal Toxicity (TGMT) agreed, at its meeting in Tokyo in 1974, that the critical organ for lead effect is hematopoietic bone marrow (TGMT, 1976). It also concluded that the critical effect of lead is adverse interference with heme synthesis. Increased concentrations of heme intermediates in blood or urine were defined as critical effects, whereas decreased activity of *delta*-aminolevulinic dehydratase (ALAD) in red blood cells was defined as a subcritical effect. The Task Group also considered the possibility that continued research may develop methods for measuring neurochemical or neurophysiologic disturbances which underlie nervous system manifestations occurring prior to the known effect upon heme synthesis. At that time (1974), it was not possible to state whether or not nervous disorders may precede critical hematologic disturbances in particular circumstances, but since then more data have accumulated.

## IV. DOSE-EFFECT AND DOSE-RESPONSE CURVES FOR HEMATOLOGIC EFFECTS

The effect of lead upon the hematopoietic system can be attributed to the combined effect of the inhibition of hemoglobin synthesis and the shortened life span of erythrocytes, which result in anemia. Various hematologic effects of lead are related to concomitant blood lead (PbB) levels insofar as published data allow the construction of dose-effect and dose-response curves. Among the effects treated in some detail are inhibition of δ-ALAD activity, accumulation of protoporphyrins (PP's) in the erythrocytes and increase in the excretion of δ-ALA in the urine. Some additional effects are discussed more briefly; these are increased excretion of coproporphyrin III (CP) in the urine, decrease of the blood hemoglobin (Hb) concentration and shortening of the erythrocyte life span. Here the data available are scanty, conflicting, or, in the case of CP, similar to the more thoroughly treated effects of ALA. Furthermore, only sparse information is available on other effects, such as inhibition of ferrochelatase, decrease of reduced glutathione and inhibition of erythrocyte $Na^+ + K^+$-ATPase for which dose dependence is essentially unknown.

### A. Inhibition of Erythrocyte Delta-aminolevulinic Acid Dehydratase Activity

The earliest effect related to the hematopoietic tissue is inhibition of the activity of ALAD in circulating erythrocytes. As far as is known, this enzyme activity is without functional importance in the mature erythrocyte, in which heme synthesis no longer occurs. However, since direct relationships between lead concentrations in bone marrow and their effects on the immature erythrocyte precursors are unknown, the easily sampled mature erythrocyte has been ascribed the role of indicating what is thought to take place in other tissues (Secchi et al., 1974).

Although ALAD is inhibited *in vitro* by a number of substances, (*e.g.*, ethylenediaminetetraacetic acid, mercury, copper, silver, manganese and cobalt) (Hernberg and Nikkanen 1972) and transiently *in vivo* by ethanol (Moore et al., 1971), the inhibitory action of all these substances is less than that of lead. Furthermore, with the exception of ethyl alcohol, significant concomitant exposure to other inhibitory agents is so rare that such confounding is unlikely to interfere with dose-effect and dose-response relationships. Hence, for practical purposes, the inhibitory action of lead on ALAD activity can be regarded as nearly specific.

There are numerous studies proving the existence of a close negative linear relationship between PbB and the logarithm of ALAD activity (Hernberg et al., 1970; Haeger-Aronsen et al., 1971, Schaller et al., 1971; Granick et al., 1973; Tola, 1973; Tola et al., 1973; Sakurai et al., 1974; Alessio et al., 1977). When high lead levels are included and when the precision and accuracy of both methods are good, the correlation coefficients are as high as −0.8 to −0.9. Figure 1 shows the correlation found in one such study (Tola, 1973); in the 1,147 persons examined, the noneffect level seemed to lie between 10 and 20 μg Pb/100 ml blood. Other authors have reached the same conclusion (Haeger-Aronsen et al., 1971; Sakurai et al., 1974).

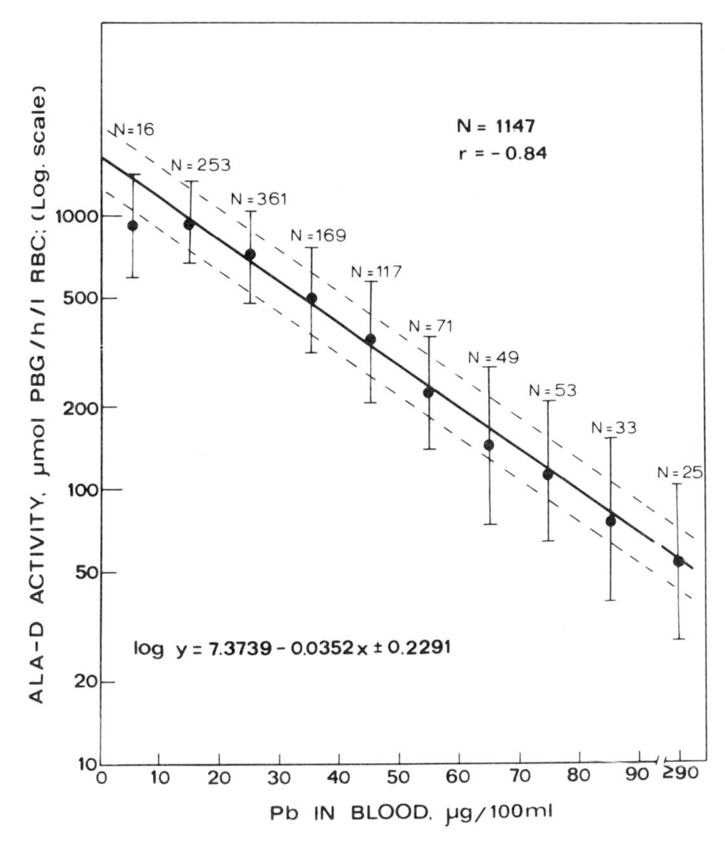


**Figure 1.** Relationship between the inhibition of erythrocyte ALAD activity and PbB. Note the logarithmic scale on the abscissa (Tola, 1973).

However, in a study on the Finnish general population, a statistically significant correlation was obtained between ALAD activity and PbB values of less than 10 µg/100 ml (PbB mean value, 8.4 µg/100 ml); this result creates some doubt about the existence of a noneffect level (Nordman and Hernberg, 1975). It may well be that the noneffect level found in most studies is merely an artifact reflecting some analytic imprecision and inaccuracy in the determination of PbB, which is particularly problematic at low levels. The interindividual variation of ALAD activity also tends to add scatter to the upper end of the regression curve.

From Figure 1 and from similar curves published elsewhere, it can be seen that a minor average depression (approximately 30%) of ALAD activity occurs at a PbB level of approximately 30 µg/100 ml. At around 50 µg Pb/100 ml blood, the average reduction of ALAD activity is about 70%; an average reduction of 90% occurs at PbB levels in the magnitude of 75 to

80 μg/100 ml. Although the correlation between PbB and ALAD is high in groups of subjects, the predictability of ALAD for PbB has rather wide confidence limits for individuals (Hernberg et al., 1970). The wide range is probably due to interindividual variations in ALAD activity and to methodologic considerations, rather than to a time lag between ALAD inhibition and commencement of exposure, as is the case with protoporphyrin accumulation in red blood cells. It has been shown that the PbB—ALAD relationship is roughly the same, at least for adults in industry, regardless of whether the exposure is long term and steady state or sudden or has been terminated (Tola 1972; TGMT, 1976).

The interindividual variation in ALAD activity can be decreased if the proportion of the enzyme activity inhibited by lead is reactivated by dithiothreitol and the ratio of activity with versus without enzyme activation is used instead of absolute values (Granick et al., 1973). Similar results have been obtained by the heating of the hemolysate at 60° for 5 min (Tomokuni and Kawanishi, 1975). However, these and similar studies, while of great theoretical interest, do not solve any practical problems, since the greatest advantage of the ALAD assay over the determination of PbB has been its simplicity and good reproducibility in different laboratories (Berlin et al., 1974b). The treatment of the hemolysate with heat or dithiothreitol complicates the assay so much that these advantages are lost.

The dose-response relationships for PbB and ALAD inhibition can be calculated from the dose-effect curve. In principle, any point on the curve can be used as a cut-off point between "effect +" and "effect −". Zielhuis (1975) calculated dose-response curves for adults and children from individual data provided by the Institutes of Occupational Health in Helsinki, Finland, and Erlangen-Nürnberg, Federal Republic of Germany. He used cut-off points of 40 and 70% inhibition of the activity found in individuals having PbB's of $\leq 15$ μg/100 ml (Fig. 2.) In adults the $<5\%$ response levels for 40 and 70% inhibition were 15 to 20 and 25 to 30 μg/100 ml, respectively; in children it was slightly lower, 4 to 10 μg/100 ml for 40% inhibition and 20 to 25 μg/100 ml for 70% inhibiton.

It can be stated that the existence of close negative dose-effects and dose-response relationships between PbB and ALAD activity are now so well established that studies failing to reveal such relationships can be disregarded on methodologic grounds. ALAD activity is almost completely inhibited at PbB levels in excess of 70 to 80 μg/100 ml, but the question of whether or not a threshold exists for the inhibition is still open. However, as already stated, ALAD inhibition was considered to be only a subcritical effect by the Task Group on Metal Toxicity (TGMT, 1976). Another expert group which met in Amsterdam in 1976 came to the same conclusion (Zielhuis, 1977).

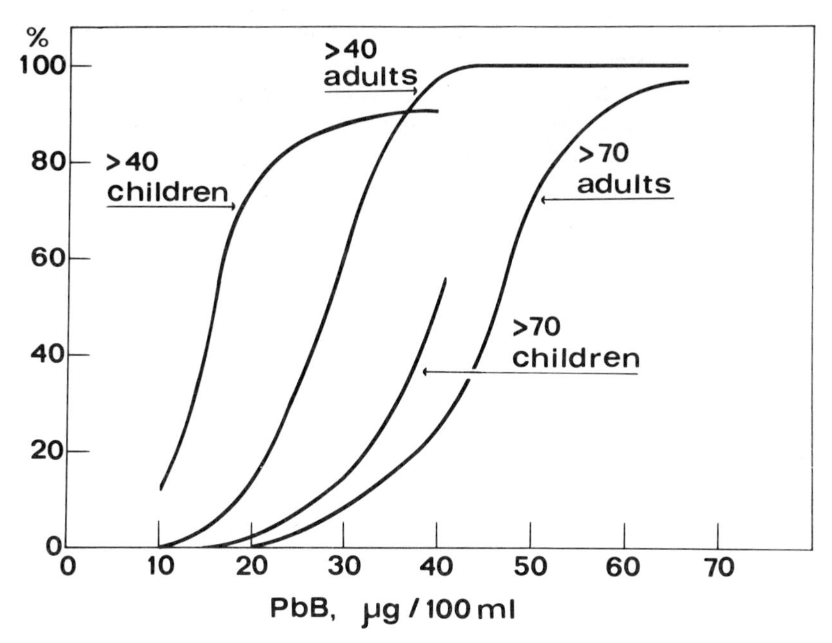


**Figure 2.** Effect of PbB concentrations on the inhibition of ALAD activity (>40% and >70%) in adults and children (Zielhuis, 1975).

## B. Accumulation of Protoporphyrin IX in Erythrocytes

As a result of the inhibition of ferrochelatase (heme synthetase) activity in bone marrow, excessive accumulation of the substrate, protoporphyrin, takes place in the erythrocytes. Some of the analytic methods measure the protoporphyrin IX (PP) concentration, while others assess total free erythrocyte porphyrins (FEP). According to recent studies, FEP's are in fact not "free," but exist as zinc protoporphyrin IX (ZPP). This complex can readily be detected by fluorometry in dilute whole blood (Lamola and Jawane, 1974; WHO, 1977, Blumberg et al., 1977).

Several studies have shown a close relationship between PbB and erythrocyte porphyrin content (Haeger-Aronsen, 1971; Sassa et al., 1973; Piomelli et al., 1973; Stuik, 1974; Chisolm et al., 1974; Roels et al., 1975; Tomokuni and Ogata, 1976, Joselow and Flores, 1977; Eisinger et al.; Haeger-Aronsen and Schütz, 1978). Although different methods yield slightly different results, it is not crucial in principle whether FEP, ZPP or PP IX is measured, since about 90% of excess porphyrins caused by lead exposure are composed of protoporphyrin IX (Baloh, 1974). However, it is important to remember that close correlations are found only under steady state conditions of exposure. This is an understandable phenomenon, since

PP is formed in the mitochondria during the differentiation of the erythrocyte in the hematopoietic bone marrow and the conversion of PP to heme requires an undisturbed ferrochelatase function. This enzyme activity is inhibited by lead. Because the mitochondria are lost when the erythrocyte matures, the content of PP in circulating red blood cells represents what has accumulated during maturation and, consequently, describes an effect taking place in bone marrow, not in the peripheral blood. Since the normal average life span of an erythrocyte is about 120 to 130 days, the exchange of the entire population requires that length of time. Consequently, the PP content of erythrocytes cannot reflect actual exposure conditions until the whole population has been exchanged. Therefore, there is a time lag in the increase of erythrocyte PP as compared to the rise of PbB which weakens the correlation between the two parameters during the first 4 months of exposure (Sassa et al., 1973; Cools et al., 1976; WHO, 1977).

Other reports have also shown that the PP content of erythrocytes remains increased out of proportion to current PbB levels after the termination of "unnatural" (occupational or accidental) exposure (*e.g.*, Rubino et al., 1958; de Bruin, 1971; Alessio et al., 1976), probably because the lead stores in the hematopoietic bone marrow decrease more slowly than the circulating lead. These considerations explain well the fact that close correlations between PbB and erythrocyte PPs exist only under steady state conditions; the following recapitulation is restricted to such a situation.

Figure 3 shows the regression curve between FEP and PbB in 201 adult males occupationally exposed for at least 1 year (Alessio et al., 1976). The authors gave a normal mean value for FEP of 29 μg/100 ml erythrocytes and a mean + 2 SD of 46 μg/100 ml. These values correspond to a mean PbB level of 32 μg/100 ml and a mean + 2 SD of 49 μg/100 ml; they reflect the unusually high values for general populations repeatedly measured in Milan. Figure 3 shows that the increase in FEP beyond the upper normal limits (*i.e.*, 46 μg/100 ml erythrocytes) began at PbB levels of about 45 μg/100 ml. However, even below that value there was a PbB—FEP relationship, starting at a PbB concentration of about 30 to 35 μg/100 ml. Within the PbB range of 50–90 μg/100 ml, FEP increased exponentially to a level of 240 μg/100 ml. Thereafter, the curve tended to be asymptotic.

Other authors have, in principle, obtained similar results, although methodologic differences render a comparison of absolute PP values difficult. However, the asymptomatic slope of the curve may be due to the method used (*i.e.* incomplete extraction of PP at higher levels), since other authors using more modern methods have not found this phenomenon (Tomokuni, 1975; Tomokuni and Ogata, 1975; Joselow and Flores, 1977).

As already stated, methodologic differences complicate the comparison of PP values from different studies. Most investigations indicate, however, that the concentration of PP in erythrocytes starts to increase at PbB levels

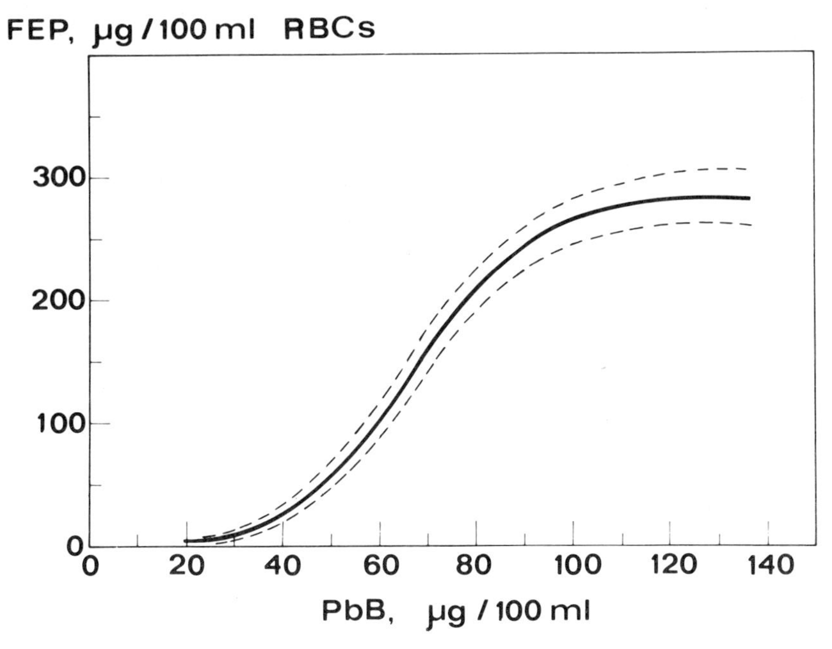


**Figure 3.** Relationship between free erythrocyte porphyrins (FEP) and PbB in 201 adult males (r = 0.90). The regression function is nonlinear (Alessio et al., 1976a).

of about 30 to 40 μg/100 ml in men (Stuik, 1974; Roels et al., 1975; Zielhuis, 1975; TGMT, 1976; Alessio et al., 1976; Joselow and Flores, 1977). In women and children, the increase starts at lower levels (*i.e.*, around 25 to 30 μg/100 ml) (Sassa et al., 1973, Stuik, 1974; Roels et al., 1975; Zielhuis, 1975; Alessio et al., 1977) and the slope is steeper than for men (Alessio et al., 1977). The difference is probably explained by lower iron stores in females and children in men. Three recent studies (Alessio et al., 1978, Roels et al., 1979, Cavalleri et al., in press) cast some doubt on these thresholds, however. According to the results, the PP-concentration increases linearly without any threshold at PbB levels much below 35 μg/100 ml.

With individual data provided by Roels, Schaller and Piomelli as a basis, Zielhuis (1975) calculated dose-response curves for PbB levels and the increase in erythrocyte PP in males, females and children. He could not use absolute values because of methodologic differences; instead, he made the cut-off level the PP/FEB level not exceeded by that of about 95% of the subjects with a PbB concentration of less than 20 μg/100 ml. Figure 4 shows the percentage of subjects with increased PP/FEP levels with increasing PbB concentrations. There was an increased response of PP/FEB in females as compared to males. The data for females were limited in this study, but later publications have corroborated this assumption (Alessio et al., 1977).

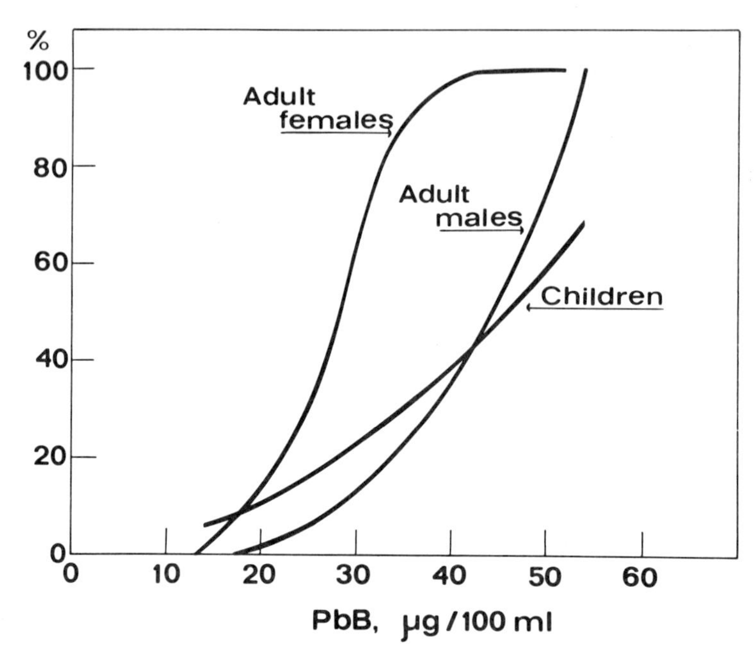


**Figure 4.** Dose-response curve between PbB and FEP. The cut-off point for "response" is a PP/FEP level not exceeded by that of 95% of the subjects with a PbB concentration of ≤20 μg/100 ml (Zielhuis, 1975).

According to the calculations of Zielhuis, the 5% response level for the increase of PP occurs at a PbB concentration of about 20 to 25 μg/100 ml for adult females and children and 25 to 30 μg/100 ml for adult males.

## C. Increased Excretion of Delta-aminolevulinic Acid in Urine

Measurement of the urinary excretion of delta-aminolevulinic acid (ALAU) has long been used as a biologic test to monitor occupationally exposed workers (Haeger-Aronsen, 1960; Hernberg, 1975 and 1976; TGMT, 1976; WHO, 1977). Several studies have shown that an increased concentration of ALAU correlates with concomitant PbB levels; however, a threshold occurred for the increase in the mean ALA excretion at a mean PbB level of about 35 to 45 μg/100 ml (Selander and Cramér, 1970; Tola et al., 1973; Hernberg, 1975; TGMT, 1976) below which no correlation existed. Even at PbB levels in excess of 40 μg/100 ml, the correlation was not so close as for ALAD and PP; correlation coefficients in the range of 0.5 to 0.7 are usually reported (TGMT, 1976). Increasing PbB levels above the threshold value of about 40 μg/100 ml (in males) were accompanied by rapidly increasing ALAU values, but the confidence limits for the regression line were wide. According to Roels et al. (1975), females may be more sensitive than males with respect to this effect, but a correlation between PbB and ALAU occurred for females even in the PbB range of 20 to 50 μg/100 ml.

Using ALAU concentrations of >5 and >10 mg/l as cut-off points, Zielhuis (1975) calculated dose-response relationships for PbB and ALAU in male workers (Fig. 5). According to his calculations, the 5% response level for ALAU >5 and >10 occurred at PbB concentrations of about 30 to 40 μg/100 ml and 40 to 50 μg/100 ml, respectively. At PbB levels between 60 and 70 μg/100 ml, 88% of the subjects had an ALAU concentration above 5 mg/l; 50% a value above 10 mg/1.

## D. Increased Excretion of Coproporphyrin in Urine

The pattern of coproporphyrin excretion in urine (CPU) parallels that of ALAU, although the latter is more specific for lead (Tola et al., 1973; Zielhuis, 1975; TGMT, 1976). The noneffect level is approximately 35 to 40 μg Pb/100 ml blood.

## E. Decreased Hemoglobin (Hb) Concentration

There is conflict in the literature regarding whether or not a decrease in the Hb level has a dose-response relationship to PbB. Several cross-sectional

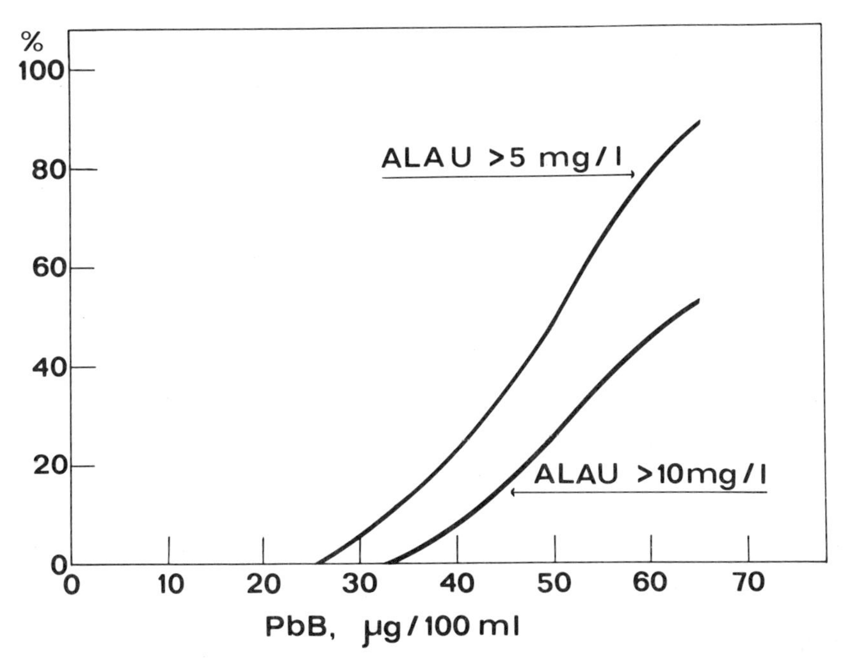


**Figure 5.** Dose-response curves for PbB and the proportion of subjects with a urinary ALA excretion of ≤5 and ≤10 mg/l, respectively (Zielhuis, 1975).

studies have failed to show such an association (Williams, 1966; Cooper et al., 1973; WHO, 1977). On the other hand, it is a well known fact that anemia occurs in clinical lead poisoning. The failure to find a relationship between PbB and Hb values in cross-sectional studies may reflect the non-specificity of the Hb parameter, as well as the great interindividual variations in its sensitivity. When Tola et al. (1973) carried out serial Hb measurements and used the difference between initial and final Hb values 3 to 4 months later instead of absolute values, a statistically significant decrement in the Hb levels of workers who were new to occupational exposure was shown. The mean PbB level at the end of the follow-up was about 50 $\mu g/100$ ml. This approach to analysis is certainly more sensitive than traditional cross-sectional studies.

In children, significant negative correlations between PbB and Hb have been reported even with cross-sectional approaches (Betts et al., 1973, and Peuschel et al., 1972, in the PbB range of 37 to 60 $\mu g/100$ ml and 40 to 130 $\mu g/100$ ml, respectively). Hence it is quite possible that at least for some groups of children a reduction in Hb may occur at a PbB level of approximately 40 $\mu g/100$ ml (WHO, 1977).

### F. Shortened Erythrocyte Life Span

Several studies have shown that erythrocyte survival time is shortened in lead poisoning, but so far only one attempt has been made to correlate the degree of shortening with PbB levels. Using $^3$H-di-isofluorophosphonate (DFP), it was possible to demonstrate a statistically significant negative correlation ($r = -0.6$) between the PbB level and erythrocyte survival time (Hernberg et al., 1967). However, the PbB levels were rather high (Fig. 6). Although only 3 of the 17 subjects studied had clinical signs of poisoning, the lowest PbB levels were around 60 $\mu g/100$ ml. In these subjects, the mean erythrocyte survival time was about 110 days, which is slightly shorter than the average normal life span of 120 to 130 days. Four unexposed subjects were also examined and their values were 106, 120, 126 and 132 days, respectively. Although not conclusive, these data suggest that the non-effect level for the shortening of the erythrocyte life span may be below 60 $\mu g$ Pb/100 ml blood (Fig. 6).

### G. Other Hematologic Effects

Many other hematologic effects occur after excessive lead exposure. They include reticulocytosis, basophilia, depression of erythrocyte $Na^+ + K^+$-ATPase activity and decreases in the reduced glutathione content of the blood (Zielhuis, 1975). Secchi et al. (1973) found a dose-response relationship between decrease in $Na^+ + K^+$-ATPase activity and PbB levels. The correlation coefficient between PbB and reduced enzyme activity was not

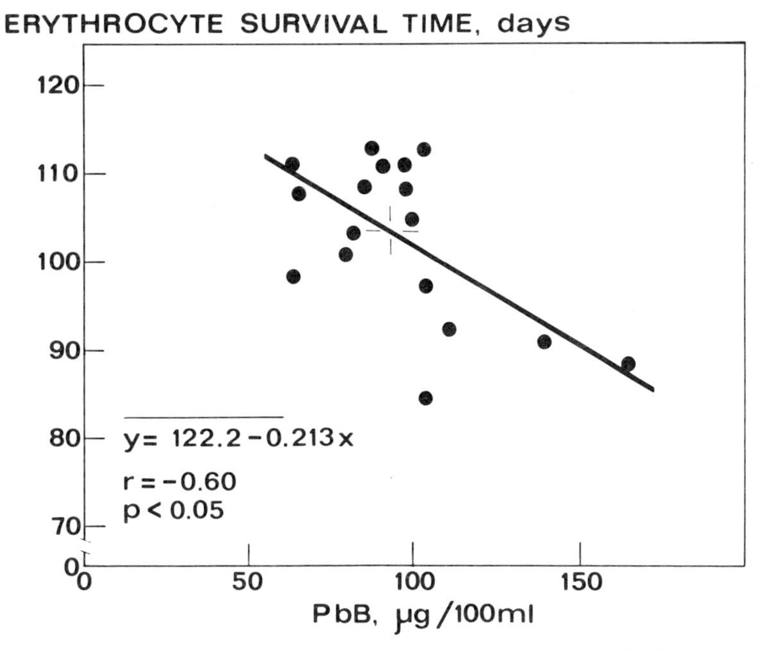


**Figure 6.** Correlation between PbB and erythrocyte survival times in 17 lead workers (Hernberg et al., 1967).

higher than 0.37, but, at PbB levels of ≤30 μg/100 ml, 50% of the subjects showed moderate inhibition.

Roels et al. (1975) and Howard (1978) reported a dose-effect relationship between PbB and reduced glutathione. The correlation coefficients were 0.42 and 0.59, respectively. In both studies the noneffect level was rather low, ranging between 20 and 30 μg Pb/100 ml blood. This finding is interesting, since it has been postulated that the inhibition of ALAD activity may partly be due to a lack of available free —SH radicals (Howard, 1978).

Figure 11 summarizes the most important hematologic effects and responses in relation to the PbB level. The earliest hematologic effect of lead is partial inhibition of ALAD, which is measurable at PbB levels below 10 to 20 μg/100 ml and is perhaps without threshold. Within the PbB range of 20 to 30 μg/100 ml, the inhibition of erythrocyte $Na^+ + K^+$-ATPase and a decrease in reduced glutathione concentration have been suggested. Erythrocyte protoporphyrins show the first signs of increasing at PbB levels of about 30 to 40 μg/100 ml in men and 25 to 30 μg/100 ml in women and children, while increased urinary ALA and CP excretion starts at PbB levels of about 40 μg/100 ml. The erythrocyte life span probable begins to decrease below a PbB level of 60 μg/100 ml, but this assumption is based on very few data. The Hb level has been shown to decrease at PbB level of the order of 50 μg/100 ml in a longitudinal study, but cross-sectional investigations have not revealed any decrement even at PbB levels as high as 80 μg/100 ml. In children, however, the Hb concentration probable falls earlier. Anemia is usually connected with clinical lead poisoning and becomes common at PbB levels in excess of 80 to 100 μg/100 ml.

## V. DOSE-EFFECT AND DOSE-RESPONSE CURVES FOR NEUROLOGIC EFFECTS

Lead impairs both the central and the peripheral nervous system. Impairment of the central nervous system manifests itself as encephalopathy of different intensity; such a disorder may leave irreversible sequelae. The symptoms and signs vary from severe neurologic defects to subtle psychological or behavioral changes. Peripheral nervous impairment ranges from paresis to slight functional defects that can only be detectable by sensitive electrophysiologic techniques. Because dose-effect and dose-response relationships mainly serve the purpose of defining health criteria, the lower exposure range is emphasized and the results of accidental or otherwise exceptionally heavy exposure are mentioned only in passing in this section.

Relating neurologic effects to blood lead (PbB) level is more difficult than relating hematologic effects, because useful data are sparse and the PbB level measured in connection with the examination of the nervous system is usually far from representative of the one which actually caused the effect. This lack of representativeness is especially so for behavioral disorders in adolescents whose relevant exposure took place years earlier. Moreover, peripheral nervous system damage may also persist for a long time. In such situations, the current PbB level cannot be a good indicator of the "dose" that caused the effect. Some other measure, perhaps one which reveals past exposure, would better serve this purpose. Several studies have, in fact, utilized the lead content of hair and deciduous teeth or the urinary excretion of lead after chelation. Such observations, while important for revealing causal connections between lead exposure and various nonspecific symptoms and signs, are difficult to relate to exposure intensities or PbB levels. Neither safety norms nor monitoring programs can be based on the measurement of the lead content in hair, teeth or bone; even measuring the provoked excretion of lead in urine is too cumbersome for monitoring purposes. For obvious reasons, prevention must focus on controlling current, not past, exposure. Hence it is operational to use the PbB level as the dose indicator even for neurologic effects, in spite of severe conceptual shortcomings. These considerations also explain why close relationships between current PbB levels and neurologic effects can be expected to occur only when no major changes in the exposure intensity have taken place over time.

Unfortunately, there are few publications providing information (PbB levels) of past exposure; whenever such data are available, the neurologic effects have been related to them rather than to current PbB levels.

## A. Peripheral Nervous System

It is a well known fact that peripheral paresis occurs in severe lead poisoning. The PbB levels required to produce this manifestation are not well defined and considerable interindividual variation probably occurs (Hernberg 1975). Lead palsy is rare today.

Of much more interest are the electrophysiologically detectable functional abnormalities that occur in the peripheral nerves, even in the absence of clinical neurologic signs (Sessa et al., 1965; Catton et al., 1970; Seppäläinen and Hernberg 1972). However, in the first studies of this kind, most of the PbB levels greatly exceed the "safe" limits of 70 or 80 μg/100 ml that were recommended at the time and many of the subjects had symptoms and signs of lead poisoning. The most important neurophysiologic abnormalities recorded consisted of slowing of nervous motor conduction velocity, especially that of the slower fibers, and electromyographic abnormalities, such as fibrillations and a diminished number of motor units in maximal contraction.

More recently, neurophysiologic studies have been extended to include subjects without symptoms of poisoning and with PbB levels below 70 or 80 μg/100 ml. The results indicate that slight functional impairment occurs even in this exposure category.

Seppäläinen et al. (1975) studied 26 storage battery workers whose PbB levels had been monitored during the entire exposure period (13 months to 17 years). The criterion for being included in the series was that no single PbB value had ever exceeded 70 μg/100 ml. The PbB levels had ranged mostly between 35 and 60 μg/100 ml and occasionally between 20 and 70 μg/100 ml. In the study, the mean PbB level was 40 ± 8.9 μg/100 ml. Maximum conduction velocity (MCV) of the median, ulnar, deep peroneal and posterior tibial nerves, sensory conduction velocity (SCV) of the median and ulnar nerves and conduction velocity of the slower motor fibers (CVSF) of the ulnar nerve were measured. Needle electromyograph (EMG) of three to seven muscles was also performed on 11 workers with abnormal or borderline conduction velocities. The results of the velocity measurements were compared to those of age- and sex-matched referents.

In comparison with the referents, the exposed subjects showed statistically significant lower MCV's of the median and ulnar nerves and, especially, lower CVSF's of the ulnar nerve. The SCV's also were slower than in the reference group. Of the 11 EMG's made, 9 were abnormal. Denervation potentials (fibrillations) were found in five cases; the other abnormalities consisted of an abnormally long duration of motor unit potentials and fewer motor units in maximal contraction. All these findings are compatible with peripheral neurogenic lesions. Figure 7 shows a comparison of the CVSF's of patients with lead poisoning in another study (Seppäläinen and Hernberg, 1972) with the workers whose PbB levels had never exceeded 70 $\mu g/100$ ml in this study and the two different referent groups. As can be seen, the lowest average CVSF was found in the poisoned group, while those with a maximal PbB level below 70 $\mu g/100$ ml came next. Both reference groups had significantly better results. It was not possible to establish a dose-effect relationship for the CVSF within the exposed group with a PbB of $<70$ $\mu g/100$ ml, but on a group basis such a relationship was present (Fig. 7.) A dose-response relationship was also indicated. If 40 m/s is considered as the

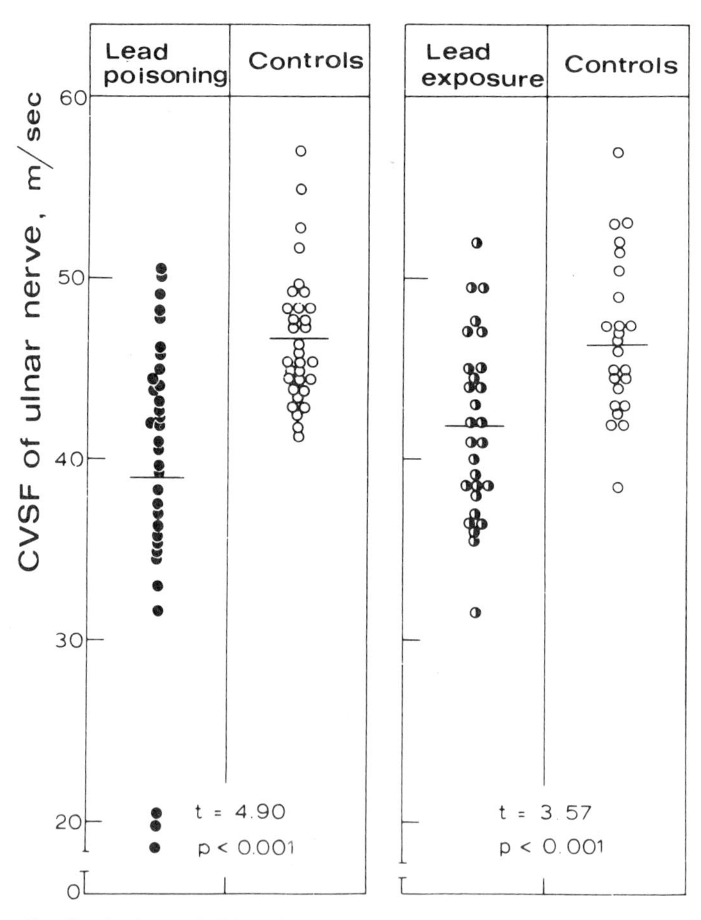


**Figure 7.** Conduction velocities of the slower motor fibers (CVSF) of the ulnar nerve in patients with lead poisoning, workers exposed to lead (PbB never $>70$ $\mu g/100$ ml) and referents (Seppäläinen et al., 1975).

lower normal limit, only 1 of the 54 referents, or 2%, had an abnormal value, whereas 10 of the 26 (40%) exposed (PbB <70 μg/100 ml) and 15 of the 33 (45%) poisoned workers had abnormally slow conduction. EMG's were not made for the referents, but the frequency of pathologic findings was 15 out of 32 (47%) in the poisoned group and at least 9 out of 26 (35%) in the exposed group. (EMG's were only performed for those of the latter with abnormal or borderline conduction velocities, and hence represent a conservative estimate.)

Abbritti et al. (1977) also recently found a similar pattern for EMG's but unfortunately measured only the MCV of the ulnar, femoral and peroneal nerves. Studies were conducted on 118 exposed subjects, 57 of whom had current PbB levels in excess of 80 μg/100 ml. Of those 59 subjects with PbB levels between 41 and 80 μg/100 ml, 30 (51%) had a pathologic EMG, while 32 of the subjects (65%) with PbB levels between 81 and 120 μg/100 ml and all 8 of those with a PbB level of ≥121 μg/100 ml had an abnormal EMG. In this comparison only those with normal MCV were considered. When a combination of reduced MCV and abnormal EMG was considered, 2% of those with PbB levels between 41 and 80 μg/100 ml and 12% of those with PbB levels between 81 and 120 μg/100 ml showed both findings. These reports provided no data for studying dose-effect and dose-response relationships in sufficient detail, because the exposure range was too narrow and there were too few measurements from subjects with low exposure.

A recent study (Araki and Honma 1976) has revealed statistically significant dose-effect relationships between the current PbB levels and the MCVS of both the median and the posterior tibial nerve, as well as the mixed nerve conduction velocity (MNCV) of the median nerve. This relationship between the PbB level and the MCV of the median nerve in 38 subjects is seen on Figure 8. If 50 m/s is taken as the lowest normal value, the dose-response relationship suggests a nonresponse level in the range of 31 to 50 μg/100 ml (Fig. 8).

Recently, Seppäläinen et al. (1979) made a more detailed study of dose-effect and dose-response data for conduction velocities. They examined 61 storage battery plant and mechanical workshop workers whose PbB levels had never exceeded 70 μg/ml, 17 workers whose PbB level had exceeded 70 μg/100 ml and 34 referents without any present or past occupational exposure to lead. The PbB level of the lead workers had been regularly monitored during their entire exposure history. The mean PbB level of the referent group was 10.5 μg/100 ml (SD 3.5 μg/100 ml) at the time of the examination. This study once more confirmed the presence of group differences between workers with low lead exposure and nonexposed referents. When only workers whose PbB levels had always been below 70 μg/100 ml were compared to the referents, statistically significant dif-

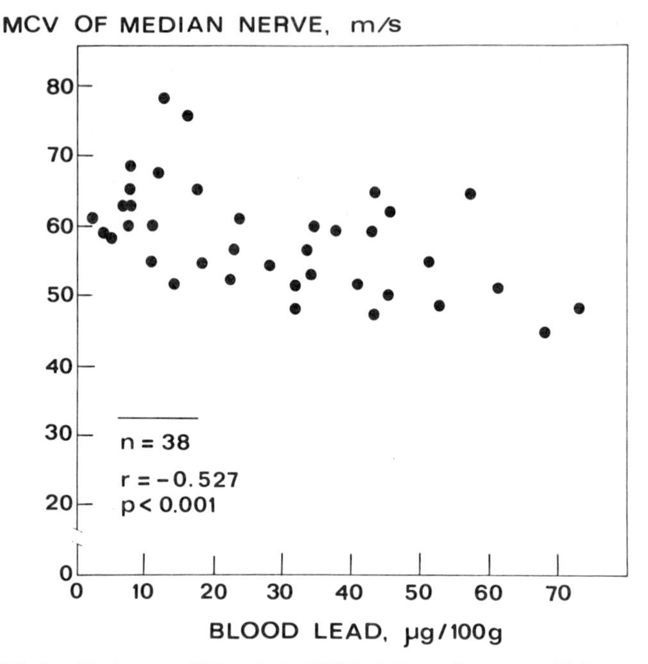


**Figure 8.** Relationship between PbB and the MCV of the median nerve. PbB measured at the neurophysiologic examination (Araki and Honma, 1976).

ferences were found for the MCV of the posterior tibial nerve, the SCV and the distal SCV (dSCV) of the median nerve and the CVSF of the ulnar nerve. Within the entire study population, statistically significant correlations were also found between the maximal PbB level and several measures of the median nerve, *i.e.*, the MCV, the SCV, the sDCV and the distal latency. In addition, the CVSF and the SCV of the ulnar nerve and the MCV of the posterior tibial nerve correlated with the maximal PbB level. Figure 9 shows the correlation between the maximal PbB concentration and the SCV of the median nerve. The current and time-weighted average PbB level also correlated to measures of nervous function but this was to be expected since there were high intercorrelations between the different PbB indicators.

Seppäläinen and her coworkers (1979) also made an attempt to define the noneffect level for impairment of nervous function by dividing the exposed group into subcategories according to the maximal PbB level and by testing the differences of the means by the student's t test (Table 1). The CVSF of the ulnar nerve was the most sensitive indicator; the group with a maximal PbB level between 40 and 49 $\mu$g/100 ml differed significantly from the referents. The category whose maximal PbB concentration was between

50 and 59 μg/100 ml showed four significant differences; the same differences also appeared in the PbB 60–69 μg/100 ml exposure category. The highest exposure category was not so consistent; however, many of these subjects had been exposed much less for several years (Table 1).

Table 2 shows the dose-response relationship for the same exposure categories with two different measures of response. When one pathologic finding is used as the criterion, the nonresponse level seems to be in the range of 40 to 49 μg/100 ml. When the requirement is two or more pathologic findings, the level is approximately 50 to 60 μg/100 ml.

The results of this study indicate that nerve conduction impairment is induced in some subjects at PbB levels of 40 to 50 μg/100 ml. The number of subjects was rather small, however, and a definite conclusion cannot yet be drawn. It is extremely difficult to find workers who have been regularly monitored *and* who have never had high PbB levels. The lack of such subjects partly explains why so few studies have been published and why those published have such small populations. No studies have been devoted to such dose-response relationships in children.

Quite recently, Baloh et al. (1979) published a study on lead smelter workers. Relatively few differences in neurologic function were demonstrated, and the authors could not confirm the earlier reports on nerve con-

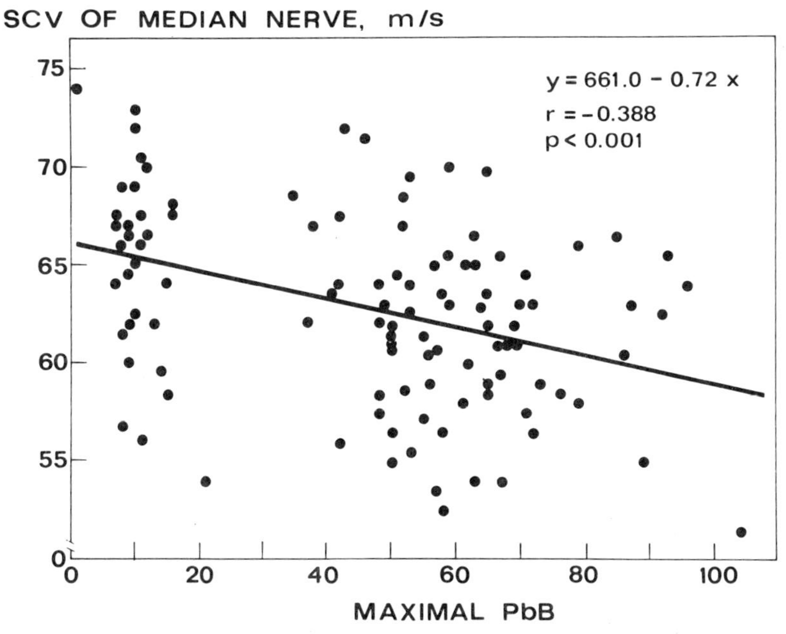


**Figure 9.** Relationship between the maximal PbB and the SCV of the median nerve (Seppäläinen, 1979).

**Table 1.** The p-values of differences in the conduction velocities of lead workers in various exposure categories (according to the maximal PbB level, μg/100 ml) and 34 nonexposed referents. Only velocities with at least one significant difference are shown (modified from Seppäläinen et al. 1979).

| Nerve and conduction velocity | | PbB Levels (μg/100 ml) | | | |
|---|---|---|---|---|---|
| | | PbB 40–49 (N = 11) | PbB 50–59 (N = 28) | PbB 60–69 (N = 19) | PbB > 70 (N = 17) |
| | | p | p | p | p |
| Median nerve: | MCV | ns* | ns | ns | <0.02 |
| | SCV | ns | <0.005 | <0.01 | <0.005 |
| | dSCV | ns | <0.02 | <0.01 | <0.05 |
| Ulnar nerve: | CVSF | <0.025 | <0.05 | <0.02 | ns |
| | SCV | ns | ns | ns | <0.05 |
| Posterior tibial nerve: | MCV | ns | <0.05 | <0.025 | ns |

* ns, not significant

duction slowing. Instead, the study showed impaired accuracy and reaction time of saccadic eye movements. The PbB levels of the exposed group had mostly been in the range of 60–80 μg/100 ml, while those of the controls showed an average as high as 22 μg/100 ml at examination.

With allowance for the uncertainties for small populations, the available information on dose-effect and dose-response relationships for peripheral nervous damage may be summarized as follows.

1. Dose-effect relationships for some conduction velocities (MCV, CVSF, SCV), especially of arm nerves, have been reported in the PbB range of

**Table 2.** Conduction velocity scores (equal to number of abnormal* conduction velocities) for lead workers in various exposure categories (according to the highest PbB recorded) and the percentage of those with abnormal scores (Seppäläinen et al., 1979)

| Maximal PbB (μg/100 ml) | N | Scores* 0 | 1 | 2 | 3 | 4 | Score ≥ 1 (%) | Score ≥ 2 (%) |
|---|---|---|---|---|---|---|---|---|
| Referents (Current mean PbB 11) | 34 | 33 | | 1 | | | 3 | 3 |
| <40 | 3 | 2 | 1 | | | | — | — |
| 40–49 | 11 | 8 | 3 | | | | 27 | 0 |
| 50–59 | 28 | 19 | 5 | 2 | 1 | 1 | 32 | 14 |
| 60–69 | 19 | 11 | 5 | 2 | | 1 | 42 | 16 |
| >70 | 17 | 8 | 7 | 1 | 1 | | 53 | 12 |

* Conduction velocity scores equal to number of abnormal conduction velocities, defined as being lower than the mean minus two SD of the laboratory's normal population, for each specific nerve.

20 to 70 μg/100 ml. The noneffect level has not yet been clearly defined, but it may be in the 40 to 50 μg/100 ml range.

2. The nonresponse level also appears to be somewhere in the PbB range of 40 to 50 μg/100 ml.
3. PbB levels in excess of 50 μg/100 ml reveal an increasing frequency of abnormal conduction velocities among occupationally exposed workers.
4. Rather high frequencies of EMG abnormalities have been reported among workers whose maximal PbB has been above 50 μg/100 ml. The frequency probably increases at higher levels of exposure (PbB $>80$ μg/100 ml).
5. The length of time that the PbB concentration must remain above a particular level to cause impairment has not yet been established. The effects of comparatively high PbB levels of short duration and those of moderately increased PbB levels persisting for longer periods of time have not been compared.
6. Neurologic manifestations such as paresis do not generally occur at PbB levels below 80 μg/100 ml and the nonresponse level for these effects is probably considerably higher.

## B. Central Nervous System

While there is no doubt that heavy lead exposure causes encephalopathy, some controversy still prevails as to whether lower intensities of exposure can cause subtle changes in central nervous functions (WHO, 1977). It has also been postulated that infants and young children are more sensitive to the toxic effects of lead than adults, partly because the developing brain is more vulnerable than the mature central nervous system, another reason is the empirical observation that encephalopathy is more commonly associated with childhood poisoning than adult poisoning. On the other hand, it may well be that childhood poisoning is related to higher exposure intensities; these higher intensities may explain the higher incidence of encephalopathy.

Central nervous impairment can be studied with a variety of clinical neurological methods. To study subtle effects, however, the selection is narrower. The most widely used methods for this purpose consist of different psychologic tests and symptom questionnaires.

Establishing the *causative* PbB level for central nervous symptoms and signs is probably even more complicated by the time factor than determining the corresponding level for peripheral effects. Many of the central nervous effects found at the time of examination are sequelae of past episodes of acute poisoning. In fact, the association between the PbB level and neurologic status may in such cases create a false impression concerning the level of lead exposure when the damage was initiated (*i.e.*, the actual PbB level responsible for the symptoms was probably much higher than the one measured at the examination (WHO, 1977).

Another difficulty in studying central nervous effects, especially subtle ones such as hyperactivity, the reduction of psychological performance, etc., is the nonspecific nature of these manifestations. All studies should be able to control a great number of selective and confounding factors, but such control is not always possible. Most of the studies published on subtle central nervous effects have, in fact, been criticized for not having solved this problem (WHO, 1977). Childhood lead poisoning has been a particular problem in this respect.

As long as there is severe disagreement on whether or not subtle behavioral abnormalities in children can be related to past lead exposure, it is not justifiable to construct dose-effect and dose-response relationships from positive studies only and neglect completely those which have failed to reveal any such correlations. Besides, even the positive studies do not usually relate the findings to PbB levels, but rather relate to excretion of lead in the urine provoked with EDTA or penicillamine or to dentine or hair levels of lead. Because of these uncertainties, the dose-effect and dose-response relationships between PbB levels and central nervous dysfunction in children cannot be described as yet. The WHO criteria document on lead (WHO, 1977) assumes that the probability of noticeable brain dysfunction increases in children once the PbB level reaches approximately 50 $\mu$g/100 ml and that the probability of acute or chronic encephalopathy increases at a level of about 60 $\mu$g/100 ml. This view is more of an educated guess than a statement of fact.

The accessibility of past PbB data from regular monitoring procedures makes assessment somewhat easier in the case of occupationally exposed workers. It may be difficult, however, to find enough workers who have been monitored throughout the entire length of exposure and who represent different intensities of exposure, particularly the lower range.

Thus far, four research groups have published results on the psychological performance of lead workers with PbB levels below 80 $\mu$g/100 ml. All the groups found some impairment, but the average performance differences between the exposed workers and the referents were inconsistent. This inconsistency was probably partly due to differences in the performance levels of the referent groups. Finding a referent group that corresponds to the exposed group with respect to primary performance level and has the motivation to cooperate is difficult. Therefore, intragroup comparisons are generally more meaningful. This also appears to be the case for lead workers, since the results of the four research groups become more consistent when the performances (as measured by similar tests) are related to the PbB levels within the exposed groups.

Repko et al. (1974) studied 316 workers with an average PbB level of 63 $\pm$ 22 $\mu$g/100 ml and 112 referents. The performances of the lead workers were not inferior to those of the referents, but within the exposed group the

PbB level was significantly related to muscular strength and endurance, tremor and visual-motor functions (*i.e.*, eye-hand coordination). The same relationship was found for auditory threshold levels. Later, Repko and his collaborators (1978) reported another study on 42 storage battery workers and 18 referents. The range of the PbB levels in these workers was 12 to 79 μg/100 ml. The results partly corroborated the findings of the earlier study.

Hänninen et al. (1978) found a relationship between lead uptake and impaired psychological performance among 49 workers whose highest PbB had never exceeded 70 μg/100 ml. Their average PbB level was 32 ± 11 μg/100 ml. The test procedures comprised subtasks of the Wechsler Adult Intelligence Scale (WAIS) and Wechsler Memory Scale (WMS) and tasks for visual-motor performance. The comparison of mean performances with those of 24 nonexposed referents did not yield statistically significant differences, although most of the performances were inferior in the lead-exposed group. However, the test results of the exposed workers were related both to the actual PbB, the highest PbB level ever recorded and the time-weighted average PbB. Of these, the time-weighted average PbB correlated best with impaired performance. Hence it seems as if the psychological effects found were due to the accumulated uptake of lead, rather than to current exposure levels.

The performances that were most affected by lead were dependent on visual intelligence and visual-motor functions (Fig. 10). The two tests that appeared to be the most sensitive indicators of lead-induced impairment were Block Design (BD) for visual intelligence and the Santa Ana Dexterity Test (SA) for visual-motor ability. In the Santa Ana test, "poor" results (one standard deviation below the "normal" average) did not occur at all among those 22 lead workers whose PbB levels had never exceeded 50 μg/100 ml. Among the 27 workers whose maximal PbB levels had been between 51 and 70 μg/100 ml, "poor" results in this test occurred in 7 cases (26%). These findings indicated that in the Santa Ana test the nonresponse level for PbB is approximately 50 μg/100 ml.

Valciukas et al. (1978) also found a negative relationship between PbB levels and some psychological performances, but they used a less extensive test procedure than Hänninen et al. (1978). The study population comprised 99 secondary lead smelter workers whose current PbB levels were 17% less than 40, 61% between 40 and 59, 20% between 60 and 79 and 1% more than 80 μg/100 ml. The results were analyzed for the exposed group and also compared with those of 36 nonexposed referents. The four tests used in this study were all for visual intelligence and visual-motor ability. Three of them, the Block Design (BD), Digit Symbols (DSy) and Embedded Figures (EF), yielded significant differences between the exposed group and the referents; furthermore they displayed significant correlations with the PbB level within the exposed group. The age-corrected scores for BD and EF

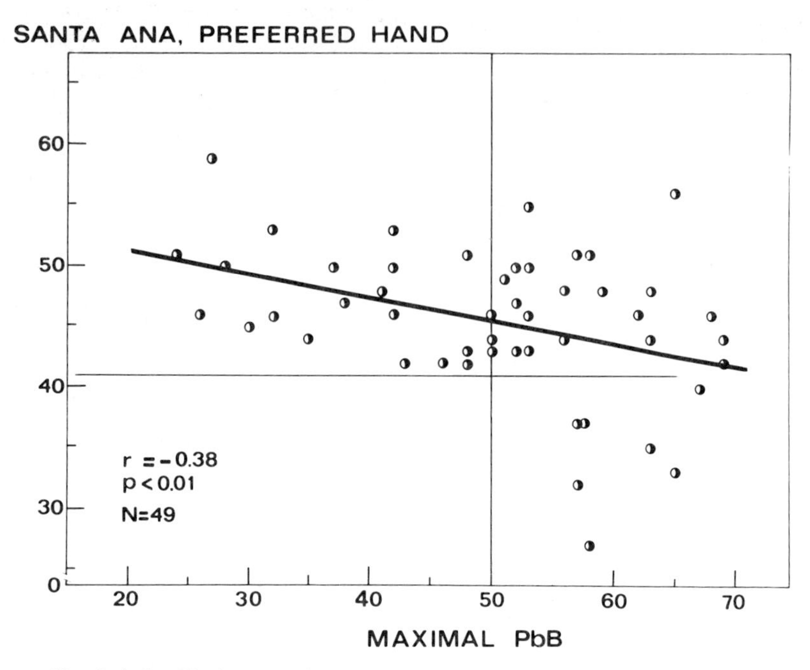


**Figure 10.** Relationship between the maximal PbB ever recorded and performance in the Santa Ana Dexterity Test (Hänninen et al., 1978).

correlated with the PbB level and these correlations were improved when the erythrocyte zinc protoporphyrin (ZPP) content was used as indirect dose index in place of PbB. (ZPP is rather an effect indicator than a dose indicator.) The DSy test also correlated with ZPP levels. The fact that the correlations were better for ZPP than PbB indicates that earlier, not current, PbB levels were more relevant. In contrast to the first two studies mentioned (Repko et al., 1978; Hänninen et al., 1978), the dexterity test (SA) did not show any relationship to PbB level in this investigation. According to the authors, the reason may be that they used only one trial for measuring each variable and this procedure decreases the reliability of the method.

Grandjean et al. (1978) found impaired performance in a wide area of psychological functions among 42 exposed workers in comparison with 22 referents. The PbB levels of the exposed workers ranged from 0.6 to 4.2 μmol/l (12 to 82 μg/100 ml), with a median value of 2.2 μmol/l (45 μg/100 ml). The test procedures included all subtasks of WAIS and additional tests for memory and cognitive functions. Impaired performances covered both verbal and visual intelligence and memory functions; for most tests the decrease was significantly correlated with indices of lead exposure (PbB and ZPP) within the exposed group. The wider range of psychological impairments, when compared with those found in the other subjects (Hänninen et al., 1978), may be explained by a higher mean PbB value (32 versus 45 μg/100 ml) and by the higher upper limit of the range of PbB (69 versus 82 μg/100 ml). Unfortunately, neither Grandjean et al. (1978) nor Valciukas et al. (1978) reported individual values, so the thresholds for effect and response cannot be evaluated. It should be stressed that the aforementioned research groups carried out an intercomparison program for PbB analyses in 1976; the results were compatible since all used comparable methodologies.

Effects on the central nervous system can also appear as subjective symptoms reflecting disturbances of sleep, mood, memory, attention, etc. Repko et al. (1974) reported that the lead-exposed workers exhibited significantly more hostility, depression and general dysphoria than the nonexposed referents. Hänninen et al. (1979) inquired into subjective symptoms by means of a questionnaire, with groups subdivided into those with maximal PbB levels below 50 μg/100 ml and those with a level between 50 and 69 μg/100 ml. Several symptoms were more prevalent in the higher exposure category, including fatigue, sleep disturbances, forgetfulness, absentmindedness, restlessness, paresthesia of the upper limbs, weakness of the lower limbs and difficulties in walking in the dark. Although many more differences were found when the entire group was compared with the reference group, it is prudent to give more emphasis to those differences which showed a dose-related pattern among the exposed workers because of possible systematic differences in attitudes between the exposed and nonexposed subjects. Corresponding results have been reported by Lilis et al. (1977), who found an excess of central nervous symptoms and pain in the muscles and joints of workers whose PbB did not exceed 80 μg/100 ml. The frequency of these symptoms was related to ZPP levels.

In conclusion four different research groups have found some impairment in the psychological performance of workers exposed to lead; three groups discovered an excess of subjective symptoms among workers exposed to comparatively low intensities of lead (below 80 μg/100 ml; sometimes below 60 μg/100 ml). Tests measuring visual-motor functions and visual intelligence seem to be the most sensitive indicators of these effects, even at lower PbB levels (below 70 or even 60 μg/100 ml), assuming that sufficiently reliable methods are used. Impairment of a wider area of psychological functions seems to be detectable when the exposure range is somewhat higher (up to 80 μg/100 ml). Subjective central nervous symptoms have been found to increase significantly at the same PbB levels.

The nonresponse level for central nervous impairment is approximately 50 μg/100 ml; the time-weighted average exposure is a better dose indicator than actual PbB levels. Current information does not yet allow the establishment of dose-response relationships for chronic central nervous effects, particularly for childhood exposure. There is no doubt that heavy exposure to lead (usually ingestion) causes encephalopathy in children and later affects psychological performance, but is is not yet possible to describe the relationship between lower PbB levels and subtle central nervous effects.

## VI. DOSE-EFFECT AND DOSE-RESPONSE RELATIONSHIPS FOR OTHER EFFECTS

Lead also causes damage to other organ systems, but it usually requires higher exposure intensities. The renal system, the gastrointestinal tract, possibly the cardiovascular system, the liver, the gonads and other endocrine organs may be affected. (WHO, 1977). Chromosome aberrations have also been reported in some studies, while others have given negative results in this respect (WHO, 1977; Zielhuis, 1977).

### A. Renal Effects

The dose-effect and dose-response relationships of lead on the renal system are poorly understood. Goyer and Rhyne (1973) suggested that blood lead (PbB) levels in the range of 40–80 $\mu$g/100 ml are associated with inclusion bodies in the renal tubular epithelium. Fanconi's syndrome may develop among exposed children at PbB levels above 150 $\mu$g/100 ml (Chisolm and Leany, 1962). In occupationally exposed workers, kidney function impairment may occur at PbB levels of 100 to 120 $\mu$g/100 ml or even higher, but the duration of exposure must probably be long (TGMT, 1976). Lilis et al. (1968) found 18 cases of clinical chronic nephropathy (arteriosclerotic changes, interstitial fibrosis, glomerular atrophy and hyaline degeneration of the vessels) among 102 cases of lead poisoning. The PbB levels at the time of diagnosis ranged from 42 to 141 $\mu$g/100 ml, but more than likely the levels had been much higher in the past. Nephropathy in adults has also been connected with lead poisoning contracted in childhood (Emmerson, 1963; WHO, 1977). the development of chronic nephropathy probably requires very high exposure levels and hence would rarely occur in industrial situations.

### B. Gastrointestinal Symptoms

Typical gastrointestinal symptoms caused by lead exposure include constipation or diarrhea, epigastric pain, nausea, indigestion and anorexia (Hernberg, 1975). Mild symptoms may occur in the PbB range of 50 to 70 $\mu$g/100 ml (Hänninen et al., 1979); more pronounced symptoms become common at PbB levels above 80 $\mu$g/100 ml. The likelihood of more severe manifestations increases with rising PbB levels; colic becomes relatively common when the PbB exceeds 130 to 150 $\mu$g/100 ml, although wide individual variations occur in both directions (Hernberg 1976).

### C. Chromosomal Aberrations

The literature is controversial in regard to chromosomal aberrations in subjects exposed to lead (WHO, 1977). However, a recent Swedish study established a dose-response relationship between PbB level (mean of several determinations at different periods of time) and chromosomal aberrations in 26 lead-exposed smelter workers (Nordenson et al., 1978). An increased risk of chromosomal damage was found at a PbB level of 25 μg/100 ml, and the frequency of gaps, chromosome aberrations and chromatide aberrations increased as the mean PbB level of three subgroups increased from 22.5 μg/100 ml in the "low exposure" group, to 39 μg/100 ml in the "medium exposure" group and to 65 μg/100 ml in the "high exposure" group. Only chromosome aberrations were statistically increased (Nordenson et al., 1978). The merits of this study lie in the longitudinal exposure data and the competent cytogenetic techniques used. Unfortunately, the authors did not discuss the possibility of concurrent exposure to other metals (*e.g.*, arsenic) which may also have occurred in the lead smelter.

## VII. SUMMARY AND CONCLUSIONS

The effects of lead have already been related to PbB levels. A comparative summary of the various dose-effects and dose-response relationships of lead is shown in Table 3 and reveals the noneffect PbB levels established by a WHO expert group (WHO, 1977) for a variety of hematologic and neurologic effects. Other attempts have been made to combine dose-effect and

**Table 3.** No detected effect levels in relation to PbB (μg Pb/100 ml of blood) (WHO, 1977)

| No Detected Effect Level (μg/100 ml) | Effect | Population |
|---|---|---|
| $<10$ | Erythrocyte ALAD inhibition | Adults, children |
| 20–25 | FEP | Children |
| 20–30 | FEP | Adult, female |
| 25–35 | FEP | Adult, male |
| 30–40 | Erythrocyte ATPase inhibition | General |
| 40 | ALA excretion in urine | Adults, children |
| 40 | CP excretion in urine | Adults |
| 40 | Decrease in blood Hb | Children |
| 40–50 | Peripheral neuropathy | Adults |
| 50 | Decrease in blood Hb | Adults |
| 50–60 | Minimal brain dysfunction | Children |
| 60–70 | Minimal brain dysfunction | Adults |
| 60–70 | Encephalopathy | Children |
| $>80$ | Encephalopathy | Adults |

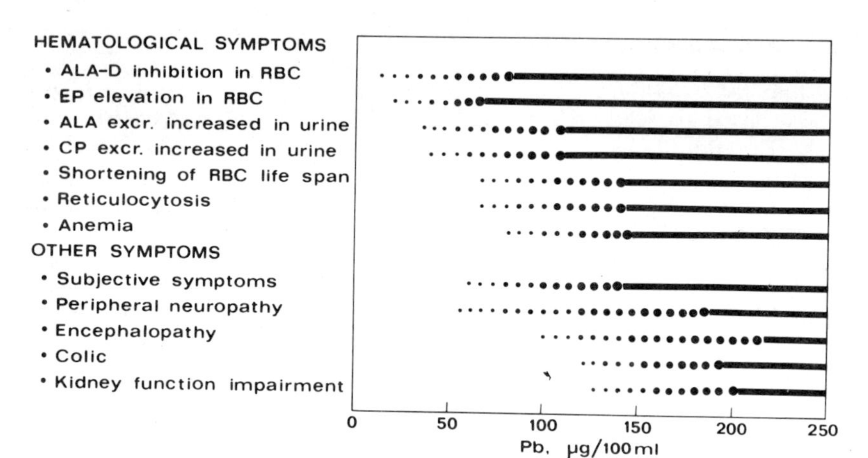


**Figure 11.** Summary of relationship between PbB levels and the onset of some toxic effects. For explanation, see text (Hernberg, 1976).

dose-response relationships through the correlation of both the intensity and the frequency of different symptoms and signs with concomitant PbB levels (Hernberg 1976) (Fig. 11). There are wide variations in individual sensitivity and no definite cut-off points can be defined. It should be evident that there is no clear-cut delineation between those levels of lead producing overt pathological signs and those producing symptoms of toxicity.

## ACKNOWLEDGMENTS

I wish to thank Helena Hänninen, Ph.L., Claës-Henrik Nordman, M.D., Anna Maria Seppäläinen, M.D., and Sakari Tola, M.D., for the long and fruitful collaboration which has resulted in many of the data cited in this chapter; their constructive criticism of this manuscript is also gratefully acknowledged.

## REFERENCES

Abbritti, G., Perticoni, G.F., Colangeli, C., Curradi, F., Cianchetti, C., Siracusa, A., and Morucci, P. (1977). *Med. Lavoro 68:*412.

Alessio, L., Bertazzi, P. A., Monelli, O., and Foà, V. (1976). *Int. Arch. Occup. Environ. Health 37:*89.

Alessio, L., Bertazzi, P.A., Monelli, O., and Toffoletto, F. (1976). *Int. Arch. Occup. Environ. Health 38:*77.

Alessio, L., Bertazzi, P.A., Toffoletto, F., and Foà, V. (1976). *Int. Arch. Occup. Environ. Health 37:*73.

Alessio, L., Castoldi, M.R., Buratti, M., Maroni, M., and Bertazzi, P.A. (1977). *Int. Arch. Occup. Environ. Health 40:*283.

Alessio, L., Castoldi, M.R., Buratti, M., Calzaferri, G., Odone, R., and Juana, C. (1978). *Med. Lavoro 69:*563.

Araki, S., and Honma, T. (1976). *Scand. J. Work Environ. Health 4:*225.
Baloh, R.W. (1974). *Arch. Environ. Health 28:*198.
Baloh, R.W., Spivey, G.H., Brown, C.P., Morgan, D., Campion, D., Browdy, B.L., Valentine, J.L., Gonick, H.C., Massey, F.J., and Culver, B.D. (1979) *J. Occup. Med. 21:*490.
Berlin, A., Lauwerys, R., Buchet, J.P., Roels, H., Del Castilho, P., and Smeets, J. (1974a). In "Recent Advances in the Assessment of the Health Effects of Environmental Pollution," International Symposium, Proceedings, Paris, June 24–28, 1974. Luxembourg, Commission of the European Communities, 1975.
Berlin, A., Schaller, K.H., and Smeets, J. (1974b). In "Recent Advances in the Assessment of the Health Effects of Environmental Pollution," International Symposium, Proceedings, Paris, June 24–28, 1974. Luxembourg, Commission of the European Communities, 1975.
Betts, P.R., Astley, R., and Raine, D.N. (1973). *Br. Med. J. I:*402.
Blumberg, W.E., Eisinger, J., Lamola, A.A., and Zuckerman, D.M. (1977). *J. Lab. Clin. Med. 89:*712.
Catton, M.J., Harrison, M.J.G., Fullerton, P.M., and Kazantzis, G. (1970). *Br. Med. J. 2:*80.
Cavalleri, A., Baruffini, A., Minoia, C., and Bianco, L. (1979). *Environm. Res.*, in press.
Chisolm, J.J., Jr., and Leany, N.B. (1962). *J. Pediat. 60:*1.
Chisolm, J.J., Jr., Mellits, E.D., Keil, J.E., and Barrett, M.D. (1974). *Environ. Health perspect. Exp. Issue 7:*7.
Cools, A., Sallé, H.J.A., Verberk, M.M., and Zielhuis, R.L. (1976). *2:*129.
Cooper, W.C., Tabershaw, I.R., and Nelson, K.W. (1973). In "Environmental health aspects of lead." International Symposium, Amsterdam, 1972. Luxembourg, Commission of the European Communities, 1973.
De Bruin, A. (1971). *Arch. Environ. Health 23:*249.
Donovan, D.I., Vought, V.M., and Rakow, A.B. (1971). *Arch. Environ. Health 23:*111.
Emmerson, B.T. (1963). *Austr. Ann. Med. 12:*310.
Goyer, R.A., and Rhyne, B.C. (1973). *Int. Rev. Exp. Path. 12:*1.
Grandjean, P., Arnvig, E., and Beckmann, J. (1978). *Scand. J. Work Environ. Health 4:*295.
Granick, J.L., Sassa, S., Granick, S., Levere, R.D., and Kappas, A. (1973). *Biochem. Med. 8:*149.
Haeger-Aronsen, B. (1960). *Scand. J. Clin. Lab. Invest. 12: suppl.* 47.
Haeger-Aronsen, B. (1971). *Br. J. Ind. Med. 28:*52.
Haeger-Aronsen, B., and Schütz, A. (1978). *Läkartidningen 75:*3427.
Haeger-Aronsen, B., Abdulla, M., and Fristedt, B.I. (1971). *Arch. Environ. Health 23:*440.
Hänninen, H., Hernberg, S., Mantere, P., Vesanto, R., and Jalkanen, M. (1978). *J. Occup. Med. 20:*683.
Hänninen, H., Mantere, P., Hernberg, S., Seppäläinen, A.M., and Kock, B. (1979). *Neurotoxicology*, in print.
Hernberg, S. (1975). C. Zenz (Ed.), "Toxic metals and their compounds: Lead," In "Occupational Medicine: Principles and Practical Applications." Year Book Medical Publishers, Chicago, Ill. p. 715.
Hernberg, S. (1976). Meeting of the Subcommittee on the Toxicology of Metals, Tokyo, November, 18–23, 1974. G.F. Nordberg, (Ed.), p. 404. Elsevier Scientific Publishing Company, Amsterdam.

Hernberg, S., and Nikkanen, J. (1972). *Pracov. Lék. 24:*77.
Hernberg, S., Nikkanen, J., Mellin, G., and Lilius, H. (1970). *Arch. Environ. Health 21:*140.
Hernberg, S., Nurminen, M., and Hasan, J. (1967). *Environ. Res. 1:*247.
Howard, J.K. (1978). *J. Toxicol. Environ. Health 4:*51.
Joselow, M.M., and Flores, J. (1977). *Am. Ind. Hyg. Assoc. J. 38:*63.
Lamola, A.A., and Yamane, T. (1974). *Science 186:*4167.
Lilis, R., Fischbein, A., Diamond, S., Anderson, H.A., and Selikoff, I.J. (1977). *Arch. Environ. Health 32:*256.
Moore, M.R., Beattie, A.D., Thompson, G.G., and Goldberg, A. (1971). *Clin. Sci. 40:*81.
Nordenson, I., Beckman, G., Beckman, L., and Nordström, S. (1978). *Hereditas 88:*263.
Nordman, C-H., and Hernberg, S. (1975). *Scand. J. Work Environ. Health 1:*219.
Peuschel, S. M., Kopito, L., and Schwachmann, H. (1972). *J. Amer. Med. Ass. 222:*462.
Piomelli, S., Davidow, B., Guinee, V.F., Young, P., and Gay, G. (1973). *Pediatrics 51:*254.
Rabinowitz, M.B., Wetherill, G.W., and Kopple, J.D. (1973). *Science 182:*4113.
Repko, J.D., Corum, C.R., Jones, P.D., and Garcia, L.S., Jr., (1978). NIOSH Research Report, HEW Publication No. 78-128.
Repko, J.D., Morgan, B.B., and Nicholson, J.A. (1974). Interim Technical Report N:o ITR-74-27. HEW Contract No. HSM 99-72-123.
Roels, H.A., Buchet, J.P., and Lauwerys, R.R. (1974). *Int. Arch. Occup. Environ. Health 3:*277.
Roels, H.A., Lauwerys, R.R., and Buchet, J.P. (1975). *Int. Arch. Occup. Environ. Health. 34:*97.
Roels, H., Balis-Jacques, M.N., Buchert, J-P., and Lauwerys, R. (1979). *J. Occup. Med.*, in press.
Rubino, G.F., Pagliardi, E., Prato, V., and Giangrandi, E. (1958). *Br. J. Haematol. 4:*103.
Sakurai, H., Tugita, M., and Tsuchiya, K. (1974). *Arch. Environ. Health 29:*157.
Sassa, S., Granick, J.L., Granick, S., Kappas, A., and Levere, R.R. (1973). *Biochem. Med. 8:*135.
Schaller, K.H., Mache, K., Haas, T., Mache, W., and Valentin, H. (1971). EC Meeting, Erlangen.
Secchi, G.C., and Alessio, L. (1975). *Med. Lavoro 65:*293.
Secchi, G.C., Alessio, L., and Cambiaghi, G. (1973). *Arch. Environ. Health 27:*399.
Selander, S., and Cramér, K. (1970). *Br. J. Ind. Med. 27:*28.
Seppäläinen, A.M., and Hernberg, S. (1972). *Br. J. Ind. Med. 29:*443.
Seppäläinen, A.M., Hernberg, S., and Kock, B. (1979). *Neurotoxicology*, In print.
Seppäläinen, A.M., Tola, S., Hernberg, S., and Kock, B. (1975). *Arch. Environ. Health 30:*180.
Sessa, T., Ferraro, E., and Colucci d'Amato, C. (1965). *Folia Med. 48:*658.
Stuik, E.J. (1974). *Int. Arch. Occup. Environ. Health 33:*83.
Task Group on Metal Accumulation (1973). *Environ. Physiol. Biochem. 3:*65.
Task Group on Metal Toxicity (1976). Proc. International Meeting on the Toxicology of Metals, Tokyo, Nov. 18–23, 1974. G.F. Nordberg (Ed). Elsevier Scientific Publishing Company, Amsterdam.
Tola, S. (1972). *Work-Environ.-Health 9:*66.
Tola, S. (1973). *Work-Environ.-Health 10:*26.

Tola, S., Hernberg, S., Asp, S., and Nikkanen, J. (1973). *Br. J. Ind. Med. 30:*134.
Tomokuni, K. (1975). *Ind. Health 13:*197.
Tomokuni, K., and Kawanishi, T. (1975). *Arch. Toxicol. 34:*253.
Tomokuni, K., and Ogata, M. (1975). *Ind. Health 13:*31.
Tomokuni, K., and Ogata, M. (1976). *Arch. Toxicol. 35:*239.
Valciukas, J.A., Lilis, R., Eisinger, J., Blumberg, W.E., Fischbein, A., and Selikoff, J. (1978). *Int. Arch. Occup. Environ. Health 41:*217.
Williams, M.K. (1966). *Br. J. Ind. Med. 23:*105.
World Health Organization (1977). Environmental Health Criteria 3. Lead. WHO Geneva.
Zielhuis, R.L. (1975). *Int. Arch. Occup. Health 35:*1.
Zielhuis, R.L. (1977). *Int. Arch. Occup. Environ. Health 39:*59.

# Neurophysiological Effects of Lead

*G. P. Cooper and C. D. Sigwart*

TABLE OF CONTENTS



## I. INTRODUCTION

Although lead has long been known to produce harmful neurological effects in man and animals, only within the past decade or so have substantial research efforts been initiated which may eventually provide a reasonably good understanding of the neurotoxicity of this heavy metal. A chief concern of much current research is that the nervous system may still be at serious risk at levels of lead exposure previously considered safe (about 80 $\mu$g Pb/100 ml blood). There are now numerous independent lines of evidence, albeit by no means conclusive, to support this point of view. There is little doubt that clinical and behavioral studies have been the most influential in provoking and maintaining our concern, particularly with respect to the consequences of low-level lead exposure in developing organisms. Yet, in no area of this research is the evidence consistent enough or conclusive enough to make reliable conclusions concerning (a) the seriousness of observed effects, (b) relative permanence of effects, (c) "safe" levels of exposure or (d) mechanisms of action.

Electrophysiological recording and stimulating techniques have traditionally been one of the most useful and easily-applied set of methods for quantitative functional assessment of neurons and neuronal systems. Yet, surprisingly, they have been used very sparingly and unevenly in studies of lead toxicity, having been applied most extensively in examinations of the effects of lead on peripheral nerve conduction velocity and peripheral synaptic transmission. Most studies of peripheral nerve conduction in animals and humans have been done under chronic exposure conditions while most of the work on synaptic effects has been characterized by acute *in vitro* or *in vivo* exposures. There have been remarkably few systematic electrophysiological studies of central nervous system functions or sensory processes in animals or humans. Routine electroencephalographic studies in humans have yielded little information of value from either a diagnostic or heuristic standpoint.

## II. EFFECTS OF LEAD ON PERIPHERAL NERVE

A classic, but inconstant, sign of chronic inorganic lead intoxication in adult humans is a peripheral motor neuropathy which, if severe, can result in distal muscular weakness, particularly of the extensor muscles, which may be characterized by "wrist drop" or "ankle drop" (Aub *et al.*, 1925). Its pathogenesis is still not entirely understood. However, since Gombault's (1873, 1880) work with humans and guinea pigs, *chronic* lead exposure has been known to be capable of producing segmental demyelination and axonal (Wallerian) degeneration in peripheral nerves, although such effects are highly variable within a species and also seem to be quite species-dependent. In man, the most prominent feature reported in cases of clinically apparant lead poisoning is axonal degeneration (Gombault, 1873; Dejerine, 1879; Fiaschi *et al.*, 1976; Buchthal and Behse, 1979).

Few effects are seen in the baboon (Hopkins, 1970; Hopkins and Dayan, 1974) while in guinea pigs (Gombault, 1880; Fullerton, 1966) and rats (Lampert and Schochet, 1968; Schlaepfer, 1969; Krigman *et al.*, 1977; Ohnishi, *et al.*, 1977; Dyck *et al.*, 1977; Brashear *et al.*, 1978) segmental demyelination is prominent with some axonal degeneration also present. In contrast, rabbits show little or no segmental demyelination, the principal feature being axonal degeneration (Key, 1924; de Villaverde, 1930). It should be realized that differences in histopathology observed between species may not represent true species difference but, rather, could to some degree be a consequence of differences in lead exposure dose and exposure duration.

Since either severe axonal degeneration or demyelination may be expected to produce decrements in nerve conduction velocity, electrophysiological measures of conduction velocity might be expected to provide a rapid and non-destructive means of evaluating potential peripheral neuropathy in both man and animals. Helmholtz (1850) was the first to experimentally determine motor nerve conduction velocities. The basic procedures he used with the frog nerve-muscle preparation are still those used today, although electromyographic (EMG) recording of muscle response (Hodes *et al.*, 1948) has replaced kymographic recording of the mechanical response of muscles which Helmholtz had, perforce, to use. By determining the latency of muscle response to electrical stimulation of two different points along a nerve, the conduction velocity (v) is simply;

$$v(m/sec) = \frac{d(mm)}{t_1(msec) - t_2(msec)}$$

where $t_1$ and $t_2$ are the latencies of muscular response to stimulation of the nerve at proximal and distal sites, respectively and d is the distance between the two points of stimulation. More elaborate procedures have been developed for clinical measurements of the conduction velocities of slow motor neurons (Hopf, 1962; Seppalainen and Hernberg 1972) and sensory nerves (Dawson and Scott, 1949; Magladery and McDougal, 1950). For general presentations of procedure as well as discussions of sources of error and variation the reader may consult Kaeser (1970). Very few studies of the effects of lead on nerve conduction velocity have been done in animals. However the results available are in good accord with correlated neuropathological observations. For instance, chronic lead poisoning in baboons produced only minimal histological evidence of damage to peripheral nerve and no decrease in nerve conduction velocity (Hopkins, 1970); Hopkins and Dayan, 1974). In other studies, Fullerton (1966) exposed guinea pigs chronically to lead acetate in quantities sufficient to produce weight loss, anemia, and convulsions in many animals. In 17 of 40 lead poisoned animals tested, the conduction velocity of the posterior tibial nerve was reduced; however in all animals which showed a marked reduction in conduction velocity, segmental demyelination was also found. Figures 1 and 2 show, respectively, electromyographic recordings from a control guinea pig and from three lead poisoned guinea pigs. As can be seen, lead exposure not only produced large decreases in nerve conduction velocity, but also resulted in irregularities of EMG response and a dispersal of the EMG wave-form over time, which might be a consequence of selective effects of lead on larger diameter nerve fibers.

### A. Studies in Human Adults

Recent studies of peripheral nerve conduction velocity in adult humans exposed to lead have been devoted largely to an assessment of the potential health effects of relatively low-level lead absorption in persons exposed to lead in the work place. Many of these studies were done on workers with no obvious, clinically apparent, neurological impairment and they represent serious attempts to re-evaluate our standards for acceptable levels of lead exposure.

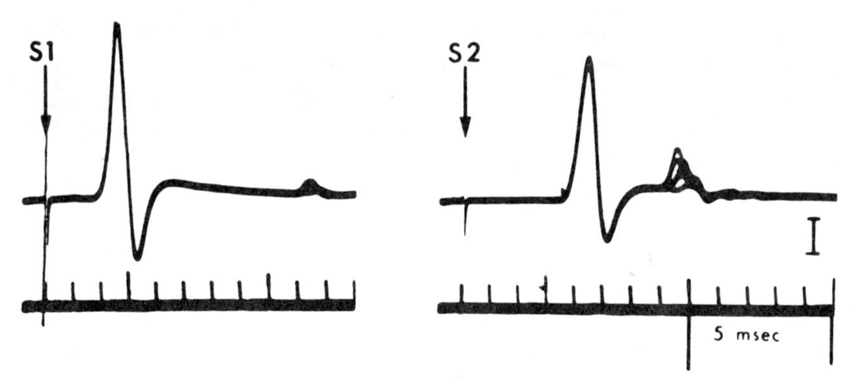


**Figure 1.** Muscle action potentials from a control guinea pig produced by supramaximal stimulation of the tibial nerve at the ankle (S1) and the sciatic nerve in the thigh (S2). Five consecutive traces are superimposed. Calibration: 1 mV. The conduction velocity was 47 m/sec (from Fullerton, *J. Neuropath. Exp. Neurol.*, 1966).

Sessa *et al.* (1965) reported a reduction in the maximal motor conduction velocity (MCV) of the ulnar and radial nerves in twenty patients with lead poisoning who had no clinical signs of neurological impairmant. However, no control subjects were used in the study so their conclusions may be questioned. In another study Catton *et al.*, (1970) recorded MCVs of the lateral popliteal nerve in 19 men occupationally exposed to lead, 13 of whom had blood lead values of 80 $\mu$g/100 ml or more. There was no difference in maximal conduction velocity, but when the amplitude of the muscle action potential evoked by stimulation of the nerve at the knee was expressed as a percentage of the response evoked by stimulation at the ankle, lead exposed subjects had significantly lower values than those of controls. The authors interpreted this to indicate that the conduction velocity of some of the fibers had decreased, resulting in a greater dispersal of nerve action potentials over time and a consequent decrease in the peak amplitude of the muscle response.

In five out of six workers exposed occupationally to lead and having clinical signs and symptoms of lead poisoning, Feldman *et al.* (1977) found reduced peroneal MCV. Importantly, cessation of lead exposure and chelation therapy resulted in a return of the MCV toward normal. In addition to peripheral nerve damage, these workers had a wide variety of signs or symptoms attributed to CNS involvement. Feldman *et al.* (1977) emphasize that excessive exposure to lead can produce effects at all levels of the nervous system including motor neuron disease, peripheral neuropathy and encephalopathy, the manifestations of which are quite variable from individual to individual and also depend on the magnitude and duration of exposure. In the same publication Feldman *et al.* (1977) also reported a significant decrease of mean peroneal nerve MCV in 13 workers with increased lead absorption and a mean blood lead concentration of 79.5 μg/100 ml. None had physical clinical signs of lead poisoning but most did have various subjective symptoms.

Fiaschi *et al.* (1974), in a study of 49 lead workers, reported reductions in MCV of the median and sciatic-popliteal nerves, and an even greater reduction of median sensory nerve velocity. In those persons who displayed EMG abnormalities, conduction velocities were slower than those without such abnormalities. Median nerve biopsy of a subject with EMG abnormalities and slowed nerve conduction showed that the total number of nerve fibers was reduced, as was the proportion of larger diameter fibers.

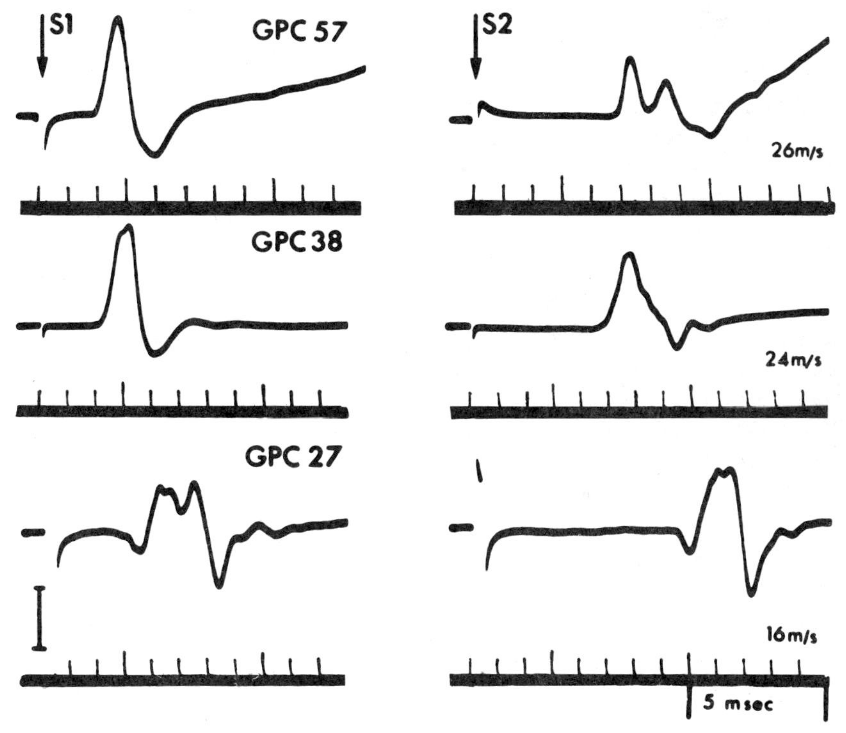


**Figure 2.** Muscle action potentials evoked by supramaximal nerve stimulation at ankle (S1) and in thigh (S2) in 3 lead-poisoned animals. Calibration mark = 2 mV for GPC 57, 1 mV for GPC 38, 500 μV for GPC 27. Conduction velocity in meters/sec given for each animal. The upper pair of records was obtained from guinea-pig GPC 57, aged 15 weeks at the beginning of the experiment and given 0,5 g/kg lead acetate 3 times a week for 19 weeks. The animal had lost no weight and showed no clinical paralysis. The maximal conduction velocity between thigh and ankle was 26 m/sec. In the case of GPC 38, the action potentials shown were recorded after 14 weeks of intermittent lead dosage. At this time, the animal was 17 weeks old; its growth was retarded, but it did not become paralyzed. Guinea-pig GPC 27, aged 24 weeks was given lead acetate intermittently in a dose of 0.5 to 1 g/kg for 22 weeks. At this time, the animal had an abnormal posture and partial paralysis of the hind limbs and it can be seen that conduction velocity was 16 m/sec (from Fullerton, *J. Neuropath. Exp. Neurol.*, 1966).

The most extensive studies of conduction velocity in people with documented chronic low-level lead intoxication have been done by Seppalainen and coworkers in Finland. In one study (Seppalainen and Hernberg, 1972), 39 lead workers and 39 age matched control subjects were studied. Most of the lead workers had been diagnosed as having lead poisoning and most had blood lead values greater than 80 $\mu g/100$ ml. None showed any signs of clinical neurological impairment, but EMG recordings revealed abnormalities such as fibrillation and diminished numbers of motor units in 24 men. In another study, 26 lead workers who had been exposed chronically to lead and whose blood lead values had never exceeded 70 $\mu g/100$ ml were examined (Seppalainen *et al.*, 1975). Lead-exposed subjects in both studies had decreased maximal conduction velocities of the median and ulnar nerves and particularly of the more slowly conducting fibers of the ulnar nerve. In neither study was there any clear relationship between the electrophysiological findings and other indices of lead poisoning. Figure 3 summarizes the determinations of conduction velocity of the slow fibers of the ulnar nerve made in these two studies. Though the mean difference between the lead-exposed and control subjects in conduction velocity is statistically significant, there is great overlap in the range of velocities and the relatively few values for Pb-exposed subjects fall below the range of control subjects. This is characteristic of the results of almost all studies of nerve conduction velocity done in people having no obvious signs of neurological impairment.

Since histological studies (Fiaschi *et al.*, 1974; Buchthal and Behse, 1979; Fullerton, 1966) have shown that demyelination or axonal degeneration occurs principally in larger diameter fibers, at least in cases of frank lead poisoning, it is somewhat surprising that the smaller diameter, more slowly conducting fibers should have been found by Seppalainen *et al.* (1975) to be more sensitive to lead. Indeed, certain other workers are either skeptical of the validity of techniques used for such measurements (Buchthal and Behse, 1979) or have failed to find decreases in the conduction velocity of the slower conducting fibers (Repco *et al.*, 1978). However, in all fairness, most studies have not included attempts to measure the conduction velocity of slower fibers, in part no doubt because the techniques involved are more complex than those used in routine MCV measurements and the recordings obtained are also more difficult to evaluate.

Repko *et al* (1978) reported statistically significant reductions in maximal MCV of the peroneal, posterior tibial, median and ulnar nerves and in ulnar nerve sensory conduction velocity in 85 lead workers whose mean blood lead concentration was 46 $\mu g/100$ ml. The mean blood lead value for 55 control subjects was 18 $\mu g/100$ ml. In this study, the MCV of the slow fibers of the ulnar nerve was determined but was not significantly reduced as compared with control values. Also, no significant correlation

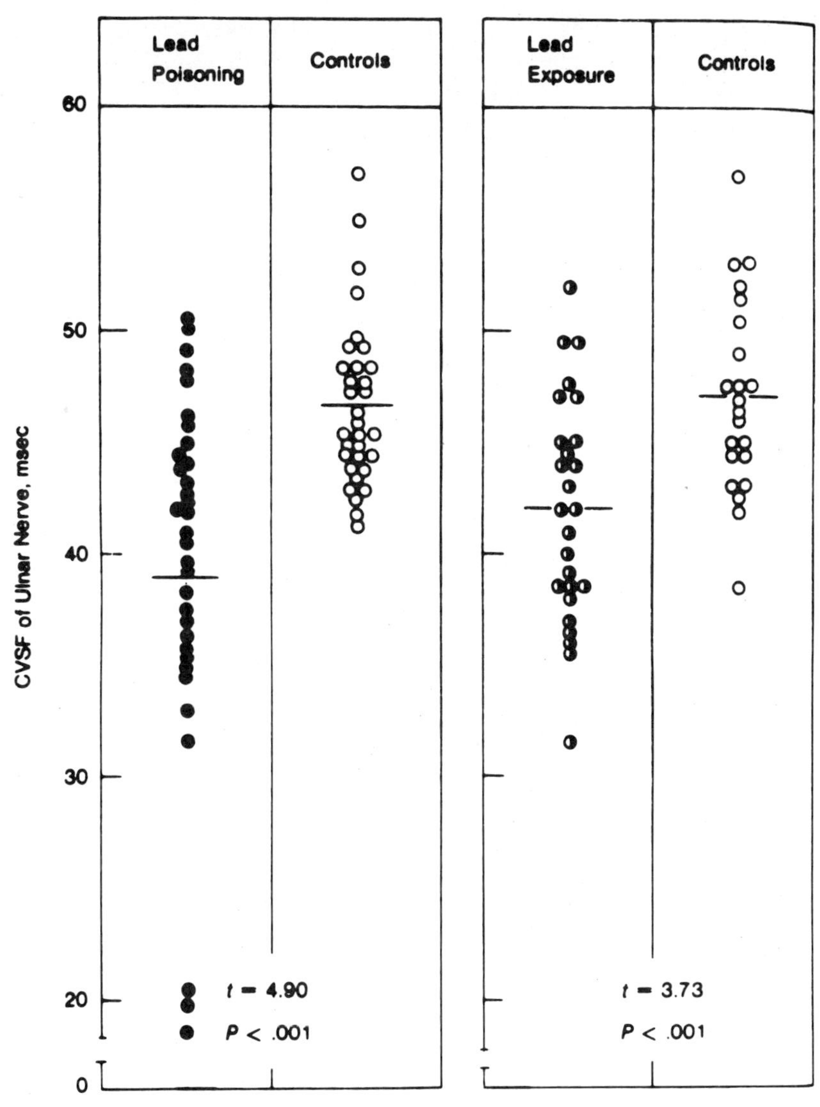


**Figure 3.** Conduction velocity of the slower motor fibers (CVSF) of the ulnar nerve in patients with lead poisoning and their controls and similar results in a series of lead workers and their controls. Most of the lead-poisoned subjects represented in the left panel had blood lead concentrations of 80 μg/100 ml or more while the lead workers represented in the right panel all had blood lead concentrations of 70 μg/100 ml or less (from Seppalainen *et al.*, *Arch. Environ. Health*, 1975).

was obtained between conduction velocity measurements and indices of lead exposure such as blood lead concentration.

Araki and Honma (1976) studied 39 lead workers whose blood lead concentration at the time of testing ranged from 2 to 73 μg/100 ml, with a mean value of 29 μg/100 ml. They found, in those workers whose blood lead concentrations ranged from 29 – 77 μg/100 ml., significant reductions in MCV of the median and posterior tibial nerves but no significant differences in sensory nerve conduction velocity or mixed nerve (sensory and motor) velocities. Unlike other studies published thus far, the MCV of lead workers in this study was significantly negatively correlated with blood lead concentrations, erythrocyte ALAD activity and 24-hour lead excretion following intravenous administration of Ca-EDTA. It is to be emphasized again, however, that very few of the MCV values of lead workers fell below the range recorded for control subjects.

Buchthal and Behse (1979) compared the maximal MCV and the sensory conduction velocities of peroneal, median and sural nerves in twenty clinically asymptomatic lead workers in age-matched controls. Seventeen of the lead workers had been exposed to lead for more than one year and blood lead values of 70 μg/100 ml. or more had been recorded for all but one worker within the year preceding examination. The average motor and sensory conduction velocities of nerves in both upper and lower limbs were significantly reduced in lead workers but the mean difference was slight. In no individual case were values obtained which were less than "borderline" nor was there any relationship between electrophysiological measures and blood lead values or exposure duration. Sural nerve biopsies revealed no evidence of axonal degeneration or demyelination; some evidence for increased incidence of remyelinization was found. Although the extent of lead exposure in these workers was apparently similar to those of workers examined in several other studies, important differences were noted. For instance, no EMG abnormalities resembling those reported by Seppalainen and Hernberg (1972) were seen and rarely was there any reduction in the ratio of the amplitude of EMG response evoked by peroneal nerve stimulation at knee and ankle as reported by Catton *et al.* (1970). The slight changes in average conduction velocities coupled with the absence of significant histological changes led to the conclusion that the health of these lead workers had not been comprimised. In the absence of observable changes in axons or myelin, the authors suggest that slight changes in conduction velocity might result from the effects of lead on nerve membranes.

## B. Studies in Children

Peripheral neuropathy has been considered to be more common in adults than in children but, judging from recent studies, it appears that many cases of neuropathy in children simply have gone unrecognized, particularly in those children having encephalopathy. In any event, few cases of peripheral neuropathy in children have been published (Seto and Freeman, 1964).

In one study, Feldman and coworkers (1973) measured peroneal nerve MCVs in 24 children with increased lead absorption and in a group of control children. The lead exposed group either had blood lead concentrations of 40 $\mu g/100$ ml. or more, increased urinary lead excretion provoked by EDTA, or radiographic evidence of increased lead absorption. Nerve conduction velocities were determined at 3 to 76 months after the cessation of lead exposure. The mean peroneal nerve MCV of 42.33 m/sec was significantly different from the value of 52.78 m/sec determined in control children of similar age. In another study, Feldman *et al.* (1977) determined the peroneal nerve MCV in 26 children suspected of having increased lead absorption. Ten of these children had MCVs one standard deviation or more below the mean of a control group. The mean blood lead concentration of these ten children was not significantly different from those children without slowed MCVs. However, the mean urinary excretion of lead provoked by chelation therapy was more than three times that expected in control children, thus confirming the occurrence of previous excessive lead exposure. Therefore, the results of this study suggest that MCV measurements can be used as one predictor of chronic increased lead absorption in children.

In a very ambitious study of 202 children 5 to 9 years old who lived near a lead smelter, a significant negative correlation was found between blood lead levels and peroneal MCV and between free erythrocyte protoporphyrin (FEP) and peroncal MCV (Landrigan *et al.*, 1976). None of the children with increased lead absorption had overt signs of neurotoxicity nor were any of their MCVs considered pathological. It is important to note that 99% of the children living within 1.6 km of the smelter had blood lead concentrations of 40 $\mu g/100$ ml. or more and 22% had concentrations equal to or greater than 80 $\mu g/100$ ml.

There have been several recent reports of peripheral neuropathy associated with concomitant lead poisoning and sickle cell disease (Anku and Harris, 1974; Erenberg *et al.*, 1974). Most of these children exhibited classical signs of wrist drop and/or foot drop and in those cases where measurements of peripheral nerve conduction velocity were less than normal. As Erenberg *et al.* (1974) point out, peripheral neuropathy in these cases might have been precipitated by an interaction between an elevated lead level and a Zn deficiency which has been documented in sickle cell disease.

## C. Summary

Although there is still some disagreement between researchers as to methods, results and interpretations of data, the combined weight of evidence from investigations of nerve conduction velocity in humans exposed to lead suggest that the level of lead exposure once considered to be safe may, in fact, not be. The major reasons that only tentative conclusions are possible are (a) nerve conduction velocity measurements have only rarely been found to correlate with any other index of lead exposure and (b) the range of nerve conduction velocities in humans is so great, even when measured by a single investigator, that in almost all studies most nerve conduction velocities of lead-exposed subjects are well within the range of normal values. As Seppalainen *et al.* (1975) have remarked, such measurements are of little diagnostic value in the individual case, but statistically do indicate the need for reducing lead exposure levels. Nerve conduction velocities should be determined in known or suspected cases of lead poisoning to more fully document possible neurological effects and to monitor the process of recovery.

Studies involving chronic low level lead exposure in animals are needed to more precisely define the conditions under which peripheral neuropathy develops and its relationship to other indices of exposure.

## III. EFFECTS OF LEAD ON SYNAPTIC TRANSMISSION

Many drugs, diseases, and toxic agents alter the transmission of excitation betwen neurons or between neurons and muscle or glands. These synaptic junctions represent areas of low "safety factor", particularly in the central nervous system where the excitability of a neuron is a delicate function of hundreds of separate synaptic inputs, both excitatory and inhibitory. Most electrophysiological studies of the effects of lead on synaptic transmission have been done on peripheral synapses. However, though peripheral chemical synaptic systems are considerably less complex than most central ones, their general functional characteristics are quite similar and it is assumed that effects which are observed in peripheral systems are also likely to occur in the CNS. Perhaps a more serious limitation on evaluating the toxicological significance of such studies is that most have been done under acute, *in vitro* exposure conditions. As may be appreciated, a disturbance in synaptic transmission may result from alterations in (a) propagation of presynaptic nerve impulses, (b) synthesis and storage of transmitter, (c) release of transmitter substance, (d) postsynaptic receptor response to transmitter, (e) breakdown or re-uptake of transmitter or (f) postsynaptic membrane excitability. Therefore, several different lines of evidence are required to fully establish the mode of action of any agent which alters or blocks synaptic transmission.

### A. Peripheral Synapses

Lead, like many other heavy polyvalent cations, has profound effects on chemically mediated synaptic transmission. In the peripheral nervous system, low concentrations of lead, manganese, cadmium, cobalt, magnesium, lanthanum and mercury, depress or block synaptic transmission. In cholinergic systems such as the frog skeletal neuromuscular junction (Manalis and Cooper, 1973), the cat and frog sympathetic ganglion (Kostial and Vouk, 1957; Kober and Cooper, 1976) and the rat phrenic nerve-diaphragm neuromuscular junction (Silbergeld *et al.*, 1974) lead reduces the postsynaptic response to presynaptic nerve stimulation without appreciably affecting the post-synaptic response to applied acetylcholine. These results clearly indicate that the synaptic blockade observed is not primarily a consequence of an effect of lead on post-synaptic membranes.

Interestingly, lanthanum (DeBassio *et al.*, 1971), mercury (Manalis and Cooper, 1975), and lead (Manalis and Cooper, 1973) actually increase *spontaneous* transmitter release in the resting (unstimulated) frog skeletal neuromuscular junction as indicated by increases in miniature endplate potential frequency; this effect usually occurs at somewhat higher metal concentrations than those required for a reduction in the synchronous release of transmitter evoked by nerve stimulation. The opposite effects of lead on evoked and spontaneous transmitter release are strikingly apparant in Figures 4 and 5. The opposite effects of lead and several other metal ions on evoked and spontaneous transmitter release may be construed as support for the hypothesis that spontaneous and phasic transmitter release are subserved by separate, though not necessarily independent, mechanisms (del Castillo and Katz, 1954).

In most cases the reduction in synaptic transmission is partially or completely reversible by increasing the calcium concentration of the bathing

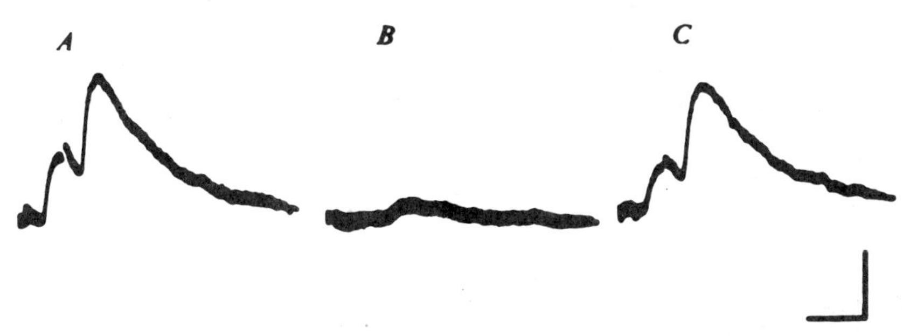


**Figure 4.** Effect of lead on the end-plate potential. Neuromuscular transmission was blocked by bathing the nerve-muscle preparation in 0.8 mM $Ca^{2+}$/5.5 mM $Mg^{2+}$-Ringer solution, thereby making it possible to record subthreshold end-plate potentials. Facilitated end-plate potentials were recorded intracellularly by stimulating the motor nerve with double shocks at a rate of 3 $s^{-1}$. A, Responses to nerve stimulation while the preparation was bathed in the control Ringer solution. B, Effect of a 1 min exposure to 0.01 mM $Pb^{2+}$ added to the modified Ringer solution. Notice that 0.01 mM $Pb^{2+}$ blocked the unfacilitated endplate potential and reduced the facilitated one. C, Responses 10 min after the preparation was washed with the control Ringer solution. Vertical calibration: 2 mV. Horizontal calibration: 10 ms (from Manalis and Cooper, *Nature*, 1973).

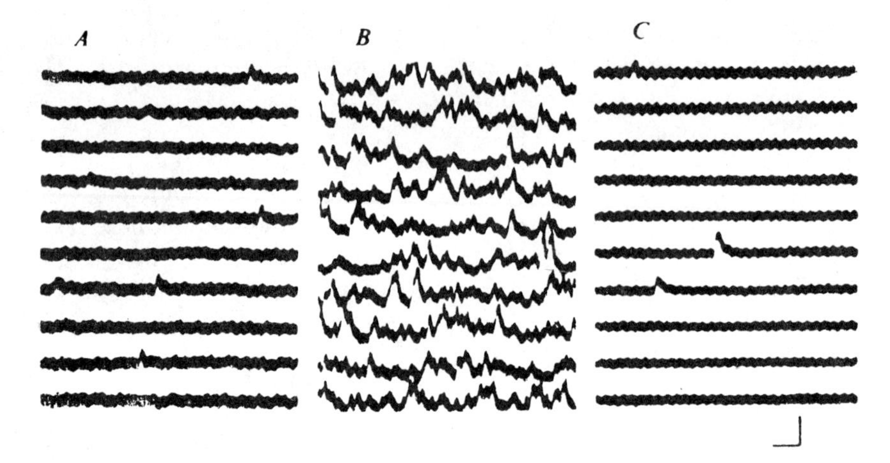


**Figure 5.** Effect of lead on the frequency of miniature endplate potentials. A, Control: the preparation was bathed in normal Ringer solution. B, Five minutes after exposing the preparation to 0.10 mM $Pb^{2+}$-Ringer solution: the frequency increased greatly while the amplitudes of the miniature end-plate potentials were essentially unchanged. C, Five minutes after washing the preparation with normal Ringer solution: the frequency was no longer elevated. Oscillograms A–C were taken from the same neuromuscular junction. Vertical calibration: 1 mV. Horizontal calibration: 50 ms (from Manalis and Cooper, *Nature*, 1973).

medium. In the case of magnesium, manganese, cobalt, and lead the antagonism has been shown to be competitive in nature (Dodge and Rahamimoff, 1967; Weakly, 1973; Balnave and Gage, 1973; Kober and Cooper, 1976; Manalis *et al.*, 1976). This competitive interaction of lead and calcium on synaptic transmission is illustrated in Figure 6. The amplitude of response recorded fron the *in vitro* bullfrog sympathetic ganglion, evoked by presynaptic nerve stimulation, can be seen in Figure 6A to be a positive function of calcium concentration. Addition of lead to the bathing medium shifts the response to the right with little change in slope. Replotting the data on modified Lineweaver-Burk coordinates (Dodge and Rahaminoff, 1967) results in the curves shown in Figure 6B. These curves intercept the ordinate at a common point, which is interpreted to indicate a competitive antagonism of calcium by lead. Since calcium is required for normal transmitter release (Katz, 1969) it therefore was concluded that lead interferes with calcium-mediated transmitter release. This conclusion has been given further support by the results of experiments in which the uptake of $^{45}Ca$ by presynaptic nerve terminals was measured in bullfrog sympathetic ganglia. As shown in Figure 7, lead reduces presynaptic $^{45}Ca$ uptake in unstimulated ganglia and in ganglia in which the preganglionic nerve was stimulated at a rate of 6/sec. Thus, it appears that lead blocks synaptic transmission by interfering with the action potential evoked influx of calcium into presynaptic nerve terminals which, in turn, results in a reduction of transmitter release.

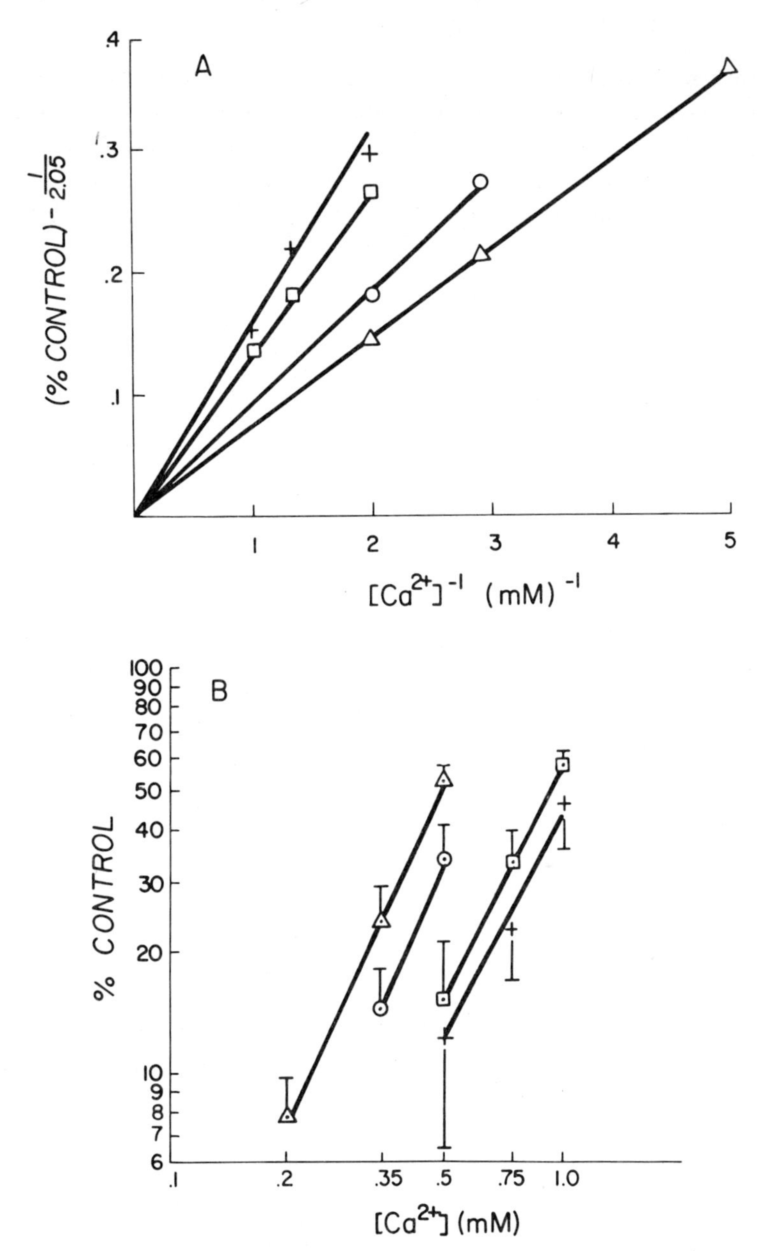


**Figure 6.** a, A log-log plot of postganglionic response amplitude against $Ca^{2+}$ concentration for 0, 5, 10, and 20 $\mu$M $Pb^{2+}$. The average slope of the lines is 2.05. Each point is the average of at least 5 determinations. The standard error of the mean is shown with each point. b, The same data plotted on modified Lineweaver-Burk coordinates. Control Ringer contained lll mM NaCl, 2.5 mM KCl, 2.0 mM CaCl and 4 mM Tris buffer. All straight lines were determined by least-squares regression. Concentration of $Pb^{2+}$ in modified Ringer's solution: $\Delta$, 0 $\mu$M; O, 5 $\mu$M; □, 10 $\mu$M; +, 20 $\mu$M; (from Kober and Cooper, *Nature*, 1976).

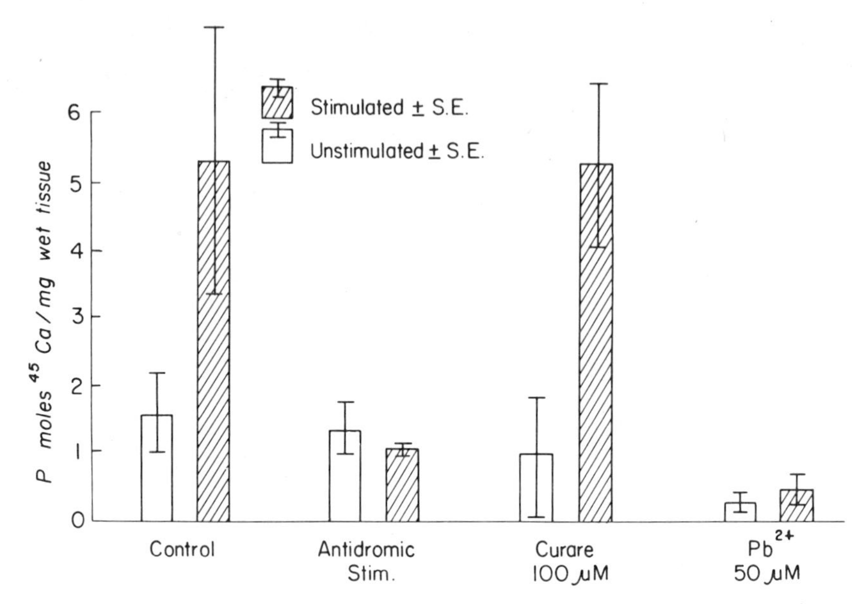


**Figure 7.** The amount of $^{45}Ca$ taken up by ganglia under the following conditions: a, unstimulated ganglia and ganglia activated by preganglionic nerve stimulation, both in normal Ringer's solution; b, unstimulated ganglia and ganglia activated by stimulation of the postsynaptic nerve, both in normal Ringer's solution; c, unstimulated ganglia and preganglionically stimulated ganglia in Ringer's solution containing 100 mM (+)-tubocurarine; d, unstimulated and preganglionically-stimulated ganglia in Ringer's solution containing 50 $\mu$M $Pb^{2+}$. 50 $\mu$M $Pb^{2+}$ blocks 100% of the postganglionic response (from Kober and Cooper, *Nature*, 1976).

Peripheral adrenergic synapses have generally received considerably less attention than cholinergic ones, but their basic characteristics are similar. For example, calcium is required for transmitter release at presynaptic nerve terminals (Kirpekar and Misu, 1967; Kuriyama, 1964) and magnesium competitively inhibits the action of calcium (Boullin, 1967). The effects of the heavier polyvalent cations have been little studied although the data available agree with observations made on cholinergic synapses. For instance, manganese, cobalt, nickel and lanthanum reduce the release of norepinephrine, evoked by splenic nerve stimulation, from the perfused cat spleen, an effect which could be partly reversed by increasing the calcium concentration of the perfusion medium (Kirpekar *et al.*, 1970; 1972).

In recent experiments, both inorganic lead and cadmium have been shown to reduce the constrictor response of *in vitro* rabbit saphenous arteries produced by sympathetic nerve stimulation (Cooper and Steinberg, 1976). Because responses to applied norepinephrine and to direct muscle stimulation remained relatively unaffected and since responses could be restored by increasing the calcium concentration of the bathing medium, it was concluded that lead and cadmium had reduced the release of norepinephrine from presynaptic nerve terminals. Thus, it appears that the mechanism of synaptic blockade by these metals is similar in adrenergic and cholinergic systems.

### B. Central Nervous System Synapses

Although much less is known about the actions of calcium and other cations in the central nervous system, calcium does appear to be required for successful transmitter release as it is for peripheral synapses (Kuno, 1971; Rubin, 1970). However, electrophysiological studies of the effects of cations other than calcium and magnesium are rare. Rozear *et al.* (1971) found that cadmium, cobalt, nickel, manganese, zinc, beryllium, and lead decrease the excitability of neurons in the cerebral cortex and brainstem of cats. Nickel and presumably the other metals, depressed excitability by increasing the threshold level for depolarization. In agreement with this is the report by Loop and Cooper (1974) that lead, injected directly into the spinal cord of cats, increases the firing threshold of motor neurons without appreciably affecting the excitatory postsynaptic potential (EPSP).

Electrophysiological evidence fo disturbances in transmitter release under conditions of chronic lead exposure is lacking, although there is evidence from neuro-chemical studies that cholinergic transmission is depressed (Carroll *et al.*, 1977) as it is in the periphery. Almost all other evidence available concerning the effects of lead on central synaptic processes derive from neurochemical determinations of the concentration and turnover rates of chemical transmitter agents, their precursors and metabolic breakdown products, and of enzymes involved in the synthesis and breakdown of transmitter substances.

In summary, it appears that lead and most of the heavy metals which have been studied cause a depression of the calcium-mediated evoked release of transmitter at peripheral synapses and possibly at some CNS synapses, particularly under acute or high-level exposure conditions. Whether such effects play any role in determining the neurotoxicological manifestations of chronic, low-level lead exposure is quite another question which, unfortunately, cannot be answered at this time.

## IV. EFFECTS OF LEAD ON THE CENTRAL NERVOUS SYSTEM

### A. Electroencephalographic Studies

Lead intoxication produces a wide spectrum of neurological symptoms and signs attributable to effects on the CNS, ranging from headache and chronic fatigue to seizures, coma and death. Acute exposure often leads to a distinct encephalopathy whose diagnosis is clinically difficult to differentiate from encephalopathies resulting from other causes. Confirmation of lead encephalopathy requires several lines of related evidence, as indicated by Whitfield *et al.* (1972). Electroencephalographic (EEG) recordings have been of little value in diagnosis and have been mainly used, when used at all, to provide additional confirmatory evidence for the presence of a diffuse encephalopathy. In adults, acute encephalopathy is now rare due to reasonable industrial hygiene practices and, in the United States, most recent cases of frank lead encephalopathy have resulted from the consumption of lead-contaminated "moonshine" whiskey (Whitfield *et al.*, 1972; Morris *et al.*, 1964).

EEG recordings in adults generally have been characterized by diffuse, irregular, patterns; slowing of the alpha (8–13 Hz) rhythm; and increases in the amount of slow wave activity in the theta (4–8 Hz) and delta ($<4$ Hz) range (Simpson *et al.*, 1964; Cosic *et al.*, 1967; Corsi and Picotti, 1965). Evidence of focal spiking associated with seizures has also been reported (Morris *et al.*, 1964). Chelation therapy appears to reverse these EEG abnormalities even though other symptoms of CNS dysfunction may persist (Simpson *et al.*, 1964; Cosic *et al.*, 1967).

Studies of the EEG in children suffering from lead encephalopathy have also shown that slow, persisting, dysrhythmic activity predominates, accompanied by focal paroxysmal discharges (Thurston *et al.*, 1955; Smith *et al.*, 1963). No significant difference from control subjects was seen in the EEG's of children with increased lead absorption (blood lead concentrations $\geq 60\ \mu g/100$ ml) but who were free of encephalopathy (Smith *et al.*, 1963).

EEG abnormalities similar to those seen in humans have been observed in a number of mammals in experimental studies of lead toxicity. Zook *et al.*(1972), in a study of 236 dogs diagnosed as having lead poisoning, recorded EEG's in eight animals which had neurological signs. EEG abnormalities consisted of increased amounts of irregular, high voltage activity in the delta range. Clinical improvement was accompanied by return of the EEG toward normal.

Tetraethyl lead (TEL) toxicity in rabbits was investigated by Morelli *et al.* (1957a; 1957b). As in clinical observations on humans, they noted great individual variation in neurological signs and in EEG abnormalities. In adult rabbits exposed to TEL in a dose of 1 mg/kg/day, EEG changes were apparent after 20 to 30 days exposure and recovery was almost complete within 60 days after cessation of exposure. When 10 mg/kg/day of TEL was given, EEG abnormalities appeared from the 10th day of exposure, often increasing until death. Animals surviving a 20 day exposure had persisting EEG abnormalities 70 days later. The abnormalities were characterized by reduced frequency and amplitude, and, in advanced intoxication, diffuse dysrhythmic activity and occasional psychomotor paroxysms were seen. The persisting EEG abnormalities were associated with ataxia, paresis, and lethargy.

In a study on rats, Saito (1973) compared the effects of leaded (TEL) and unleaded gasoline. After 10 days of TEL exposure at 7.5 mg/kg/day, the exposed animals had EEG's characterized by increased theta wave (4–8 Hz) activity and somewhat increased alpha wave (8–13 Hz) activity. These animals were hyperexcitable, but showed no gross deficiencies at that level of intoxication.

Most EEG studies have been done in animals and humans under conditions of high-level lead exposure associated with clear signs of neurological impairment including lead encephalopathy. Only a few studies have explored the potential effects of low-level lead exposure. Chronic exposure of rats to lead acetate has been shown to produce disturbance in the rapid eye-movement (REM) phase of sleep (Xintaras *et al.*, 1967) and daily exposure of infant monkeys, beginning at 1 day of age, to 500 $\mu$g of lead/kg body weight/day resulted in a shift in the EEG power spectrum toward high frequencies (Willes *et al.*, 1977). Saito and Abe (1965) reported that lead smelter workers having a mean blood lead concentration of 79 $\mu$g/100 ml and other evidence of increased lead absorption, had EEG's characterized generally by increased activity in the theta (4–8 Hz) range and decreased activity at higher frequencies.

## B. Effects on the Visual System

Even though pathological changes in the optic nerve, retina, retinal blood vessels, and extraocular muscles have been reported in humans and animals, such effects are by no means a regular feature of lead poisoning (Grant, 1974; Brown, 1974). In a thorough study of rabbits exposed to lead acetate for 1½ to 23 months (500 mg lead acetate/100 g rabbit chow diet), Brown (1974) found only moderate pathological changes in the pigment epithelium and no changes in the electroretinogram or electrooculogram, suggesting that no significant functional effects on the eye had been caused by lead.

In a study of Rhesus monkeys reared on a diet containing lead acetate for the first year of life, Busnell *et al.* (1977) tested visual discrimination ability beginning eighteen months after termination of lead treatment. Visual discrimination performance of monkeys whose blood lead concentrations had been maintained at an average value of 55 $\mu$g/100 ml was no different from that of control monkeys. However, in monkeys having a mean blood lead concentration of 85 $\mu$g/100 ml during the first year of life, visual discrimination was severely impaired at low levels of stimulus luminance. Similar scotopic visual deficits have been found in monkeys exposed to methyl mercury (Evans *et al.*, 1975), which apparently resulted from damage to the cerebral cortex. On the other hand the possibility that lead and other heavy metals may affect scotopic vision by a direct effect on retinal rods has been suggested by the results of a study of Fox and Sillman

(1979) on the isolated bullfrog retina. In this study low concentrations of lead, cadmium, and mercury were found to depress the amplitude of the rod receptor potential without altering the cone response.

Clinical studies of lead toxicity have included reports of deficits in visual learning and perception and in visual-motor performance in children (Byers and Lord, 1943; Perlstein and Attala, 1966; Thurston *et al.*, 1955; Bradley and Baungartner, 1958), and in lead workers whose blood lead values had never exceeded 70 $\mu g/100$ ml (Haenninen *et al.*, 1978). Deficits in visual discrimination learning have also been reported to result from lead exposure in lambs (Carson *et al.*, 1974) and in rats (Brown, 1975; Sobotka *et al.*, 1974; Driscoll and Stegner, 1976; Hastings *et al.*, 1979). Although tests of visual threshold or acuity have not been done in such studies, it is unlikely that the effects on learning are a consequence of sensory deficits but, rather, result from effects of lead on the CNS.

That central visual function is impaired by relatively low levels of lead exposure has been shown in studies of the visual evoked response (VER) in rats (Fox *et al.*, 1977; 1979). In these experiments, rat pups were exposed from birth to weaning to dams whose drinking water contained lead acetate. At 21 days of age (weaning), the mean blood and brain lead concentrations of these pups was 65 and 53 $\mu g/100$ g respectively. At 90 days of age, blood and brain lead concentrations were not significantly different from control animals. As shown in Figure 8, the latencies of the first four VER waves ($P_1$, $N_1$, $P_2$, $N_2$) decreased in both control and experimental rats during the suckling period. However, on any given day the response latencies were always significantly longer in lead-treated animals. Furthermore, at 90 days of age, when blood and brain lead concentrations were normal, the VER's of lead treated animals were still significantly greater than those of control animals. In the adult rats, CNS recoverability was assessed by recording the VER to pairs of light flashes, the two flashes of a pair being separated by a variable time interval. In lead-exposed rats the latency of the $P_1$ wave in response to the second flash of a pair was significantly increased at inter-flash intervals up to 175 msec, indicating a reduced ability of the CNS to recover from the effects of the first flash. Explanations for the dalayed maturation of the VER, the persisting increase in VER latency, and the reduced recoverability observed in these studies must remain speculative at this time. Cerebral cytoarchitectural changes such as hypomyelination, reduction in size of the dendritic field and number of synapses per neuron, and reduction of neuron size which occur in neonatal rats exposed to lead (Krigman and Hogan, 1974; Krigman *et al.*, 1974) may account for both the delayed maturation and persisting alteration of the VER. It is also possible that delays and/or permanent alterations in the biochemical systems involved in synaptic transmission occur, although such effects have not been demonstrated.

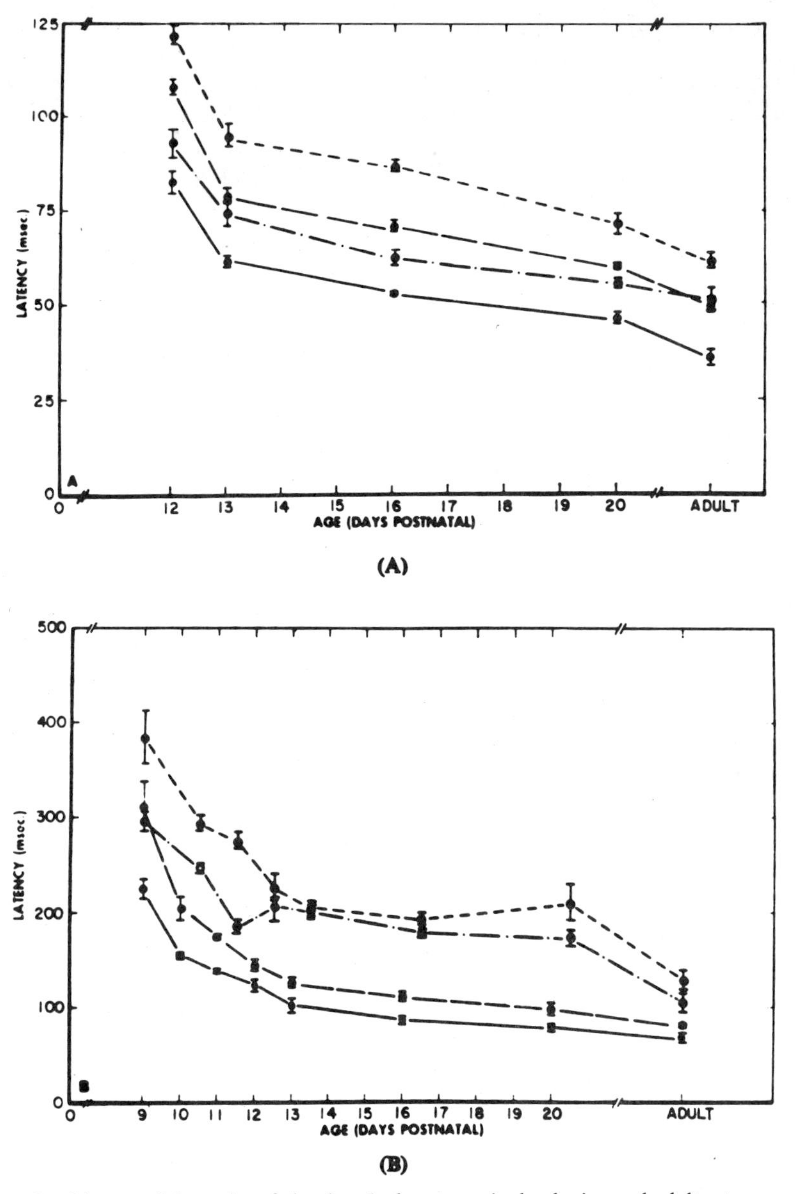


**Figure 8.** Mean peak latencies of visual evoked response in developing and adult rats exposed only during suckling to dams consuming 0.2% lead acetate solution or tap water. (A) Mean ± SE latencies of the first positive wave ($P_1$) and the first negative wave ($N_1$) of the visual evoked response. ——, $P_1$ Control; ●— —●, $P_1$ Lead-treated; ●—·—●, $N_1$ Control; ●---● $N_1$ Lead-treated. (B) Mean ± SE latencies of the second positive wave ($P_2$) and the second negative wave ($N_2$) of the visual evoked response. ——, $P_2$, Control; ●— —●, $P_2$ Lead-treated; ●—·—●, $N_2$ Control; ●---● $N_2$ Lead treated (from Fox *et al.*, *Toxicol. Appl. Pharmacol.*, 1977).

## V. SUMMARY AND CONCLUSIONS

Measurements of peripheral nerve conduction velocities in human adults and children currently constitute the only body of electrophysiological data extensive enough or consistent enough to contribute significantly to attempts to reevaluate the potential neurological effects of low-level lead exposure. Although considerable differences between studies exist with respect to methods and results, the overall conclusion that must be drawn is that peripheral nerve conduction velocities are reduced at levels of lead absorption which produces no clinically apparent neurological impairment. These data, in conjunction with other behavioral and clinical findings, suggest that current exposure standards should be considerably reduced.

Few systematic animal studies of the effects of long term lead exposure on peripheral nerve conduction velocity have been done. Correlated neurophysiological, neuropathological, and neurochemical studies in animals chronically exposed to lead are needed to gain a better understanding of the effects and mechanisms of action of lead on the peripheral nervous system.

Experiments concerned with the acute effects of lead on synaptic transmission in neuromuscular junctions and sympathetic ganglia have demonstrated that lead blocks synaptic transmission by competitively interfering with calcium in the release of transmitter agents from presynaptic nerve terminals. While it is not yet known whether this interference with the action of calcium accounts for any of the neurological problems encountered in chronic lead toxicity, it is probable that a mechanism of this sort, involving other trace metals important in the action of enzymes, does play a significant role in determining lead toxicity.

Routine electroencephalographic recordings in humans and animals with lead poisoning have not contributed very much to improvements in diagnosis, treatment, or understanding of lead encephalopathy, much less the more subtle effects of lead on the CNS. However, considerable advances in this field could be expected from truly sophisticated EEG studies and, particularly, from the application of more specific and quantitative electrophysiological procedures such as evoked response methodology.

## ACKNOWLEDGEMENTS

The author's work was supported in part by U.S. National Institute of Environmental Health Sciences Grants ES00159, ES00649, ES05042 and ES01494 and by U.S. Environmental Protection Agency Contract 68-03-0429.

## REFERENCES

Anku, V.D. and Harris, J.W. (1974). *Pediatrics 85:*337.
Araki, S. and Honma, T. (1976). *Scand. J. Work Environ. Health 4:*225.

Aub, J.C., Fairhall, L.T., Minot, A.S. and Reznikoff, P. (1925) *Medicine 4:* 1.
Balnave, R.J. and Gage, P.W. (1973). *Brit. J. Pharmacol. 47:*339.
Boullin, D.J. (1967). *J. Physiol. London 189:*85.
Bradley, J.E. and Baumgartner, R.J. (1958). *J. Pediatr 53:*311.
Brown, D.R. (1975). *Toxicol. Appl. Pharmacol. 32:*628.
Brown, D.V.L. (1974). *Trans. Am. Opth. Soc. 72:*404
Buchthal, F. and Behse, F. (1979). *Brit. J. Ind. Med. 36:*135.
Bushnell, P.J., Bowman, R.E., Allen, J.R. and Marlar, R.J. (1977). *Science 196:*333.
Byers, R.K. and Lord, E.E. (1943). *Amer. J. Dis. Child 60:*471.
Carrol, P.T., Silbergeld, E.K. and Goldberg, A.M. (1977). *Biochem. Pharmacol. 26:*397.
Carson, T., Van Gelder, G.A., Karas, G.C. and Buch, W.B. (1974) *Arch. Environ. Health 29:*154.
Catton, M.J., Harrison, M.J.G., Fullerton, P.M. and Kazantzis, G., (1970). *Brit. Med. J. 2:*80.
Cooper, G.P. and Steinberg, D. (1976). *Amer. J. Physiol.* In press.
Corsi, G.G. and Picotti G. (1965). *Folia Medica 48:*856.
Cosic, V. Kapor, G., Kusic, R., Kop, P. and Marenic, S. (1967). *Lij. Vjes. 89:*19.
Dawson, G.D. and Scott, J.W. (1949). *Neurol. Neurosurg. Psychiat. 12:*259.
DeBassio, W.A., Schnitzler, R.M. and Parsons, R.L. (1971) *J. Neurobiol. 2:*263.
Dejerine, J. (1879). *C.R. Soc. Biol.* (Paris) *7:*11.
del Castillo, J. and Katz, B. (1954). *J. Physiol. 124:*553.
de Villaverde, J.M. (1930). *Trab. hab. Invest. Biol. Univ. Madr. 26:*163.
Dodge, F.A. and Rahaminoff, R. (1967). *J. Physiol. London 193:*419.
Drissoll, J.W. and Stegner, S.E. (1976). *Pharmacol. Biochem. Behav. 4:*411.
Dyck, P.C., Obrien, P.C. and Ohnoshi, A. (1977). *J. Neuropathol. Exp. Neurol. 36:*570.
Erenberg, G., Rinsler, S.S. and Fish, B.G. (1974). *Pediatrics 54:*438.
Evans, H.L., Laties, V.G. and Weiss, B. (1975). *Fed. Proc. 34:*1858.
Feldman, R.G., Aladdow, J., Kopito, L. and Schwachman, H. (1973) *Amer. J. Dis. Child. 125:*39.
Feldman, R.G., Hayes, M.K. Tounes, R. and Aldrich, F.D. (1977) *Arch. Neurol. 34:* 481.
Ferraro, A. and Hernandez, R. (1932). *J. Psychiat. Quart. 6:*121 and 319.
Fiaschi, A., DeGrandis, D. and Ferrari, G. (1974). *Rev. Pat. Nerv. Ment. 95:*914.
Fox, D.A., Lewkowski, J.P. and Cooper, G.P. (1979). *Neurobehav. Toxicol.* In Press.
Fox, D.A. Lewkowski, J.P. and Cooper, G.P. (1977). *Toxicol. Appl. Pharmacol. 40:*449.
Fox, D.A. and Sillman, A.J. (1979). *Science.* In Press.
Fullerton, P. (1966). *J. Neuropathol. Exper. Neurol. 25:*214.
Gombault, A. (1880). *Arch. Neurol. (Paris). 1:*11.
Gombault, A. (1873). *Arch. Physiol. Norm. Path. 5:* 592.
Grant, W.M. (1974). "Toxicology of the Eye", 2nd ed. Charles C Thomas, Pub. Springfield, Ill.
Haenninen, H., Hernberg, S., Mantere, P., Vesanto, R. and Jalkanen, M. (1978). *J. Occup. Med. 20:*683.
Hastings, L., Cooper, G.P., Bornschein, R.L. and Michaelson, I.A. (1979). *Pharmacol. Biochem. Behav.* In Press.
Helmholtz, Arch. (1850). *Anat. Physiol.* p. 277.
Hodes, R., Larrabee, M.G. and German, W. (1948). *Arch. Neurol. Psychiat. (Chicago). 60:*340.

Hopf, H.C. (1962). *Dtsch. Z. Nervenheilk. 183:*579.
Hopkins, A.P. and Dayan, A.D. (1974). *Brit. J. Ind. Med. 11:*128.
Hopkins, A.P. (1970). *Brit. J. Ind. Med. 27:*130.
Kaeser, H.E. (1970). Vinken, P.J. and Bruyn, G.W. (Eds.), In "Handbook of Clinical Neurology," Vol. 7, North Holland Press, Amsterdam, p. 116.
Katz, B. (1969). *The Release of Neural Transmitter Substances.* C.C. Thomas, Springfield, Ill.
Key, J.A., (1924). *Amer. J. Physiol. 70:*86.
Kirpekar, S.M., Dixon, W. and Prat, J.C. (1970). *J. Pharmacol. Exp. Thera. 174:*72.
Kirpekar, S.M., Prat, J.C., Puig, M. and Wakade, A.R. (1972). *J. Physiol. London 221:*601.
Kirpekar, S.M. and Misu, Y. (1967). *J. Physiol. London 188:*219.
Kober, T.E., and Cooper, G.P. (1976). *Nature 262:*704.
Kostial, K. and Vouk, V.B. (1957). *Brit. J. Pharm. 12:*219.
Krigman, M.R., Druse, M.J., Traylor, T.D., Wilson, M.H., Newel, L.R., and Hogan, E.L. (1974). *J. Neuropathol. Exp. Neurol. 33:*58.
Krigman, M.R. and Hogan, E.L. (1974). *Environ. Health Perspect 7:*187.
Krigman, M.R., Kopp, V.J., Bendeich, E.G. and Brashear, C.W. 1977. *J. Neuropathol. Exp. Neurol. 16:*610.
Kuno, M. (1971). *Physiol. Rev. 51:*647.
Lampert, P.W. and Schochet, S.S. (1968). *J. Neuropathol. Exp. Neurol. 27:*527.
Landrigan, P.J., Baker, E.L., Jr., Feldman, R.G., Cox, D.H., Eden, K.V., Orenstein, W.A., Mather, J.A., Yankel, A.J., and Von Lendern, I.H. (1976). *J. Pediatrics 89:*904.
Loop, W.C. and Cooper, G.P. (1974). *Fed. Proc. 33:*341.
Magladery, J.W. and McDougal, D.B., Jr. (1950). *Bull. Johns Hopk. Hosp. 86:*265.
Manalis, R.S. and Cooper, G.P. (1975). *Nature 257:*690.
Manalis, R.S., Cooper, G.P. (1973). *Nature 243:*354.
Manalis, R.S., Cooper, G.P. and Pomeroy, S.L. (1976). *Neuroscience Ab. 2:*715.
Morelli, A., Guiliani, V. and Serra, C. (1957a). *Lavoro Umano 9:*433.
Morelli, A., Guiliani, V. and Serra, C. (1957b). *Lavoro Umano 9:*545.
Morris, C.E., Heyman, A. and Pozefsky, T. (1964). *Neurology 14:*493.
Ohnishi, A., Schilling, K. Brimijoin, W.S., Lambert, E.H., Fairbanks, V.G., and Dyck, P.J. (1977). *J. Neuropathol. Exp. Neurol. 36:*499.
Perlstein, M.A., and Attala, R. (1966). *Clin. Pediatr. 5:*292.
Repco, J.D., Corum, C.R., Jones, P.D. and Garcia, L.S., Jr., (1978). "The effects of inorganic lead on behavioral and nurological function:. U.S. Dept. of Health Education and Welfare (NIOSH) publication No. 78–128. Washington, D.C. U.S. Govt. Printing Office.
Rozear, M.R., DeGroof, R. and Somjen, G. (1971). *J. Pharmacol. Exp. Ther. 176:*109.
Rubin, R.P. (1970). *Pharmacol. Rev. 51:*389–428.
Sabotka, T.J., Brodie, R.E. and Cook, M.P. (1975). *Toxicology 5:*175.
Saito, K. (1973). *Brit. J. Ind. Med. 30:*352.
Saito, K. and Abe, S. (1965). *Japaan J. Ind. Health 7:*20.
Schlaepfer, W.W., (1969). *J. Neuropathol. Exp. Neurol. 28:*401.
Seppalainen, A.M. and S. Hernberg (1972). *Brit. J. Indus. Med. 29:*443.
Seppalainen, A.M., Tola, S., Hernberg, S. and Kock, B. (1975) *Arch. Environ. Health 30:*180.
Sessa, J. Ferrari, E. and Colucci d'Amato, C. (1965). *Folia Med. Napoli 48:*658.
Seto, D.S.Y. and Freeman, J.M. (1964). *Amer. J. Dis. Child. 107:*337.

Silbergeld, E.K., Fales, J.T. and Goldberg, A.M. (1974). *Nature 247:49.*
*Simpson, J.A., Seaton, D.A. and Adams, S.F. (1964). J. Neurol. Neurosurg. Psychiat. 27:*536.
Smith, H.D., Bachner, R.L., Carney, T. and Majors, W.J. (1963). *Amer. J. Dis. Child. 105:*609.
Thurston, D.L., Middelkamp and Mason, E. (1955). *J. Pediatr. 47:*413.
Weakly, J.N. (1973). *J. Physiol. London. 234:*597.
Willes, R.F. (1977). *Toxicol. Appl. Pharmacol. 41:*207.
Whitfield, C.L., Ch'ien, L.T. and Whitehead, J.D. (1972). *Amer. J. Med. 52:*289.
Xintaras, C., Sobecki, M.F. and Ulrich, C.E. (1967). *Toxicol. Appl. Pharmacol. 10:*384.
Zook, B.C., Carpenter, J.L. and Roberts, R.M. (1972). *Amer. J. Vet. Res. 33:*891.

# Nutrient-Lead Interactions

*Kathryn R. Mahaffey*

**TABLE OF CONTENTS**



## I. INTRODUCTION

Nutritional status is known to alter the susceptibility of humans and animals to disease. The role of nutrition in altering the disease process has been observed for both infectious and chemically-induced diseases. Nutrition alters both the likelihood of an individual contracting the disease and the severity of the outcome of the disease. General mechanisms involved in nutrient-toxicant interactions and a wide range of examples of the role of nutrition in infectious and environmental diseases have been reviewed recently (Mahaffey and Vanderveen, 1979).

Fifty to one hundred years ago, milk was recommended as a supplement to the diet of lead industry workers to protect them against lead poisoning. In the 1920's, 1930's and early 1940's, experiments were performed with diets containing varying levels of calcium and phosphorus in an attempt to increase excretion of lead from the body (Sobel et al., 1940). With the advent of chelation therapy (Hardy et al., 1954), more effective means of deleading patients became available, and interest in nutrition and lead declined for some years. In the 1970's, prevention of lead toxicity, rather than treatment, has been emphasized. The related role of nutrition as part of preventive medicine in reducing the accumulation of the body burden of lead has received increasing attention. As techniques for studying nutrient-lead interactions have become more sophisticated, the physiologic mechanisms through which these interactions occur are elucidated. More recently, an additional area has received emphasis; this is the identification of the effects of lead on nutrient metabolism.

## II. PHYSIOLOGIC MECHANISMS

Physiologic effects on lead toxicity due to nutrition occur at different stages in the development of lead intoxication. At the very earliest stage of intoxication, absorption of lead from the gastrointestinal tract may be greatly influenced by the dietary intake of nutritionally significant compounds. For example, lead absorption is known to be altered by dietary calcium (Meredith et al., 1977), phosphorus (Quarterman and Morrison, 1975), iron (Ragan, 1977), fat (Barltrop and Khoo, 1975), and vitamin D (Smith et al., 1978; Barton et al., 1978a). In addition to changing the absorption of lead, nutrients also can influence resistance to lead toxicity in several ways. The first is by sequestering lead in body depots, specifically bone mineral; combined in the mineral phase of bone, lead is relatively unavailable, for adverse effects on metabolism would occur if it were deposited in organs such as liver and kidney. Under conditions of low calcium diet, where bone formation was lower or bone was resorbed to provide calcium, the lead content of kidney and blood increased dramatically (Mahaffey-Six and Goyer, 1970). Although bone lead content was increased several fold over that observed with a nutritionally adequate diet, the lead concentration in kidney increased to a far greater extent (Table 1). Second, the concentration of mineral nutrients such as copper or iron can markedly alter the effects of lead on the activity of cellular enzymes. For example, the enzyme delta-aminolevulinic acid dehydrogenase (ALAD), which is inhibited by lead, requires zinc for activity. As shown in rats, zinc had an antagonistic effect on ALAD inhibition produced by lead (Meredith et al., 1974). *In vitro* lead inhibits ALAD activity over a range of lead concentrations. Addition of zinc to incubation media containing lead activated the enzyme; the degree of activation was proportional to zinc concentration above threshold zinc concentration (Border et al., 1976). Cellular synthesis of protein is inhibited by lead and a portion of this inhibition could be overcome by adding higher concentrations of iron to the incubation media (Borsook et al., 1957).

Certain nutrients and lead affect the same organ systems or parameters of toxicity, although they alter different enzymes or different steps in

**Table 1.** Effect of dietary calcium deficiency on blood, kidney and femur Pb concentration in rats

| | Blood Pb (μg/dl) | Kidney Pb (μg/g)* | Femur Pb (μg/g)* |
|---|---|---|---|
| Normal Ca†† | < 10‡ | 2.6 ± 1.2 | 2.2 ± 1.0 |
| Low Ca ¶ | < 10 | 4.4 ± 0.6 | 9.7 ± 2.2 |
| Normal Ca + Pb** | 50 ± 10 | 29.6 ± 7.0 | 202.0 ± 22.2 |
| Low Ca + Pb | 180 ± 15 | 691.0 ± 203.0 | 225.0 ± 15.2 |

* Wet weight

† Represents 0.7% of diet

‡ Mean ± SEM

¶ Represents 0.1% of diet

** Pb in drinking water; 200 ppm Pb added as lead acetate

(Mahaffey-Six and Goyer, 1970)

biochemical synthesis. Maines and Kappas (1977) have reviewed a number of their own investigations on the role of metals as regulators of heme metabolism. Various metals stimulate both synthesis and degradation of metalloprotein. In the hematopoietic system, anemias develop as a result of deficiencies of cobalt, copper, iron, fluorine and vitamin E, as well as with lead toxicity. Generally, the anemia becomes more severe when nutrient deficiency and lead toxicity are combined, although these compounds affect different steps in the hematopoietic process.

More complicated parameters of toxicity such as behavior also can be altered by both nutritional deficiency and exposures to lead. In experimental animals, behavioral changes attributed to specific toxicity of lead may be influenced in part by nutrition. This subject is discussed in some detail in Section VII.

## III. ANIMAL STUDIES OF LEAD/NUTRIENT INTERACTIONS

### A. Major Nutrients

The overall importance of major nutrients on the absorption of lead was demonstrated by the observation that approximately 10 times as much lead was absorbed when it was consumed between meals as when it was consumed with meals (Rabinowitz et al., 1975). Quarterman et al. (1976a) found that food restriction always increased lead retention in rats fed diets containing 200 or 400 mg lead/kg for 3 or 6 weeks. Of the major nutrients, variations in dietary protein and fat content have been most thoroughly studied to date. Baernstein and Grand (1942) found that mortality was lower and weight loss was diminished in rats fed lead chloride with increasing protein levels in the diet (6 to 13 to 20%). Gontaea and coworkers

(1964) observed that doubling the protein content of the diet from 9 to 18% in the rat lessened susceptibility to lead intoxication as judged by the lead content of liver, kidney and blood. Der et al. (1974) reported that rats fed 4% protein diets had higher blood lead concentrations than animals fed 20% protein diets when lead was injected subcutaneously. These data are difficult to interpret because both groups of rats were injected with 100 $\mu$g of lead acetate daily for 40 days; however, rats receiving diets containing 4% protein were severely growth-retarded. Although equal quantities of lead were injected, the dose per unit body weight was markedly different. In these three studies, the comparison has been made between deficient and adequate levels of dietary protein. Barltrop and Khoo (1975) found that high as well as low protein diets fed for short periods of time elevated the lead content of several tissues. Barton et al. (1978a) reported significant increases in lead absorption by rats when lead was administered in a 1 mM solution of methionine.

One of the most significant dietary factors found to influence absorption of lead is fat. Barltrop and Khoo (1976) observed that lead absorption is dependent upon both the quantity and type of dietary fat. Increasing the corn oil content of the diet from 5 to 40% resulted in 7- to 14-fold increases in lead content of several tissues. However, decreasing fat from 5 to 0% of the diet did not affect tissue lead content. Quarterman et al. (1977) have also observed that the degree of lead absorption varied with the type of dietary fat and showed that lecithin, when mixed with bile salts and, to a smaller extent, choline, increased lead uptake. However, the importance of small amounts of phospholipids in stimulating lead absorption remains in doubt. In 10-week studies, Ku et al. (1978) found that for both young and mature rats ingesting either lead acetate or a complex of lead with phospholipids resulted in similar concentrations of lead in femur, kidney, liver and brain. Interestingly, Buck (1970) has observed that lead in greases and oils is far more toxic to cattle than elemental lead or lead salts.

The carbohydrate component of the diet, especially complex carbohydrates, can also alter the toxicity of lead. Carr et al. (1969) reported that, in 7- to 8-week-old rats the standard laboratory diet of which contained 10% alginate, lead absorption was reduced. Kostial et al. (1971) indicated that alginates reduced lead absorption from the gastrointestinal tract of newborn rats. However, in three human volunteers, addition of alginate to the diet did not affect lead absorption (Harrison et al., 1969). The importance of variation in type of relatively digestible carbohydrates on lead absorption in monogastric animals remains largely uninvestigated. Dietary lactose is thought to stimulate intestinal absorption of calcium and magnesium by infants (Kobayaski et al., 1975), but Barltrop and Khoo (1975) reported that addition of 10% lactose to the control diet used in their studies had no effect on lead absorption in rats.

## B. Minerals

**1. Calcium** Awareness of metabolic relationships between calcium and lead has existed for many years. In 1926, Aub stated that the "lead stream" follows the "calcium stream." Early investigation indicated that vitamin D, calcium and phosphorus influenced deposition of lead in soft tissue and bone (Sobel et al., 1940). Shields and Mitchell (1941) observed that lowering the calcium and/or phosphorus intake could increase bone or soft tissue lead at low levels of lead intake (32 ppm). These workers also reported that the influence of calcium intake of lead metabolism was seen primarily at low levels of calcium intake; increasing dietary calcium to high levels, such as 1.1% of the diet, did not decrease tissue stores of lead below that found on a normal calcium diet. Lederer and Bing (1940) observed that lowered dietary calcium increased deposition of lead in the femur when lead acetate was fed in the diet, but did not affect femur lead content when lead acetate was administered by daily intraperitoneal injection.

Many of the early studies were carried out with the hope that manipulation of dietary calcium would be preventive or useful in treatment of occupational lead poisoning. In fact, it was customary at that time in Great Britain and to a lesser extent in the United States to provide workmen in lead industries with free milk while they were working (Hunter, 1975). One difficulty in interpreting many of the older papers arises from the emphasis on the dietary calcium/phosphorus ratio, which resulted in the investigators changing more than one nutrient in the same experiment, *e.g.*, calcium, phosphorus and vitamin D. Interest in the calcium-lead relationship may have declined because of the advent of clinically useful chelating agents, *e.g.*, ethylene diamine tetraacetic acid (EDTA), for treating lead poisoning around 1950 (Hardy et al., 1954).

Several decades later, the role of lowered dietary calcium in increasing tissue lead content and the biochemical and morphologic changes associated with lead toxicity were reported (Mahaffey-Six and Goyer, 1970). These studies provided a new awareness that low dietary calcium greatly enhanced the susceptibility and pathologic effects of a particular dose of lead. A low calcium diet resulted in approximately a fourfold increase in blood lead content in rats ingesting equivalent amounts of lead (Fig. 1). Tissue lead concentrations were greatly elevated with low calcium diets (Tables 1 and 2); however, the increase of lead in soft tissue was much greater than for bone. Urinary excretion of delta-aminolevulinic acid (ALA) was markedly increased by low calcium diets in lead-treated animals, as were morphologic changes in the proximal renal tubules. The magnitude of the calcium effect was illustrated by the observation that rats ingesting water containing 12 ppm lead on a low calcium diet (0.1%) had concentrations of tissue lead similar to those of rats receiving water containing 200 ppm lead and a

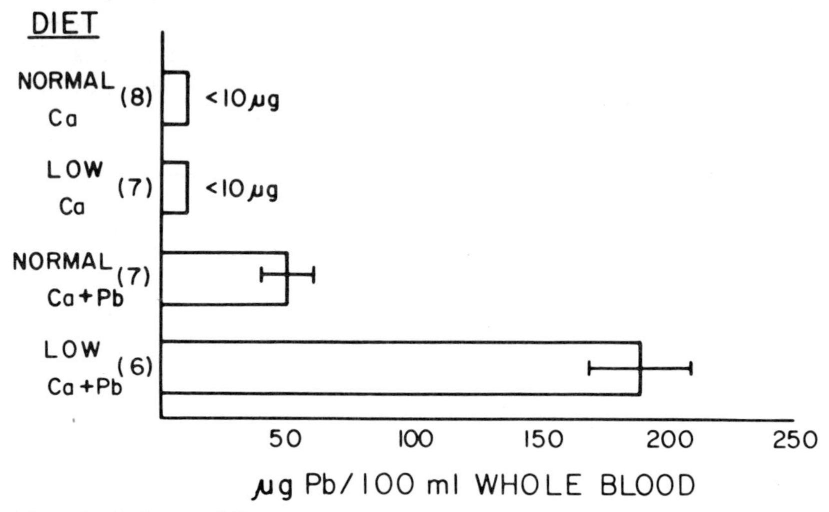


**Figure 1.** Influence of dietary calcium on blood lead levels (Mahaffey-Six and Goyer, 1970).

normal calcium diet (0.7%) (Mahaffey et al., 1973). Reduction of dietary calcium to approximately 40 to 50% of the dietary requirement was necessary before increases in tissue lead content occurred (Mahaffey et al., 1975). These levels approximate the physiologic requirement for calcium in the rat. Possibly, part of the interaction between low calcium diet and lead occurs only when physiologic mechanisms adapting the body to a low calcium diet become activated.

Comparison of concentrations of lead that produce nuclear inclusion bodies and hematologic disorders with normal and low calcium diets is

**Table 2.** Renal and femur lead content of rats fed normal (0.7%) and low (0.1%) calcium diets and varying concentrations of lead

| Pb in drinking water (μg/ml) | Renal Pb (μg/g)* | | Femur Pb (μg/gm) | |
|---|---|---|---|---|
| | Normal Ca diet | Low Ca diet | Normal Ca diet | Low Ca diet |
| | 1.0 ± 0.1† | 3.6 ± 0.5 | 4.8 ± 1.1 | 50 ± 13 |
| 3 | 1.3 ± 0.1 | 6.6 ± 1.4 | 5.2 ± 0.5 | 63 ± 9 |
| 12 | 1.9 ± 0.1 | 19.6 ± 2.0 | 16.8 ± 2.8 | 240 ± 14 |
| 48 | 5.1 ± 0.4 | 154 ± 51 | 65.6 ± 7.7 | 430 ± 47 |
| 96 | 6.9 ± 0.7 | 629 ± 170 | 100.2 ± 10.9 | 708 ± 109 |
| 200 | 21.3 ± 3.8 | 942 ± 362 | 236.5 ± 25.4 | 922 ± 188 |
| 400 | 20.4 ± 1.9 | | 230.2 ± 22.1 | |

* Fresh weight

† Mean ± 1 SEM

(Mahaffey et al., 1973)

shown in Table 3. Similar effects of a low calcium diet on susceptibility to lead toxicity in the rat were reported by Quarterman and Morrison (1975). Confirmation of the role of calcium in increasing toxicity of a particular dose of lead has been reported for the dog (Calvery et al., 1938), pig (Hsu et al., 1973), horse (Willoughby et al., 1972a) and lamb (Morrison et al., 1977).

Because low dietary calcium influenced the toxicity of ingested but not injected lead (Lederer and Bing, 1940; Quarterman and Morrison, 1975), a similar influence of calcium on gastrointestinal absorption of lead appeared likely. Specific studies of the gastrointestinal absorption of lead have subsequently shown this to be a more complex situation than was first envisioned. The most complete work published to date on this subject is the study by Barton et al. (1978a). Prior conditioning of their rats by low or high calcium diets did not significantly alter lead absorption *in vivo* using ligated intestinal loop techniques (Table 4). However, when the concentration of calcium used in the intubation media was varied within physiologic ranges, lead absorption decreased with increasing calcium concentration (Fig. 2). Similar results indicating that decreasing concentrations of calcium in the intubation media, but not calcium deficiency *per se*, increased lead absorption in rats and chicks have been observed by Mahaffey and coworkers (1979) during studies on the role of vitamin D in lead absorption. Reduced lead uptake from ligated gut loops following increased levels of calcium in intubation media have been reported by Barltrop and Khoo (1976). Meredith et al. (1977) have shown that oral calcium administered immediately before lead was highly effective in decreasing lead absorption in rats.

These studies do not discount enhanced gastrointestinal absorption of lead as a component in the higher retention of lead by animals on calcium-deficient diets, because the physiologic equivalent of lowered concentrations

**Table 3.** Minimal concentration of lead in drinking water of rats fed a low calcium diet that produce various lead-induced changes

| | Minimal toxic dose ($\mu$g/ml) | |
|---|---|---|
| | Low Ca (0.1%) diet | Normal Ca (0.7%) diet |
| Intranuclear inclusion bodies* | 12 | 200 |
| Increased urinary excretion of ALA | 12 | 200 |
| Increased kidney weight/ body weight ratio | 3 | 200 |

* In renal proximal tubule lining cell

(Mahaffey, 1974)

**Table 4.** Influence of dietary calcium content on lead absorption

| | Calcium-deficient diets | | Normal diet | High calcium diets | |
|---|---|---|---|---|---|
| Dietary calcium content (% of diet) | 0.23 | 0.34 | 0.46 | 0.92 | 1.38 |
| Lead absorption (%) | 19.6 ± 3.6 | 19.2 ± 3.7 | 21.8 ± 2.4 | 25.4 ± 3.0 | 25.7 ± 3.8 |
| p value | ns* | ns | | ns | ns |

* ns, nonsignificant change at the 0.05 level of probability
(Barton et al., 1978a)

of calcium in intubation media would probably occur during ingestion of calcium-deficient diets. However, increased absorption of lead is probably not the only mechanism for increases in tissue lead content associated with low calcium diets. Barton et al. (1978a) observed that excretion of injected lead in rats on calcium-deficient diets was slower than in animals receiving normal calcium diets. Animals receiving diets containing higher than normal calcium showed lead retention similar to those rats on normal calcium diets. The pattern of these changes again illustrates that the major calcium effects on lead metabolism occur when dietary calcium is deficient.

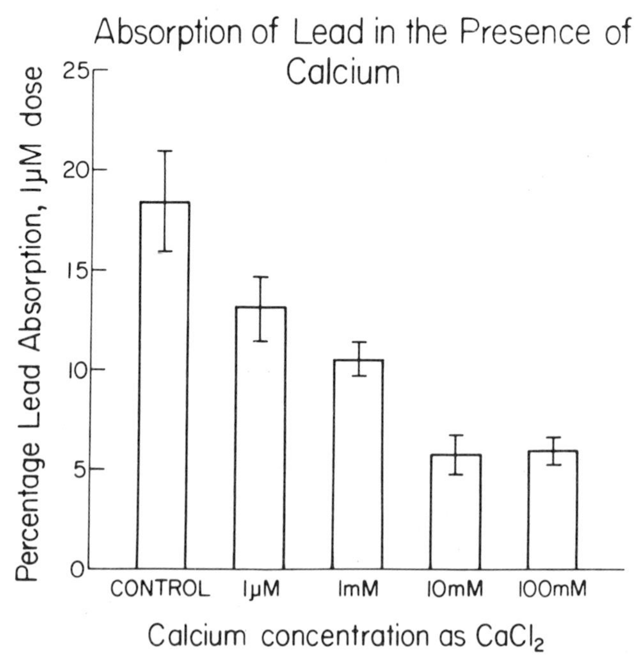


**Figure 2.** Effect of variation of calcium chloride in the intubation media on the absorption of a single dose of radiolabeled lead chloride (Barton et al., 1978a).

Little influence of calcium on lead metabolism is observed if elevated levels of calcium are compared with normal dietary calcium intake. In an editorial accompanying the report of Barton et al. (1978a), Goyer (1978) commented that Barton's data invite speculation that the kidney may be the site for the low calcium effect. Mobilization of lead to soft tissue, including the kidney, accompanied by increased metal-protein complex formation (*e.g.*, intranuclear inclusion bodies) may account in some degree for the increased retention of lead accompanying low dietary calcium.

Yamamoto et al. (1974) demonstrated that lead decreased $^{47}Ca$ radioactivity in bone and increased $^{47}Ca$ radioactivity in liver without altering $^{47}Ca$ radioactivity in blood. The authors indicated that lead mobilized calcium from the bone. However, these results must be interpreted with caution, because the $^{47}Ca$ and $Pb^{2+}$ were administered in single ip injections and the lead doses were extremely high. The relevance of this experimental model to physiologic conditions accompanying long term inorganic lead poisoning in humans is probably remote.

Calcium and lead interactions can also be observed in bone metabolism using both *in vitro* and *in vivo* techniques. Recent studies of bone organ culture (Rosen and Wexler, 1977) have demonstrated that low media levels of calcium, with or without parathyroid hormone or 1,25-dihydroxy vitamin $D_3$, enhanced the release of radioactive lead from bone explant to the medium. In addition to the effects of calcium on bone lead metabolism, lead affected the bone metabolism of calcium. The lead line resulting at the metaphysis of bone was suggested some years ago as being due to an abnormal calcification of cartilage. Hsu et al. (1973) observed inclusion bodies in multinucleated giant cells in bone and indicated that the osteoclasts became poisoned during osteolysis of bone containing large amounts of lead. Eisenstein and Kawanoue (1975) demonstrated that these giant cells were osteoclasts and chondroclasts and interpreted the lead line to represent lead induced inability of cartilage-resorbing cells to degrade bone mineral.

In neural tissue, lead and calcium interactions are also observed. Lead appears to block presynaptic neural transmission by reducing the release of acetylcholine. Using *in vitro* techniques, this lead effect was reversed by adding calcium to the intubation medium (Kostial and Vouk, 1957; Silbergeld, 1973).

Using rat heart mitochondria, Parr and Harris (1976) have demonstrated that very low concentrations of $Pb^{2+}$ decreased the capacity of heart mitochondria to remove $Ca^{2+}$ from the incubation medium. Below 0.5 nmol $Pb^{2+}$/mg protein, this effect had the characteristics of competitive inhibition. Addition of 1 nmol $Pb^{2+}$/mg protein resulted in almost complete inhibition of $Ca^{2+}$ sequestering. Although these results were based almost entirely on *in vitro* work, the authors calculated that 0.5 nmol of $Pb^{2+}$/mg protein is equivalent to between 3 and 9 ppm $Pb^{2+}$ in whole tissue. Persons without unusual exposure to lead have much lower concentrations of cardiac lead; however, in persons with unusually high lead exposure, the aorta showed the highest lead concentrations of any soft tissue, in the region of 2 ppm wet weight (Barry, 1978). Cardiac complications are considered an unusual sequela of lead exposure in humans, although a few cases have been reported. For example, Freeman (1973) reported myocarditis in a 3-year-old child with severe lead poisoning; cardiac status returned to normal within 4 days of chelation therapy with ethylenediamine tetracetic acid for reduction of the body burden of lead. Others have reported cardiac changes accompanying lead intoxication (Kline, 1960; Read and Williams, 1952; Myerson and Eisenhauer, 1963).

**2. Phosphorus** As with calcium, changes in dietary phosphorus alter susceptibility to lead toxicity. Phosphorus-deficient diets resulted in higher tissue concentrations of lead at equivalent levels of lead exposure in rats (Sobel et al., 1940; Shields and Mitchell, 1941; Quarterman and Morrison, 1975). Tissue accumulation of lead was no lower when dietary phosphorus was increased above the nutritional requirement for phosphorus than with a normal phosphorus diet (Shields and Mitchell, 1941). Stated another way, phosphorus-deficient diets increase tissue lead accumulation, but increasing dietary phosphorus to levels above those recommended as nutritional requirements is not protective against lead toxicity. This effect of phosphorus occurred at low, adequate and high concentrations of dietary calcium (Shields and Mitchell, 1941; Lederer and Bing, 1940); when low dietary calcium and phosphorus were combined, tissue lead retention was much greater than that observed when the diet was low in only one of these nutrients (Shields and Mitchell, 1941). Supplementation of a low phosphate diet with phosphate decreased the quantity of lead in the blood and femur of lambs, but did not significantly reduce lead concentration in liver and kidney (Morrison et al., 1974). Retention of ip injected lead was not influenced by level of dietary phosphorus in rats (Quarterman and Morrison, 1975), suggesting that phosphorus affected lead absorption. Similar observations with calcium have invited the speculation that calcium-deficient diets resulted in increased lead absorption. However, experimental observations have shown these Ca-Pb interactions to be more complex (Barton et al., 1978a).

Overall, the effects of phosphorus on lead retention appear to be of smaller magnitude than those of calcium. For example, low dietary phosphorus resulted in smaller increases in carcass lead content than did lowering dietary calcium (Shields and Mitchell, 1941). Quarterman and Morrison (1975) observed that low dietary calcium resulted in large increases in kidney, liver and erythrocyte lead content, while low dietary phosphorus decreased only liver lead content. With regard to the effects of these two nutrients in humans, the intake of dietary calcium in both children and adults is frequently lower than recommended, while inadequate dietary phosphorus is quite uncommon among persons ingesting a Western-type diet (Section VI).

**3. Iron** Iron deficiency, like deficiencies of various other minerals, increased susceptibility to lead toxicity in the rat (Mahaffey-Six and Goyer, 1972). Iron-deficient rats have increased concentrations of lead in kidney and bone (Table 5) as compared to rats ingesting equivalent quantities of water containing 200 ppm lead and an iron-adequate diet. In contrast to calcium deficiency, iron deficiency does not result in a redistribution of lead to nonosseous tissue (Table 6). The degree of iron deficiency associated with increased body burden of lead need not be severe. Experimental animals with increased lead burden showed only a slight decrease in hematocrit; however, liver iron stores were greatly reduced. Although lead toxicity and iron deficiency act at different steps in the heme biosynthetic pathway, the degree of anemia produced by the combination of iron-deficient diet and lead intoxication is greater than that resulting from either alone. Measurement of various intermediate compounds in heme synthesis illustrates both independent and combined effects of iron and lead. For example, erythrocyte protoporphyrins are increased moderately in iron deficiency but are greatly increased by lead intoxication. Generally, lead intoxication results in far greater increases in free erythrocyte protoporphyrins than does iron deficiency (Piomelli, 1977). Urinary excretion of ALA is not increased by iron deficiency alone, but is elevated by lead toxicity (Table 5). Simultaneous occurrence of these conditions results in higher ALA excretions than those produced by lead alone; however, this may be due to an increased body burden of lead in iron deficiency.

Increased absorption of lead in iron-deficient animals was thought to occur because it has been known that iron-deficient diets increased the

**Table 5.** Hematocrit, urinary ALAD and renal and femur Pb content of rats fed normal and low Fe diets with and without Pb

| Dietary group | N | Hematocrit* | Urinary ALAD (μg/24 hr) | Renal Pb (μg/g wet tissue) | Femur Pb (μg/g wet tissue) |
|---|---|---|---|---|---|
| Normal Fe | 9 | 45.7 ± 0.6 | 20 ± 5 | 1.0 ± 0.1 | 5.6 ± 1.4 |
| Low Fe | 8 | 42.6 ± 1.0 | 22 ± 5 | 1.9 ± 0.4 | 10.6 ± 3.0 |
| Normal Fe + Pb† | 8 | 44.2 ± 0.9 | 180 ± 25 | 14.5 ± 1.6 | 75.2 ± 13.1 |
| Low Fe + Pb† | 8 | 37.8 ± 0.9 | 355 ± 50 | 38.7 ± 4.8 | 225.2 ± 15.2 |

* Mean ± 1 SEM

† Drinking water containing 200 μg Pb/ml water

(Mahaffex-Six and Goyer, 1972)

**Table 6.** Comparison of tissue concentrations of lead in rats fed diets deficient in calcium (LCa) and iron (LFe) and nutritionally adequate diets (NCa, NFe)

| Diet | Pb (μg/g wet tissue) | |
|---|---|---|
| | Bone | Kidney |
| No Pb | | |
| NCa, NFe | 2.2 ± 1.0* | 2.6 ± 1.2 |
| LFe | 10.6 ± 3.0 | 1.9 ± 0.4 |
| LCa | 9.7 ± 2.2 | 4.4 ± 0.6 |
| 200 μg Pb/ml $H_2O$ | | |
| NCa, NFe | 74 ± 12 | 22 ± 4.3 |
| LFe | 225 ± 15 | 28.7 ± 4.8 |
| LCa | 202 ± 202 | 691 ± 203 |

* Mean ± 1 SEM

(Mahaffey-Six and Goyer, 1972)

absorption of certain metals, including manganese (Diez-Edward et al., 1968; Pollock et al., 1965), cobalt (Pollock et al., 1965; Valberg et al., 1969) and zinc (Pollock et al., 1965), but not cesium, mercury, calcium or copper (Pollock et al., 1965). Recently, using retention of $^{210}Pb$ as an estimate of gastrointestinal absorption, Ragan (1977) demonstrated sixfold increases in tissue lead in rats when body-iron stores were reduced, but before frank iron deficiency developed. Using *in situ*, ligated gut loop techniques, Barton et al. (1978b) have demonstrated that iron deficiency secondary to bleeding and iron-deficient diets increase lead absorption and that iron loading rats decreased lead absorption (Fig. 3). These support and provide some physiologic explanations for the observations of Mahaffey et al. (1978) that increasing dietary iron concentrations to higher levels than those recommended for nutritional purposes decreased kidney, femur and blood lead levels in rats fed varying concentrations of iron for 10 weeks (Table 7). Barton et al. (1978b) also determined whole body retention of a single iv injection of $^{210}Pb$ over a period of 4 weeks and found that rats retained similar quantities of lead whether or not they were iron deficient, normal or iron loaded (Fig. 4). This demonstrated that body iron status does not alter lead excretion.

Lowered tissue iron levels in iron-deficient animals can alter effects of lead on enzyme systems. Waxman and Rabinowitz (1966), using incubation of reticulocytes, reported that lead inhibited hemoglobin synthesis as a specific antagonist of heme synthesis and as a general heavy metal inhibitor of cell metabolism. Addition of ferrous iron protected both functions against the action of lead. Effects of lead on iron metabolism are discussed in Section V.

**Table 7.** Influence of dietary iron and lead content on weight gain, tissue lead and iron concentration in rats[a]

| | Deionized water | | | | | 200 $\mu g$ Pb/ml water | | | | |
|---|---|---|---|---|---|---|---|---|---|---|
| Dietary Fe (mg/kg) | Weight gain (g) | Blood Pb ($\mu g$/dl) | Femur Pb ($\mu g$/femur)* | Renal Pb ($\mu g$/kidney)* | Liver Fe (mg/kg)* | Weight gain (g) | Blood Pb ($\mu g$/dl)* | Femur Pb ($\mu g$/femur)* | Renal Pb ($\mu g$/kidney)*† | Liver Fe (mg/kg)* |
| 10 | $251 \pm 7$‡ | $5.9 \pm 0.05$ | $3.11 \pm 0.26$ | $1.37 \pm 0.18$ | $44.7 \pm 1.9$ | $172 \pm 8$ | $159 \pm 8$ | $339 \pm 14$ | 311(254,382) | $53.6 \pm 3.7$ |
| 15 | $269 \pm 9$ | $5.8 \pm 1.1$ | $2.67 \pm 0.50$ | $0.99 \pm 0.16$ | $118.0 \pm 13.4$ | $189 \pm 7$ | $101 \pm 6$ | $265 \pm 17$ | 165(128,211) | $115.4 \pm 11.7$ |
| 25 | $274 \pm 11$ | $4.3 \pm 0.5$ | $2.20 \pm 0.27$ | $1.03 \pm 0.09$ | $136.9 \pm 10.1$ | $225 \pm 6$ | $100 \pm 8$ | $209 \pm 15$ | 58(47,72) | $156.6 \pm 10.3$ |
| 40 | $272 \pm 11$ | $9.7 \pm 1.8$ | $1.04 \pm 0.14$ | $1.05 \pm 0.14$ | $149.7 \pm 6.2$ | $247 \pm 10$ | $113 \pm 11$ | $209 \pm 22$ | 44(33,61) | $144.1 \pm 6.6$ |
| 50 | $263 \pm 11$ | $4.7 \pm 0.6$ | $1.95 \pm 0.36$ | $1.19 \pm 0.17$ | $157.4 \pm 9.2$ | $196 \pm 10$ | $110 \pm 7$ | $203 \pm 16$ | 91(72,115) | $174.6 \pm 12.2$ |
| 120 | $261 \pm 14$ | $5.8 \pm 1.0$ | $0.91 \pm 0.08$ | $0.85 \pm 0.10$ | $147.9 \pm 7.0$ | $250 \pm 8$ | $76 \pm 7$ | $187 \pm 23$ | 26(19,36) | $160.0 \pm 6.9$ |

* Fe effect, $p < 0.0001$

† Analysis of renal Pb content of Pb groups was performed on log-transformed data and the antilogs of the means and SEM (which are not necessarily symmetric about the means because of the transformation) are listed here

‡ Mean $\pm$ 1 SEM

(Mahaffey et al., 1978)

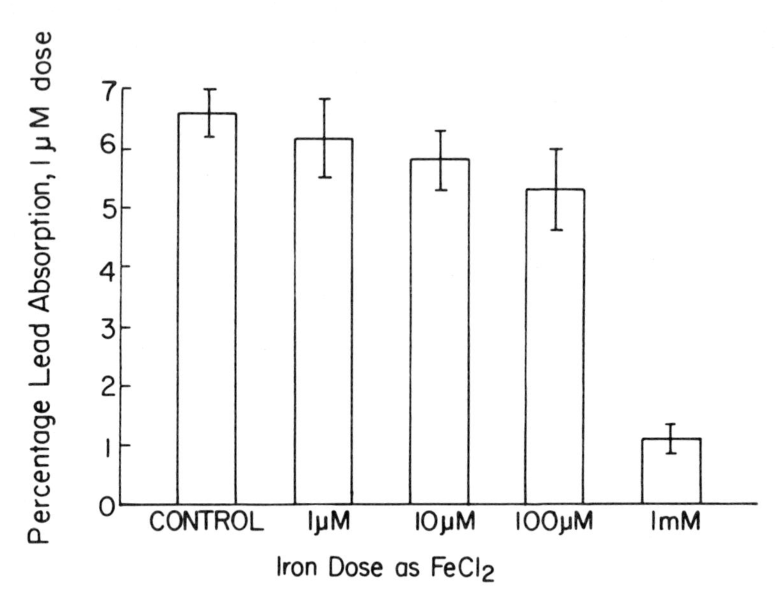


**Figure 3.** Effect of variation of iron chloride in the intubation media on the absorption of radiolabeled lead chloride (Barton et al., 1978b).

**4. Zinc** Cerklewski and Forbes (1976a), who investigated the effects of dietary zinc (8, 35 and 200 ppm) on the toxicity of lead (0, 50 and 200 ppm), observed that, as dietary zinc content increased, tissue lead levels and lead toxicity decreased. Because varying concentrations of dietary zinc did not influence the toxicity of injected lead, the investigators concluded that zinc exerts its effect on lead at the level of the gastrointestinal tract. Also supporting this conclusion was a decline in apparent absorption of lead with increasing dietary zinc concentration. Zinc concentration in tissues can alter the inhibition of aminolevulinic acid dehydratase activity produced by lead. Unfortunately, variation in tissue zinc levels produced by these changes in dietary zinc were not reported by Cerklewski and Forbes. Using *in vivo* studies in rats, zinc was shown to have an antagonistic effect on lead-induced ALAD inhibition (Meredith et al., 1974). *In vitro* lead inhibited ALAD activity over a wide range of lead concentrations. Addition of zinc to incubation media containing lead activated the enzyme; the degree of activation was proportional to zinc concentration above a threshold zinc concentration (Border et al., 1976). The interaction of lead and zinc on ALAD activity can also be observed in feeding experiments. Finelli et al. (1975) have reported that erythrocyte ALAD activity of rats fed diets containing either 2.5 or 40 ppm zinc was significantly lower in the rats on the low zinc diet. Addition of 200 ppm lead to the drinking water reduced ALAD activity to zero in the low zinc group and to near zero in the zinc supplemented group. Following removal of the lead exposure, erythrocyte ALAD activity returned to normal values more quickly in rats receiving the higher dietary zinc level than in animals receiving the low zinc diet.

Effects of feeding highly elevated levels of zinc were observed in the studies of Hsu et al. (1975), in which the combined influence of high levels of dietary zinc (4000 ppm) and lead (1000 ppm), when compared to lead alone, increased lead levels in blood, soft tissue and bone and enhanced the toxic effects of lead, as shown by reduced weight gain and by severe clinical signs of lead toxicity in Yorkshire pigs. Dietary zinc concentrations used by Hsu et al. (1975) were much greater than those used by Cerklewski and Forbes (1976a). In contrast to the results of Hsu et al. (1975), were those of Willoughby et al. (1972b), who reported that toxic amounts of zinc might prevent development of clinical signs of experimentally induced lead poisoning in the young growing horse. Using weight gain as an overall index of toxicity, pigs tolerated 4000 ppm zinc and 1000 ppm lead either alone or in combination on a normal calcium diet, but only zinc or lead alone on a low calcium diet (Hsu et al., 1975).

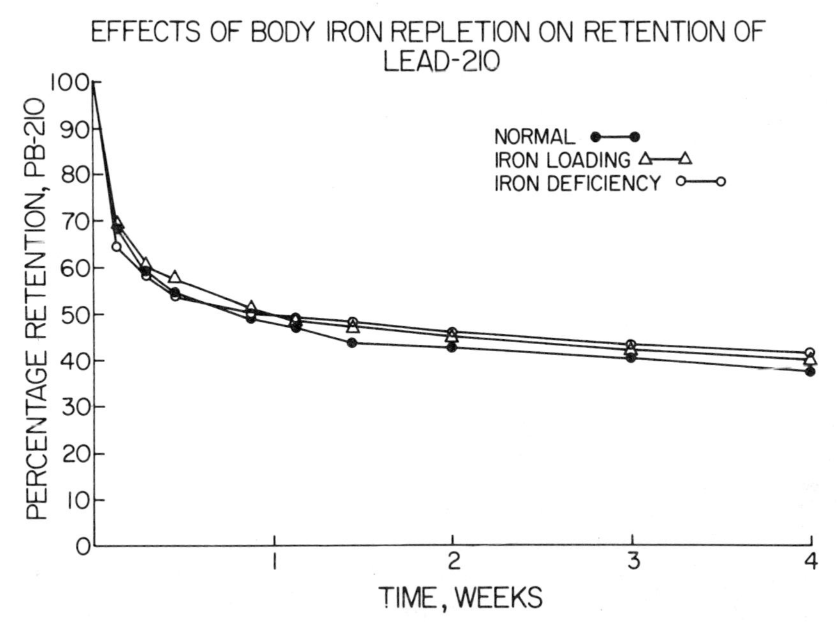


**Figure 4.** Comparison of the retention of $^{210}$Pb by normal, iron-loaded and iron-deficient rats (Barton et al., 1978b).

Overall, the role of zinc in susceptibility to lead poisoning is not entirely clear and is likely related to whether or not dietary zinc levels used in the experiments produced zinc deficiency or zinc toxicity. Similar changes associated with deficiency or excess appear to exist with selenium. Investigation of the role of zinc deficiency on susceptibility to lead toxicity is of practical importance to humans. Methods of assessing zinc nutriture in humans include estimates of dietary zinc content and evaluation of hair and plasma zinc levels. Reduced hair and plasma zinc levels, which are biochemical changes associated with zinc deficiency, have been reported among low-income children (Hambridge, 1977) and dietary levels of zinc have been found to be inadequate (Wolfe et al., 1977).

**5. Copper** Cerklewski and Forbes (1977) observed that increasing dietary copper (1, 5 and 20 ppm) increased the severity of lead toxicity (200 ppm) in rats, as shown by increasing excretion of urinary ALA and tibia lead content. Small differences in tissue lead content were observed when copper was increased from 5 to 20 ppm, although urinary ALA excretion decreased when copper was increased. The experimental period used was 4 weeks. During this time, hemoglobin level had fallen dramatically, with serum ceruloplasmin and liver copper concentrations about one-eighth the levels observed on a 5 ppm copper diet. By contrast, Klauder and Petering (1977) reported that increasing dietary copper (0.5 and 8.5 ppm) in rats for either an 8- or 12-week period decreased the severity of intoxication produced by consuming drinking water containing 500 ppm lead. Comparison of these two studies is difficult; differing results are possibly related to the severity of the copper deficiency produced. When dietary iron was adequate (40 ppm), average hemoglobin concentrations were above 12 g/100 ml of whole blood in animals fed 0.5 ppm copper for either 8 or 12 weeks (Klauder and Petering, 1977). The animals in the experiments of Cerklewski and Forbes appear to be more severely copper-deficient on the basis of hemoglobin concentrations, but not when serum ceruloplasmin levels are compared. After 4 weeks of a diet containing 1 ppm copper, hemoglobin concentrations averaged approximately 8 g/100 ml of whole blood despite adequate iron (35 ppm) (Cerklewski and Forbes, 1977). However, at the end of 12 weeks of copper-deficient diet, the plasma ceruloplasmin values of Klauder and Petering's animals were lower than those reported by Cerklewski and Forbes. Because the length of the experimental period and other conditions differed between these studies, explanations about differences in their conclusions cannot be made with certainty.

Klauder and Petering (1977) suggest that the induction of anemia by lead may be related to copper. They noted that many characteristics of anemia due to lead poisoning are similar to those of copper deficiency and postulated that lead may induce copper deficiency and thus interfere with iron metabolism and utilization. Tephly et al. (1978) indicated that *in vitro* inhibition of ferrochelatase activity produced by lead could be restored to near control levels by increasing the copper content of the reaction mixture. Addition of excess ferrous iron did not reverse the lead inhibition. Tephly et al. (1978) suggested that lead inhibits ferrochelatase activity, not by competing for the iron binding sites on the enzyme, but by attacking an adjacent site, specifically the one occupied by copper in the presence of reduced glutathione.

**6. Selenium** Although selenium has been reported as protective against the toxicity of mercury (Ganther et al., 1972) and cadmium (Parizek et al., 1971), the effects of selenium on lead metabolism are less marked. Cerklewski and Forbes (1976b) reported that selenium was mildly protective against the toxic effects of 200 ppm lead, but only up to concentrations of 0.50 ppm dietary selenium with tissue lead concentrations and urinary ALA excretion as the criteria of toxicity. Selenium at 1.0 ppm of the diet appeared to exacerbate lead intoxication. Qualitatively similar results, *e.g.*, higher levels of selenium resulting in slight decreases in tissue lead content, were reported for the Japanese quail by Stone and Soares (1976), although these investigators fed only two levels of selenium, 0 and 1 ppm. Levander et al. (1977a) compared the relative effects in rats of vitamin E and 0.5 ppm selenium on the development of splenomegaly, anemia and mechanical fragility of erythrocytes and concluded that vitamin E was more protective than selenium against these lead-induced effects.

Diets producing selenium toxicity appear to decrease the toxicity of lead. Cerklewski and Forbes (1976b) observed that addition of selenium to the diet decreased kidney lead concentrations. Levander et al. (1977a) reported that excess levels of selenium (2.5 and 5.0 ppm) in the diet of vitamin E-deficient rats depressed the splenomegaly and anemia of lead poisoning; however, these levels of selenium are toxic. Rastogi et al. (1976) regard selenium and lead as mutually detoxifying agents. However, they observed only slight effects of selenium on lead-induced changes in body weight, kidney weight/body weight ratio, tissue lead content and liver cytochrome P-450 activity. In their study, the groups treated with selenium and lead were substantially more comparable to animals receiving lead alone than to control animals. An exception to the above was the return to near control values for ALAD activity of blood, liver and kidney in rats treated with lead plus selenium compared to those receiving lead alone. Overall, the effects of selenium on lead metabolism and toxicity are slight.

**7. Other Metals** Lead balance studies in dogs receiving a synthetic cation-free diet (no detectable calcium or magnesium) for a 15-day period showed that addition of magnesium to the diet decreased apparent lead absorption from 26.5 ± 1.7% to 8.6 ± 1.4% of ingested lead (Fine et al.,

1976). Blood lead concentration was unaffected by the added magnesium. The ability of animals to maintain serum calcium and magnesium concentrations, despite dietary deprivation of these nutrients, was shown by these studies (Table 8). Morrison et al. (1975a) have shown that the addition of sulfur to a low sulfur diet of sheep greatly reduced tissue lead content. Because ruminant animals including sheep can metabolize sulfur to sulfide or sulfur-amino acids, Quarterman et al. (1976b) investigated the effect of feeding varying levels of methionine and cystine to rats; he observed that cystine lowered blood lead but increased carcass lead, while methionine lowered both carcass and blood lead when compared to rats receiving lead alone. Mahaffey et al. (1976a) fed diets containing high levels of fluorine and 200 ppm lead to rats for 10 weeks. Neither fluorine nor lead were toxic alone, but the combination produced high mortality and substantial weight loss in survivors. Fluorine administration produced significantly higher concentrations of lead in blood and femur, but not kidney, of lead-exposed rats. This increase in lead retention did not account for the sharp increase in toxicity observed when lead and fluorine were fed together.

### C. Vitamins

**1. Vitamin D** Vitamin D may increase the adverse effects of a particular level of lead exposure. Addition of vitamin D to the diet of rats receiving lead increased the deposition of lead in bone (Sobel et al. 1938). Several years later, Rapaport and Rubin (1941) administered cod liver oil to rats or exposed rats to sunlight and observed increases in blood lead concentrations. Both of these treatments should have increased the serum level of vitamin D metabolites, although no measurements of vitamin D were performed in either of these studies. The investigations were carried out in an attempt to explain the increased incidence of cases of lead toxicity among children during the summer months. Because vitamin D stimulates calcium absorption and low calcium diets increase susceptibility to lead toxicity, the hypothesis was developed that vitamin D stimulated lead absorption from the gastrointestinal tract, possibly through an active transport mechanism. Using *in vitro* everted gut sac and *in vivo* techniques, vitamin D was shown to stimulate lead absorption in the distal small intestine of rats (Smith et al., 1978 and 1.25 dihydroxycholecalciferol was

**Table 8.** Blood chemistry values of dogs fed a cation-free ration containing 10 ppm Pb for a 15-day period

| | Control | Mg fed | p value |
|---|---|---|---|
| Whole Blood Pb (μg/dl) | $53.8 \pm 7.4$ | $51.5 \pm 3.5$ | ns |
| Serum Ca (mg/dl) | $10.5 \pm 0.2$ | $10.7 \pm 0.2$ | ns |
| Serum Pb (mg/dl) | $8.5 \pm 1.6$ | $8.5 \pm 0.9$ | ns |
| Serum Mg (meg/l) | $1.40 \pm 0.03$ | $1.41 \pm 0.03$ | ns |

(Fine et al., 1976)

shown to be the active metabolite involved in stimulating lead absorption (Mahaffey et al. 1979). The distal small intestine was the major site of lead absorption, whereas the greatest stimulation of calcium transport by vitamin D was found in the duodenum. It appears, therefore, that lead and calcium do not share a common pathway for absorption as is often assumed. Vitamin D stimulates phosphate absorption in the distal small intestine, suggesting that lead absorption may be related to phosphate transport in some manner. Although lead absorption was increased when vitamin D intake was raised from deficient to adequate levels, high doses of vitamin D did not markedly enhance the absorption of lead above that achieved with physiologic doses of the vitamin (Table 9). Barton et al. (1978a) indicated that lead is bound to both vitamin D-dependent calcium binding protein and a higher molecular weight mucosal protein which is probably not vitamin D-dependent. Based on the relative binding of radiolabeled lead to these protein fractions, these authors suggest that more lead is bound to the fraction thought not to be induced by vitamin D. The relative physiologic importance of those two fractions remains to be established; however, it is clear that vitamin D does affect lead absorption. The mechanism through which vitamin D affects lead absorption is not clear at this time. The magnitude of the vitamin D effect on lead requires further evaluation in longer term studies such as those performed for calcium and iron (Mahaffey-Six and Goyer, 1972; Mahaffey et al., 1973).

**2. Vitamins A and K** To the best of this author's knowledge, there are no publications on interactions of lead and vitamins A or K in the English language.

**3. Vitamin E** In 1954, DeRosa reported that adding vitamin E to the diets of lead-intoxicated rabbits decreased lead-induced coproporphyrinuria and anemia. Further evidence that vitamin E affected lead-induced anemia was provided about 20 years later when Bartlett et al. (1974) reported that vitamin E was protective against anemia in rabbits fed a basal ration plus lead for a period of 12 weeks. The work suggested that vitamin E had a stimulatory effect on heme synthesis, apparently through its action of delta-aminolevulinic acid synthetase. Although the effects on anemia were statistically significant, the differences were not large. However, added vitamin E resulted in a marked and significant reduction of basophilic stippling. Beginning in 1975, Levander and various coworkers published a series of papers establishing that lead poisoning 1) caused a more pronounced anemia, splenomegaly and increased red blood cell fragility in vitamin E-deficient rats and 2) caused a high rate of reticulocytosis. They indicated that these vitamin E-related differences were not due to further impairment in heme synthesis (Table 10). Levander et al. (1977b) have investigated the mechanisms accounting for increased destruction of red blood cells. They observed that 1) the mechanical fragility of red blood

**Table 9.** Effect of increasing doses of vitamin D on lead acetate absorption *in vivo* by the rat. Six rats per group were fed a 1.2% calcium and 0.3% phosphorus diet for 4 weeks

| Cholecalciferol dose | Lead acetate absorption |
|---|---|
| ($\mu$g/day) | (%) |
| 0 | 16 ± 2 |
| 6.25 | 31 ± 6 |
| 12.50 | 37 ± 8 |
| 25.0 | 49 ± 4 |
| 100.0 | 45 ± 5 |

(Smith et al., 1978)

cells from lead-poisoned, vitamin E-deficient rats during *in vitro* testing was greater than that resulting from either condition alone and 2) the differences in mechanical fragility were due to marked alterations in the morphology of the red blood cells from vitamin E-deficient rats, whether the rats were receiving lead or not. As a result of vitamin E deficiency, cells changed from a normal discocytic shape to the highly abnormal stomatopherocytic shape (Levander et al., 1977b). This change affected primarily the filterability of the old population of red blood cells. Additional description and discussion of the mechanisms by which lead and vitamin E affect the red blood cell membrane are reported by Levander (1979).

**4. Vitamin B complex** The effect, if any, of various members of vitamin B-complex on lead toxicity is not well established. Anemia is an accompaniment of deficiencies of a number of B-vitamins. The several existing studies on interactions between B-vitamins and lead have emphasized measurement of heme metabolites. Nicotinic acid treatment during lead poisoning reduced coproporphyrinuria, free erythrocyte protoporphyrin and urinary ALA (Kao and Forbes, 1973).

Although Pecora et al. (1966) suggested that nicotinic acid was protective against lead poisoning, Kao and Forbes (1973) found that a tenfold

**Table 10.** Effects of vitamin E deficiency on the toxic response to lead in rats. Mean values of four to six animals ± SE

| Diet | Lead in water (ppm) | Weight gain (g) | Hematocrit value (% by volume) | Mechanical fragility index |
|---|---|---|---|---|
| + Vitamin E | 0 | 195 ± 8 | 41.0 ± 0.6 | 4.7 ± 0.4 |
| − Vitamin E | 0 | 177 ± 7 | 41.0 ± 0.4 | 3.7 ± 0.3 |
| + Vitamin E | 250 | 143 ± 7 | 36.8 ± 0.7 | 3.4 ± 0.3 |
| − Vitamin E | 250 | 124 ± 4 | 33.4 ± 1.0 | 16.0 ± 2.7 |

(Levander et al., 1975)

increase in the nicotinic acid content of the diet fed for 23 days did not influence the effects of lead on blood heme levels, *in vitro* heme synthesis or urinary ALA. Unfortunately, tissue lead levels were not reported.

**5. Ascorbic acid** Ascorbic acid has been suggested as a nutrient influencing the toxicity of a particular level of exposure to lead, although experimental data are sparse and contradictory. Ascorbic acid is of clinical interest because it is present in both deficient and excess levels in the diet. Ascorbate is one of four nutrients commonly consumed at less than recommended levels in the diet. Conversely, some individuals have very high intakes of ascorbic acid because of the popularity of ingesting high levels of vitamin C for a variety of illnesses.

Holmes et al. (1939) observed reduction of basophilic stippling and reduced morphologic evidence of anemia among painters and lead-exposed industrial workers when treated with supplements of ascorbic acid. Similar amelioration by ascorbic acid of lead-induced hematologic changes occurred in guinea pigs (Pal et al., 1975). However, this improvement in hematology does not appear to be due to a decreased tissue burden of lead. In 1940, Pillemer et al., while studying the effect of lead on vitamin C metabolism, indicated that ascorbic acid did not affect tissue levels of lead. However, the number of animals used in these studies was very small. Mahaffey and Banks (1974), using guinea pigs and larger numbers of animals, found that varying the ascorbic acid intake from deficient to normal to excessive levels for a period of 6 weeks produced relatively little effect on kidney and liver lead concentrations. Bone lead concentrations were not measured in this experiment. Barton et al. (1978b) have demonstrated that administration of radiolabeled lead in a 1 mM ascorbic acid solution increased lead absorption above that of controls. This effect was removed by addition of 1 mM $FeCl_2$ to the intubation solution containing 1 mM ascorbic acid. These studies were performed using isolated gut loops in rats. Guinea pigs, like humans, require a dietary source of ascorbic acid, whereas the rat synthesizes its own ascorbate. Recently, it has been suggested that ascorbic acid is a useful adjunct to chelation therapy in the rat and promotes removal of lead from the central nervous system tissue, compared to chelation with EDTA alone (Goyer and Cherian, 1978). However, McNiff et al. (1978) reported that feeding diets containing 1% ascorbic acid for 24 hr did not significantly alter urinary or fecal excretion of lead by adult rats injected intravenously with lead acetate at the level of 1 mg lead/kg body weight.

## IV. NUTRITION-LEAD INTERACTIONS IN HUMANS

There are fewer studies on the role of nutrition in susceptibility to lead in humans than on effects of dietary variables on susceptibility of animals to lead toxicity. Available information for humans is of two types, the balance study data and epidemiologic surveys on nutrient intake among normal and lead-burdened children. Ideally, demonstration of nutritional effects in humans evaluates the impact of nutritional changes over relatively long periods of time. Sorrell et al. (1977) and Mahaffey et al. (1976b), using case-control epidemiologic studies, have established that dietary calcium intake was lower in children having elevated blood lead concentrations than in children with normal blood lead levels. Further, Sorrell et al. (1977) observed a significant negative correlation between blood lead level and dietary calcium intake. Environmental lead exposure was not documented in these studies; however, the control children came from similar socioeconomic and cultural backgrounds and lived in similar neighborhoods. In some cases the children were siblings. Mahaffey et al. (1976b) observed lower dietary phosphorus intake in lead-burdened children than in controls.

Using balance study techniques, Ziegler et al. (1978) demonstrated that not only did young children absorb a greater percentage of ingested lead than did adults, but that lead absorption and retention were inversely correlated with the calcium content of the diet. When dietary calcium level was higher, lead absorption decreased. Interestingly, the various calcium levels used in this study were all greater than those recommended by the National Research Council as adequate for human infants and toddlers.

Iron deficiency is considered to be important to the development of lead toxicity because animal studies have indicated that iron deficiency increased tissue lead retention. Children from the same socioeconomic groups who are at highest risk of elevated lead exposure are the most likely to be iron-deficient. However, the studies of Sorrell et al. (1977), Mahaffey et al. (1976b) and Mooty et al. (1975) have not shown differences in dietary iron intake between normal and lead-intoxicated children. Watson et al. (1958) reported that children with lead toxicity had lower serum iron and iron-binding capacity than children with uncomplicated nutritional anemia. However, this study involved small numbers of subjects and was poorly controlled. Angle et al. (1976) found that administration of oral iron to children with mild iron deficiency appeared to increase lead absorption. However, these investigators were observing effects of increasing levels of dietary iron for children having elevated blood lead concentrations at the start of iron therapy. Accordingly, it is not surprising that iron therapy was not associated with decreases in blood lead concentration, as the role of iron appears to be in the prevention of accumulation of body burden of lead rather than stimulation of lead excretion (Figs. 3 and 4).

The role of nutrients other than calcium in human susceptibility to lead toxicity is not well documented. Although there may be specific metabolic interactions between lead and nutrients, the nutritional effects on human susceptibility to lead toxicity are likely part of a broad sociologic and cultural pattern. It is suggested that the pattern of families being unable to cope with their multiple financial, social, cultural and emotional problems may influence the development of pica and lead poisoning. It is likely that these same factors interfere with the ability of the family to provide diets that contain adequate levels of nutrients.

## V. EFFECTS OF LEAD ON METABOLISM OF NUTRIENTS

Relatively little investigation has been directed to the effects of lead on absorption or metabolism of nutrients. In the broadest sense, the adverse effects of lead on nutrient metabolism were revealed by failure of lead-treated rats to gain weight comparable to control animals, although food intakes were similar (Mahaffey and Goyer, 1972 and 1973).

Addition of lead acetate to the diet for 7 weeks decreased the absorption of sodium, glucose and the amino acids glycine, lysine and phenylalanine from the gastrointestinal tracts of rats (Wapnir et al., 1977). *In vivo* techniques were used to estimate absorption. Sodium absorption in lead-exposed animals was reduced to less than half the transport level for control animals. In lead-treated animals, glucose absorption decreased between 10 and 31%; reduction in absorption of amino acids averaged approximately 20%. Intestinal mucosal ($Na^+K^+$)-ATPase was concomitantly lower in lead-intoxicated rats; however, the following enzyme concentrations were not altered in the intestinal mucosa of lead-poisoned rats: fructose-1,6-diphosphatase, succinic dehydrogenase, pyruvate kinase and tryptophan hydroxylase.

Lead-induced changes in calcium metabolism have also been observed. Doses of lead acetate greater than 20 $\mu$g/day decreased the transfer of calcium across the dueodenal wall in rats, as determined by everted gut sac methods (Gruden, 1975). Doses of lead up to 200 $\mu$g/day continued to reduce calcium transfer; however, above 200 $\mu$g/day no further decreases occurred. Similar results were seen for whole body retention of $^{47}Ca$ when measured 3 days after po doses of $^{47}Ca$. Animals used in these experiments were not on chronic lead administration, but received lead for only 7 days before measurement of $^{47}Ca$ retention. The diets fed to the animals in the experiments of Gruden contained high levels of calcium (1.1% of diet) and approximately the recommended levels of phosphorus (0.65% of diet).

Lead also may alter serum calcium and phosphorus concentrations. Serum calcium was reduced and serum phosphorus was moderately elevated to statistically significant levels ($p = 0.01$ and 0.001, respectively) by lead in rats fed 200 ppm lead for 70 days (Mahaffey-Six and Goyer, 1970). Kidney calcium content was dramatically increased ($p = 0.002$) when lead was fed with a low calcium diet. Because renal lead content and the percentage of rats with renal intranuclear inclusion bodies were both increased, the elevation in renal calcium content may simply be secondary to generalized renal damage.

Addition of lead to the diet causes varying effects on tissue levels of a number of nutritionally required metals. Cerklewski and Forbes (1977) found that 200 ppm lead did not consistently increase or decrease liver copper concentration or serum ceruloplasmin levels when 0, 5 or 200 ppm copper diets were fed. A similar absence of influence of lead on tissue copper content and on ceruloplasmin was observed by Klauder and Petering (1977). Lead, however, consistently lowered kidney selenium levels (Cerklewski and Forbes, 1976b). Kidney selenium was considered a good indicator of selenium status of the rat, as renal selenium content increased with increasing concentrations of selenium in the diet.

Liver iron was unaffected, but kidney iron was significantly lowered by the addition of 200 ppm lead to both iron-deficient and iron-adequate diets (Mahaffey-Six and Goyer, 1972). Liver iron concentration is usually considered a reliable index of nutritional status for iron. Exposure to lead is likely to produce other alterations in iron metabolism. Decreases in heme synthesis reduce the utilization of iron, whereas the hemolytic component of lead-induced anemia could be associated with increases in iron entering the reticulo-endothelial system. Wibowo et al. (1977) found a negative correlation between serum ferritin and whole blood lead concentration in male, but not female, adults. Kochen and Greener (1975a) reported binding of lead by serum transferrin, while Kochen and Greener (1975b) and Russell (1970) observed binding of lead by ferritin in liver, kidney and spleen. Lead has a variety of effects on metabolism of iron and on enzymes related to iron metabolism. Saturation of total iron-binding capacity markedly suppressed uptake of plasma lead by red blood cells and increased uptake of lead by liver (Kochen and Greener, 1975a). Lead does not interfere with incorporation of iron into red cell precursors; however, transfer of nonheme iron into hemoglobin was nearly completely blocked (Lothe and Falbe-Hansen, 1963). These authors suggested that the actual blockage of iron transport by lead occurred between the ferritin fraction and iron compounds in the mitochondria. Lead toxicity causes a sideroblastic anemia. Jensen et al. (1965), using electron microscopy, have shown that in lead intoxication red cell precursors contain many grossly swollen mitochondria with ferritin particles and ferruginous miscelles accumulating in the mitochondria as well as scattered in the cytoplasm. Jensen et al. (1965) also demonstrated that nonheme iron is associated with cellular basophilic stippling induced by lead. Sassa (1978), in a review of effects of lead on heme metabolism, notes that lead interferes with utilization of iron for heme formation in the mitochondria, producing excessive accumulation of iron and destruction of the mitochondria. Borova et al. (1973) and Ponka and Neuwirt (1975) indicate that accumulation of iron in the mitochondria is due to increased uptake of iron from the cytoplasm as well as decreased iron utilization.

Combined feeding of high concentrations of lead (1000 ppm) and zinc (4000 ppm) to Yorkshire pigs reduced soft tissue zinc content (Hsu et al., 1975). However, the zinc content of bone was actually elevated when the diet was low in calcium (Table 11). Changes in bone zinc concentrations are not simply parallel to alteration in bone mineral content. Bone mineral content is reduced by either zinc or lead, with the greatest reduction accompanying high dietary levels of both metals.

Lead also alters the metabolism of vitamins. Pal et al. (1975) investigated ascorbic acid metabolism in lead-intoxicated rats. The rat, unlike humans or other primates, can synthesize ascorbic acid *in vivo* and does not require a dietary source of ascorbate. In liver ascorbic acid, synthesis was stimulated by lead intoxication. Kidney concentrations of ascorbic acid were reduced from 19.8 ± 1.2 to 12.6 ± 1.0 mg/100 g of tissue in lead-intoxicated rats. This decrease appeared to be due to inhibition of the renal enzymes L-gulonate dehydrogenase and decarboxylase, as L-xylulose was also reduced in renal tissue of lead-poisoned rats.

## VI. NUTRITIONAL DEFICIENCIES IN THE UNITED STATES POPULATION

Although the prevalence of classic diseases of protein and vitamin deficiencies is low in the United States, Canada and Western European countries, ingestion of diets containing less than adequate amounts of nutrients is not infrequent among young children, pregnant or lactating women and elderly individuals. Circumstances associated with inadequate dietary intake usually are combinations resulting either from high nutrient requirements relative to total caloric requirements, as for pregnant or lactating women and elderly persons, or from particular patterns of food consumption, as with young children who reject meat, or with various racial groups, that

**Table 11.** Effects of variation in dietary zinc, lead and calcium on femur zinc concentration, femur ash percentage and femur ash content

| Diet | Femur zinc (ppm) | Femur ash (%) | Femur (mg ash/cc bone) |
|---|---|---|---|
| 0.7% Ca | 403* | 19.023 | 227.67 |
| 0.7% Ca + Zn† | 1113 | 15.927 | 187.33 |
| 0.7% Ca + Zn† + Pb‡ | 1360 | 14.467 | 170.33 |
| 1.1% Ca | 290 | 24.477 | 310.67 |
| 1.1% Ca + Zn† | 573 | 25.487 | 324.67 |
| 1.1% Ca + Zn† + Pb‡ | 633 | 23.127 | 291.67 |

* Mean
† In diet, 4000 ppm
‡ In diet, 1000 ppm
(Hsu et al., 1975)

limit consumption of dairy products. In all these groups, low income greatly increases the likelihood of inadequate nutritional intake. One result of limited purchasing ability is often a reduction in the types and variety of foods purchased, so that achievement of a nutritionally balanced diet becomes more difficult.

The nutrients most commonly found to be ingested in less than recommended quantities in the United States are iron, calcium, zinc, ascorbic acid, folic acid and vitamin A. The specific nutrients in shortest supply vary with age, sex, socioeconomic status and geographic location of the population surveyed. A number of excellent reviews have been published on the methodology for carrying out evaluation of nutritional status and dietary intake of nutrients.

### A. Nutritional Status of Infants and Children

Within the past decade, three major surveys of nutritional status of infants and children have been conducted in the United States; populations evaluated and characteristics are shown in Table 12. A number of regional and local studies of nutritional status in children have also been reported and are summarized in a general review by Owen and Lippman (1977), from which Table 13 is derived. This review provides detailed information on the nutritional status of children. Discussion is limited to dietary and biochemical data, because nutrient lead interactions occur with very mild nutritional deficiences.

The nutrient most commonly deficient in the diets of infants and young children is iron. Iron deficiency and iron-deficiency anemia are the most common nutritional deficiencies of infants and young children, especially those from low income groups. The highest prevalence of iron deficiency occurs in children under 5 years of age, particularly among children under 2 years. Quality and quantity of protein intakes for children are usually ade-

**Table 12.** National surveys of nutritional status in infants and children

| Survey | Dates | Population |
|---|---|---|
| Preschool Nutrition Survey | 1968–1970 | Cross-sectional; 3400 children between 1 and 6 years of age; 36 states and District of Columbia |
| Ten State Nutrition Survey | 1968–1970 | 10 states; heavy sampling of poor and near-poor income groups; 3700 children less than 6 years of age |
| Health Assessment and Nutrition Evaluation Survey I | 1971–1974 | Sample population selected as representative of United States as a whole; 3500 children between 1 and 18 years in 1971–1972 half sample |

quate. Overall, children tend to have adequate serum vitamin A levels, although in all three national surveys black children had consistently lower mean or median plasma vitamin A levels than did white children (Owen and Lippman, 1977). Spanish-American children in the southcentral and southwestern states had a greater prevalence of unacceptably low plasma vitamin A concentrations than either whites or blacks. The percentages of children with unacceptably low plasma retinol (vitamin A) concentrations were 1.5, 10 and 30 to 50% for white, black, and Spanish-American children, respectively (Owen and Lippman, 1977). In all three national surveys, approximately one-third of lower income children reported dietary intakes of ascorbic acid less than one-half of Recommended Dietary Allowances for this vitamin. Although median calcium intakes were adequate, between 20 and 30% of black children and 10 to 15% of white children consumed diets containing less than 400 mg calcium/day, which is about half the Recommended Dietary Allowance for these age groups. Local surveys indicate that less than adequate zinc intakes and biochemical changes indicative of zinc deficiency occur in children from low income groups (Hambridge, 1977).

### B. Nutritional Status of Adults

Both the Ten State Nutrition Survey (1968–1970) and the first Health Assessment and Nutrition Evaluation Survey (1971–1972, HANES I) indicated small but consistent clinical evidence of nutritional deficiencies. The most prevalent finding in the Ten State Survey was anemia. As many as one-fifth of the women of child-bearing age had hemoglobin levels considered to be low; the highest incidence of anemia was found among black women, followed by Spanish-American women and then white women. Women past child-bearing age and adult men had a very low prevalence of anemia. Symptoms of vitamin C deficiency were found in all adult populations, especially in those over 60 years of age. Black populations had a higher prevalence than either Spanish-Americans or whites for ages below 60 years, but differences were not significant above age 60, where about 9% of all groups were deficient. According to the HANES I survey, a low rate of vitamin A deficiency (less than 2%) existed for the entire population of adults except for blacks over 60 years of age. In this group, the deficiency rate was nearly 9%. The HANES I survey also showed that niacin deficiency is still evident, primarily among black populations; however, the symptoms were those of only marginal deficiencies. Similarly, thiamine deficiencies were frequently found among black populations, but again the symptoms were those indicating only a marginal deficiency.

Many older populations showed signs of deficiencies which occurred during infancy and childhood, but are likely no longer a problem. Vitamin D was the most prominent example; bowlegs and knock-knees were frequently encountered. All adult populations showed evidence of calcium-

**Table 13.** Selected regional and local studies of nutritional status of children

| Sample | Location | Findings and comments |
|---|---|---|
| 170 Mexican-American children, 0–66 months of age | San Ysidro, California | Some children with low intake of energy, vitamin C, niacin and iron; vitamin A intakes good; plasma vitamin A concentrations normal; iron deficiency prevalent; some youngsters had low plasma vitamin C. |
| 41 Black inner city preschoolers | Philadelphia | Comparable intake of iron by day care (0.7 mg/kg/day) and by non-day care (0.5 mg/kg/day). Anemia more prevalent among non-day care than among day care children. |
| 50 Black children of low income | Nashville | Approximately 10% of children were anemic; 15% had low plasma vitamin A. |
| 281 Mixed racial children, 2–3 years of age | Honolulu | Intakes of calcium, vitamin A, ascorbic acid and riboflavin low for some groups (inversely related to income); between 5 and 10% were anemic. |
| 168 Mexican-American children, 6–9 years of age | Coachella Valley California | 52% had hemoglobin <10 g/dl. |
| 178 Black preschoolers | South Carolina | 25% anemia (hemoglobin <10 g/dl); 50% iron deficient (serum iron <40 μg/dl) |
| 115 White children, 2–6 years of age | Minnesota | Dietary survey only; some low intakes of iron and vitamin C. |
| 109 Black and white children, 4 months-5 years of age | Michigan | 10% with severe iron-deficiency anemia; 13% mild iron-deficiency anemia; 23% with nonanemic iron deficiency. |
| 250 Mexican-American and White children, 0–17 years of age | Denver | Evidence of low zinc stores among 45% of children under 4 years of age (based on hair zinc determinations). |
| 843 Eskimos, 2–6 years of age | Alaska | Low levels of intake of calories, calcium and vitamin C among some children; protein intakes were generally high. |
| 70 White children, 0–18 months of age | Seattle | More than half of the nonanemic (hemoglobin >11 g/dl) infants studied had iron deficiency (transferrin saturation <15%). |
| 36 Black children, 4–10 months of age | South Carolina | Approximately 10% of infants had low levels of albumin and plasma vitamin C. |
| 60 Families | Iowa and North Carolina | Dietary survey only; adequate intake of protein, calcium and vitamins C, $B_1$ and $B_2$, as well as iron. |

*continued*

**Table 13.**—*continued*

| Sample | Location | Findings and comments |
|---|---|---|
| 113 Indian families | Fort Belknap, Montana | Some children with low intakes of vitamins A and C, as well as of calcium. Approximately one-third of children had low levels of hemoglobin, plasma vitamin A and erythrocyte riboflavin. |
| 40 White children | Nebraska | Urinary thiamin and riboflavin excretion varied with income; approximately 10% of children had low levels of hemoglobin. |
| 160 Mexican-American children, 0–9 years of age | Lower Rio Grande Valley, Texas | Clinical signs of vitamin A and vitamin D deficiencies; one-third of children with low level of vitamin A in plasma; 10% anemic. |
| 100 Low income children, 0–14 years of age | New York City | No clinical signs of riboflavin deficiency, although 10% had biochemical evidence of deficiency. |
| 386 White children, 3–10 years of age | Tennessee | Many children had low intakes of iron (6–18%), vitamin C (20%), energy (30%) and vitamin A (50%). |

(Owen and Lippman, 1977).

phosphorus imbalance; however, the prevalence of this condition was considerably higher in black women than in other adult groups. Finally, some evidence of protein deficiency was also found for blacks among adult populations studied in HANES I; the largest percentage, approximately 80%, was reported in the 45- to 59-year-old-group.

### C. Unsurveyed Nutrients

Clinical studies for deficiencies of certain nutrients have not been done. The prevalence of deficiencies of nutrients such as copper, zinc or vitamin E in the general population is not known. Frank deficiency of these nutrients is thought to be exceedingly rare in persons not having metabolic defects or receiving parenteral nutrition. However, biochemical changes indicating less than optimal nutritional status have been reported for these nutrients. For example, Levander et al. (1975) cite unpublished data of R. Hepner showing that 2.5 to 6.1% of clinically well inner-city children between 3 and 14 years of age had serum alpha-tocopherol levels below 0.5 mg/100 ml, the usual criterion of low vitamin E intake.

## VII. INFLUENCE OF NUTRITION ON HEALTH EFFECTS ATTRIBUTED TO LEAD TOXICITY

High concentrations of dietary lead may interfere with the metabolism of a variety of nutrients, as discussed in Section V. In a general way, animals fed high lead diets show this effect as failure to gain weight when compared with control animals. Feeding high lead diets may also reduce food consumption by experimental animals, producing impaired weight gain (Morrison et al., 1975b; Maker et al., 1975). Water consumption usually parallels food intake. These changes in food and water intake determine the quantity of lead consumed. If lead is administered at a set quantity rather than on a body weight basis, failure to gain weight will influence the dose ($\mu$g/kg body weight) the animal receives.

Interactions between lead and various nutrients can affect the activity of a specific enzyme, *e.g.*, ALAD, or an organ system, *e.g.*, heme synthesis. In addition to these specific examples, gross failure of animals to grow or loss of body weight also complicates interpretation of changes produced by lead. An example of this occurs in studies of behavioral or neurologic disorders attributed to lead. Nutritional deficiency, usually of sufficient severity to result in weight loss or growth retardation, produced behavioral abnormalities in several species, including humans, dogs, pigs and rodents (Barnes et al., 1967; Dobbing, 1967; Stewart and Platt, 1967). Deficiencies of zinc (Halas and Sandstead, 1975), or iron (Pollitt and Leibel, 1976) produced behavioral abnormalities in animals fed these diets.

Turning to studies of behavioral or neurologic changes induced by high lead diets, examples of the general effects of lead can be observed. Interactions between lead, nutrition and behavioral and psychologic changes are reviewed by Michaelson in this book. Pentschew and Garro (1966) fed high concentrations of lead (4% lead carbonate) to lactating rat dams and evaluated neurologic effects in young rats suckled by mothers fed high lead diets. Specifically, 90% of the young animals developed paraplegia within 23 to 29 days of feeding. Rosenblum and Johnson (1968) performed similar studies in the mouse with comparable results. However, these investigators noted that the general condition of suckling mice of mothers fed high lead diets was so abnormal that it could be suggested that the neurobiologic changes observed were remote expressions of lead effects and were more directly related to some fundamental alteration(s) in body metabolism. These authors suggested that "lead in the maternal diet may have caused a more complex change in the composition of maternal milk than the simple addition of lead to the product. Hence, the alteration in neonatal development and structures may reflect dietary abnormalities other than the presence of lead."

When exposure to high concentrations of lead produces either weight loss or failure to gain weight, separating the response attributed to lead into a component due to nutrition and a component due to lead is very difficult. However, a few careful investigators have attempted to evaluate the nutritional effects. Maker et al. (1975) have studied growth and food and water consumption, as well as pathological and behavioral parameters, in two strains of mice fed high concentrations of lead. Dams reduced their food consumption in response to diets containing 0.5% lead or more. At 30 days of age, litters fed diets containing 0.08% or 0.16% of lead or more had significantly lower body and brain weights than did controls. At 30 days of age, body weights of pups on the 0.4% lead diet averaged only two-thirds of normal and their brain weights were 10% below controls. Food consumption of the pups was not directly studied and quantity of milk consumed was not known. Neither the C57 nor Swiss-Webster mice would accept drinking water containing 0.5% lead while lactating; those animals receiving water containing 0.5% lead sharply reduced their food intake. In young of these litters, growth was retarded to the same degree as in young of dams receiving similar concentrations of lead in the diet. Reduction of number of pups per litter from six to three did not improve weight gain; this indicated either that the effects were not due to reduced milk production, as suggested by the authors, or that the nutritional and/or metabolic effects were so severe that simply reducing litter size did not alleviate the effect of lead on the quality or quantity of milk produced or the metabolic effects of lead on growth in mouse pups. Pups of control dams pair-fed so that their food intakes were equal to those of dams exposed to lead had body weights intermediate between pups of *ad libitum* fed controls and lead-exposed dams. This indicated that food restriction of the dam is an important factor in the reduced growth of lead-exposed pups, but that other factors were also important. Lead exposure may reduce the efficiency of conversion of nutrients ingested from the diet into nutrients present in milk, may result in abnormal milk composition or may have direct metabolic action on the pups.

It is unlikely that nutritional deficiencies can account for all the neuropathologic changes observed in lead-exposed animals, because comparison with severely underfed animals reveals changes in the lead-exposed animals not present in the nutritionally deprived. Such analysis of primary and secondary effects of high lead exposures might be of minimal importance if the lead-related effects were produced in animals fed levels of lead low enough to result in no reduction in total body weight or brain weight of the pups. However, when Hastings et al. (1977) fed diets containing sufficiently low concentrations of lead to rat dams so that growth of the pups was not retarded, these pups did not show hyperactivity and aggressiveness or impaired ability to discriminate visually, which has been reported to be due to lead. Analysis of blood lead concentrations of the pups showed blood lead concentrations of 35 to 50 μg/dl whole blood. Considerable differences of opinion exist concerning the neurologic-behavioral consequences of elevation of blood lead of this magnitude in humans. The studies of Hastings et al. (1977) are of substantial interest and, if repeated, will further establish the association between behavioral changes, lead and nutrition.

## VIII. SUMMARY AND CONCLUSIONS

Several generalizations about interactions of lead and nutrition can be made.

1. Increases in lead toxicity and tissue retention of lead follow ingestion of diets inadequate in certain nutrients, *e.g.*, calcium, phosphorus, iron.
2. Diets containing nutrients at higher levels than recommended on a nutritional basis, *e.g.*, elevated levels of calcium or vitamin D, usually do not decrease lead toxicity. An exception to this generalization appears to be iron. Ingestion of high levels of many nutrients may result in toxicity from the nutrient, *e.g.*, selenium, zinc or copper. Ingestion of high levels of other nutrients may result in increased tissue deposition of lead, *e.g.*, fat or protein.
3. The role of nutrition is to prevent the accumulation of an elevated body burden of lead rather than to treat lead intoxication.

Improvement of nutritional status and control of environmental exposure to lead should be seen as separate public health goals. Ideally, standards on environmental exposure to lead should protect nutritionally deprived persons as well as individuals having optimal nutritional status. However, dose-response information on human lead toxicity, expecially for children, is not precisely documented and the nutritional status of children for whom case reports exist usually is not known. At the present time, environmental standards which protect nutritionally deprived persons are at best speculative.

Clinically, lead toxicity and ingestion of diets inadequate in such nutrients as iron and calcium occur in the same population group: economically poor preschool children. The economic and familial circumstances associated with lead toxicity are also liable to increase the likelihood of a child ingesting an inadequate diet. Ingestion of an inadequate diet will increase the adverse effects of lead exposure.

Improvement of nutritional status is in no way a substitute for decreasing environmental lead exposure; however, the effectiveness of environmental lead control can be optimized if children are well nourished. Further, optimal nutritional status is protective against a variety of toxic substances.

## REFERENCES

Angle, C.R., Stelmark, K.L., and McIntire, M.S. (1976). D.D. Hemphill, (Ed.), In "Trace Substances in Environmental Health," vol. IX, p. 377. University of Missouri.

Aub, J.C., Fairhall, L.T., Minot, A.S., and Reznikoff P. (1926). "Lead Poisoning; Medicine Monographs, Vol. 7. Williams & Wilkins Co., Baltimore, Md.
Baernstein, H.D., and Grand, J.A. (1942). *J. Pharm. Exp. Therap. 74:*18.
Barltrop, D., and Khoo, H.E. (1975). *Postgrad. Med. J. 51:*795.
Barltrop, D., and Khoo, H.E. (1976). *Sci. Total Environ. 6:*265.
Barnes, R.J., Moore, A.U., Reid, I.M., and Pond, W.G. (1967). N.S. Scrimshaw and J.E. Gordon (Eds), In "Malnutrition, Learning and Behavior," p. 203. MIT Press, Cambridge, Mass.
Barry, P.S.I. (1978). J. Nriagu (Ed.), In "Biogeochemistry of Lead," p. 124. Elsevier/North Holland Biomedical Press, Amsterdam.
Bartlett, R.S., Rousseau, J.E., Frier, H.I., and Hall, R.C. (1974). *J. Nutr. 104:*1637.
Barton, J.C., Conrad, M.E., Harrison, L., and Nuby, S. (1978a). *J. Lab. Clin. Med 91:*366.
Barton, J.C., Conrad, M.E., Nuby, S., and Harrison, L. (1978b). *J. Lab. Clin. Med. 92:*536.
Border, E.A., Cantrell, A.C., and Kilroe-Smith, T.A. (1976). *Brit. J. Indust. Med. 33:*85.
Borová, J., Poňka, P., and Neuwirt, Jr. (1973). *Biochem. Biophys. Acta 320:*143.
Borsook, H., Fischer, E.H., and Keighley, G. (1957). *J. Biol. Chem. 229:*1059.
Buck, W.B. (1970). *J. Amer. Veterin. Med. Associ. 156:*1468.
Calvery, H.O., Laug, E.P., and Morris. H.J. (1938). *J. Pharm. Exper. Therap. 64:*365.
Carr, T.E.F., Nolan, J., and Durkovic, A., (1969). *Nature 224:* 1115.
Cerklewski, F.L., and Forbes, R.M. (1976a). *J. Nutr. 106:*689.
Cerklewski, F.L., and Forbes, R.M. (1976b). *J. Nutr. 106:*778.
Cerklewski, F.L., and Forbes, R.M. (1977). *J. Nutr. 107:*143
Der, R., Fahim, Z., Hilderbrand, D., and Fahim, M. (1974). *Res. Communic. Chem. Pathol. Pharmacol. 9:*723.
DeRosa, R. (1954). *Acta Vitaminol. 8:*167.
Diez-Edwald, M.-Weintraub, L.R., and Crosby, W.H. (1968). *Proc. Soc. Exp. Biol. Med. 129:*448.
Dobbing, J. (1967). N.S. Scrimshaw and J.E. Gordon (Ed.), In "Malnutrition, Learning and Behavior," p. 203. MIT Press, Cambridge, Mass.
Eisenstein, R., and Kawanoue, S. (1975). *Amer. J. Pathol. 80:*309.
Fine, B.P., Barth, A., Shaffet, A., and Lavenhar, M.A. (1976). *Environ. Res. 12:*224.
Finelli, V.N., Klauder, D.S., Karaffa, M.A., and Petering, H.G. (1975). *Biochem. Biophys. Res. Commun. 65:*303.
Freeman, R. (1973). *Clin. Toxicol. Bull. 3:*75.
Ganther, H.E., Goudie, C., Sunde, M.L., Kopecky, M.I. Wagner, P., Oh, S.-H, and Hoekstra, W.G. (1972). *Science 175:*1122.
Gontzea, I., Sutzesco, P., Cocora, D., and Lungu, D. (1964). *Archives Sci. Physiol. 18:*211.
Goyer, R.A., and Cherian, M.G. (1978). *Fed. Prac. 37:*938.
Goyer, R.A. (1978). *J. Lab. Clin. Med. 91:*363.
Gruden, N. (1975). *Toxicol. 5:*163.
Halas, E.S., and Sandstead, H.H. (1975). *Pediat. Res. 9:*94.
Hambridge, K.M. (1977). *Pediat. Clin. N. Amer. 24:*95.
Hardy, H.L., Elkins, H.B., Ruotolo, BPW, Quinby, J., and Baker, W.H. (1954). *J. Amer. Med. Assoc. 154:*1171.
Harrison, G.E., Carr, T.E.F., Sutton, A., and Humphries, E.R., (1969). *Nature 224:*1115.

Hastings, L., Cooper, G.P. Bornschein, R.L., and Michaelson, I.A. (1977), *Pharmacol. Biochem. Behavior. 7:*37.
Holmes, H.N., Campbell, K., and Amberg, E.J. (1939). *J. Lab. Clin. Med. 24:*1119.
Hsu, F.S., Krook, L., Pond, W.G., and Duncan, J.R. (1975). *J. Nutr. 105:*112.
Hsu, F.S., Krook, L., Shively, J.N., and Duncan, J.R. (1973). *Science 181:*447.
Hunter, D. (1975). "The Diseases of Occupations," Ed. 5, p. 275. The English Universities Press, Ltd., London.
Jensen, W.N., Moreno, G.D., and Bessis, M. (1965). *Blood 25:*933.
Kao, R.L.C., and Forbes, R.M. (1973). *Arch. Environ Health 23:*31.
Klauder, D.S., and Petering, H.G. (1977). *J. Nutr. 107:*1779.
Kline, T.S. (1960). *Amer. J. Diseases of Child. 99:*48.
Kobayaski, A., Kawai, S., Ohbe, Y., and Nagashima, Y. (1975). *Amer. J. Clin. Nutr. 28:*681.
Kochen, J., and Greener, Y. (1975a). *Pediatr. Res. 9:*323.
Kochen, J., and Greener, Y. (1975b). *Pediatr. Res. 9:*323.
Kostial, K., and Vouk, V.B. (1957).*Brit. J. Pharmacol. Chemothera. 12:*219.
Kostial, J., Simonovic, I., and Pisonic, M. (1971). *Nature (London) 233:*564.
Ku, Y., Alverez, H.G., and Mahaffey, K.R. (1978). *Bull. Environ. Sci. Technol. 20:*561.
Lederer, L.B., and Bing, F.C. (1940). *JAMA 114:*2457.
Levander, O.A. (1979). Environ. Health Perspective. In press.
Levander, O.A., Morris, V.C., and Ferretti, R.J. (1977a). *J. Nutr. 107:*378.
Levander, O.A., Fisher, M., Morris, V.C., and Ferretti, R.J. (1977b). *J. Nutr. 107:*1828.
Levander, O.A., Morris, V.C., Higgs, D.J., and Ferretti, R.J. (1975). *J. Nutr. 105:*1481.
Lothe, K., and Falbe-Hansen, I. (1963). *Clin. Sci. 24:*47.
Mahaffey, K.R. (1974). *Environ. Health Perspect. 7:*107.
Mahaffey, K.R., and Banks, T.A. (1974). *Fed. Proc. 32:*267.
Mahaffey, K.R., and Vanderveen, J.E. (1979). *Environ. Health Perspect.* In press.
Mahaffey, K.R., Banks, T.A., Stone, C.L., Capar, S.G., Compton, J.F., and Gubkin, M.H. (1975). In "International Conference on Heavy Metals in the Environment," Symposium Proceedings, Vol. 3, p. 155. Toronto, Canada.
Mahaffey, K.R., Goyer, R.A., and Haseman, J. (1973). *J. Lab. Clin. Med. 82:*92.
Mahaffey, K.R., Smith, C. Tanako, Y., and DeLuca, H.F. (1979). *Fed. Proc. 38:*384.
Mahaffey, K.R., Stone, C.L., and Fowler, B.A., (1976a). *Fed. Proc. 35:*256.
Mahaffey, K.R., Stone, C.L., Banks, T.A., and Reed, J. (1978). M. Kirchgessner (Ed.), In "Proceedings of the Third International Symposium on Trace Element Metabolism in Man and Animals," p. 584. Arbeitskreis fur Tierernahrung-Storschung Weihenstephan.
Mahaffey, K.R., Treloar, S., Banks, T.A., Peacock, B.J., and Parekh, L.E. (1976b). *J. Nutr. 107.*
Mahaffey-Six, K., and Goyer, R.A. (1970). *J. Lab. Clin. Med. 76:*933.
Mahaffey-Six, K., and Goyer, R. A. (1972). *J. Lab. Clin. Med. 79:*128.
Maines, M.D., and Kappas, A. (1977). *Science 198:*1215.
Maker, H.S., Lehrer, G.M., and Silides, D.J. (1975). *Environ. Res 10:*76.
McNiff, E.G., Cheng, L.K., Woodfield, H.C., and Fung, H.L. (1978). *Res. Commun. Chem. Pathol. Pharmacol. 20*(1):131.
Meredith, P.A., Moore, M.R., and Goldberg, A. (1974). *Biochem. Soc. Trans. 2:*1243.

Meredith, P.A., Moore, M.R., and Goldberg, A. (1977). *Biochem. J. 166:*531.
Mooty, J., Ferrand, C.F., and Harris, P. (1975). *Pediatr. 55:*636.
Morrison, J.N., Quarterman, J., and Humphries, W.R. (1974). *Proc. Nurtr. Soc. 33:*88A.
Morrison, J.N., Quarterman, J., and Humphries. W.R. (1977). *J. Comp. Pathol. 87:*417.
Morrison, J.N., Quarterman, J., Humphries, W.R., and Mills, C.F., (1975a). *Proc. Nutr. Soc. 34:*77A.
Morrison, J.H., Olton, D.S., Goldberg, A.M., and Silbergeld, E.K. (1975b). *Develop. Psychobiol, 8:*389.
Myerson, R.M., and Eisenhauer, J.H. (1963). *Amer. J. Cardiol. 11:*409.
National Center for Health Statistics, Advanced Data Number 46. (1979). U.S. Department of Health Education and Welfare.
Owen, G., and Lippman, G. (1977). *USA Pediatr. Clin. N. Amer. 24:*211.
Pal, D.R., Chatterjee, J., and Chatterjee, G.C. (1975). *Inter. J. Vit. Nutr. Res. 45:* 429.
Parizek, J., Ostudalova, I., Kalouskova, J., Babicky, A., and Benes, J., (1971). W. Mertz and W.E. Cornatzer (Eds.), In "Newer Trace Elements in Nutrition," p. 85. Marcel Dekker, New York.
Parr, D.R., and Harris, E.J. (1976). *Biochem. J. 158:*289.
Pecora, L. Silvestron, A., and Brancaccio, A. (1966). *Panminerva Medica 8:*284 (Kettering Abstr. 42, 1967).
Pentschew, A., and Garro, F. (1966). *Acta Neuropathol. 6:*266.
Pillemer, L., Seifter, J. Keuhn, A.O., and Ecker, E.C. (1940). *Amer. J. Med. Sci. 200:*322.
Piomelli, S. (1977). *Clin. Chem. 23:*264.
Pollitt, E., and Leibel, R.L. (1976). *J. Pediatr. 88:*372.
Pollock, S., George, I.N., Rita, R.C., Kaufman, R.M., and Crosby, W.H. (1965). *J. Clin. Invest. 44:*1470.
Poňka, P., and Neuwirt, G. (1975). *New Engl. J. Med. 293:*406.
Quarterman, J., and Morrison, J.N. (1975). *Brit. J. Nutr. 34:*351.
Quarterman, J., Morrison, J.N., and Humphrics, W.R. (1976a). *Environ. Res. 12:*180.
Quarterman, J., Humphries, W.R., and Morrison, J.N. (1976b). *Proc. Nutr. Soc. 35:*33A.
Quarterman, J., Morrison, J.N., and Humphries. W.R. (1977). *Proc. Nutr. Soc.:* 104A.
Rabinowitz, M., Wetherill, G., and Koyple, J. (1975). In "Proceedings," p. 361. Ninth Annual Conference on Trace Substances in Environmental Health, University of Missouri, Columbia, Mo.
Ragan, H.A. (1977). *J. Lab. Clin. Med. 90:*700.
Rapaport, M., and Rubin, M.D. (1941). *Amer. J. Dis. Child. 61:*245.
Rastogi, S.C., Calusen, J., and Shivastava, (1976). *Toxicology 6:*377.
Read, J.L., and Williams, J.P. (1952). *Amer. Heart J.* 44:797.
Rosen, J.R., and Wexler, E.E. (1977). *Biochem. Pharmacol. 26:*650.
Rosenblum, W.I., and Johnson, M.G. (1968). *Arch. Pathol. 85:*640.
Russell, J.A. (1970). *Bull. Environ. Contamin. Toxicol. 5:*115.
Sassa, S. (1978). G.V.R. Born, O. Eichler, A. Farah, H. Herken, and A.D. Welch, (Eds.), In "Handbook of Experimental Pharmacology," Vol. 44. F. DeMatteis and W.N. Aldridge (Eds.), In"Heme and Hemoproteins." Springer-Verlag, New York.

Shields, J.B., and Mitchell, H.H. (1941). *J. Nutr. 21:*541.
Silbergeld, E. (1973). *Fed. Proc. 32:*275.
Smith, C.M., DeLuca, H.F., Tanaka, Y., and Mahaffey, K.R. (1978). *J. Nutr. 108:*843.
Sobel, A.E., Wexler, I.B., Petrovsky, D.D., and Kramer, B. (1938). *Proc. Soc. Exp. Biol Med. 38:*435.
Sobel, A.E., Yusha, H. Peters, D.D., and Kramer, B. (1940). *J. Biol. Chem. 132:*239.
Sorrell, M., Rosen, J.F., and Roginsky, M.R. (1977). *Arch. Environ. Hlth. 32:*160.
Stewart, R.J.C., and Platt, B.S. (1967). N.S. Scrimshaw and J.E. Borden (Eds.), In "Malnutrition, Learning and Behavior," p. 168. MIT Press, Cambridge, Mass.
Stone, C.L., and Soares, J.H. (1976). *Poultry Sci. 55:*341.
Ten State Nutrition Survey, 1968–1970. (1972). V. Dietary, U.S. Dept. HEW Atlanta, Georgia.
Tephly, T.R., Wagner, G. Sedman, R., and Piper, W. (1978). *Fed. Proc. 37:*35.
Valberg, L.S., Ludwig, J., and Olatunbosun, D. (1969). *Gastroenterology 56:*241.
Wapnir, R.A., Exeni, R.A., McVicar, M., and Lifshitz, F. (1977). *Pediatr. Res. 11:*153.
Watson, J., Decker, C., and Lichtman, H.C. (1958). *Pediatrics 21:*40.
Waxman, H.S., and Rabinovitz, M. (1966). *Biochem. Biophys. Acta 129:*369.
Wibowo. A.A.E., P. Del Castilho, Herber, R.F.M., and Zielhuis, R.L. (1977). *Int. Arch. Occup. Environ. Hlth. 39:*113.
Willoughby, R.A., Thirapatsakun, T., and McSherry, B.J. (1972a). *Amer. J. Vet. Res. 33:*1165.
Willoughby, R.A., MacDonald, E., Mesherry, B.J., and Brown. G. (1972b). *Can. J. Comp. Med. 36:*348.
Wolfe, W. R., Holden, J., and Greene, F.E. (1977). *Fed. Proc. 36:*1175.
Yamamoto, T., Yamaguchi, M., and Suketa, Y. (1974). *Toxicol. Appl. Pharmacol. 27:*204.
Ziegler, E.E., Edwards, B.B., Jensen, R.L., Mahaffey K.R., and Fomon, S.J. (1978). *Pediatr. Res. 12:*2934

# Lead and Other Metals: A Hypothesis of Interaction

*J. Julian Chisolm, Jr.*

**TABLE OF CONTENTS**



## I. INTRODUCTION

Over 50 years ago, Aub et al. (1926) suggested in their classic monograph that the "lead stream" follows the "calcium stream." Even before that, occupational physicians had surmised that milk might have a protective effect in lead workers, so that it was common practice in Great Britain for lead workers to be provided each day with free milk (Hunter, 1978). About 40 years ago, Sobel and others reported that both calcium and phosphorus in adequate amounts reduced the absorption and retention of lead. More recently, Kostial and Vouk (1957) found in the perfused superior cervical ganglion of the cat that lead (5 to 40 $\mu$M/l) blocked ganglionic transmission and reduced the output of acetylcholine. Calcium (10 mM/l) relieved the block caused by lead and restored the output of acetylcholine. These observations have lain fallow until the past decade, during which interest in environmental overexposure to lead has stimulated a renewal of experimental study of interrelationships between lead and calcium.

About a decade ago, Willoughby et al. (1972) fed toxic amounts of zinc and lead separately and in combination to young foals. They were stimulated to carry out this experiment because they and other veterinarians had observed that young foals grazing on contaminated grass in the vicinity of lead-zinc smelters developed zinc poisoning (lameness and "arthritis-like lesions"), but did not develop the clinical syndrome of lead poisoning in the horse (pharyngeal and laryngeal paralysis). In separate experiments, they produced zinc and lead poisoning; however, when the same toxic amounts of zinc and lead were fed together, symptoms of zinc poisoning, but not of lead poisoning, developed, even though the amounts of lead in the tissues were the same as those achieved in the lead-poisoned foals fed excess lead alone. Subsequent experiments in rats (Willoughby and Thawley, 1975) confirmed their initial impression that zinc appears to have a protective effect against the toxicity of lead. It is now known that $\delta$-aminolevulinic acid dehydratase (ALAD) is a zinc-dependent enzyme (Cheh and Neilands, 1973) and that inhibition of this enzyme by lead can be reversed *in vivo* with nontoxic amounts of zinc (Haeger-Aronsen et al., 1976). While this may not account fully for the clinical observations in young foals, it does provide an explanation for at least one type of interaction between lead and zinc. Parenthetically, it may be noted in the original description of ALAD that zinc in high concentration was a more potent inhibitor of ALAD *in vitro* than lead (Gibson et al. 1955).

Bovine pancreatic carboxypeptidase is a zinc-metalloenzyme with both peptidase and esterase activity. When the zinc moiety is replaced with mercury, cadmium or lead, some esterase activity is retained; however, peptidase activity *in vitro* is completely abolished (Coleman and Vallee, 1961). Ulmer and Vallee (1968) found that lead markedly inhibited growth of Rhodopseudomonas spheroides in cultures deficient in iron and that the inhibitory effect could be largely reversed by increasing the iron content of the medium from 1 to 10 $\mu$M. Heme synthesis was also reduced in iron-deficient cultures by manganese and cobalt. the effects of these metals were additive and could be reversed by increasing the iron supplement. They postulated that these metals caused a "conditioned iron deficiency" through their antagonism of its utilization. Trace amounts of chromium are apparently protective against the adverse effects of lead on growth and mortality in rats (Schroeder and Tipton, 1968).

Environmental data collected during the past decade show quite clearly that some populations at high risk for lead toxicity are overexposed not only to lead, but also to other metals (Landrigan et al., 1975; Creason et al., 1975; Dorn et al., 1978). Children residing near ore smelters have been overexposed not only to lead, but also to zinc, copper, cadmium and arsenic. Not only are the concentrations of lead and other toxic heavy metals apparently higher in urban soil and dust than in suburban soil and dust, but also the ratios of essential and nonessential trace elements may vary in such soils and dusts. Workers in the lead trades may be exposed to varying combinations of lead, silver, zinc, antimony and tellurium. Some populations at high risk for lead toxicity may have inadequate dietary intakes of iron, calcium and possibly zinc. These epidemiologic observations point quite clearly to the need for a better understanding of how metals interact in the body.

Mechanisms through which metals may interact have been reviewed by several authors (Vallee and Ulmer, 1972; Nordberg et al., 1978; Magos and Webb, 1978). Metals may interact through synergism, antagonism or joint independent action. They can compete for metal-binding proteins, as well as induce formation of metal-binding proteins. In this chapter, advances in the field of metal-metal interactions involving lead during the past 10 years will be considered; however, other factors which influence the metabolism and toxicity of lead will not be considered, including age, dietary protein, dietary fat, interactions with metalloids (selenium), hormones and vitamins.

## II. EXPERIMENTAL STUDIES

### A. Interactions Affecting Absorption, Retention, Distribution and Excretion of Lead

The effects of calcium (Ca), magnesium (Mg), iron (Fe), zinc (Zn), copper (Cu), cadmium (Cd) and arsenic (As) on the absorption, retention, internal distribution and excretion of lead have been studied by a number of investigators during the past decade. Generally, these studies have taken the form of long term feeding experiments of 10 to 12 weeks in duration, in which growing rats are fed lead (usually in drinking water) and semipurified diets containing the recommended dietary allowances for all minerals and other components except for the metal under study. Usually, a single metal has been varied. Often, some of lead's known adverse metabolic effects are measured and the lead content of the tissues is determined at the end of the experiment. What is actually measured is the retention and distribution of lead within the tissues, although effects of the various dietary manipulations on the absorption of lead are usually inferred. Such experiments are intended as a model for chronic low level overexposure to lead in man.

Using the chronic experimental design just described, Six and Goyer (1970) compared the effects of low (0.1%) and high (0.7%) calcium diet on the absorption, retention and distribution of a constant oral dose of lead (200 $\mu$g Pb/ml of drinking water). This oral dose of lead was the highest amount which in combination with a normal calcium diet had not previously produced evidence of toxicity due to lead. At the deficient dietary level of calcium (0.1%), bone weight was significantly reduced, the lead content of the kidney was increased over 20-fold and the lead content of bone was increased threefold. In the presence of normal calcium diet, the blood lead level was increased from control values with no added lead ($<10$ $\mu$g Pb/dl whole blood) to approximately 50 $\mu$g Pb/dl whole blood, while in the presence of deficient dietary calcium the same oral dose of lead raised the blood lead level to approximately 180 μg Pb/dl whole blood. Fecal lead output was reduced and urinary lead output was increased among rats receiving the low calcium diet, in comparison with those receiving a normal calcium diet. Thus, the experiments indicated that low calcium diet not only enhanced the absorption of lead, but also raised kidney lead and blood lead to a greater extent than it did the level of bone lead. Mahaffey et al. (1973) next extended this experiment by varying the concentration of lead in the drinking water (Table 1). In the presence of normal dietary calcium, 200 μg Pb/ml in drinking water produced a renal lead content of approximately 20 μg/g and the occurrence of inclusions, perhaps the most sensitive histologic indicator of increased lead absorption. However, only 12 μg Pb/ml of drinking water in the presence of low calcium diet was required to produce the same kidney lead content and appearance of inclusions. Subsequently, in similar experiments, Mahaffey et al. (1975) held oral lead dosage constant at 200 μg Pb/ml drinking water and varied dietary calcium stepwise at 0.1, 0.2, 0.3, 0.4 or 0.5% Ca. It was found that the effect of dietary Ca intake on lead metabolism was not linear. Marked increases in femur and kidney lead content were found only at the 0.1% dietary calcium level, a percent which does not meet the physiologic requirement of the rat for calcium. These studies do not take into account the effects of vitamin D, parathyroid hormone and thyrocalcitonin, which also affect both calcium and lead absorption and distribution. Throughout, evidence of inhibition of heme synthesis and the renal effects of lead were apparently related to the concentrations of lead in the affected tissues.

Recently, Barton et al. (1978a), using a different experimental design, have reported findings which extend considerably our understanding of the effects of calcium on the absorption and retention of lead. Male albino rats weighing 200 to 225 g were prepared on calcium-deficient, normal and calcium-replete diets. Just prior to the experiment, food was removed overnight, but water was permitted. Labeled lead ($^{203}Pb$) was injected into isolated intestinal segments. Animals were sacrificed 4 hr later and total body radioactivity was measured. These experiments demonstrated no *direct* effect of calcium-deficient, normal and calcium-replete diets on the percentage of lead absorbed. However, when both calcium and lead were injected intraduodenally a dose-related increase in the percentage of lead absorbed was observed as the concentration of calcium was reduced from 10 mM to no added calcium. Long term experiments with $^{210}Pb$ also showed that the calcium-deficient diet increased lead retention and, by inference, decreased lead excretion. They also isolated two intestinal mucosal calcium-binding proteins, one a low molecular weight vitamin D-induced calcium-binding protein and the other a higher molecular weight calcium-binding protein. While the higher molecular weight protein had a greater affinity for lead

**Table 1.** Kidney lead content and incidence of intranuclear inclusions in relation to dietary calcium and oral lead administration (Mahaffey et al., 1973)

| Dietary group | N | Lead ($\mu$g Pb/g kidney, wet weight)* | Inclusions† | Dietary group | N | Lead ($\mu$g Pb/g kidney, wet weight)* | Inclusions† |
|---|---|---|---|---|---|---|---|
| Normal Ca‡ | 7 | 1.0 ± 0.1 | 0/7 | Low Ca | 5 | 3.6 ± 0.5 | 0/5 |
| Normal Ca + 3¶ | 5 | 1.3 ± 0.1 | 0/5 | Low Ca + 3¶ | 5 | 6.6 ± 1.4 | 0/5 |
| Normal Ca + 12 | 6 | 1.9 ± 0.1 | 0/6 | Low Ca + 12 | 7 | 19.6 ± 2.0 | 5/7 |
| Normal Ca + 48 | 6 | 5.1 ± 0.4 | 0/6 | Low Ca + 48 | 6 | 154 ± 51 | 5/6 |
| Normal Ca + 96 | 6 | 6.9 ± 0.7 | 0/5 | Low Ca + 96 | 5 | 629 ± 170 | 5/5 |
| Normal Ca + 200 | 6 | 21.3 ± 3.8 | 5/6 | Low Ca + 200 | 5 | 942 ± 362 | 5/5 |
| Normal Ca + 400 | 5 | 20.4 ± 1.9 | 5/5 | | | | |

* No significant differences found in renal wet weight

† Intranuclear inclusions in proximal renal tubular cells characteristic of plumbism

‡ 0.7%

¶ $\mu$g Pb/ml drinking water

than did the vitamin D-induced protein, the addition of calcium reduced the binding of lead by both protein fractions (Table 2). These studies indicate that lead and calcium, when present simultaneously in intestinal contents, compete for binding sites on absorptive proteins. Of note was the observation that calcium-replete diets did not promote increased lead elimination, suggesting that the mechanism for the excretion of lead, unlike that for the excretion of calcium is probably quite limited.

Noting that lead poisoning in young children is often associated with iron deficiency anemia, Six and Goyer (1972) carried out chronic feeding studies in rats receiving a purified diet containing adequate iron (25 ppm) and other nutrients and a diet deficient only in iron (5 ppm). As in the case of calcium deficiency, iron deficiency resulted in an increased retention of lead in liver, kidney and bone and increased urinary lead output. Biochemical parameters of lead toxicity were greater in the iron-deficient rats fed lead than in rats receiving adequate diets and lead. Barltrop and Khoo (1975) in 48-hr absorption studies in the rat were able to show an increase in lead absorption and retention in association with diets deficient in calcium and magnesium, but were not able to show any effect on lead absorption with diets deficient in iron, copper and zinc.

Recently, Conrad and Barton (1978) gave rats iron-deficient and iron-supplemented diets for 3 weeks prior to injection of labeled iron and lead into isolated duodenal loops. The iron-deficient rats absorbed significantly more lead from the test dose ($21.3 \pm 2.3\%$) than did the iron-supplemented rats ($11.2 \pm 1.4\%$; $p < 0.01$). Iron-loaded rats absorbed significantly less lead ($7.71 \pm 1.2\%$; $p < 0.05\%$). When lead and iron were injected together, iron decreased the absorption of lead in a dose-related manner, but lead had no effect on the absorption of iron. In more recent studies in rats, Barton, et al (1978b) have shown that addition of iron to test doses of lead markedly reduces the amount of lead bound by a high molecular weight intestinal mucosal protein with an affinity for both iron and lead. Mahaffey et al. (1978) have also reported that iron-loaded rats absorb less lead in chronic feeding experiments than do rats receiving the recommended dietary

**Table 2.** *In vivo* labeling of duodenal heat-stable mucosal proteins from the rat expressed as percentage of radioisotope recovered in crude mucosal homogenates (Barton et al., 1978a)

| | Higher molecular weight protein | | Vitamin D-induced protein | |
|---|---|---|---|---|
| Isotopic label | Calcium | Lead | Calcium | Lead |
| $^{45}Ca$ only | 0.5 | | 2.6 | |
| $^{203}Pb$ only | 0 | 35.9 | | 8.0 |
| $^{45}Ca + ^{203}Pb$ | 0.1 | 14.3 | 0.8 | 1.0 |

allowance for iron. It is of interest that Halliday *et al.* (1976) have isolated three proteins from intestinal mucosal cells of the rat which bind iron, only one of which was shown immunochemically to be ferritin.

In earlier studies, toxic dietary doses of zinc were found to be protective against lead toxicity in the young horse (Willoughby et al., 1972). Cerklewski and Forbes (1976) studied relationships between nontoxic dietary levels of zinc and marginally toxic dietary levels of lead. In chronic feeding experiments, they administered 8, 35 and 200 ppm zinc and 0, 50 and 200 ppm lead to young weanling rats. Again, in rats receiving 200 ppm Pb in the diet, progressive increase in dietary zinc decreased the lead content of blood, liver, kidney and tibia; the decrease in organ lead content was associated with a decrease in the urinary excretion of ALA and decrease in accumulation of free erythrocyte porphyrins. Evans et al. (1975) have identified a zinc-binding substance in the gastrointestinal tract of the rat. They reported a low molecular weight zinc-binding substance in the gastrointestinal tract of the rat. Also, a low molecular weight zinc-binding ligand was found in the pancreas of the rat and in pancreatic secretions from the dog. The uptake of $^{65}Zn$ from everted intestinal segments was markedly increased in the presence of the zinc-binding ligand from pancreatic secretion. While long term combined lead and zinc feeding studies suggest the possibility that there may be competition between zinc and lead for ligands which facilitate absorption, no studies on this or other aspects of possible zinc-lead interactions at the absorptive level have been reported.

Long term feeding studies with lead (200 ppm), cadmium (50 ppm) and arsenic as sodium arsenate (50 ppm) singly and in various combinations in rats have been reported (Mahaffey and Fowler, 1977). With lead alone, an average blood lead level of 45 $\mu$g Pb/dl whole blood was found at the end of 10 weeks. When cadmium was added to lead, there was a significant reduction in blood lead to 32 $\mu$g Pb/dl whole blood ($p < 0.001$). Retention of lead in tissues was also significantly reduced, as was the occurrence of lead-induced intranuclear inclusion bodies in renal tubular cells (Mahaffey, personal communication). Inorganic arsenic reduced blood lead, but to a lesser degree; lead plus organic arsenic had a negligible effect on blood lead level. These experiments strongly suggest that excess cadmium reduces absorption of lead. Der et al. (1976) reported that lead and cadmium together have a synergistic effect more damaging to rat testes than higher levels of either lead or cadmium when fed alone. No tissue concentrations of lead or cadmium were reported in their studies.

The studies just cited have indeed demonstrated that deficiencies of certain essential minerals do increase the absorption and retention of lead. In the case of cadmium, this toxic metal apparently decreases lead absorption. Competition for absorptive proteins has been identified as the mechanism by which calcium and iron can alter lead absorption. It would be fruitful to extend the approach used to study interactions between lead, iron and calcium at the absorptive level to other metals.

### B. Interactions Affecting Heme Synthesis

There is abundant evidence that lead partially inhibits at least two enzymes in the biosynthetic pathway for heme formation on all organs so far tested (Nordberg, 1976). The principle enzymes inhibited are two sulfhydryl-dependent enzymes, δ-aminolevulinic acid dehydratase (ALAD) and ferrochelatase. When heme formation is reduced, there is also a compensatory increase in the activity and/or formation of the first enzyme in the pathway, δ-aminolevulinic acid synthetase (ALAS). These enzymatic effects are associated with a pattern of accumulation and excretion of heme intermediates which is pathognomonic for lead in man; these are increased urinary excretion of ALA, normal or nearly normal excretion of porphobilinogen and uroporphyrin, increased excretion in urine of coproporphyrin and increased accumulation of zinc protoporphyrin in circulating erythrocytes. The increased urinary excretion of ALA and increase in erythrocyte zinc protoporphyrin are particularly sensitive biochemical indicators of early lead effect.

Abdulla and Haeger-Aronsen (1971) reported that zinc enhanced the activity of ALAD from human erythrocytes both *in vitro* and *in vivo*. Cheh and Neilands (1973) demonstrated that beef liver ALAD is a zinc-dependent enzyme. Haeger-Aronsen, et al. (1976) subsequently showed that inhibition of ALAD by lead could be reversed *in vivo* in rabbits by zinc. Thawley et al. (1978) administered lead po with and without zinc, which was given either po or by ip injection. The increase in urinary ALA characteristic of lead was largely suppressed by both po and ip injection of zinc. These authors also concluded that the antagonistic effect of zinc was not exclusively at the absorptive level, but rather that interaction between lead and zinc also takes place internally. Labbe and Finch (1977) have suggested that iron deficiency in rats may increase the activity of ALAD in erythrocytes. Immediately following injection of iron (Imferon), they found a threefold increase in erythrocyte ALAD activity in rats receiving an iron-deficient diet. The initial increase in ALAD activity was followed by a progressive decline thereafter. Although they postulated that there might be a direct stimulating action of iron on ALAD, others (Battistini et al., 1971) have found that ALAD activity is also related to age in erythrocytes, with the highest activity being found in the youngest cells. Since reticulocytes were not measured in this study, this possibility, which might also explain the findings, cannot be excluded.

It is strongly suspected that there may be an interaction between iron and lead at the ferrochelatase step in heme biosynthesis, which is the step in which ferrous iron is inserted into protoporphyrin to form heme. While lead inhibits relatively crude preparations of this enzyme *in vitro*, it is not certain that this is the major effect of lead *in vivo*. Lead also impairs the transport of iron into the reticulocyte; accumulation of ferric iron as ferritin in blood cells is a characteristic of lead poisoning in the iron-sufficient subject. At this point, lead may act by inhibiting the reduction of iron from the ferric to the ferrous state, the preferred substrate of ferrochelatase. These questions, however, remain to be elucidated (Granick et al., 1978).

Fowler and Mahaffey (1978) have studied the effects of lead, cadmium and arsenic singly and in various combinations, with emphasis on alterations in heme biosynthesis. The pattern of urinary porphyrin excretion served as the assessment criterion. As expected, increased dietary lead alone produced increased urinary ALA, uroporphyrin and coproporphyrin. Dietary cadmium caused a modest increase in urinary level of uroporphyrin only. When lead and cadmium were fed together, the pattern of excretion of these metabolites was altered so that urinary ALA was decreased while uroporphyrin was markedly increased, with little change in urinary coproporphyrin in comparison with lead alone. The combination of lead and inorganic arsenic had little effect on the excretion of ALA, but greatly magnified excretion of coproporphyrin and uroporphyrin. Although tissue concentrations are not reported, these findings suggest that metabolic interactions among lead, cadmium and arsenic may occur.

The studies of Klauder and Petering (1975) reveal some of the complexities of trace metal interactions involving lead, copper, iron and zinc. These workers fed rats for 12 weeks on diets deficient (6 ppm) or optimal (40 ppm) in iron and deficient (0.5 ppm) or optimal (8.5 ppm) in copper with or without added (500 ppm) lead, together with adequate (20 ppm) or excess (140 ppm) zinc. Various combinations were fed. Some of their results are shown in Tables 3, 4 and 5. Combined dietary deficiency of copper and iron magnified kidney lead content almost 20-fold, but iron deficiency was the major factor in increasing kidney lead content, while excess zinc had no demonstrable effect (Table 3). In the same animals, on the other hand, the

**Table 3.** Kidney lead (Klauder and Petering, 1975)

| Dietary composition (ppm) | | | | Kidney lead |
|---|---|---|---|---|
| Zinc | Copper | Iron | Lead | ($\mu$g/g dry weight)* |
| 20 | 0.5 | 6 | 500 | 1910[a] |
| 20 | 8.5 | 6 | 500 | 1490[b] |
| 20 | 0.5 | 40 | 500 | 126[a] |
| 20 | 8.5 | 40 | 500 | 111[b] |
| 140 | 0.5 | 40 | 500 | 112 |
| 140 | 8.5 | 40 | 500 | 90 |

* Geometric means given. Matched superscript letters indicate significance at $p < 0.01$. All control values (no Pb added) were 1 to 2 $\mu$g/g

**Table 4.** Hemoglobin (Klauder and Petering, 1975)

| Dietary composition (ppm) | | | Hemoglobin (g/100 ml blood)† | |
|---|---|---|---|---|
| Zinc | Copper | Iron | Lead (0 ppm)†† | Lead (500 ppm)†† |
| 20 | 0.5 | 6 | 5.2[a,b,c] | 3.1[b,d,e] |
| 20 | 8.5 | 6 | 8.2[a,f,g] | 6.5[d,f,h] |
| 20 | 0.5 | 40 | 12.3[c,j,i*] | 8.8[e,j,k,l] |
| 20 | 8.5 | 40 | 13.9[g,i*] | 13.2[h,k] |
| 140 | 0.5 | 40 | 11.9[m,o] | 7.1[l,o,p] |
| 140 | 8.5 | 40 | 13.6[m] | 13.4[p] |

† Arithmetic means given. Matched letters indicate significance at $p < 0.01$; asterisk shows significance at $p < 0.05$

†† Dietary addition

data revealed that anemia could be corrected by iron alone in the presence of copper deficiency if lead was absent. However, in the presence of excess lead, correction of both iron and copper deficiency was necessary to restore hemoglobin to control levels. Excess zinc appeared to have an effect additive to that of lead in the presence of copper deficiency, perhaps because of the antagonistic effect of zinc on copper metabolism (Table 4). Heart cytochrome c oxidase was adversely affected by lead and zinc, but not by lead alone; this effect was further intensified by copper deficiency (Table 5). Thus, interactions affecting heart cytochrome c oxidase activity involve zinc, copper and lead, but not iron. Their data indicate that interactions among these metals can differ in different tissues of the *same* animal. Using the same basic experimental design, Klauder and Petering (1977) have presented data which suggest that a lead-copper interaction may be the principle factor responsible for lead's antagonistic effect on hematopoiesis, as lead may interfere with the copper-dependent mobilization and utilization of

**Table 5.** Heart cytochrome oxidase (Klauder and Petering, 1975)

| Dietary composition (ppm) | | | Heart cytochrome oxidase $k \times 10^4$ mg protein† | |
|---|---|---|---|---|
| Zinc | Copper | Iron | Lead (0 ppm)†† | Lead (500 ppm)†† |
| 20 | 0.5 | 6 | 12.9[a,b*] | 8.8[c,d*] |
| 20 | 8.5 | 6 | 39.1[a] | 41.8[c] |
| 20 | 0.5 | 40 | 9.9[d] | 7.3[e] |
| 20 | 8.5 | 40 | 39.0[d] | 38.5[e,f*] |
| 140 | 0.5 | 40 | 11.1[g] | 9.3[h] |
| 140 | 8.5 | 40 | 33.2[g] | 24.9[f,h*] |

† Geometric means given. Matched superscript letters indicate significance at $p < 0.01$; asterisk shows significance at $p < 0.05$

†† Dietary addition

iron stores. These experiments represent an extension of earlier work by this group (Murthy et al., 1975), in which interactions among copper, zinc, lead and cadmium were studied. The data indicated that lead and cadmium interfered with copper metabolism in at least an additive manner.

Evidence is accumulating which indicates that the levels of heme in various tissues is influenced by the activities of two inductible enzymes, ALAS and heme oxygenase; metals may act as regulators of heme metabolism through their influence on these enzymes (Maines and Kappas, 1977a). In particular, these workers have studied the microsomal hemoprotein cytochrome P-450. Although cobalt is perhaps the most potent inductor of hepatic heme oxygenase, the enzyme which catalyzes the initial degradation of heme, chromium, iron, nickel, copper, zinc, cadmium and mercury also induce heme oxygenase activity (Maines and Kappas, 1976). Lead appears to induce heme oxidase activity in kidney, but not in heart. With the finding that nickel and platinum, metals which do not form metalloporphyrins in biologic systems, have an influence on the regulation of heme metabolism in liver and kidney, these workers have offered the provocative suggestion that the iron atom itself is the proximately active regulator of ALAS and heme oxygenase. The iron tetrapyrrole complex, rather than being the regulator, may function primarily as a means of transport of the metal to the regulatory sites in the cells (Maines and Kappas, 1977b and 1978). While this groups of workers has so far studied the effects of several metals singly in various tissues (principally heart, liver, kidney), the data suggest that theirs might be a most fruitful experimental system in which to study interactions among trace metals.

### C. Interactions in Other Tissues

Most of the recent work on trace metal interactions involving lead has been concentrated in the areas of absorption, distribution and heme synthesis. The possibility of interactions involving other tissues and metabolic pathways has received little attention. In lead-poisoned suckling rats, it has been reported that zinc and copper concentrations in brain are depressed at day 25 (Michaelson and Sauerhoff, 1973). Silbergeld (1977) has extended the earlier observations of Kostial and Vouk (1957). She used synaptosomal preparations from the caudate nucleus or whole forebrain of adult rats and found that lead-induced inhibition of high affinity choline uptake was mimicked by reduced *in vitro* calcium concentrations. This finding is explainable by a lead-calcium interaction. However, inhibition of dopamine uptake by lead was not reversed by addition of calcium. Ulmer and Vallee (1968), noting that lead is also bound by nucleotides, amino acids and sulfhydryl groups, suggested that interactions between lead and other metals involving these metabolites might well be occurring; however, these areas have received scant attention.

## III. CLINICAL CONSIDERATIONS

### A. Clinical Investigations

In man, major emphasis has traditionally been placed on the study of single etiologic agents. Not surprisingly, human data on possible interactions between lead and other metals are quite limited. In an epidemiologic study of 121 children with PbB ranging from 17 to 160 $\mu$g Pb/dl whole blood, Sorrell et al. (1977) have found significantly lower serum calcium and serum 25-hydroxyvitamin D (25-OHD) in children with PbB $>60$ $\mu$g/dl whole blood (Table 6). They also found a significant inverse correlation ($r = -0.82$, $p < 0.001$, $N = 49$) between reported daily calcium intake and blood lead concentration. In metabolic 3-day balance studies in healthy infants less than 2 years of age, Ziegler et al. (1978) found a weak, but statistically significant, inverse relationship between measured dietary calcium intake and lead retention. These studies, in conjunction with the abundant data in animals, leave little doubt that dietary calcium has a most important influence on lead retention. The findings take on further significance in view of the fact that nutritional surveys indicate that a high proportion of 2- to 3-year-old children in low income groups, the groups also most highly exposed to lead, do not receive adequate dietary intakes of calcium and iron (Table 7).

The concentration of protoporphyrin in circulating erythrocytes is increased 10-fold or more in lead poisoning. This substance persists throughout the life span of the erythrocyte in which it is present as the metalloporphyrin, zinc protoporphyrin, although it is often measured as "free" erythrocyte protoporphyrin (FEP). Statistically highly significant dose-effect relationships between erythrocyte protoporphyrin and blood lead concentration have been found in numerous recent studies (WHO, 1977).

**Table 6.** Serum concentrations of calcium and 25-OHD and whole blood lead in young children

| | Group I (17–29)* | Group II (30–59)* | Group III (60–160)* |
|---|---|---|---|
| | N = 40 | N = 35 | N = 49 |
| Blood Lead, $\mu$g/dl whole blood | 23 ± 1.0† | 48 ± 1.0 | 84 ± 5.0 |
| Calcium, mg/dl | 9.85 ± 0.16 | 9.96 ± 0.15 | 9.08 ± 0.14‡ |
| 25-OHD, ng/ml, | | | |
| Overall | 32 ± 1 (40) | 30 ± 1 (35) | 18 ± 1‡ (49) |
| November–January | 30 ± 1 (12) | 31 ± 1 (10) | 18 ± 1‡ (15) |
| June–August | 33 ± 1 (20) | 31 ± 1 (16) | 20 ± 1‡ (20) |

* Numbers indicate range of blood Pb in $\mu$g/dl

† Values presented as mean ± SEM

‡ Significantly different from Groups I and II, $p < 0.001$

**Table 7.** Cumulative percentage distribution of iron and calcium intakes for children 24 to 36 months of age for low income ratio states (Ten State Nutrition Survey, 1968–1970) (Mahaffey, 1974)

| Iron | | Calcium | |
|---|---|---|---|
| Intake (mg/day) | Children (%) | Intake (mg/day) | Children (%) |
| 3.9 | 24.1 | 200 | 13.7 |
| 7.9 | 67.3 | 400 | 28.9 |
| 11.9 | 90.9 | 600 | 50.1 |
| 15.0 | 98.0 | 800* | 65.9 |

* Recommended daily allowances for children 2 to 3 years old, 1968 (NAS); however, recommended daily allowance for infants (500 to 600 mg) and children (400 to 500 mg) (Calcium, 1978)

Roels et al. (1975) reported that the blood lead threshold for the FEP response is somewhat lower in adult females and children (PbB approximately 25 to 30 μg Pb/dl) than it is in adult males (PbB approximately 40 μg Pb/dl whole blood). Whitehead (1978) pointed out in his statistical critique of these data that three groups of statistical outlyers could be identified among the adult females. If these "atypical female responders" were excluded, there was no statistically significant difference between the responses of the remaining females and the entire adult male group.

Hormonal differences and iron deficiency have been suggested as the factors responsible for the apparent difference between the erythrocyte protoporphyrin responses of males and females. Roels et al. (1978) administered testosterone, estradiol and progesterone to castrated male and female rats. They found an interaction of sex hormones with the FEP response to lead in adult female rats, but not in male rats. These were perhaps not the best choice of steroids, inasmuch as studies on steroid induction of porphyrin synthesis in chick embryo liver indicate that the primary endocrine secretions (testosterone, estradiol, progesterone) are only weak inducers of porphyrin synthesis, while their metabolites (etiocholanolone, pregnanediol, 11-ketopregnanolone and pregnanetriol) are among the most potent porphyrin-inducing endogenous steroids. Conversely, the glucuronide conjugates of these metabolites are totally devoid of porphyrin-inducing activity (Kappas et al., 1968).

In man, erythrocyte protoporphyrin is elevated in the latent stage of iron deficiency and it is elevated to an even greater degree in iron deficiency anemia. In children with PbB $<30$ μg/dl whole blood, there is an inverse relationship between elevation of erythrocyte protoporphyrin and reduced iron saturation (serum iron/total iron binding capacity) and reduced serum ferritin concentration (Stockman et al., 1975; Koller et al., 1978). Indeed, the degree of increase in erythrocyte protoporphyrin may be just as high in iron deficiency anemia as it is in lead poisoning (Table 8) (Campbell et al., 1978). These workers and Sassa et al. (1973) point out that simultaneous measurement of erythrocyte protoporphyrin and ALAD activity could help to differentiate the relative contributions of iron deficiency and increased lead absorption to the FEP response. Clearly, the erythrocyte protoporphyrin elevation can reflect either iron deficiency or increased lead absorption. Indeed, erythrocyte protoporphyrin levels may be grossly elevated if both conditions are present. In some studies of relationships between lead and erythrocyte protoporphyin, anemia has been excluded by design (Alessio et al., 1976). Lilis et al. (1978) measured erythrocyte protoporphyrin, blood lead, serum iron and total iron binding capacity in a group of lead-exposed workers. Although highly significant correlation was found between blood lead level and erythrocyte protoporphyrin, no evidence of an interaction between iron and lead was found. However, serum iron and total iron binding capacity were within the normal range in the lead workers in comparison with the control population. If such an interaction is demonstrable in man in studies in which only blood is sampled, experimental data indicate that not only blood lead, erythrocyte protoporphyrin, serum iron, total iron binding capacity and serum ferritin, but also serum copper (or ceruloplasmin) and ALAD activity should also be measured (Klauder and Petering, 1975). Even so, the persistence of erythrocyte protoporphyrin throughout the life span of the circulating erythrocytes and the difficulties in locating populations whose iron, copper, lead and hormonal status are stable may make such studies extremely difficult. Certainly, longitudinal rather than cross-sectional studies will be needed.

Chisolm et al. (1978) have used a different approach in young children in whom a steady metabolic state is unlikely, due to growth and other factors. They have used the calcium disodium EDTA mobilization test, one of the standard tests used in the diagnosis of lead poisoning, as a "chemical biopsy." In this test, calcium disodium EDTA is injected parenterally and the amount of metal mobilized into urine during the subsequent 24 hr is measured. This drug enhances, not only the excretion of lead, but also the excretion of zinc, cadmium, manganese and iron. The 24-hr output of ALA in urine is taken as the indicator of biologic effect. When the chelatable zinc-lead and iron-lead ratios are plotted against urinary ALA, significant curvilinear relationships are found (Fig. 1). The data are best fitted by an intersecting-lines statistical approach. Increase in ALA output above the normal range is associated with reduced chelatable zinc-lead and iron-lead ratios. These data have provided the first evidence in humans for zinc-lead and iron-lead interactions. The zinc-lead interaction is readily explained by an interaction between lead and zinc on the activity of ALAD. How an iron-lead interaction is mediated at this stage of heme biosynthesis is not

**Table 8.** Erythrocyte protoporphyrin, ALA concentrations and ALAD dehydratase activity in normal subjects and subjects with various forms of anaemia (Campbell et al., 1978)

| Diagnosis | N | Blood lead ± SD (μmol/l) | Hgb ± SD (g/dl) | Serum iron ± SD (μmol/l) | EPP* + SD (nmol/l) | ALA† ± SD (μmol/l) | ALAD†† ± SD (mmol ALA produced/min/ml min/ml RBC) |
|---|---|---|---|---|---|---|---|
| Controls | 37 | 1.0 ± 0.3 | >12 g/dl (f)<br>>13 g/dl (m) | | 0–637<br>(range) | 3.09 ± 0.71 | 27.06 ± 5.49 |
| Iron deficiency anaemia | 31 | 1.2 ± 0.4 | 8.8 ± 2.4 | 4.9 ± 2.9 | 5050 ± 3853 | 4.64 ± 1.82<br>$p < 0.001$¶ | 37.34 ± 19.08<br>$p < 0.005$ |
| Megaloblastic anaemia | 12 | 1.4 ± 0.6 | 8.8 ± 2.2 | 26.7 ± 17.0 | 703 ± 250 | 3.81 ± 1.41<br>$p < 0.01$ | 41.43 ± 27.20<br>$p < 0.005$ |
| Secondary anaemia | 22 | 0.8 ± 0.6 | 9.1 ± 1.9 | 15.6 ± 11.3 | 833 ± 513 | 4.82 ± 2.49<br>$p < 0.001$ | 27.75 ± 13.16<br>ns |
| Lead poisoning | 12 | 4.3 ± 2.0 | 11.0 ± 1.6 | | 6050 ± 3200 | 6.64 ± 0.62<br>$p < 0.001$ | 5.28 ± 3.41<br>$p < 0.001$ |

* EPP, erythrocyte protoporphyrin

† ALA, delta-aminolaevulinic acid

†† ALAD, delta-aminolaevulinic acid dehydrase

¶ P values are with respect to normal controls and are established by Student's unpaired t test

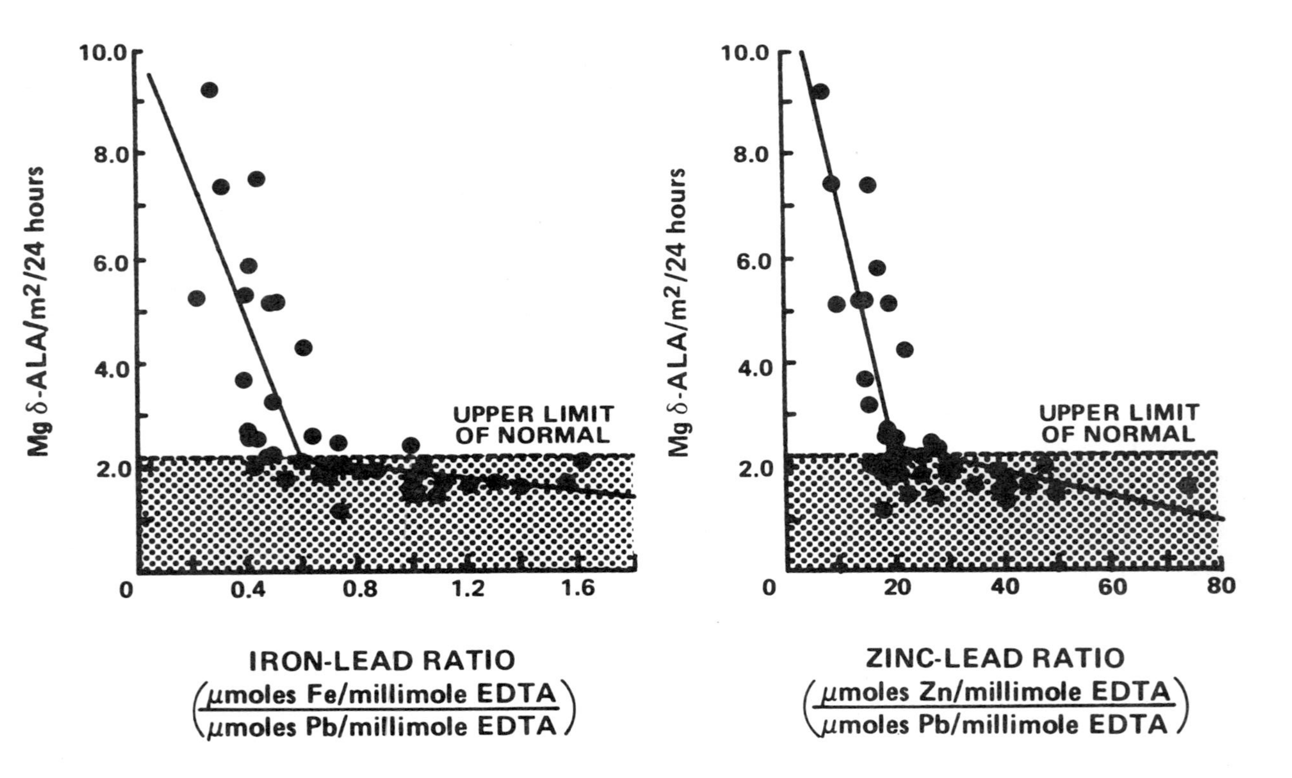


**Figure 1.** Relationship between chelatable lead-zinc ratio (left hand panel) and chelatable iron-lead ratio (right hand panel) and ALAU in 45 children. Highly statistically significant curvilinear relationships are found between these parameters; however, the data are best fitted by an intersecting-lines technique, as shown above: (Chisolm et al., 1978).

entirely clear, but may result from a compensatory increase in the activity of ALA synthetase secondary to reduced heme formation.

### B. Clinical Implications of Experimental Data

Children and workers overexposed to lead may also be overexposed to other trace metals, both essential and nonessential. Conversely, inadequate dietary intake of iron and calcium (Table 7) are most prevalent in child population groups most likely to be overexposed to lead in the United States and possibly elsewhere. There is a strong suspicion that dietary intake of zinc in American children may be marginal (Zinc, 1978). Furthermore, the bioavailability of zinc in cow's milk is less than that of zinc in human milk (Hurley et al. (1979). Only in recent years has prepared infant formula in the United States been supplemented with zinc. The experimental data in animals cited above clearly indicate the importance of interactions between lead and various other metals which enhance lead toxicity. Such interactions may well be of clinical importance from another viewpoint. Neurobehavioral deficits which may subsequently impede the progress of children during the early school years have been attributed not only to low level lead toxicity (Needleman, et al., 1979), but also to iron deficiency (Pollit and Leibel, 1976). Experimental studies in which lead and iron are examined separately give reasonable support to this hypothesis; however, clinical studies, in which lead and iron status have been evaluated separately, have given less certain results. Experimentally, neurobehavioral deficits have been reported in zinc-deficient neonatal animals. The combined effects of excess lead and deficient iron and zinc may have an important bearing on neurobehavioral development in young nutritionally-deprived children overexposed to lead.

Interaction among trace metal imbalances and genetic factors have received little attention in clinical investigation. Sickle cell disease (hemoglobin-type SS) is associated with secondary zinc deficiency, owing to increased urinary loss of zinc. It is of more than passing interest that 7 of the 15 reported cases of peripheral neuropathy due to lead in children have occurred in children with sickle cell disease. Inheritance of deficiency of erythrocyte ALAD activity as an autosomal dominant trait has been identified in a pedigree (Labbe, personal communication). How frequent this is and what the effect of lead-zinc interaction might be in such persons is unknown. Acrodermatitis enteropathica is still another inherited disorder associated with zinc deficiency. Another genetically determined condition which could, in theory, increase susceptibility to lead is intestinal lactase deficiency (Kretchmer, 1972). Lactase-deficient individuals may be intolerant of milk and voluntarily avoid milk and milk products, thereby reducing their dietary intake of calcium.

Certain therapeutic maneuvers in clinical practice may also disturb trace metal metabolism. For example, calcium disodium EDTA, one of the most common drugs used to treat lead poisoning, causes a very substantial diuresis of zinc, equivalent on the average in children to one-half of the recommended daily dietary allowance for zinc (Chisolm, unpublished observations). D-penicillamine, another drug used in the treatment of lead poisoning, mobilizes not only lead, mercury, arsenic and gold, but also copper, iron and zinc. These points have received only scant attention in the clinical area. Disturbances in trace metal metabolism have occurred in association with hemodialysis. Solutions used for total parenteral alimentation in severe burns and other disorders have not always been adequate in trace metal content, although progress in this area is being made.

## IV. SUMMARY AND CONCLUSIONS

During the past decade, there has been a substantial resurgence of interest in trace metal interactions. Most experimental work has taken the form of chronic feeding studies in which the experimenter manipulates the concentration of calcium, zinc, iron, copper, cadmium and lead in semipurified diets fed to rats. Such studies have demonstrated that deficient (sometimes, extremely deficient) intakes of calcium, iron, copper and zinc enhance absorption and retention of lead and affect the distribution of lead within the organs. Simultaneous excessive intake of cadmium and lead depresses lead absorption. Although such studies provide only limited insight into the mechanisms through which lead and these other metals interact, they are relevant to the human situation. In particular, workers and certain groups of young children, can have chronic mixed overexposure to lead, cadmium, copper, zinc and arsenic. It is in the same groups that deficient dietary intakes of calcium and iron may be most prevalent. Marginal dietary deficiency of zinc may also coexist (Zinc, AAP, 1978; Zinc, Lancet, 1973). In the vicinity of ore smelters, 50% or more of the lead, cadmium, copper and zinc emitted is found in particulates of respirable size, so that animal feeding experiments are not entirely comparable to human exposures (Landrigan, 1975; and Dorn et al., 1976). Even so, the animal data make one wonder if simultaneous overexposure to copper, zinc and lead might have modified the toxicity of lead in children so exposed. Studies on the possible neurodevelopmental effects of dietary iron and zinc deficiency have not as yet considered the question of interactions between these and other metals.

Studies concerning the mechanisms through which lead may interact with other metals are still quite limited. Magos and Webb (1978) list possible mechanisms through which metals may interact; these are direct chemical reaction between two metals, competition for carriers, metabolic interference, induction of protein-binding sites, morphologic factors and synergistic or antagonistic effects. There is now evidence that lead competes with calcium and iron for absorptive ligands in the intestinal tract. It is strongly suspected that lead, copper and zinc may interact in the intestine through competition for proteins which facilitate absorption, although such proteins have not as yet been isolated and identified. Attempts to identify a low molecular weight protein which could bind lead in the tissues and blood have so far been unsuccessful, although a preliminary study indicates that such a protein(s) may, in fact, exist (Raghavan et al., 1978). ALAD is a zinc-dependent enzyme. Interaction between lead and zinc on this enzyme represent an antagonistic type of interaction. How iron and lead interact in the biosynthetic pathway for heme formation remains unclear. It is suspected, however, that a three-way interaction among iron, copper and lead may be involved. Although a beginning had been made, it is clear that much remains to be learned about the mechanisms involved.

Studies have been greatly facilitated by the development within the past decade of extraordinarily sensitive micro methods for measuring trace metals. Even so, interlaboratory studies reveal that substantial improvements in both precision and accuracy in such determinations must still be made (Lauwerys et al., 1975). The use of certified reference materials obtainable from the United States National Bureau of Standards or comparable materials certified as to their metal content can facilitate such improvement. Their use should be mandatory.

To date, only one clinical study has provided evidence for interaction between iron and lead and zinc and lead. In this study, it was found that the urinary output of ALA is apparently a function of the chelatable iron-lead and zinc-lead ratios. Although a number of studies suggest that the level of zinc protoporphyrin in circulating erythrocytes is influenced by interaction among iron-lead and possibly copper in the bone marrow, such interaction has not as yet been demonstrated in human studies. Experimental data have several other clinical implications which are yet to be addressed in clinical investigation. Such genetic conditions as sickle cell disease, inborn deficiency of ALA dehydratase activity and acrodermititis enteropathica may well influence lead-zinc interactions. Intestinal lactase deficiency and other disorders associated with malabsorption could well play a role in altering the absorption of a number of trace metals. Among these and possibly other clinical entities, it would not be surprising to find groups which are particularly susceptible to lead toxicity.

The traditional approach in both experimental and clinical investigation has been to study the effects of one metal at a time without regard to the possible influence of trace metal imbalances. It is rather striking that excess of lead and deficiency of iron and zinc appear to have rather similar adverse effects on neurobehavioral development. Although nutritional survey data suggest that such a combination may occur in the same population, the possibility of interaction has not been considered in any clinical study to date. Murthy et al. (1975) have put the matter forcefully and succinctly. "The importance of considering the possible combined toxic effects of environmental chemicals cannot be overemphasized, even though such studies are complex and fraught with experimental and interpretive difficulties. Unless experimental studies of this nature are undertaken we shall continue to debate unrealistic and simplistic concepts about the toxicity of one chemical at a time."

## REFERENCES


Abdulla, M., and Haeger-Aronsen, B. (1971). *Enzyme 12:*708.

Alessio, L., Bertazzi, P.A., Toffoletto, F., and Foa, V. (1976). *Int. Arch. Occup. Environ. Hlth. 37:*73.

Aub, J.C., Fairhall, L.T., Minot, A.S., and Reznikoff, P. (1926). In "Medicine Monographs," Vol. 7, The Williams & Wilkins Co., Baltimore, p. 1.

Barltrop, D., and Khoo, H.E. (1975). *Postgrad. Med. J. 51:*795.

Barton, J.C., Conrad, M.E., Harrison, L., and Nuby, S. (1978a). *J. Lab. Clin. Med. 91:*366.

Barton, J.C., Conrad, M.E., Nuby, S., and Harrison, L. (1978b). *J. Lab. Clin. Med. 92:*536.

Battistini, V., Morrow, J. J., Ginsburg, D., Thompson, G., Moore, M.R., and Goldberg, A. (1971). *Brit. J. Haematol. 20:*177.

Calcium requirements in infancy and childhood. American Academy of Pediatrics Committee on Nutrition. (1978). *Ped. 62:*826.

Campbell, B.C., Meredith, P.A., Moore, M.R., and Goldberg, A. (1978). *Brit. J. Haem. 40:*397.

Cerklewski, F.L., and Forbes, R.M. (1976). *J. Nutr. 106:*689.

Cheh, A., and Neilands, J.B. (1973). *Biochem. Biophys. Res. Com. 55:*1060.

Chisolm, J.J., Jr., Mellits, E.D., and Barrett, M.B. Presented at Second International Symposium on Environmental Lead Research, University of Cincinnati, Cincinnati, Ohio. Dec 5–7, 1978.

Coleman, J.E., and Vallee, B.L. (1961). *J. Biol. Chem. 23:*2244.

Conrad, M.E., and Barton, J.C. (1978). *Gastroenter. 74:*731.

Creason, J.P., Hinners, T.A., Bumgarner, J.E., and Pinkerton, C. (1975). *Clin. Chem. 21:*603.

Der, R., Fahim, Z., Yousef, M., and Fahim, M. (1976). In "Trace Substances in Environmental Health-X," D.D. Hemphill (Ed.), p. 505. University of Missouri, Columbia.

Dorn, C.R., Pierce, II, J.O., Phillips, P.E., and Chase, G.R. (1976). *Atmospheric Environ. 10:*443.

Evans, G.W., Grace, C.I., and Votava, H.J. (1975). *Amer. J. Physiol. 228:*501.

Fowler, B.A., and Mahaffey, K.R. (1978). *Environ. Health Perspec. 25:*87.

Gibson, K.D., Neuberger, A., and Scott, J.J. (1955). *Biochem. J. 61:*618.

Granick, J.L., Sassa, S., and Kappas, A. (1978). O. Bodansky and A.L. Latner (Eds.), In "Advances in Clinical Chemistry," Vol. 20, p. 287. Academic Press, New York.

Haeger-Aronsen, B., Schütz, A., and Abdulla, M. (1976). *Arch. Environ. Health 31:*215.

Halliday, J.W., Powell, L.W., and Mack, U. (1976). *Brit. J. Haem. 34:*237.
Hunter, D. (1978). "The Disease of Occupation," 6th edition, Hodder and Stoughton, London. p. 248.
Hurley, L.S., Lonnerdal, B., and Stanislowski, A.G. (1979). *Lancet 1:*677.
Kappas, A., Song, C.S., Levere, R.D., Sachson, R.A., and Granick, S. (1968). *Proc. Natl. Acad. Sci. 61:*509.
Klauder, D.S., and Petering, H.G. (1975). *Environ. Health Perspec. 12:*77.
Klauder, D.S., and Petering, H.G. (1977). *J. Nutr. 107:*1779.
Koller, M.-E., Romslo, I., Finne, P.H., Brockmeier, F., and Tyssebotn, I. (1978). *Acta Paediatr. Scand. 67:*361.
Kostial, K., and Vouk, V.B. (1957). *Brit. J. Pharmacol. Chemother. 12:*219.
Kretchmer, N. (1976). In "Human Physiology and the Environment in Health and Disease," Readings from *Scientific American*, p. 47.
Labbe, R.F., and Finch, C.A. (1977). *Biochem. Med. 18:*323.
Landrigan, P.J., Gehlbach, S.H., Rosenblum, B.F., Shoults, J.M., Candelaria, R.M., Barthel, W.F., Liddle, J.A., Smrek, A.L., Staehling, N.W., and Sanders, J.F. (1975). *New Engl. J. Med. 292:*123.
Lauwerys, R., Buchet, J.P., Roels, H., Berlin, A., and Smeets, J. (1975). *Clin. Chem. 21:*551.
Lilis, R., Eisinger, J., Blumberg, W., Fischbein, A., and Selikoff, I.J. (1978). *Environ. Health Perspect. 25:*97.
Magos, L., and Webb, M. (1978). *Environ. Health Perspect. 25:*151.
Mahaffey, K.R. (1974). *Environ. Health Perspect. 7:*107.
Mahaffey, K.R., and Fowler, B.A. (1977). *Environ. Health Perspect. 19:*165.
Mahaffey, K.R., Goyer, R., and Haseman, J.K. (1973). *J. Lab. Clin. Med. 82:*92.
Mahaffey, K.R., Stone, C.L., Banks, T.A., and Reed, G. (1978). M. Kirchgessner (Ed.), In "Proceedings of the Third International Symposium on Trace Element Metabolism in Man and Animals," p. 584. Freising, West Germany.
Mahaffey, K.R., Banks, T.A., Stone, C.L., Capar, S.G., Compton, J.F., and Gubkin, M.H. (1975). In "International Conference on Heavy Metals in the Environment," Symposium Proceedings, Vol. 3, Toronto, Ontario. p. 155.
Maines, M.D., and Kappas, A. (1976). *Annals Clin. Res. 8*(suppl. 17):39.
Maines, M.D., and Kappas, A. (1977a). *Proc. Natl. Acad. Sci. 74:*1875.
Maines, M.D., and Kappas, A. (1977b). *Science 198:*1215.
Maines, M.D., and Kappas, A. (1978). *J. Biol. Chem. 253:*2321.
Michaelson, I.A., and Sauerhoff, M.W. (1973). *Life Sciences 13:*417.
Murthy, L., Highhouse, S., Levin, L., Menden, E.E., and Petering, H.G. (1975). In "Trace Substances in Environmental Health-IX," D.D. Hemphill (Ed.), p. 395. University of Missouri, Columbia.
Needleman, H.L., Gunnoe, C., Leviton, A., Reed, R., Peresie, H., Maher, C., and Barrett, P. (1979). *New Engl. J. Med. 300:*689.
Nordberg, G.F. (1976). "Effects and Dose-Response Relationships of Toxic Metals." Elsevier Scientific Publishing Co., Amsterdam.
Nordberg, G.F., Fowler, B.A., Friberg, L., Jernelov, A., Nelson, N., Piscator, M., Sandstead, H.H., Vostal, J., and Vouk, V.B. (1978). *Environ. Health Perspec. 25:*3.
Pollitt, E. and Leibel, R.L. (1976). *J. Ped. 88:*372.
Raghavan, S.R.V., Culver, B.D., and Gonick, H.C. Presented at Second International Symposium on Environmental Lead Research, University of Cincinnati, Cincinnati, Ohio. Dec 5–7, 1978.
Roels, H.A., Lauwerys, R.R., Buchet, J.P., and Vrelust, M.-T. (1975). *Int. Arch. Arbeitsmed. 34:*97.

Roels, H.A., Buchet, J.P., Bernard, A., Hubermont, G., Lauwerys, R.R., and Masson, P. (1978). *Environ. Health Persp. 25:*91.
Sassa, S., Granick, J.L., Granick, S., Kappas, A., and Levere, D. (1973). *Biochem. Med. 8:*135.
Schroeder, H.A., and Tipton, I.H. (1968). *Arch. Environ. Health 17:*965.
Silbergeld, E.K. (1977). *Life Sciences 20:*309.
Six, K.M., and Goyer, R.A. (1970). *J. Lab. Clin. Med. 76:*933.
Six, K.M., and Goyer, R.A. (1972). *J. Lab. Clin. Med. 79:*128.
Sorrell, M., Rosen, J.F., and Roginsky, M. (1977). *Arch. Env. Health 32:*160.
Stockman, J.A., III, Weiner, L.S., Simon, G.E., Stuart, M.J., and Oski, F.A. (1975). *J. Lab. Clin. Med. 85:*113.
Thawley, D.G., Pratt, S.E., and Selby, L.A. (1978). *Environ. Res. 15:*218.
Ulmer, D.D., and Vallee, B.L. (1968). In "Trace Substances in Environmental Health-II," D.D. Hemphill (Ed.), p. 7. University of Missouri, Columbia.
Vallee, B.L., and Ulmer, D.D. (1972). E.E. Snell (Ed.), In "Annual Review of Biochemistry," Vol. 41, Annual Reviews, Inc., Palo Alto, Calif. p. 91.
Whitehead, J. (1978). Report to World Health Organization, Geneva.
Willoughby, R.A., and Thawley, D.G. (1975). In "International Conference on Heavy Metals in the Environment," Symposium Proceedings, Vol. 3, Toronto, Ontario. p. 143.
Willoughby, R.A., MacDonald, E., McSherry, B.J., and Brown, G. (1972). *Canadian J. Compar. Med. 36:*348.
World Health Organization. (1977). "Environmental Health Criteria," Vol. 3, "Lead." Geneva, Switzerland.
Ziegler, E.E., Edwards, B.B., Jensen, R.L., Mahaffey, K.R., and Fomon, S.J. (1978). *Ped. Res. 12:*29.
Zinc deficiency in man. (1973). *Lancet 1:*299.
Zinc. American Academy of Pediatrics, Committee on Nutrition. (1978). *Ped. 62:*408.

# Lead in the Atmosphere And Its Effect on Lead in Humans

*Jerome O. Nriagu*

TABLE OF CONTENTS



## I. INTRODUCTION

The question of airborne lead prophylaxis has drawn more emotional debate than any other aspect of lead in the environment. On the one hand, there are people who maintain that no lead level is "safe" and point to the potential insidious health repercussions of low level lead concentrations (Patterson, 1965; Schroeder, 1975; Bryce-Smith et al., 1979). On the other hand, some people firmly assert that the present levels of lead emission do not constitute any danger to the health of any section of the public (Occupational Safety and Health Administration, 1978). The conflicting conclusions have remained unresolved because there are no epidemiologic data showing unambiguous relationships between ambient lead concentrations, elevated body burdens of lead and frequency of lead-related disorders or diseases (Bathea and Bathea, 1975).

This chapter summarizes what is known about lead in ambient air and discusses the contribution of airborne lead to the human body burden of lead. It has drawn extensively on two recent comprehensive reviews on lead in the atmosphere (Environmental Protection Agency, 1977; Nriagu, 1978a).

## II. SOURCES OF LEAD IN THE ATMOSPHERE

Lead in the atmosphere comes from a wide variety of natural and anthropogenic sources. Worldwide emission of lead from natural sources is estimated to be 24,500 tons, or $2.45 \times 10^{10}$ g (Table 1). Contributions from windblown dusts account for about 65% of the natural lead emissions. About 6% of airborne lead from natural sources is genetically linked to plant exudates and other natural hydrocarbons. The remaining 28% comes from forest fires, volcanogenic (26%) and meteoritic particles. The contributions of lead from radioactive decay (Jaworowski and Kownacka, 1976) and seasalt sprays are rather insignificant (Table 1).

The release of lead from natural sources is small (about 5%) when compared to technogenic emissions, which amounted to $4.5 \times 10^{10}$ g during the 1974–1975 period (Table 2). About 61% of the annual technogenic emissions come from the combustion of leaded gasoline. The other major contributors of airborne lead include the production of steel and base metals (23%), the mining and smelting of lead (8%) and the nonautomative burning of fossil fuels (5%).

The source of pollutant lead in the atmosphere varies considerably from place to place and from time to time. For example, about 88% of the lead released annually into the atmosphere in the United States is believed to come from gasoline combustion, with a remaining 10% attributable to the various stationary sources (Table 2). Nriagu (1978a), however, argues that the stationary sources account for about 40% of the pollutant lead emissions outside of the United States.

Irrespective of the current emission rates, the relative importance of each industrial source has changed with time. From the onset of the industrial revolution to the early part of the 20th Century, the smelting of various metals and the use of coal were the paramount sources of atmospheric lead. Since the 1920's, rapid increases in atmospheric lead emission have been associated with the widespread use of lead additives as antiknock agents in automobile gasoline. Historical changes in the flux of

**Table 1.** Worldwide emissions of lead from natural sources (Nriagu, 1979)

| Source | Lead emission ($\times 10^6$ kg yr$^{-1}$) | Total emission (%) |
|---|---|---|
| Windblown dusts | 16 | 65.3 |
| Forest fires | 0.5 | 2.0 |
| Volcanogenic particles | 6.4 | 26.1 |
| Vegetation | 1.6 | 6.5 |
| Seasalt sprays | 0.02 | 0 |
| Total | 24.5 | |

**Table 2.** Worldwide antrhopogenic emissions of lead during 1975 (Nriagu, 1979)

| Source | Lead emission ($\times 10^6$ kg $yr^{-1}$) | Total emission (%) |
|---|---|---|
| Mining, nonferrous metals | 8.2 | 1.8 |
| Primary nonferrous metal production | | |
| Copper | 27 | 6.0 |
| Lead | 31 | 6.9 |
| Nickel | 2.5 | 0.6 |
| Zinc | 16 | 3.6 |
| Secondary nonferrous metal production | 0.8 | 0.2 |
| Iron and steel production | 50 | 11.1 |
| Industrial applications | 7.4 | 1.6 |
| Coal combustion | 14 | 3.1 |
| Oil (including gasoline) combustion | 273 | 61 |
| Wood combustion | 4.5 | 1.0 |
| Waste incineration | 8.9 | 2.0 |
| Phosphate fertilizers | 0.1 | 0.02 |
| Miscellaneous | 5.9 | 1.3 |
| Total | 449 | |

lead into the atmosphere have now been well documented from the analysis of polar ice layers (Murozumi et al., 1969; Jaworowski, 1968), the dating of sediment profiles (Chow et al., 1973; Edgington and Robbins, 1976), the analysis of lead in tree rings (Hutchinson, 1972), analysis of soils (Page and Ganje, 1970; Ewing and Pearson, 1974), peat profiles (Lee and Tallis, 1973) and moss specimens (Ruhling and Tyler, 1968; Robinson et al., 1973).

Table 3 shows the historical flux rates and the all-time emissions of lead into the atmosphere. It is estimated that over 20 million tons of lead have been transmitted via the atmosphere to the various ecosystems. Between 1900 and 1970, the flux of lead into the atmosphere increased at

**Table 3.** All-time worldwide consumption and emission of lead (Nriagu, 1979)

| Period | Lead consumption ($\times 10^9$ kg) | Lead emission ($\times 10^9$ kg) |
|---|---|---|
| Pre–1900 | 80 | 3.52 |
| 1901–1910 | 10.7 | 0.47 |
| 1911–1920 | 11.2 | 0.49 |
| 1921–1930 | 14.2 | 1.12 |
| 1931–1940 | 14.6 | 1.64 |
| 1941–1950 | 14.9 | 1.67 |
| 1951–1960 | 24 | 2.69 |
| 1961–1970 | 33 | 3.70 |
| 1971–1980 | 38 | 4.26 |
| Total, all-time | 241 | 19.6 |

the rate of about 62 $\times$ $10^6$ kg/decade. Today, the cumulative atmospheric input exceeds the natural lead reservoirs in many ecosystems, while the rate of deposition exceeds the lead being cycled annually in the living biota and flora (Nriagu, 1978b).

## III. AIRBORNE LEAD LEVELS

Synoptic surveys and long term monitoring programs have now generated a large volume of data on the distribution of lead in the atmosphere. Ambient monitoring of airborne lead under various meteorologic conditions is required to develop feasible strategies for maintaining health air quality standards for lead. The data on lead concentrations at remote locations have provided insight into the global atmospheric circulation patterns, in addition to constituting the baseline to which other polluted atmospheres can be referred. In many countries, environmental impact statements now require the determination of the original air quality (including the lead levels) prior to any project development that may affect airborne lead concentrations.

Measurements of airborne lead levels have been made at such remote locations as the South Pole (Zoller et al., 1974), Antarctica (Duce, 1972), Thule in Greenland (HASL, 1975), the middle of the oceans (Egorov et al., 1970) and remote mountain tops (Chow et al., 1973; Adams et al., 1977). The lowest lead concentration at a continental location was the 0.23 ng $m^{-3}$ measured at the Cape of Desire (Egorov et al., 1970). Generally, the atmospheric lead levels in remote continental areas fall within the range of 0.5 to 10 ng $m^{-3}$. Lead levels at representative remote locations are shown in Table 4.

Considerably higher ambient lead levels have been reported in rural and urban areas. Between 1970 and 1974, the mean annual lead concentrations at the nonurban National Air Surveillance Network (NASN) sites ranged from 47 to 139 ng $m^{-3}$, with the majority (86%) of the reported annual averages being below 200 ng $m^{-3}$ (Table 5). Similar mean annual values have also been reported at semirural sites in Britain; they were 163, 126, 110 and 118 ng $m^{-3}$ in 1972–1973, 1974, 1975 and 1976, respectively (Cawse, 1974–1977). Fugas et al. (1973) found the average ambient lead concentrations in rural parts of South Norway to be $<50$ ng $m^{-3}$ during 1974 and 1975, whereas the lead level at two rural sites in Belgium was 174 ng $m^{-3}$ between 1972 and 1975 (Kretzschmar et al., 1977). During 1971–1972, the mean airborne lead levels at several nonurban sites in Europe were less than 500 ng $m^{-3}$, with observed daily maximums being $<1000$ ng $m^{-3}$ (WHO, 1977). These data show that, in general, the average concentrations of lead in rural air range from 50 to 500 ng $m^{-3}$.

The highest concentrations of lead in ambient air occur near highways and in densely populated areas. Annual mean lead concentrations in towns and cities in the USA range from 140 ng $m^{-3}$ in Los Almos (Tepper and Lewin, 1972) to 7500 ng $m^{-3}$ in mid-Manhattan, New York (Chow, 1973). The 1970–1974 data for the urban NASN sites are summarized in Table 5. Notice that 5% of the urban stations had lead concentrations in excess of 2,000 ng $m^{-3}$, whereas the levels at 1% of the stations exceed 3,000 ng $m^{-3}$ in 1974. As one moves away from a major city, the lead levels commonly fall from 210 ng $m^{-3}$ near the city to 96 ng $m^{-3}$ at intermediate distances from the city and 22 ng $m^{-3}$ at more remote areas (McMullen et al., 1970). The data from the National Air Pollution Surveillance (NAPS) of Environment Canada show average lead concentration in Canadian towns and cities during 1974 and 1975 to be 700 ng $m^{-3}$; the range is 200 ng $m^{-3}$ (Lethbridge, Alberta) to >2000 ng $m^{-3}$ (Montreal). The average for 28 Ontario towns and cities is 1000 ng $m^{-3}$ (Barton et al., 1975).

The average atmospheric lead concentrations in North American cities may be compared with the mean values of 300 ng $m^{-3}$ for 7 large Japanese cities in 1973 (WHO, 1977) and <1000 ng $m^{-3}$ for 10 Polish cities (Maziarka et al., 1971). Janssens and Dams (1975) found the average atmospheric levels in residential and industrial locations in Belgium to be 1800 and 1600 ng $m^{-3}$, respectively. From the data given by Ireland (1972) and Lawther et al. (1972), the mean airborne lead levels in British cities are estimated to be 800 ng $m^{-3}$. Blokker (1972) and WHO (1977) summarized the data on atmospheric lead concentrations in major European cities and found the range to be 400 to 8,000 ng $m^{-3}$. The dearth of measurements from developing countries is obvious. Branquinho and Robinson (1976) found the average lead concentration in air at six stations in Rio de Janeiro to be about 1,000 ng $m^{-3}$. Between April and September, 1974, the mean lead concentration in eight South American cities was 230 ng $m^{-3}$ (Harley, 1970; Fox and Ludwick, 1976).

In general, the average suspended lead concentration in a given district is related to the population density. In a survey of 24 USA cities during 1954–1955, the mean lead levels were found to increase from 1470 ng $m^{-3}$ in cities with populations of <0.1 million to about 2000 ng $m^{-3}$ in cities with populations of 1 to 2 million to >3000 ng $m^{-3}$ in cities with populations of over 3 million (Cholak, 1964). The observed relationship apparently reflects the increasing intensity of lead emissions from automotive and industrial sources. The relative decrease of lead aerosol concentration from urban to rural areas results from 1) dilution and mixing with cleaner air as the urban air masses move across the rural areas and 2) progressive removal of lead particulates by rainout and snowout and dry deposition.

## IV. DISPERSION OF AIRBORNE LEAD

The literature on the distribution of lead near line sources (roadways), point sources (smelters, power plants, incinerators, etc.) and area sources (cities)

**Table 4.** Atmospheric lead concentrations in remote areas

| Location | Lead concentrations ($ng/n^3$)* | Period | Reference |
|---|---|---|---|
| Novaya Zemlya, USSR | 0.23 | 1968–1969 | Egorov et al., 1970 |
| Dickson Island, USSR | 0.87 | 1968–1969 | Egorov et al., 1970 |
| Solehard, USSR | 1.26 | 1968–1969 | Egorov et al., 1970 |
| Sevastopol, USSR | 8.25 | 1968–1969 | Egorov et al., 1970 |
| Jungfraujoch, Switzerland | 8.7 (0.13–25) | Oct. 1973–Oct. 1974 | Janssens and Dams, 1975 |
| Lakselv, Norway | 5.6 (0.6–25) | Nov. 1971–Dec. 1972 | Janssens and Dams, 1975 |
| Windward Coast, Oahu, Hawaii | 3 (0.3–13) | June 1969–Feb. 1970 | Hoffman et al., 1972 |
| Windward Coast, Honolulu, Hawaii | 1.7 | Summer 1968 | Jernigan et al., 1971 |
| White Mountain, California | 8.0 (1.2–29) | 1969–1970 | Chow et al., 1972 |
| Quillayute, Washington | 9.4 (0.21–57.1) | 1974 | Fox and Ludwick, 1976 |
| The Antarctica | 0.4 | 1971 | Duce, 1972 |
| Thule, Greenland | 0.5 | Summer 1965 | Murozumi et al., 1969 |
| Thule, Greenland | <10.0 | 1967 | Harley, 1967 |
| Thule, Greenland | 12.6 (6.8–19.0) | April–Sept., 1974 | HASL, 1975 |
| Kap Tobin, Greenland | 2.5 (1.8–3.6) | April–Sept., 1974 | HASL, 1975 |
| Mauna Loa, Hawaii | 5.5 (2.6–10.1) | April–Sept., 1974 | HASL, 1975 |

| | | | |
|---|---|---|---|
| Isle de Pasque, Easter Island | 4.8 (2.8–8.0) | April–Sept., 1974 | HASL, 1975 |
| Antarctica | 3.4 | April–Sept., 1974 | HASL, 1975 |
| South Pole Station | 4.4 (1.4–7.1) | April–Sept., 1974 | HASL, 1975 |
| The South Pole | 0.63 (<0.19–1.19) | Oct. 1970 | Zoller et al., 1974 |
| N. Atlantic, varible westerlies | 0.049† | 1971–1972 | Chester et al., 1974 |
| N. Atlantic, westerlies/N.E. Trades | 0.046† | 1971–1972 | Chester et al., 1974 |
| N. Atlantic, N. E. Trades | 1.6† | 1971–1972 | Chester et al., 1974 |
| S. Atlantic, S.E. Trades | 0.18† | 1971–1972 | Chester et al., 1974 |
| S. Atlantic, S. African Coast | 1.77† | 1971–1972 | Chester et al., 1974 |
| North-central Pacific | 1.0 (0.3–1.5) | 1967 | Chow et al., 1969 |
| N. Indian, N.E. Monsoon | 0.52† | 1971–1972 | Chester et al., 1974 |
| N. Indian Ocean | 4.45 | 1968–1969 | Egorov et al., 1970 |
| S. Indian Ocean | 1.01 | 1968–1969 | Egorov et al., 1970 |
| S. Indian Ocean | 0.13† | 1971–1972 | Chester et al., 1974 |
| S. China Sea, S. W. Monsoon | 0.32† | 1971–1972 | Chester et al., 1974 |
| E. China Sea | 1.49† | 1971–1972 | Chester et al., 1974 |
| S. Japan Coast | 3.59† | 1971–1972 | Chester et al., 1974 |
| Sea of Japan | 4.45† | 1971–1972 | Chester et al., 1974 |
| Java Sea | 1.03† | 1971–1972 | Chester et al., 1974 |

* Ranges in concentration are shown in brackets

† Sand-sized particles only

**Table 5.** Cumulative frequency distributions of airborne lead measurements in the USA (Akland, 1976)

| | | | | | | | | | | | Arithmetic | | Geometric | |
|---|---|---|---|---|---|---|---|---|---|---|---|---|---|---|
| Year | N | Minimum* | 10 | 30 | 50 | 70 | 90 | 95 | 99 | Maximum | Mean | Standard Deviation | Mean | Standard Deviation |
| A. Urban areas | | | | | | | | | | | | | | |
| 1970 | 797 | LD | 0.47 | 0.75 | 1.04 | 1.37 | 2.01 | 2.59 | 4.14 | 5.83 | 1.19 | 0.80 | 0.99 | 1.84 |
| 1971 | 717 | LD | 0.42 | 0.71 | 1.01 | 1.42 | 2.21 | 2.89 | 4.38 | 6.31 | 1.23 | 0.87 | 1.00 | 1.89 |
| 1972 | 708 | LD | 0.46 | 0.72 | 0.97 | 1.25 | 1.93 | 2.57 | 3.69 | 6.88 | 1.13 | 0.78 | 0.93 | 1.87 |
| 1973 | 559 | LD | 0.35 | 0.58 | 0.77 | 1.05 | 1.62 | 2.08 | 3.03 | 5.83 | 0.92 | 0.64 | 0.76 | 1.87 |
| 1974 | 594 | 0.08 | 0.36 | 0.57 | 0.75 | 1.00 | 1.61 | 1.97 | 3.16 | 4.09 | 0.89 | 0.57 | 0.75 | 1.80 |
| B. Nonurban areas | | | | | | | | | | | | | | |
| 1970 | 124 | 0.003 | 0.003 | 0.003 | 0.003 | 0.003 | 0.267 | 0.383 | 0.628 | 1.471 | 0.088 | 0.190 | 0.040 | 3.72 |
| 1971 | 85 | 0.003 | 0.003 | 0.003 | 0.003 | 0.003 | 0.127 | 0.204 | 0.783 | 1.134 | 0.047 | 0.155 | 0.008 | 4.80 |
| 1972 | 137 | 0.007 | 0.007 | 0.007 | 0.107 | 0.166 | 0.294 | 0.392 | 0.950 | 1.048 | 0.139 | 0.169 | 0.090 | 2.59 |
| 1973 | 100 | 0.015 | 0.015 | 0.015 | 0.058 | 0.132 | 0.233 | 0.392 | 0.698 | 0.939 | 0.110 | 0.149 | 0.068 | 2.77 |
| 1974 | 79 | 0.007 | 0.007 | 0.053 | 0.087 | 0.141 | 0.221 | 0.317 | 0.496 | 0.534 | 0.111 | 0.111 | 0.083 | 2.30 |

* Concentrations are given in $\mu gm^{-3}$; LD means below the detection limit of the spectrographic method used in sample analysis.

has become quite voluminous. In general, these studies show the following:

1. The lead concentrations show diurnal and day-to-day changes which are determined primarily by variations in rate and point of emission, wind direction and the local ventilation factors. In urban areas, the maximal Pb concentration generally coincides with morning and evening traffic rush hours and the average concentrations are often higher during the day than at night (Janssens and Dams, 1975). It is equally noteworthy that periods of unusually high levels of metals in air are often associated with the development of stagnant anticyclones; Demuynck et al. (1976) recorded a 10-fold increase in airborne lead levels during one such incident in Ghent, Belgium.
2. The decrease in airborne lead concentration with distance from the emission source is characteristically curvilinear. Thus, the highest incidence of lead contamination occurs in a rather narrow zone bordering the leeside of the source. Near roadways, the lead content of the air often decreases by about 50% between 10 and 50 m from the road. On the other hand, Yankel et al. (1977) showed that the lead level decreased by more than 50% (10.3 to 4.9 $\mu g\ m^{-3}$) at a distance of 4 to 10 km from the smelter in Silver Valley, Idaho.
3. The development of vertical gradients of lead in the atmosphere has not been unambiguously established. Several reports (Schroeder, 1974; Edwards, 1975) have shown that lead concentrations at the street level are generally higher than the values at the roof level.
4. The size of airborne lead particulates decreases with the distance away from the source. For example, Daines et al. (1970) showed that 50% of the particles larger than 6.5 $\mu m$ had dropped out at a distance of about 600 m from the roadway.
5. Indoor concentrations of lead tend to be quite variable and generally amount to 30 to 70% of the concentration in the adjacent outdoor environment.
6. Marked seasonal variations in ambient lead levels are common and may be attributed to seasonal differences in the intensity of lead emissions from different sources and/or to the variability in regional ventilation factors (Fig. 1).
7. The physical and chemical characteristics of automotive lead in the air are also related to the nature of the fuel and the operating mode of the vehicles. For example, it is now generally accepted that a larger fraction of lead in automobile exhaust is released on highways than is released during street driving.
8. Representative data on lead in street dusts and soils in urban areas are shown in Table 6. It is clear that the fallout of leaded aerosols has resulted in abnormally high levels of lead in urban and roadside soils.

**Table 6.** Lead in urban dusts.

| Sample location | Lead concentration ($\mu g/g$)* | Reference |
|---|---|---|
| 77 cities in USA | | |
| commercial sites | 2413 | Natl. Acad. Sci. Report, 1972 |
| 77 cities in USA | | |
| residential sites | 1636 | Natl. Acad. Sci. Report, 1973 |
| London | | |
| Main roads, gutter | 1530 (±600) | Duggan and Williams, 1977 |
| Main roads, pavements | 1460 (±520) | ” |
| Side roads, gutter | 1030 (±450) | ” |
| Side roads, pavement | 900 (±420) | ” |
| Birmingham | | |
| Major arterial roads, gutter | 2350 (160–10000) | Archer and Barrat, 1976 |
| Subsidiary roads, gutter | 1557 (300–4600) | ” |
| Residential roads, gutter | 1050 (220–4300) | ” |
| Residential roads, pavement | 932 (200–5800) | ” |
| Industrial estate roads, gutter | 2574 (160–50000) | ” |
| Dustfall at highway interchange | 1093 (160–8650) | ” |
| External surfaces other than roads | 2240 (100–67000) | ” |
| Internal surfaces (inside buildings) | 2780 (100–280000) | ” |
| Birmingham, around a lead factory | | |
| All roadside samples | 2660 (450–19600) | Archer and Barrat, 1976 |
| Residential/industrial roads, gutter | 2900 (450–19600) | ” |
| External surfaces | 6420 (230–128000) | ” |
| Internal surfaces | 7930 (136–470000) | ” |
| Illinois (9 cities) | | |
| Residential areas | 656–3067 | Schroeder, 1974 |
| Commercial areas | 1774–3549 | |
| Indiana (8 cities) | | |
| Residential areas | 290–9972 | ” |
| Commercial areas | 942–6597 | ” |

| | | |
|---|---|---|
| Iowa (5 cities) | | |
| Residential areas | 849–2525 | ” |
| Commercial areas | 1127–3790 | ” |
| Kentucky (4 cities) | | |
| Residential areas | 1945–2702 | ” |
| Commercial areas | 1350–5089 | ” |
| Michigan (9 cities) | | |
| Residential areas | 1041–3042 | ” |
| Commercial areas | 2524–4722 | ” |
| Ohio (12 cities) | | |
| Residential areas | 206–2639 | ” |
| Commercial areas | 352–2933 | ” |
| W. Virginia (5 cities) | | |
| Residential areas | 1084–2045 | ” |
| Commercial areas | 375–6979 | ” |
| Philadelphia | | |
| Residential | 614 (293–1030) | Needleman et al., 1974 |
| Suburban | 830 (277–1517) | ” |
| Urban industrial | 3855 (929–15680) | ” |
| Classrooms | 2000 | Shapiro et al., 1973 |
| Playground | 3000 | ” |
| Window frames | 1750 | ” |
| Urbana-Champaign, Illinois | | |
| Residential | 600 | Solomon and Hartford, 1976 |
| Offices, hallways | 3380 | ” |
| Offices, no rugs | 1450 | ” |
| Offices, rugs | 2320 | ” |
| Classrooms, no rugs | 930 | ” |
| Public schools, rugs/mats | 730 | ” |
| Public schools, no rugs | 650 | ” |
| Hospitals, corridors, rooms | 360 | ” |
| Supermarkets | 490 | ” |
| Chemical labs | 11400 | ” |

* The ranges reported are shown in parentheses.

Indeed, lead in the play environment is now being implicated in the etiology of lead poisoning in urban children (Lepow et al., 1974; Jordan and Hogan, 1975; Duggan and Williams, 1977).

Any factors which affect the distribution of lead in air obviously will influence the incidence of exposure of human beings and other organisms to lead in particular environmental settings. The study of the distribution of lead in the atmosphere thus provides an insight into the pathways of lead to man.

## V. LONG TERM TRENDS IN ATMOSPHERIC LEAD LEVELS

Long term trends in ambient lead levels are likely to reflect changes in climatic patterns and/or changes in the flux of pollutant lead in the atmosphere. Such information is thus important in any attempt to maintain reasonable air quality standards for lead.

The trend in lead levels at 92 NASN urban sites form 1965 through 1974 is shown in Fig. 1. In general, urban lead concentrations increased from 1965 until 1971 and have since been declining. Between 1971 and 1974,

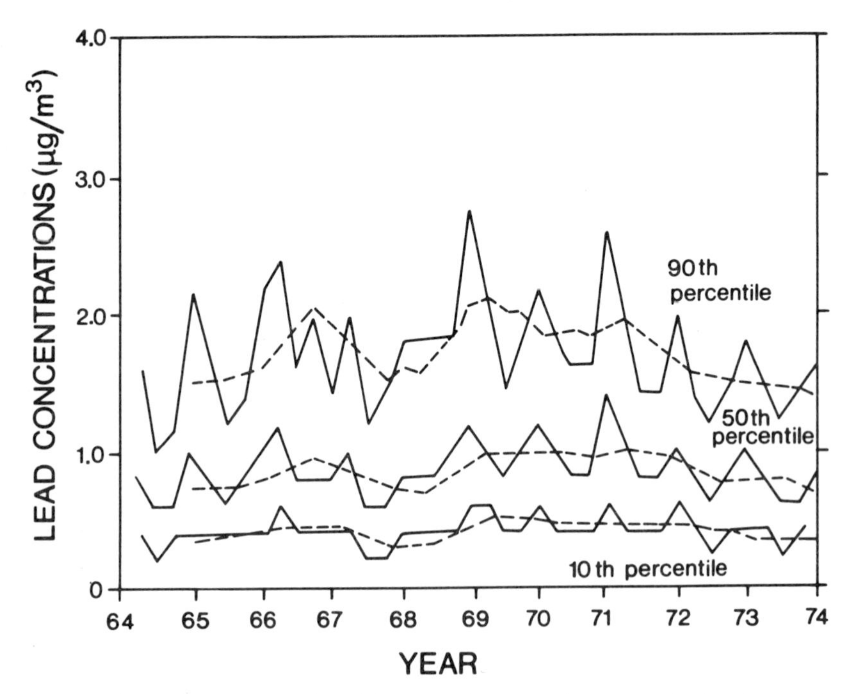


**Figure 1.** Seasonal patterns and trends in quarterly average lead concentrations in the urban areas of the USA (Faoro and McMullen, 1977).

the rate of decline as described by the 50th percentile was about 6% $yr^{-1}$. The lead content of the particulate matter also follows this trend, suggesting that the observed profile is not just a result of particulate controls, but is also related to decreases in lead emissions (Environmental Protection Agency, 1977). The decline in ambient lead levels apparently is related to the reduction in lead content of gasolines and the increasing use of low lead or non-lead gasoline which began in 1970. These lead pollution control measures seem to be enough to override the general increase in annual gasoline consumption of about 5% per year. The general reduction in particulate emissions during the same period (Council on Environmental Quality, 1977) could also have been a contributing factor in the observed downward trend in ambient lead levels.

The situation in semirural areas in the United Kingdom is somewhat different. Salmon et al. (1978) showed that, following the Clean Air Act of 1956, the rate of increase of airborne lead at Chilton was reduced to about 3% $yr^{-1}$ between 1962 and 1974. Apaprently, the effect of the reduction in particulate emissions engendered by the Clean Air Act could not over-ride the emission of lead from gasoline combustion, which increased by about 12% $yr^{-1}$ between 1960 and 1970 (Salmon et al., 1978).

It follows that the temporal trend in airborne lead levels tends to reflect the overall rate of emissions. In most developing countries, the ambient lead levels will likely be on the upswing, whereas, in countries with effective lead pollution control measures, a downward trend in the levels should be expected. It is significant in this respect that the data gathered in the USA and Europe before the early 1970's often show a general increase in airborne lead level. For example, Chow and Earl (1970) showed that the lead aerosol concentration in San Diego increased at the rate of 5% per year between 1957 and 1969. Jost et al. (1973) reported that the atmospheric lead in Frankfurt/Main showed an increase of about 10% per year between 1966 and 1970. The data published by Tepper and Levin (1972) indicate that, between 1961 and 1969, the geometric mean lead concentrations in Los Angeles, Philadelphia and Cincinnati registered increases which varied between 20 and 56%. Some data collected between 1968 and 1974 tend to show little change or marginal changes in lead concentrations with time. Between 1967 and 1974, the reported lead levels in the seven South American cities remained unchanged (Harley, 1967; Hasl, 1975). No discernable trend is apparent in mean lead levels in Canadian towns and cities from 1971 to 1975 (Environment. Canada, NAPS Reports).

The opposite trends reported prior to 1970 are probably artifacts related to changes in sampling station location and to improvements in sampling and analytical methodologies. For example, it has been reported that lead levels at Cincinnati showed a gradual downward trend in mean concentrations of lead from 5.1 $\mu g/m^3$ in 1941 to 1.4 $\mu g/m^3$ in 1962

(Cholak, 1964). The range of values was also reported to have decreased so that, after 1954, no concentration above 8.0 $\mu g/m^3$ was recorded. The notable feature of this study, however, is that the array of data during the 20-year span are not exactly comparable because of changes in sampling stations, as well as in the sampling and analytical methodologies. An analogous downward trend was observed in suburban Salt Lake City, where the average lead levels decreased from 1.4 $\mu g/m^3$ between 1944 and 1949 to 0.30 $\mu g/m^3$ during 1964–1965 (Nelson, 1970).

It is important to conclude this section by noting that the short residence times of lead aerosols in the atmosphere (from a few minutes up to three weeks; Nriagu, 1978a) would preclude an inexorable buildup of lead in the air.

## VI. THE NATURE OF AIRBORNE LEAD

The shape, size and chemical composition of lead particulates varies extensively and reflects both the source features and the past atmospheric history of the aerosols. Considerable attention is now being paid to variations in these physical attributes, which determine to a large extent the rate of absorption of particles into the blood stream.

Very little is known about the chemical composition of airborne lead aerosols. Biggins and Harrison (1978) identified lead ammonium sulfate, $(NH_4)_2SO_4 \cdot PbSO_4$, as the principal lead compound in urban air, with $PbSO_4$ and PbBrCl as minor constituents. Considering that most of the airborne lead in urban areas comes from automobiles and that lead halides (particularly PbBrCl) are the principal forms of lead from automotive sources, it follows that the automotive lead particulates react rather quickly with pollutant sulfur oxides and ammonia to become converted to $(NH_4)_2SO_4 \cdot PbSO_4$. The conversion reaction apparently involves losses of the halide ions and may be enhanced photochemically. If the atmospheric levels of sulfur oxides and ammonia are low, the lead halides may also be converted to oxides and carbonates on aging (Ter Haar and Bayard, 1971). Near the point sources of lead, the plumbiferous aerosols will likely contain lead oxides, sulfates, sulfides, silicates and even elemental lead, depending on the nature of lead products being handled (Nriagu, 1978a).

Harrison and Perry (1977) have reviewed the available information on alkyllead in urban areas. In general, high values of airborne alkyllead occur around gasoline stations (210 to 590 ng $m^{-3}$) and underground car parks (1800 2200 ng $m^{-3}$), where considerable evaporative losses of leaded gasoline are encountered. At these hot spots, alkyllead constitutes between 4 and 18% of the total airborne lead. Elevated levels (about 5% of total lead) of alkyllead also occur close to busy streets (Harrison and Laxen, 1978). Typically, the level of airborne alkyllead in urban air is in the range of $<6$

to 206 ng $m^{-3}$, or about 1 to 4% of the total lead (Harrison and Perry, 1977). By contrast, alkyllead in rural air around Lancaster constitutes between 1.5 and 33% of the total airborne lead, suggesting that a significant portion of the organolead has been derived from the natural methylation of lead in soils and wetlands (Harrison and Laxen, 1978). However, this observation needs to be confirmed by measurements in other rural areas.

A large volume of literature has been built up on the size distribution of atmospheric lead-containing particulates (Environmental Protection Agency, 1977; Nriagu, 1978a). Typically, the size distribution of most plumbiferous aerosols is approximately log-normal with over 90% of the particles being less than 1.0 $\mu$m in diameter. Particles in this size distribution are readily respired and easily absorbed into the blood stream. Furthermore, it has been shown (Linton et al., 1976) that there is a marked enrichment of lead at the surfaces of fine particles, a feature which should greatly enhance the entry of lead (in the aerosols) into the blood stream.

## VII. PATHWAYS OF AIRBORNE LEAD TO MAN

The atmosphere is the principal medium involved in the transmission of lead from stationary and mobile sources to most terrestrial ecosystems. Figure 2 depicts the potential pathways of airborne lead to man. The magnitude of lead flow along most of the pathways remains essentially unknown; the available information is summarized in reports by WHO (1977), Environmental Protection Agency (1977), Institute of Environmental Studies (1977), Bogges and Wixon (1977) and Nriagu (1978).

Figure 2 clearly illustrates the fact that lead is a multimedia pollutant. Pollutant lead from a primary source may result in secondary and tertiary contamination in the environment. For example, airborne lead contributes via fallout to the lead in soils, which in turn affects the lead concentration in the various foodchains. The focus of this chapter, however, is on direct human exposure to lead in the air, with some attention paid to exposure from airborne lead now in dirt and dust.

## VIII. EFFECT OF AIRBORNE LEAD ON BODY LEAD BURDEN

If the estimated natural flux of lead to the atmosphere (of $24.5 \times 10^9$ g/yr; Table 1) is assumed to be uniformly distributed to a height of 10 km and then turned over 40 times during the year, the estimated background lead concentration in the pre-industrial atmosphere would be 0.3 $\mu g/m^3$. This figure would imply that human activities have elevated the level of airborne lead levels in the average rural and urban areas of the USA by factors of between 500 and 4,000 (Table 5). The average atmospheric lead level at remote locations in North America is about 10 $ng/m^3$ (Chow et al., 1972),

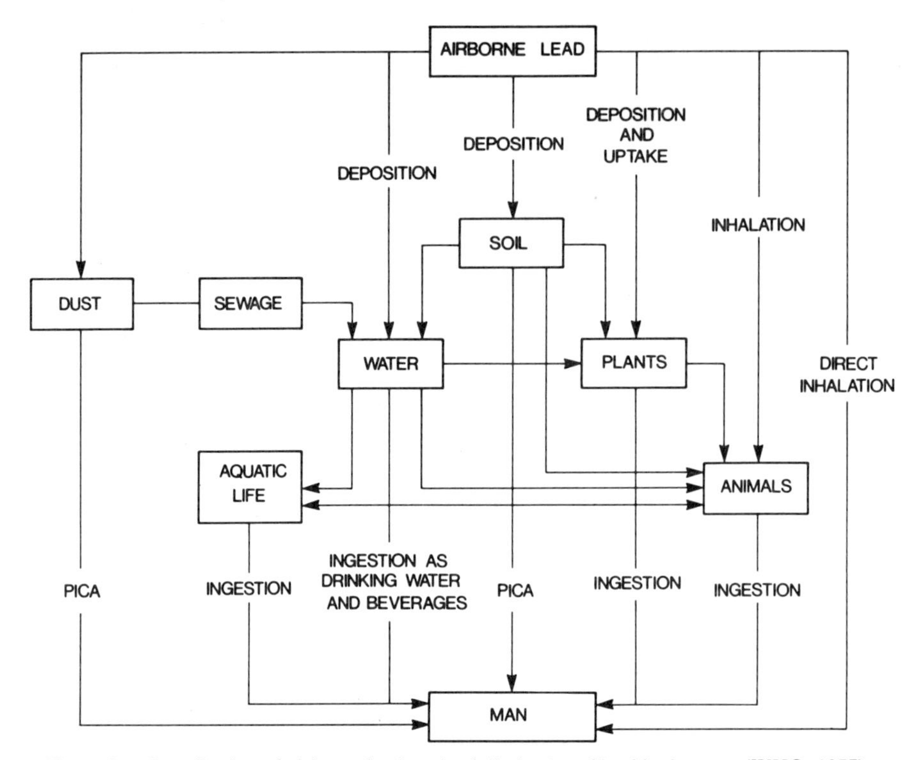


**Figure 2.** Contribution of airborne lead to the daily intake of lead by humans (WHO, 1977).

which is still about 50 times higher than the expected level in a pre-industrial atmosphere. Comparable levels of lead contamination of the atmosphere apparently occur in most countries (cf. Table 4).

The full impact of the 50- to 4,000-fold increase in the atmospheric lead levels on human populations remains to be fully evaluated. Widespread pollution of the atmosphere with lead inevitably results in extensive contamination of food and water with lead (Fig. 2). From geochemical arguments, Elias et al. (1977) estimated the average lead absorption by a mesolithic man to be under 0.3 $\mu g$/day. The most recent studies of present-day dietary lead intake (WHO, 1977) have put the daily absorption figure at 20 to 40 $\mu g$/day, or about 100 times higher than the natural background rate. Undoubtedly, atmospheric lead accounts for most of the lead in the human diet. Thus, it has been shown that 90 to 99% of the lead in sedge and voles in remote Rocky Mountains originated from smelter fumes and gasoline (Hirao and Patterson, 1974; Ellias et al., 1977).

The effect such manifold environmental contamination has had on the human body burden of lead is a concern. Since lead is retained predominantly in calcified tissues, analyses of archaeological bone finds have provided important clues on lead exposures in the past. For example, a study of prehistoric Danish bone samples showed seven times less lead than modern samples (Grandjean, 1975). A more recent study of the bones and teeth of ancient Nubians who lived 4,800 to 5,300 years ago found 10- to 100-fold less lead than in modern specimens (Grandjean, 1978). Barry (1978) has reviewed the available data on lead content of human remains from antiquity; in general, it has been shown that specimens from prehistoric cultures tend to have very low lead levels. It should be emphasized that factors pertaining particularly to analytical errors and to possible contaminating or leaching effects of soil-water can make archaeological bone finds unreliable indicators of the true skeletal lead burdens.

In their very careful study of the bones of Peruvian Indians who lived between 1,400 and 2,000 years ago, Erickson et al. (1979) obtained an average lead concentration in the bone of 0.08 $\mu g/g$ and an average atomic Pb/Ca ratio (in the same specimens) of $6 \times 10^{-8}$. The value of 0.08 $\mu g/g$ is about 300-fold less than the average level in bones of present-day North Americans, whereas the Pb/Ca ratio is about 450-fold below the level typical in bones of modern Britons and Americans. In essence, their results, along with those of many others, suggest that the present-day body burdens of lead may be orders of magnitude higher than those of the prehistoric cultures.

The preceding sections clearly show that no place on earth is free of lead pollution and that the current levels of lead in air at most rural and urban areas are 50- to 4000-fold higher than the levels in pre-industrial atmosphere. Airborne pollutant lead is now strongly implicated in the widespread contamination of human diet which has resulted in a modern daily absorption of lead (from both inhalation and ingestion) 50 to 200 times higher than that of neolithic man. Several lines of evidence are now converging to indicate that the intake of large doses of pollutant lead apparently have been responsible for the present-day body burdens of lead which may be orders of magnitude higher than those of the prehistoric population.

The growing concern about the public health implications of lead pollution stems, at least in part, from the realization that there is no toxicologically acceptable safety margin between the range of body lead levels now regarded as normal and those levels generally agreed to be capable of producing clinical poisoning in some individuals (Bryce-Smith et al., 1979). Until fairly recently, lead content of human tissues from rural areas remote from any major sources of lead pollution were regarded (quite erroneously) as "normal" (or "natural background") and the lead levels in such persons were widely believed to be safe (Kehoe et al., 1933). During the past decade, considerable progress has been made in evaluating some of the cryptic effects of lead poisoning and the definition of "acceptable" exposure has continually been revised downward (Zielhuis, 1977; David, 1978; Grandjean, 1978; Sassa, 1978; Bryce-Smith et al., 1979). It is reasonable to expect that future research will reveal other subtle manifestations of lead toxicity and that the limit of acceptable body lead burden will be further lowered. The critical question is how far downward the revision will be and whether it eventually will be determined that the typical level of lead in people today exceeds the threshold limit of the acceptable burden of lead in the body. Indeed, Patterson (1965) made the cogent observation that normal mammalian biochemical functions evolved in the presence of very low levels of lead and suggested that the grossly elevated exposures of 100 to 1,000 times the natural dose would almost certainly have adverse effects on human health.

The existence of adverse biologic effects at lower-than-normal levels of lead will certainly be difficult to demonstrate. In the first instance, there will be no suitable unexposed population that can be used as a control. Furthermore, the sought-for effects are usually ill-defined and nonspecific and include uneasiness, changes in mood, irritability, headaches, muscular weakness, gastrointestinal distress, fatigue, disorders of mentation (hyperactivity, disturbance of intelligence, learning disability, etc.) and increased susceptibility to disease. Such afflictions are typical of most environmental toxins and toxicologic tests are still to be developed to link such effects to specific doses of specific toxins. The wide variation in individual susceptibility to lead, as well as synergism or antagonism with other elements, are other factors which can mitigate against "normal" exposures to lead being recognized as an important etiologic factor in a disease condition.

Irrespective of what the true threshold value is for lead exposure or body burden of lead, there is increasing concern that the present airborne lead levels in many urban and even rural areas constitute a health hazard, particularly to children. Grossly elevated lead intake is certainly not doing them any good. The need for substantial reductions in the rate of lead discharge to the ambient air simply cannot be overemphasized.

## REFERENCES

Adams, F., Dams, R., Guzman, L., and Winchester, J.W. (1977). *Atmosph. Environ. 11*:629.

Akland, G.G. (1976). Air Quality Data for Metals 1970 through 1974 from the National Air Surveillance Networks. National Techn. Info. Service, Report No. PB 260 905. Springfield, Virginia.

Archer, A., and Barrat, R.S. (1976). *Sci. Total Environ. 6*:275.

Barry, P.S.I. (1978). J.O. Nriagu (ed.), In "The Biogeochemistry of Lead in the Environment," Part 1B, p. 97. Elsevier, Amsterdam.

Barton, S.C., Shenfeld, L., and Thomas, D.A. (1975). International Conference on Heavy Metals in the Environment, Toronto, p. C91–C93.
Bathea, R.M., and Bathea, N.J. (1975). *Residue Res. 54:*55.
Biggins, P.D.E., and Harrison, R.M. (1978). *Nature 272:*531.
Blokker, P.C. (1972). *Atmosph. Environ. 6:*1.
Boggs, W.R., and Wixon, B.G. (Eds.), (1977). "Lead in the Environment," Report No. PB 278 278, National Techn. Info. Service. Springfield, Virginia.
Branquinho, C.L., and Robinson, V.J. (1976). *Environ. Pollut. 10:*287.
Bryce-Smith, D., Mathews, J., and Stephens, R. (1979). *AMBIO 7:*192.
Cawse, P.A. (1974–1977). A Survey of Atmospheric Trace Elements in the U.K. AERE Publ. No. R-7669, R-8038, R-8393, R-8869. H.M. Stationery Office, London.
Council on Environmental Quality. (1975). Environment Quality, 1975. Council on Environmental Quality, 6th Annual Report. U.S. Government Printing Office, Washington, D.C.
Chester, R., Aston, S.R., Stoner, J.H., and Bruty, D. (1974). *J. Res. Atmosph.* p. 777.
Cholak, J. (1964). *Arch. Environ. Health 8:*126.
Chow, T.J. (1973). *Chem. in Britain 9:*258.
Chow, T.J., and Earl, J.L. (1970). *Science 169:*577.
Chow, T.J., Earl, J.L., and Synder, C.B. (1972). Science *178:*401.
Chow, T.J., Bruland, K.W., Bertine, K., Soutar, A., Koide, M., and Goldbert, E.D. (1973). *Science 181:*551.
Daines, R.H., Motto, H., and Chilko, D.M. (1970). *Environ. Sci. Technol. 4:*318.
David, O.J. (1978). Proc. Symp. on Lead Pollution—Health Effects, p. 30. The Conservation Society, London.
Demuynck, M., Rahn, K.A., Janssens, M., and Dams, R. (1976). *Atmosph. Environ. 10:*21.
Duce, R.A. (1972). B.C. Parker, (ed.), Proc. Colloquim on Conservation Problems in Antarctica, p. 27. Allen Press, Lawrence, Kansas.
Duce, R.A., Hoffman, G.L., and Zoller, W.H. (1975). *Science 187:*59.
Duggan, M.J., and Williams, S. (1977). *Sci. Total Environ. 7:*91.
Edgington, D.N., and Robbins, J.A. (1976). *Environ. Sci. Technol. 10:*266
Edward, H.W. (1975). In "Proc., Int'l. Symp. on Recent Advances in the Assessment on the Health Effects on Environmental Pollution," Vol. 3, p. 1277. Commission on European Communities, Luxembourg.
Egorov, V.V., Zhigalovskaya, T.N., and Malakhov, S.G. (1970). *J. Geophys. Res. 75:*3650.
Elias, R., Hirao, Y., and Patterson, C.C. (1977). T.C. Hutchinson (ed.), Proc. Int'l. Conf. on Heavy Metals in the Environment, Vol. 2, p. 257. Inst. of Environmental Studies, University of Toronto, Toronto.
Environmental Protection Agency. (1977). "Air Quality Criteria for Lead." U.S. Environmental Protection Agency, Office of Research and Development, Washington, D.C.
Ericson, J.E., Shirahata, H., and Patterson, C.C. (1979). Unpublished Report, Division of Geology and Planetary Sciences, California Institute of Technology, Pasadena.
Ewing, B.B., and Pearson, J.E. (1974). *Advances Environ. Sci. Technol. 3:*1.
Faoro, R.B., and McMullen, T.B. (1977). "National Trends in Trace Metals in Ambient Air, 1967–1974." U.S. Environmental Protection Agency, Office of Air

Quality Planning and Standards, Research Triangle Park, N. Carolina, Publ. No. EPA-450/1-77-003.
Fox, T.D., and Ludwick, J.D. (1976). *Atmosph. Environ. 10:*799.
Fugas, M., Wilder, B., Paukovic, R., Hrsak, J., and Steiner-Skreb, D. (1973). In "Environmental Health Aspects of Lead," p. 961. Commission of the European Communities, Amsterdam.
Grandjean, P. (1975). F. Coulston and F. Korte (Eds.), In "Environmental Quality and Safety," Vol. 2, p. 6. Academic Press, New York.
Grandjean, P. (1978). *Environ. Res. 17:*303.
Harley, J.H. (1970). *Environ. Sci. Technol. 4:*225.
Harrison, R.M., and Perry, R. (1977). *Atmosph. Environ. 11:*847.
Harrison, R.M., and Laxen, D.P.H. (1978). *Nature 275:*738.
HASL (1975). Health and Safety Laboratory Environmental Quarterly Report No. 297, Appendix B, p. B-139.
Hicks, R.M. (1972). *Chem. Biol. Interactions 5:*361.
Hirao, Y., and Patterson, C.C. (1974). *Science 184:*989.
Hoffman, G.L., Duce, R.A., and Hoffman, E.J. (1972). *J. Geophys. Res. 77:*5322.
Hutchinson, T.C. (1972). Publ. No. EH-2, Inst. Environ. Sci., University of Toronto, 27 pp.
Institute of Environmental Studies. (1977). "Environmental Contamination by Lead and Other Heavy Metals," Vols. 1 to 5. Institute of Environmental Studies, University of Illinois.
Ireland, F.E. (1972). H.A. Waldron and D. Stoffen, In "Sub-Clinical Lead Poisoning," p. 224. Academic Press, New York.
Janssens, M., and Dams, R. (1975). *Water Air Soil Pollut. 5:*97.
Jaworowski, A. (1968). *Nature (London) 217:*151.
Jaworowski, Z., and Kownacka, L. (1976). *Nature (London) 263:*303.
Jaworowski, Z., Bilkiewica, J., and Dobosz E. (1975). *Environ. Pollut. 9:*305.
Jernigan, E.L., Ray, B.J., and Duce, R.A. (1971). *Atmosph. Environ. 5:*881.
Jordan, L.D., and Hogan, D.J. (1975). *New Zealand J. Sci. 18:*253.
Jost, D., Muller, J., and Jendricke, U. (1973). In "Environmental Health Aspects of Lead," p. 941. Commission of the European Communities, Amsterdam.
Keohoe, R.A., Thamann, F., and Cholak, J. (1933). *J. Industrial Hygiene 15:*257.
Kretzschmar, J.G., Delespaul, I., Rijck, T.D., and Verduyn, G. (1977). *Atmosph. Environ. 11:*263.
Lawther, P.J., Commins, B.T., Ellison, J.M., and Biles, B. (1972). P. Hepple (Ed.), In "Lead in the Environment," p. 8. Institute of Petroleum, London.
Lee, J.A., and Tallis, J.H. (1973). *Nature (London) 245:*216.
Lepow, M.L., Bruckman, L., Rubino, R.A., Markowitz, S., Gillette, M., and Kapish, J. (1974). *Environ. Health Perspectives 7:*99.
Linton, R.W., Loh, A., Natusch, D.F.S., Evans, C.A., and Williams, P. (1976). *Science 191:*852.
Maziarka, S., Strusinski, A., and Wyzynska, H. (1971). Cited in *Chem. Abstr.*, Vol. *75.*
McMullen, T.P., Faoro, R.B., and Morgan, G.B. (1970). *J. Air Pollut. Control Assoc. 20:*369.
Murozumi, M., Chow, T.J., and Patterson, C.C. (1969). *Geochim. Cosmochim. Acta 33:*1247.
National Academy of Sciences Report. (1972). "Airborne Lead in Perspective," 330 p. National Academy of Science, Washington D.C.

National Air Pollution Surveillance Reports. (1971–76). Air Pollution Control Directorate, Ottawa, Canada, Report Series EPS-5-AP.

Needleman, H.L., Davidson, I., Sewell, E.M., and Shapiro, I.M. (1974). *New England J. Med. 290:*245.

Nelson, K.W. (1969). Presented at the Air Quality and Lead Symposium, Minneapolis, Minnesota.

Nriagu, J.O., (Ed). (1978). "The Biogeochemistry of Lead in the Environment," Parts 1A and 1B. Elsevier, Amsterdam.

Nriagu, J.O. (1978a), J.O. Nriagu (Ed.), In "The Biogeochemistry of Lead in the Environment," Part 1A, p. 137. Elsevier, Amsterdam.

Nriagu, J.O. (1978b). J.O. Nriagu (Ed.), In "The Biogeochemistry of Lead in the Environment," Part 1A, p. 1. Elsevier, Amsterdam.

Nriagu, J.O. (1979). Global inventory of natural and anthropogenic emissions of trace metals, to the atmosphere. *Nature, 279:*409.

Occupational Safety and Health Administration. (1978). "Inorganic Lead." Final Environmental Impact Statement. U.s. Dept. of Labor, Occupational Safety and Health Administration, Washington, D.C.

Page, A.L., and Ganje, T.J. (1970). *Environ. Sci. Technol. 4:*140.

Patterson, C.C. (1965). *Arch. Environ. Health 11:*344.

Robinson, J.W., Wolcott, D.K., Slevin, P.J., and Hindman, G.D. (1973). *Anal. Chim. Acta 66:*13.

Ruhling, A., and Tyler, G. (1968). *Bot. Notiser, 121:*321.

Salmon, L., Atkins, D.H.F., Fisher, E.M.R., Healy, C., and Law, D.V. (1978). *Sci. Total Environ. 9:*161.

Sassa, S. (1978). In "Handbook of Experimental Pharmacology," Vol. 44. Springer-Verlag, Berlin.

Schroeder, H.A. (1974). "The Poisons Around Us," p. 144. Indiana University Press, Bloomington.

Shapiro, I.M., Needleman, H.L., Dobkin, B., and Tuncay, O.C. (1973). *Clin. Chem. Acta, 46:*119.

Solomon, R.L., and Hartford, J.W. (1976). *Environ. Sci. Technol. 10:*773.

Tepper, L.B., and Levin, L.S. (1972). "A Survey of Air and Population Lead Levels in Selected American Communities," p. 1. Final Report, Environmental Protection Agency, Cincinnati, Ohio.

Ter Haar, G.L., and Bayard, M.A. (1971). *Nature 216:*353.

Thrane, K.E. (1978). *Atmosph. Environ. 12:*1555.

WHO (1977). "Environmental Health Criteria." 3. "Lead." World Health Organization, Geneva.

Yankee, A.J., von Lindern, I.H., and Walter, S.P. (1977). *J. Air Pollut. Control Assoc. 27:*763.

Zielhuis, R.L. (1977). *Intl. Arch. Occup. Environ. Health 39:*59.

Zoller, W.H., Gladney, E.S., and Duce, R.A. (1974). *Science 183:*198.

# Subject Index